The Construction Sector in Asian Economies

This is a collaborative project contributed to by all members of *AsiaConstruct*. It presents a comprehensive and up-to-date collection of data on the Asian construction sector forming a unique guide to construction economics in the following countries; Australia, China, India, Indonesia, Japan, Malaysia, Singapore, South Korea, Sri Lanka and Vietnam, as well as Hong Kong, a Special Administrative Region of China.

The state of the construction sector is a good indicator of the overall health of an economy. In emerging economies, activity in the construction sector signifies resources being put into infrastructure development; in developed countries it demonstrates the extent to which development is maintained and indicates the general direction of the economy, for instance in housing starts. In the last 15 years, the Asian region's construction industry has grown dramatically, and is poised to continue growing.

Significant new trends in the construction sector in Asia are: larger private sector participation: increased vertical integration and increased foreign participation. These may be attributed to both the globalisation and de-regulation of markets. While such trends present opportunities worldwide, they have also highlighted the issues affecting sustainable development of the Asian construction industries. These issues are examined in the detailed analyses of the construction sector in each country over several years. The book constitutes a unique overall collection of the construction economy performances and characteristics of a sizeable sample of Asian nations as they exist at the turn of the century. It constitutes a record of lasting value to scholars of economics.

Yat-hung Chiang is Associate Professor at the Research Centre for Construction and Real Estate Economics, Department of Building and Real Estate, The Hong Kong Polytechnic University.

Michael Anson is Emeritus Professor at the Department of Building and Real Estate, Faculty of Construction and Land Use, The Hong Kong Polytechnic University.

John Raftery is Dean of the School of Built Environment at Oxford Brookes University and formerly Chair Professor, Research Centre for Construction and Real Estate Economics, Department of Building and Real Estate, The Hong Kong Polytechnic University.

The Construction Sector in Asian Economies

Edited by

Y.H. Chiang, Michael Anson and John Raftery

Routledge
Taylor & Francis Group
LONDON AND NEW YORK

First published 2004
by Spon Press

This edition published 2013 by Routledge

2 Park Square, Milton Park, Abingdon, Oxon OX14 4RN

Simultaneously published in the USA and Canada by Routledge

52 Vanderbilt Avenue, New York, NY 10017

Routledge is an imprint of the Taylor & Francis Group, an informa business

First issued in paperback 2020

Typeset in Sabon by
Integra Software Services Pvt. Ltd, Pondicherry, India

British Library Cataloguing in Publication Data
A catalogue record for this book is available from the British Library

Library of Congress Cataloging in Publication Data
The construction sector in Asian economies / edited by John Raftery,
Y.H. Chiang and Michael Anson.
p. cm.
Includes bibliographical references and index.
1. Construction industry—Asia. 2. Asia—Economic conditions.
I. Raftery, John. II. Chiang, Y.H. III. Anson, Mike (Mike J.)
HD9715.A782 C66 2003
338.4′7624′095—dc22

2003025737

ISBN 13: 978-0-367-57815-2 (pbk)
ISBN 13: 978-0-415-28613-8 (hbk)

Contents

List of figures vii
List of tables ix
Foreword xvii

1 **Regional overview** 1
J. RAFTERY, M. ANSON, Y.H. CHIANG AND S. SHARMA

2 **Australia** 30
CONSTRUCTION INDUSTRY INSTITUTE, AUSTRALIA AND S. SHARMA

3 **China mainland** 66
MINISTRY OF WORKS, CHINA AND S. SHARMA

4 **China Hong Kong** 98
Y.H. CHIANG, EDWIN H.W. CHAN AND S. SHARMA

5 **India** 131
CONSTRUCTION INDUSTRY DEVELOPMENT COUNCIL, INDIA AND LENNON H.T. CHOY

6 **Indonesia** 164
MINISTRY OF PUBLIC WORKS, INDONESIA AND EDDIE C.M. HUI

7 **Japan** 192
RESEARCH INSTITUTE OF CONSTRUCTION AND ECONOMY, JAPAN, M. ANSON AND J. RAFTERY

8 **Malaysia** 325
CONSTRUCTION INDUSTRY DEVELOPMENT BOARD, MALAYSIA AND EDWIN H.W. CHAN

9 **Singapore** 360
BUILDING AND CONSTRUCTION AUTHORITY, SINGAPORE AND EDDIE C.M. HUI

10 **South Korea** 392
KOREA RESEARCH INSTITUTE FOR HUMAN SETTLEMENTS, J. RAFTERY AND M. ANSON

11 **Sri Lanka** 439
INSTITUTE FOR CONSTRUCTION TRAINING AND DEVELOPMENT, SRI LANKA AND LENNON H.T. CHOY

12 **Vietnam** 459
NATIONAL INSTITUTE FOR URBAN AND RURAL PLANNING, VIETNAM AND B.S. TANG

Index 488

Figures

1.1 The Asian two-tier construction market and the main participants 11
5.1 GDP growth rates – India 133
7.1 Trends in construction investment (nominal values), the number of licensed construction companies, and the number of workers in the construction industry 199
7.2 Trends in construction investment per construction company and construction worker 202
7.3 Construction orders in Asia 217
7.4 Trends in overseas construction orders received by the top-50 general contractors (1986–2001) 218
7.5 Overseas construction contracts in fiscal 2001 219
7.6 Trends in the construction investment growth rate 222
7.7 Estimate of construction investment 229
7.8 Trends in residential land prices 237
7.9 Trends in commercial land prices 237
7.10 Office vacancy rates and average advertised rents 238
7.11 Trends in average prices 239
7.12 Trends in average prices 239
7.13 Residential land prices since 1983 240
7.14 Commercial land prices since 1983 240
7.15 Number of land transactions since 1987 241
7.16 Average vacancy rate and rent for office space in Tokyo 243
7.17 Number of construction firms since 1989 246
7.18 Construction sector bankruptcies and debts since 1989 247
7.19 Trends in construction investment per construction company and construction worker 247
7.20 Trends in public-sector construction orders received by the major construction companies 249
7.21 Trends in private-sector construction orders received by the major construction companies 249
7.22 Number of workers in the construction sector since 1990 256
7.23 Registrations since March 1997 260
7.24 Structure of construction sector employment 265
7.25 Age distribution of the construction sector labour force 266

7.26	Construction material inventory and production indices in recent months	268
7.27	Contract awards for the major construction firms since 1989	270
7.28	Gross profitability of projects completed by major construction firms since 1993	271
7.29	Interest-bearing debts of the major construction firms since 1983	275
7.30	Overseas contracts for major construction firms since 1989	275
7.31	Trends in overseas construction orders received by the big 50 general contractors (1986–2001)	276
7.32	Two sets of statistics on overseas business	277
7.33	Number of employees at the major construction firms since 1990	278
7.34	Sales and general administrative expenses at the major construction firms since 1990	278
7.35	Ordinary profits of the major construction firms since 1993	279
7.36	Estimated growth of the renewal market	280
7.37	R&D expenditures at the top five construction firms since 1994	282
7.38	R&D expenditures: top five versus the rest	283
7.39	Structure of funding since 1970 for ordinary construction projects carried out by prefectural and local governments	306
7.40	Destinations of experts dispatched in FY 1996	317
7.41	Where trainees in FY 1996 came from	318
7.42	Number of foreign construction firms in Japan since 1988	320
9.1	Contracts awarded and progress payments	366
9.2	Construction demand 2002	370
9.3	Cost indices	376
9.4	Average prices of cement and ready-mixed concrete Grade 30	377
9.5	Average prices of steel bars and bricks	378
9.6	Average prices of granite and sand	379
9.7	Distribution of CONQUAS scores	382
10.1	Gross domestic product in KRW, US$	405
10.2	Gross domestic product: average increase rate	406
10.3	Gross national income in KRW, US$	407
10.4	Gross national income: average increase rate	407
10.5	Exports and imports	408
10.6	Balance of payments	409
10.7	Total external liabilities and assets	410
10.8	Foreign reserves	410
10.9	Foreign direct investment (FDI)	411
10.10	Inflation rate	412
10.11	Unemployment	413
10.12	Changes in construction investment	424

Tables

1.1	Asia-Pacific: real GDP growth	3
1.2	Asia-Pacific: macroeconomic indicators	4
1.3	Asia-Pacific: real GDP growth	6
1.4	Value-added per employee and GDP growth between 1995 and 2001	12
1.5	Construction industry in 1999	13
1.6	Construction investment in the period 1995–2000	14
2.1	National accounts in Australia between 1987 and 1993	33
2.2	Industry structure in Australia between 1989 and 2001	36
2.3	Annual trend growth in major economic and labour market variables	37
2.4	Key statistics: building and construction	39
2.5	Value of construction activity	40
2.6	Australian construction activity	41
2.7	Office market indicators in different cities in Australia	45
2.8	Residential market indicators in different cities in Australia	45
2.9	Industrial market indicators in different cities in Australia	47
2.10	Ten-year growth outlook for Australia	48
2.11	Economic growth forecast for the decade comparative to the region	49
2.12	Indexes of gross product per hour worked, by selected industry	50
2.13	Labour productivity in the 1990s	50
2.14	Unit construction cost for different sectors of building activities	51
2.15	Change in cost anticipated due to application of GST	51
2.16	An overview of fatality rate in Australian construction industry	55
2.17	Business enterprise R&D (ANZIC) in built environment sector, by segment	57
2.18	Research involving partners of Cooperative Research Centre (CRC)	58
2.19	Ongoing development projects in Australia and those in the pipeline	63
2.20	Extent of US investment in selected industries, US direct investment position in Australia on an historical cost basis, 1999	64
3.1	Chinese GDP change since the reform	69
3.2	Average annual rates of growth of selected economic indicators	70
3.3	Rates of growth of real GDP and inflation	71
3.4	Main macroeconomic indicators	72
3.5	Volume of investment in fixed assets	75

3.6 Annual construction volume 76
3.7 Total construction output in China 77
3.8 Beijing property market: key market indicators 78
3.9 Residential market in Beijing – key market indicators 79
3.10 Economic forecast for China in comparison to total Asia-Pacific region and ASIAN 83
3.11 Value-added per employee 84
3.12 Labour productivity of construction enterprises in various years 84
3.13 Level of mechanization of construction by projects 85
3.14 Indicators of construction cost in the case of state-owned construction companies 86
3.15 Average sectoral wage 86
3.16 Construction quality of construction enterprises 88
3.17 Distribution of fatal events by industries 91
4.1 Macroeconomic indicators 104
4.2 Land sales summary 108
4.3 Composition of GDP and employment 112
4.4 Value-added per employee 113
4.5 Services sub-sectors performance 113
4.6 Average growth rate of value-added 113
4.7 Unit construction cost 114
4.8 Average construction material price 114
4.9 Construction industry salaries and wages 115
4.10 Industrial and construction accidents 116
4.11 Construction accidents (1989–1998) 117
4.12 Environmental inputs to different stages of forward planning 127
5.1 Growth performance in the Five-Year Plans 133
5.2 Overview of national economy 136
5.3 Construction investment 139
5.4 Construction volume 140
5.5 Property prices and rents in New Delhi 141
5.6 Projected demand for electrical energy 144
5.7 Projected demand for other forms of energy 144
5.8 Project opportunities in ports 146
5.9 Value-added per employee 148
5.10 Physical measurement of construction productivity 149
5.11 Unit construction cost 149
5.12 Average construction material prices 150
5.13 Construction industry salaries and wages 150
5.14 Average sectoral wages 150
5.15 Safety record 152
5.16 Skill level of workforce and supervision level 153
5.17 New technology in automation, mechanization and prefabrication 154
5.18 Category of new technology in the underground structure 155
5.19 Major initiatives taken during the year 2000–2001 158
5.20 Commitments in General Agreement on Trade in Services (GATS) 160
5.21 Construction and related engineering services 161

5.22	Foreign direct investment and technical collaborations approved during August 1991 to March 2001 in consultancy services	162
6.1	Macroeconomic indicators: 1998 and 1999	168
6.2	Major demographic indicators (1995–1999)	169
6.3	Major macroeconomic indicators (A)	170
6.4	Major macroeconomic indicators (B)	171
6.5	Construction sector investments (constant prices) by construction type	173
6.6	The approved investment proposal by BKPM during the period of 1999–2000	174
6.7	Indonesian near-term economic restructuring and targets at a glance	178
6.8	The GDP projection for sectors other than construction year-on-year	179
6.9	Indonesian real GDP for Q1 2001	180
6.10	Productivity of field workers	181
6.11	Value-added per employee in the construction sector	181
6.12	The productivity data of construction workers	181
6.13	Unit construction cost	182
6.14	Price index of building and construction materials	183
6.15	Construction labour force by formal education for the year of 2000	184
6.16	Number of unskilled labourers	185
6.17	Indonesia: net capital inflows	188
6.18	Foreign investment approvals, 1996–2000	188
6.19	Top-ten sectors by value per year, 1996–2000	188
6.A.1	Volume of works finished and ready for use 1995–1997	190
6.A.2	Mandays of part-time field workers	190
7.1	Economic growth, 1983–1988	194
7.2	Status and future prospects of the eco-business market as of 2002	207
7.3	Public works orders received by foreign firms	212
7.4	Economic measures since the collapse of the bubble economy in 1992	220
7.5	Construction investment (or market) volume	221
7.6	Major macroeconomic, demographic and financial indicators	226
7.7	Main macroeconomic indicators	227
7.8	Main demographic indicators	227
7.9	Trends in construction investment	228
7.10	Construction market by type	230
7.11	Housing starts	231
7.12	Major projects programmed and underway at the time of writing	233
7.13	Construction costs	234
7.14	Average construction material price	234
7.15	Construction industry salaries and wages	235
7.16	Average sectoral wage costs	235
7.17	Officially announced land prices since 1997	236
7.18	Office vacancy rates and average advertised rents	238
7.19	Official land prices since 1994	241

7.20 Average rents and vacancy rate for office space 243
7.21 Recent policy on city, land and housing 244
7.22 Distribution and size of construction companies 245
7.23 Construction firms 248
7.24 Number of foreign corporations and foreign-capitalized Japanese corporations that have obtained construction business licenses 250
7.25 Foreign construction firms licensed to operate in Japan 250
7.26 Foreign firms holding a specified contractor's license 251
7.27 Trends in the number and value of contracts of land quality assessment firms 252
7.28 Trends in the number and value of contracts of surveying firms 252
7.29 Trends in the number and value of contracts of construction consulting firms 253
7.30 Trends in the area of designs actually implemented by architectural design firms 253
7.31 Trends in the number of registered consulting firms 253
7.32 Trends in registered licensed surveyors and assistant surveyors by fiscal year 254
7.33 Trends in the number of licensed architects 255
7.34 Domestic demand for key construction materials since 1996 255
7.35 Key construction equipment: output value and unit production since 1994 255
7.36 Trends in the number of workers and technical workers/labourers in the construction industry 257
7.37 Accident frequency rates 261
7.38 Number of workers passing the technical certification 263
7.39 Value-added per employee 264
7.40 Physical measurements of construction productivity 264
7.41 Breakdown of construction industry employees 265
7.42 Trends in number of employees by job type 266
7.43 Construction sector wages since 1991 267
7.44 Domestic demand for key construction materials since 1996 268
7.45 Key construction equipment: output value and unit production since 1994 268
7.46 Average construction material price 269
7.47 Wholesale price indices for principal construction materials since 1993 269
7.48 Sales and number of employees for major construction firms for different types of construction projects 274
7.49 Principal R&D projects recently carried out by the major construction firms 284
7.50 New construction technologies 287
7.51 Level of prefabrication 289
7.52 Trends in number of construction business license holders (1975–2002) 289
7.53 Total points on the business evaluation $(P) = 0.35X_1 + 0.10X_2 + 0.20Y + 0.20Z + 0.15W$ 291

7.54 Outline of assessment system-items upon which a construction firm is assessed 291
7.55 Implementation of different bidding systems (1996) 292
7.56 Overview of the bidding systems 293
7.57 Contract award criteria 294
7.58 Design-build contracts now under consideration 294
7.59 VE systems now under consideration 295
7.60 Principal infrastructure: targets and current levels of development 299
7.61 Long-term public works plans 301
7.62 Principle organizations in charge of infrastructure development 303
7.63 Division of funding responsibility for infrastructure development 304
7.64 General account revenues and expenditures 305
7.65 Entities in charge of infrastructure projects 307
7.66 Trans Tokyo-Bay highway project 309
7.67 Bill promoting the use of private-sector funds for the development of public facilities 310
7.68 Guidelines for a Japan-style PFI 311
7.69 Types of ODA assistance 312
7.70 ODA assistance to Asian countries 313
7.71 Flow of funds to developing nations other than ODA 313
7.72 JICA project-type technical cooperation 314
7.73 Program for development of construction technology overseas 314
7.74 Overseas construction technology cooperation program 315
7.75 Number of experts dispatched in recent years 316
7.76 Number of training courses/trainees in recent years 317
7.77 Japanese foreign direct investments during FY 1994 to FY 1996 318
7.78 Japanese foreign direct investment in recent years 319
7.79 Participation of Japanese firms in infrastructure development abroad 319
7.80 WTO government procurement agreement: requirements in the construction services sector 320
7.81 Foreign direct investment in Japan 320
7.82 Foreign direct investment in Japan 321
7.83 Construction material imports by the principal construction firms in recent years 321
7.84 Construction material imports 321
7.85 Selected construction equipment: production and domestic demand (1995) 322
7.86 Construction equipment exports (1996) 322
7.87 Overseas construction contracts awarded to Major Japanese construction firms 323
7.88 Access problems in overseas construction markets 323
7.89 Market access problems mentioned by foreign firms in Japan 324
8.1 Gross domestic product by industrial origin at constant 1978 prices (per cent share in GDP) sector/industry 1985, 1990, 1995, 2000 330
8.2 Main macroeconomic indicators 333

8.3 Growth trend of GDP and construction industry 334
8.4 Total volume and number of projects in 2000 335
8.5 Quarterly project values and numbers 335
8.6 Number and value of property transaction by sector 338
8.7 Supply of office, retail and residential space in Klang valley 339
8.8 Economic growth forecast for the decade comparative to the region 345
8.9 Growth of productivity, GDP/output and employment by sector 346
8.10 Price index for construction 348
8.11 Average construction material price 348
8.12 Construction industry salaries and wages 349
8.13 Number of accidents reported for the construction sector 352
8.14 Yearly accidents frequency in construction industry compared to industries in general 353
8.15 Development allocation for science and technology, 1991–2000 356
9.1 GDP growth in selected years 362
9.2 Main macroeconomic indicators 363
9.3 GDP growth of each sector in 2001 364
9.4 Contracts awarded by sector and type of work 368
9.5 Progress payments certified by sector and type of work 369
9.6 Historic development of property market in Singapore 373
9.7 Labour productivity growth 375
9.8 Average productivity 375
9.9 Unit construction cost 377
9.10 Construction industry salaries and wages 379
9.11 Average sectoral wages 380
9.12 Total number of certified firms 381
9.13 Yearly average CONQUAS score 381
9.14 Average buildability scores 382
9.15 Accident statistics in construction industry 383
9.16 Skill level of construction industry 384
9.17 Certifications awarded for supervisors 384
9.18 Research and development expenditure for construction 385
9.19 Major new technology developments in the construction field approved by IDS 386
9.20 Extent of design and build projects 387
9.21 Precast level 388
9.22 Major projects in the pipeline 390
10.1 Key economic indicators of the 1960s 395
10.2 Key economic indicators of the 1970s 395
10.3 Key economic indicators of the 1980s 396
10.4 Key economic indicators (1990–1996) 396
10.5 Agriculture in the national economy 398
10.6 Main macroeconomic indicators 404
10.7 The construction share of the GDP 405
10.8 Gross domestic product indices 405
10.9 Gross national income 406

10.10 Exports and imports 408
10.11 Balance of payments 408
10.12 Total external liabilities and assets 409
10.13 Foreign reserves 410
10.14 Foreign direct investment (FDI) 411
10.15 Inflation rate 412
10.16 Unemployment 413
10.17 Private investment project scale 415
10.18 Value-added product 416
10.19 Value-added per employee 417
10.20 Physical measurement of construction productivity 417
10.21 Unit construction cost 418
10.22 Price of construction materials 418
10.23 Construction industry salaries and wages 419
10.24 Average sectoral wages per month 419
10.25 Number of ISO 9000 certified construction companies 419
10.26 Accident frequency rate 420
10.27 Skill level of workforce and supervision level 420
10.28 Research and development expenditure 421
10.29 New technology in automation, mechanization and prefabrication 422
10.30 Number of ISO 14000 certified construction companies 422
10.31 Construction investment 424
10.32 Construction volume 425
10.33 Land price change 425
10.34 Average rent change 426
10.35 Average rents for office space in major cities 426
10.36 The number of bankrupt construction companies 427
10.37 Main indices in construction 427
10.38 Distribution of contractors by employment size 428
10.39 Turnover and number of employees for top-ten local construction companies 428
10.40 Main intermediate and long-term plans for infrastructure projects 431
10.41 Implementation body and burden sharing for infrastructure projects 432
10.42 Types of SOC facilities 433
10.43 SOC investment ratio in national budget 434
10.44 Sectoral SOC investment 434
10.45 Major cases of private sector capital participation during the period 1995–2001 435
10.46 Construction volume of foreign firms 436
10.47 Efforts to open markets, relax foreign capital regulations and promote foreign investment 436
10.48 The present foreign companies of general contractor license in Korea 437
10.49 Change in overseas construction 438
10.50 Top-five countries for construction export 438
11.1 Main macroeconomic indicators 442

11.2	Construction investment (or market) by type	443
11.3	Land prices in major cities	444
11.4	Average rents in major cities	444
11.5	Unit construction cost	450
11.6	Average construction material prices	450
11.7	Construction industry salaries and wages	451
12.1	GDP growth rate 1976–1995	463
12.2	Change in economic structure of Vietnam	463
12.3	Main macroeconomic indicators	465
12.4	Forecast of Vietnam's economy: some macroeconomic indicators for 2001	466
12.5	Construction investment	467
12.6	Average land rental fees of project and companies in February 2001	469
12.7	Average rent for the period between 1996 and 1999	469
12.8	Average price, typical size and share of house in Hanoi	469
12.9	Economic forecast for Vietnam in comparison to total Asia-Pacific region and ASEAN	470
12.10	Gross product per employee	471
12.11	Construction output and workers	471
12.12	Construction productivity	471
12.13	Construction cost in Hanoi	472
12.14	The residential construction cost	472
12.15	The residential construction cost in Hanoi	472
12.16	Construction material price in Hanoi	473
12.17	Average income per month of labour in state sector	473
12.18	Construction safety	474
12.19	Safety record	475
12.20	New technology in automation, mechanization and prefabrication	477
12.21	Ten steps of project development in Hanoi city	479
12.22	State investment outlays in 1999 at the then prices as of 2002 by type of component	479
12.23	State investment outlays of local government in 1999 at the then prices as of 2002 by region, province	479
12.24	Business opportunities and major projects in the pipeline	481
12.25	Top-ten sources of FDI inflows to Vietnam in 2000	485
12.26	Power demand forecast upto the year 2010	485

Foreword

This book is a unique collection of descriptions of the construction economies of 11 Asian countries plus that of Hong Kong, the Special Administrative Region of China. The countries involved are Australia, China Mainland, India, Indonesia, Japan, Malaysia, The Philippines, Singapore, South Korea, Sri Lanka and Vietnam. There is one chapter for each country relating to the years 1995–2001, preceded by an overview of the whole region in Chapter 1.

The book is a collaborative AsiaConstruct project contributed to by all the members. AsiaConstruct began as an initiative of the Research Institute of Construction Economy of Japan in 1993, which hosted the first meeting in Tokyo. The AsiaConstruct members hold an annual meeting each autumn at which each country presents its construction economy report for that year, thereby contributing to better understanding among the members of each other's characteristics. The meeting is hosted by a different country each year and a local audience is also invited to hear the country reports and contribute to discussion.

The annual compilations of the country reports are not formally published, however, and the purpose of this book is to capture for the record, during 8 years of operation, the substance of the information gathered together so painstakingly each year.

The result is an unusual and useful reference work valuable for both scholars in the economics field and for construction companies considering operations in different theatres, as well as for governments concerned with policy decisions in relation to their own construction industries.

The initial drafting of most of the chapters was carried out by staff of the Department of Building and Real Estate of the Faculty of Construction and Land Use of the Hong Kong Polytechnic University (the Department which represents China Hong Kong among the AsiaConstruct membership). Each country then became directly involved and the final product is very much a combined team effort.

Special thanks are due to Professor John Raftery one of the key drivers in the production of this book, and a pioneer in the involvement of Hong Kong with Asia-Construct, to Dr Chiang Yat-Hung, the project manager, and to Mr Sujeet Sharma who compiled much of the data from all the annual reports on a full-time basis thanks to a grant from the Hong Kong Polytechnic University.

The country reports were prepared between late 2001 and early 2002. Readers should interpret the data and information accordingly.

M. Anson
Chair Professor of Civil Engineering
The Faculty of Construction and Land Use
The Hong Kong Polytechnic University.

Note
Professor Raftery, formerly of the Department of Building and Real Estate, took up a new post in the UK at Oxford Brookes University in 2000 but has remained a central leading figure in the production of this book.

Chapter 1

Regional overview

J. Raftery, M. Anson, Y.H. Chiang and S. Sharma

1.1 Setting the scene

Following the Chinese government's announcement of its 20 billion dollars accelerated building programme upon being selected as a host for the 2008 summer Olympics, one newspaper wrote 'the news brought jubilant celebrations among the Chinese soccer fans, this also brought smiles to the lips of consultants, contractors and suppliers across geographic boundaries'. Globalization, advances in technology and changes in the structure of the global economy are presenting new challenges to the US$3.22 trillion construction industry worldwide (Reina and Tim, 2000).

Until 2002, in the last 15 years, the Asian region's construction industry had grown dramatically, as had its economy, productivity, and investment in research and development (R&D). The region's construction production was valued at over US$1000 billion in the 1990s, the number of employees running into millions. The eleven countries spent $910 billion in 1997 as construction sector investment even though the combined GDP of all ten countries (as of 2002 prices) was less than three quarters of that of the United States in the same year. Japan unequivocally tops in terms of the GDP, its construction spending, as well as the level of technology in Asia. The Asian construction market is the largest overseas market for Japanese construction companies, as about 70 per cent of the sum total of overseas construction orders in fiscal 2000 came from Asia. However several Asian countries are emulating Japan in developing their economies by increasing their capacities in high technology, and by developing a skilled labour force and better social and physical infrastructure. Resources are being devoted to high technology production, processing and distribution, including improving the quality and skills of the construction labour force.

The survey presented in this book represents the construction sector in ten participating countries of the AsiaConstruct group – Australia, the People's Republic of China (China), India, Indonesia, Japan, the Republic of Korea (South Korea), Malaysia, Singapore, Sri Lanka and Vietnam, plus Hong Kong, the Special Administrative Region of China. The AsiaConstruct Conference is modelled on EuroConstruct. However, Asia is no less, if not more, diverse than Europe in a number of social-economic and political ways. Whilst Europe has its European Currency Unit, different Asian countries have different currencies, languages, cultures, legal and government systems. Even Hong Kong is so different from its motherland, China, that its capitalist way of living is guaranteed until 2047 under the 'one country, two systems' principle. What follows is a cross-sectional study of the Asian construction sector covering a time-series beginning in the early 1990s.

1.2 The macroeconomy

1.2.1 Asia boom Asia bust

The economic landscape has changed quite dramatically over the 1990s both in Asia and internationally. The experiences in Asia in the 1990s however were mind boggling. The economic growth and subsequently the construction sector in many Asian economies have seemed to have much to do with Japan. With the benefit of hindsight, it is now clear that the Japanese economic boom of the late 1980s led to an unsustainable bubble in asset prices not only in Japan but across the Asian region as well. Mera and Renaud (2000a) attributed the massive direct and portfolio investment in these economies initially to the appreciation of the yen after the Plaza Agreement of 1985 and then to her economic stagnation after the collapse of her stock and property markets. Initially Japanese firms had to find cheaper production bases in many Asian countries, thus contributing to the economic growth of these countries. Later, not only global but also Japanese investors needed to look outside Japan for investment opportunities in Asia. Many Asian economies were showing high economic growth rates (World Bank, 1993) as a result of the direct investment and booming trade, becoming targets of global 'hot' money.

Mera and Renaud (2000a) noted the correlation between this massive influx of direct investment to Asian economies and their economic growth between 1990 and 1996, especially when compared with the stagnation of the Japanese economy. The property boom and the immense investment in infrastructure projects across the region grew out of this environment. However, when massive capital flows, themselves much the result of deregulation and globalization of capital markets, met with imprudent and even corrupt banking practices, and again with the benefit of hindsight, property busts could only be inevitable. In the middle of 1997, Thailand fought to defend her exchange rate in vain and was forced to let the Thai baht suffer a very sharp downturn. The rest, as they say, is history.

The bursting of that bubble in the late 1990s ushered in a period of uncertainty, confusion and crisis. This, however, also presented an opportunity for correction to the economic overdrive experienced in the 1990s. The economic crisis and its reverberations had continued to be the primary focus of attention for some years. The property sector, and consequently the construction sector, will continue to experience its boom and bust. Its volatility is not expected to diminish as the capital markets become more fully integrated with each other and globalized. There is a need to understand the structure of the Asian property and construction market so that a clearer background could provide better insight for a better understanding of the fundamentals. This in turn will benefit the formulation and implementation of various institutions and strategies to tame the cycles and to improve the competitiveness and productivity of the sector within individual economies.

1.2.2 Economic recovery: not a mere economic euphoria

Nevertheless, there was growing evidence that economic recovery among the crisis-hit countries had entered a more positive phase from the first or second quarter of 1999. Several countries had experienced a deep contraction in output and employment during 1998. The economies of China and India were able to sustain their surprising vigour even during the turmoil of 1998 but with few signs of a breakthrough on the domestic fiscal front or in the renewal of foreign investor confidence. In other countries experience has been mixed. Within the region, economic growth varied between countries as illustrated in Table 1.1. In countries that saw a remarkably sustained growth throughout the 1990s, recovery from the crisis was fairly rapid demonstrating the resilience of the economies and also a favourable external environment that permitted strong export-led recovery. Particularly noteworthy is the resilience of the Chinese and Indian economies and their undiluted strength that seems to have played a significant role in the rapid turnaround of the crisis in Asia.

Economic growth remained strong, averaging 7.3 per cent across the region, as domestic demand for investment and consumer goods continued to rise. The transition economies of China and Vietnam expanded most rapidly in 1996 by 9.7 and 9.0 per cent respectively. Malaysia and Indonesia followed, with growth rates of 8.2 and 7.8 per cent respectively. Policy makers of the two large economies of China and India continued to keep inflation in check by tightening money supply and price controls on some crucial products. China tamed its inflation rate to 6.1 per cent from 14.8; India managed to pare inflation to 8.5 per cent from 10.3 on average. Korea's economic recovery had been remarkable. GDP growth recorded 10.7 and 8.8 per cent in 1999 and 2000 respectively.

Table 1.2 presents macroeconomic indicators as a preview to countries' economic performance. There is a point of caution: indicators such as unemployment rate vary in definition from country to country and the numbers themselves may not be directly comparable. The historical overview suggests the effects of financial crisis were deeply entrenched in the country's employment in almost all countries. Macroeconomic developments in Asia in the 1990s had turned out to be more favourable in many ways than would have been expected at the beginning of the decade with economic growth averaging at 3–5 per cent, the unemployment levels by contrast were not so good and unemployment continued rising unabated even as the economies began to recover.

Table 1.1 Asia-Pacific: real GDP growth (annual per cent change)

Year ended December	*1995*	*1996*	*1997*	*1998*	*1999*	*2000*	*2001*
Australia	4.4	4.0	3.9	5.1	4.4	4.0	0.5
China	10.5	9.6	8.8	7.8	7.1	7.0	7.0
Hong Kong	3.9	4.5	5.0	−4.3	2.9	5.0	3.4
India	7.6	7.5	5.0	5.8	5.9	6.8	6.3
Indonesia	8.2	7.8	4.9	−13.7	−0.7	3.0	5.0
Japan	1.5	5.1	1.6	−2.5	0.7	0.9	2.0
Korea	8.9	6.8	5.0	−5.8	9.5	7.0	6.2
Malaysia	9.8	9.5	7.5	−7.5	5.3	6.5	6.0
Philippines	4.7	5.8	5.2	−5.0	3.2	4.9	4.2
Singapore	8.1	7.6	8.9	0.3	6.0	6.2	5.5
Vietnam	9.5	9.3	8.2	5.8	4.8	4.0	5.0
Total							
Asia-Pacific	7.2	7.5	5.4	1.7	4.7	5.0	5.1
OECD	2.5	3.3	3.5	2.4	2.8	2.9	2.6

Table 1.2 Asia-Pacific: macroeconomic indicators (mean value, 1995–2001)

Country	*Consumer price index*	*Inflation rate (%)*	*Unemployment rate (%)*	*Real GDP growth rate (%)*	*Current account balance (% of GDP)*	*Trade exports*	*FDI (US$ million)*
Australia							
China	–	5.3	2.17	5.2	1.08	15.03	37,160
Hong Kong	2.8						
India	8.04	7.4		6.17	−1.47	10.35	1,759
Indonesia	17.7	16.8	4.9	2.4	0.65	5.55	3,747
Japan							1,333
Korea				5.4			1,774
Malaysia	3.3	3.2	2.5	4.8	2.9	7.17	4,950
Philippines	6.9	7.04	9.8	3.7	2.17	13.66	1,558
Singapore	1.05	1.1	2.5	5.7	16.8	3.6	7,085
Sri Lanka		9.73	8.6	5.3	−4.33	6.46	–
Vietnam	–						

Current account deficits in most countries had been widened, giving rise to concern from foreign lenders and investors in the past 7 years, until 2002. However, prospects for reform remained strong in the original ASEAN economies such as the Philippines, Indonesia, Singapore and Malaysia. These nations had not only concentrated on lowering trade barriers and investment hurdles, but also agreed to establish a formal framework for financial cooperation. Envisaged first and foremost as a safety mechanism, the framework is expected to serve later as a foundation for a joint monetary market. Privatization continued to advance slowly and unevenly because the authorities in most of Asia were uncertain that the hoped-for economic benefits and efficiency gains would outweigh the negative social implications of large layoffs.

On the general foreign investment front, the increased flexibility of rules and regulations had contributed to the continuation of fairly large inflows and to the growth of intra-regional investments. China continued to attract over US$40 billion as FDI consistently every year after 1997. While foreign investor interest remains strong, the prospects for implementing many investment proposals will be determined not only by high domestic growth and favourable global market conditions, but also by improvements in the institutional framework, including the protection of property rights, the creation of regulatory agencies to supervise previously publicly owned service monopolies, a stable and transparent tax and incentive regime, and price and trade liberalization. This also applies to the other countries in the study.

One of the enduring features of the Asian economic landscape over the past decades had been the gradual implementation of the widespread programme of microeconomic reform. The major reforms include dismantling of barriers to foreign trade, financial deregulation, corporatization and privatization of the government business enterprises including new regulatory arrangements.

It was observed that such reforms initiated during the period had helped create a rather comprehensive legal framework, in parallel with the strong deployment of strategy, planning and development orientation, thereby contributing significantly to the upgrade of sector management efficiency at the macro level. Production forces had been rearranged. General corporations and strong companies had been established and consolidated for continued intensive investment in order to renovate technologies, increase production capacities, and boost competitiveness of the sector products, thus preparing the necessary prerequisites towards regional and international economic integration in the next period. In terms of structural changes, China joined the WTO thereby committing to a wide range of reforms. The benefits of these reforms will be felt in the medium and long term in the construction industry. In Australia, the implementation of GST heralded the beginning of the new set of reforms.

1.2.3 Economic outlook for the decade

Asia experienced a golden decade of uninterrupted high growth between 1986 and 1996, in the context of rapid development over the last 40 years, until 2002. A massive inflow of funds from the developed world during the sustained boom eventually led to substantial oversupply in many sectors, and to an extraordinary bust. Over the decade before the crisis, the emerging economies of Asia averaged real GDP growth rates of 8–9 per cent per year.

Table 1.3 Asia-Pacific: real GDP growth (annual per cent change)

Year ended December	*1995*	*2001–2010 (forecast)*
Australia	4.4	3.6
China	10.5	6.5
Hong Kong	3.9	4.0
India	7.6	6.5
Indonesia	8.2	4.5
Japan	1.5	3.0
Korea	8.9	5.5
Malaysia	9.8	6.5
Philippines	4.7	5.1
Singapore	8.1	5.1
Vietnam	9.5	6.0
Total Asia-Pacific	7.2	5.3
OECD	2.5	3.0

In the decade 2001–2010, this average annual growth rate is expected to fall back to 5–6 per cent per year as shown in Table 1.3. There are a few basic reasons. Structural problems, particularly in the corporate sector, will take many years to resolve. Lack of progress on reform will affect fickle investor/financial market confidence (leading to reduced inflow of funds) and financial constraints on business will linger. The rate of capital accumulation/inflow, which drove the last boom, is not expected to return to pre-crisis levels in the near future. Importantly, there was less scope for labour force growth, and productivity and technological gains. Asia increasingly competes with the developed world, rather than simply being a cheaper place to undertake labour-intensive tasks. It is rapidly 'closing the gap' and growth will trend down to developed-economy levels.

The setback of 2002 brought about by the financial crisis in 1997 had forced the governments across the region to face up to some tough issues of business and political culture, the rationalization and reform of financial sectors, and increases in foreign ownership limits. The countries which remain the strongest in the long term will be those which have removed these types of distractions from doing business, and allow the natural advantages of low costs, good infrastructure and education standards to come to the fore.

1.2.4 The property cycle

The performance of the Asian property market was drastically affected by the economic crisis of 1997/98 with recovery hinging on a revival of foreign investment in the region, which in turn, will be reliant on an improved risk/return profile and increased attractiveness of business potential. The economic indicators showed positive signs towards the recovery of the regional economy and property industry. Some property sectors have shown early recovery signs, as indicated by the increase in the Jakarta retail occupancy rate, which reached 88.2 per cent at the end of

1999, reflecting an increase of 7.9 per cent from 1998. However, property prices in Hong Kong were beset with negative equities, which showed no sign of improving. In Japan, the greying population, the uncertainty about the 'Japan Inc' and the banks' inability to extend loans have collectively caused the continuous slide in property prices and prevented the recovery of the property sector (Mera and Heikkila, 2000). These problems were structural in nature and not easy to fix. It is difficult to generalise about the current and future state of property markets. Each is at a different point in the property cycle and each will be influenced in the future by different supply/demand relationships.

Although the regional property markets operate independently of each other in supply and demand, there have been broad similarities in cyclical movements over the 1990s. In the late 1980s high construction levels occurred, as a result of rising rents amid an anticipation of boom times. The high construction levels, coupled with stagnant demand by the early 1990s, led to oversupply of varying degrees in each of these countries, and rental levels began to fall. During this period construction activity remained high as a result of the overhang from the boom, and the time lag in realising that supply was well ahead of demand up until the financial crisis in 1997. In general, while Asian economies had recovered somewhat, the construction industry was flat on its back, primarily a consequence of the construction boom of 1997. Industry analysts said that it would take 2–5 years to fill up the existing supply of office towers, upscale shopping complexes and luxury housing. Indonesia had a huge stock of empty buildings and the government had had marginal impact in turning this situation around, not only because of the political turmoil.

Prices in many Asian cities were much closer to economic fundamentals than before the crisis, reports the Asian Development Outlook, a publication of the ADB. The words 'negative equity' are sending shudders down the backs of government and property developers alike in Hong Kong as even in 1998 – well into the financial crisis – prime downtown land in Hong Kong, China and Tokyo was selling for more than \$100,000 a square metre. Prices of high-end condominium units in Malaysia and the Philippines fell by as much as 40–50 per cent in some locations from their 1997 peak values. In dollar terms, office rental rates in 12 Asian cities fell by an average of 34 per cent. Indonesia, which experienced the steepest depreciation (83 per cent), had the second sharpest drop in rents. The transaction activity in Malaysia contracted up to 47.6 per cent in value, despite a low prime office rents decline (lower than 10 per cent) in 1998. In the Philippines average rents in the CBD's best buildings discounted by as much as 35 per cent towards the end of 2000.

Market inefficiencies account for some differences between major cities rather than exchange rate movements. If rents are not allowed to fall with declining demand, either because of regulations or because landowners restrict supply, then commercial space may go unoccupied. Occupancy rates close to 50 per cent in Shanghai and Hanoi suggest that markets in those cities are not allowed to clear (ADB, 1999).

Property industry is the upstream industry of the construction sector, and the housing sector has been considered as a major economic stimulant (Mera and Heikkila, 2000, *The Economist*, 2002a). The cases of the property sector in many Asian economies have also demonstrated the close linkage between the property and the capital markets (Mera and Renaud, 2000b; *The Economist* 2002b). There is certainly much room for intra-regional cooperation in improving the governance and transparency in the capital markets and particularly the banking sector of most of the Asian economies. Improved information and market efficiency are instrumental in getting back the confidence of the investment community. Only then will the property and subsequently the construction sectors finally benefit from a return of capital inflows, businessmen and house-buyers.

1.3 The construction industry

1.3.1 The asian construction sector

Because of the massive domestic construction undertaken by Japan, Asia had always been a major construction market in terms of spending and growth. Low (1991a) identified that over a 15-year period between 1970 and 1984, Asia was the third largest region after North America and Western Europe in terms of construction spending. In addition, all the four sub-regions (West Asia, South East Asia, South Asia and East Asia) had average annual growth rates that were more than the average of all 19 sub-regions under survey. In particular, Hong Kong, Indonesia, Malaysia, Singapore and South Korea all had average annual growth that was value-added by construction of more than 20 per cent, thus putting them in the highest of the six growth categories (Low, 1991b).

Based on annual surveys conducted by *Engineering News Record*, Bon and Crosthwaite (2000) showed that in 1998, Japan had the second highest level construction spending after the US. On a per capita basis, Japan spent the most. In terms of total construction spending in that year, Japan, China, South Korea and Australia ranked second, fourth, ninth and seventeenth respectively. On a per capita basis, Japan, Singapore, Hong Kong, Australia and South Korea ranked first, sixth, ninth, fifteenth and twentieth respectively. Asia had the third largest GDP by region in 1998, after Europe and North America. However, it had the largest construction market, accounting for 35 per cent of total world construction spending in the same year, and had the largest share of GDP devoted to construction spending.

Asia was regarded as a promising international construction market in a survey of contractors in the UK (Seymour, 1987). Most of the respondents considered 'Asia, particularly South East Asia, as the market with the most potential of all developing regions'. More recently during the 1990s, Bon and Crosthwaite (2000) find that many individual Asian economies were considered to have very promising construction markets in terms of growth, profits, openness and attractiveness. China was ranked top in terms of growth and profit overwhelmingly by the respondents. Other Asian countries such as Vietnam, Malaysia, Indonesia, South Korea and India were also among the top ten fastest growing construction markets in the world. In terms of profitability, China, Japan, Indonesia, Malaysia, Vietnam, Singapore, India and Hong Kong were among the top ten. In terms of openness, China, Singapore, Hong Kong, Malaysia, Australia and South Korea were among the top ten. Last but not least, China was the most attractive construction market in the world, whilst Japan, Singapore, Malaysia, Indonesia, Hong Kong and Vietnam were also among the top ten. The surveys were conducted between 1993 and 1998. It was noted that increasing pessimism was reflected in the 1993–1998 survey regarding the Asian construction market, reflecting the impact of the Asian 'financial turmoil' on the perception of the respondents towards the prospects of many Asian countries.

Most of the economies covered in this book are 'newly industrialising countries'. The construction sector of these economies was in or near the tip of the inverted bell-shaped pattern of construction spending as share of GNP (Bon and Pietroforte, 1990). Even Japan, that Kerr (2001) called a 'construction state', had a disproportionately high volume of construction in both absolute terms and relative to her GNP when compared with other 'advanced industrial countries' (Bon and Crosthwaite, 2000; Bon and Pietroforte, 1990). Asia has been seen as the region where the construction market will remain large and continue growing at a healthy rate. Before finishing this section, it should be noted (Bon and Crosthwaite, 2000) that the region is also seen as a major exporter of construction materials and services, despite the anticipated growing demand for construction works and thus for materials and services, indicating an increasing trend of intra-regional trade.

1.3.2 Two-tier construction industry and globalization

Construction is becoming an increasingly global business, characterised by trade liberalization and the consequent increased market access offering more opportunities for firms to go global. Raftery *et al.* (1998) identified three rapidly developing trends: (i) larger private sector participation in infrastructure projects; (ii) increasing vertical integration in the packaging of construction projects; and (iii) increased foreign participation in domestic construction as early signs of Asia's increasingly consolidated (if not globalized) $1.124 trillion regional construction market led by China and Japan.[1] Such trends both at domestic and international levels, coupled with the industry's ability to respond positively to new trends, are likely to create a more fluid and open construction market within the Asian economies. Najjir *et al.* (2001) describe the social (also cultural), political, economic, legal and regulatory perspectives of globalization. Based on similar reasoning, Mitrovic (1999) attributes the globalization of markets to deregulation and integration of capital markets, formation of trading blocs and trade globalization, technological and communication improvements, and the appearance of emerging markets.

There has been more evidence of globalization of the construction market in the complex infrastructure construction sector than in the traditional building sector. As Najjir *et al.* (2001) observe, there has been a development of two rather distinct sectors in the construction industry. One is mainly the house-building sector where local contractors dominate, and the other the more complex civil engineering and infrastructure sector where big international contractors have most of the market share. Chiang *et al.* (2001) illustrate this segregation of the construction industry into two sectors with the case of Hong Kong. Local indigenous contractors and leading international contractors have captured most of the market shares in building and complex infrastructure sectors respectively.

As infrastructure projects get more complex and technologically and financially demanding, only the leading international contractors have the necessary resources to undertake them. This sector of the construction industry has to be regional, if not global, in nature. Anecdotal evidence suggests that half of the world's dredging fleet, for instance, was in Hong Kong at one time during the construction of the new airport. Only an international market could provide enough job opportunities for the large contractors to move up the learning curve so that they could reap the economy of scale and to recover the investment in plant and machinery, and in research and development. The situation is not so different to that of the automobile industry (Bjornsson, 1998).

As described by Carrillo and Heavey (2000), countries with large populations, high growth expectations and poor infrastructure provide good potential markets for construction services. Many UK contractors see the need for 'globalization'ing their services, as they see their domestic markets having become mature and saturated. There have been attempts by UK contractors to venture into Eastern Europe through both acquisitions and in a non-acquisitive manner. They were exploring outside their home even while many major UK contractors were being acquired by other international contractors. This follows the trend towards a two-tier market made up of large international contractors and small specialty firms (Ball, 1988; ENR, 1995; Carrillo, 1998). The Asian scene is very similar to the European.

This globalization of the construction industry is also illustrated by Stumpf (2000) in his study covering the UK construction industry between 1981 and 1996. Medium-sized regional building contractors had the poorest survival rates, return on capital employed and staff productivity. They became smaller and their market shares were eroded by their larger counterparts especially during recessions. Also drawing conclusions from earlier works (Seymour, 1987; Ball, 1988; Hillebrandt and Cannon, 1996), Stumpf (2000) finds that the top contractors were able to gain market share because their superior financial strength had responded better to the increasing demand for private finance and complex infrastructure projects. On the other hand, the smaller contractors were able to exploit the niche markets with their specialities. That left the medium-sized contractors somehow stuck in the diseconomy of 'step functions' when they went through periods of over-capacity as they crossed a step function 'threshold' (Hillebrandt and Cannon, 1995).

Based on ENR surveys, Male and Mitrovic (1999) find that in 1994 and 1995, Japanese and European contractors were major global competitors. In the early 1980s, it was the US contractors who were the most conspicuous in the international construction market (Seymour, 1987). However, since the late 1980s, Japanese contractors have replaced the Americans to play a leading role in East Asia, having benefited from Japanese manufacturing and real estate investment in many host Asian countries. There was also a growing presence of Chinese contractors in international construction services. In Singapore, local contractors have faced the industry's physical constraints and stiff competition from their foreign counterparts (mainly Japanese and Korean) who dominate the large and complex sector of the industry (Ofori and Chan, 2000). Many other Asian economies such as Hong Kong have faced a similar situation. Trade in international construction services is thus biased towards that between developed and developing countries.

Yet, there is some differentiation in the services offered by contractors from different nations. For example, contractors from the US used to excel at the supply of proprietary technologies (Drewer, 1990) and power and process plant construction services (Seymour, 1987), and Korean contractors at labour supply (Seymour, 1987; Drewer, 1990; Porter, 1990). However, as Porter (1990) argues strategy based on low-cost disciplined labour is not sustainable for providing financial and technical advantages, because the labour advantage is easily, and indeed was, imitated and nullified by multinational contractors sourcing cheap labour pools locally in the host countries. The implication was that Japanese contractors were expected to dominate the Asian infrastructure construction market for some time to come. They have a unique edge in human, technical and financial resources, which Seymour (1987) describes as the dominating success factor for UK contractors operating in the Middle East region in the early 1980s. This was the advantage they had over indigenous Middle East contractors or other international contractors working in the region. Such strengths brought them the 'reputation', regarded by most of the contractors and their clients alike, who responded to Seymour's survey, as a success factor of securing work in the Middle East.

Figure 1.1 illustrates the two-tier construction sector in the AsiaConstruct economies, and the dominant players in each of the tiers.

Weddikkara and Devapriya (2001) identify some global trends that give rise to institutional and policy reforms. Their study is based on the Sri Lankan construction industry. The reforms that they suggest include corporate development (increasing the technological and managerial capabilities of the contracting firms), institutional reforms and capacity building, financial resources development, and material and manpower development. These suggestions respond largely to the social-economic, political and legal sources of globalization as described by Najjir *et al.* (2001). The subsequent chapters in this book will provide the basic framework upon which further studies on the formulation and implementation of institutional reforms and corporate strategies could be made. Take the case of Hong Kong. Having been the freest economy in the world, Hong Kong has not only been reacting to, but indeed leading, many of the globalization trends such as private finance initiative, 'fast tracking' projects, the facilitation of free trade and the tariff concession. Hong Kong is better prepared perhaps both to lead and to respond more swiftly to these trends than perhaps any other country in the region. Firms doing or contemplating construction or its related business in Hong Kong have to bear this in mind when designing their business strategies.

	Indonesia*, The Philippines*, Sri Lanka*, Vietnam*	Australia, China, Malaysia, Hong Kong, India, Singapore	Japan, South Korea
Complex building and infrastructure construction**	Contractors from Japan and other AICs	Contractors from Japan and other AICs	Indigenous local contractors + Other AIC contractors
Traditional building and construction	Indigenous local contractors + Contractors from other NICs	Indigenous local contractors	Indigenous local contractors

Figure 1.1 The Asian two-tier construction market and the main participants.

Notes

* Countries importing capital and/or technology for their traditional building and construction more often than others.

** Includes manufacturing/process plants, power and design, and tunnels and bridges.

1.3.3 Construction investment and productivity

Productivity

A process of privatization, or rather re-privatization, in the construction industry is under way in China and India. Profit is the measure of success, and calls for a three-pronged approach combining added value, cost control and productivity improvement. Since productivity gains involve obtaining more output for the same input an element of productivity gains is also inherent in respect of value-added and cost control. In effect, adding value to the product is a means of improving the productivity of capital.

It is observed that value-added per employee in the construction industry grew at a phenomenal rate in China (8.5 per cent), Indonesia, Vietnam and Australia in the period between 1995 and 2000 as shown in Table 1.4. At the same time, Hong Kong, the Philippines and Singapore had seen negative average value-added growth, largely an outcome of the Asian financial crisis hitting their economies the hardest. In certain ways, the productivity trend in some countries (China, India, Vietnam) appeared to be consistent with GDP growth, while the sharp downturn in the Philippines, Singapore and Hong Kong reflected the bitter truth of a construction industry still reeling even while the general economy was picking up. Construction productivity had been rising in less developed economies and sliding in developed counterparts like Japan, which saw construction value-added per employee slump to −6.5 per cent. This is perhaps not unexpected, as the construction industry had been deployed by the Japanese government in the last decade as an economic stimulant and to generate employment opportunities. Construction productivity however is also a factor of the level of technology used and understandably construction productivity in the developing countries of Asia is at lower levels.

Construction investment

Construction is a multibillion dollar sector, with a work force of 1.2 billion serving the region. The extraordinary near-decade-long boom in construction had come to a temporary halt with an 11 per cent decline over 1998–2001. The biggest and most attractive projects for private investment have largely been done: namely new Airports in Hong Kong, Malaysia and South Korea, and construction related to the

Table 1.4 Value-added per employee and GDP growth between 1995 and 2001

Countries	*Average % GDP growth*	*Average value-added per employee*
China	8.51	8.5 (1995–1997)
Hong Kong	2.73	–
India	6.7	3.91
Malaysia	4.6	1.86
Philippines	3.5	−3.9
Singapore	5.8	−3.3
Vietnam	6.5	7.6

Sydney Olympics for example. Most infrastructure construction programmes of most countries launched in the late 1990s, have been completed and new government projects have not been brought forward fast enough to replace them. Table 1.5 shows, in 1999, that Japan led the region by a huge margin (US$4,910) in terms of construction investment per capita with Singapore, South Korea and Hong Kong following each other closely. As such, share of construction in the GDP ranged between 3.6 per cent in the Philippines and 11.5 per cent in Korea. Sri Lanka shows a low figure of 1.6 per cent which probably does not account for the large share of informal construction work in the country. This should be true with other countries such as Indonesia, Vietnam, India and the Philippines where a bulk of informal construction work remains undocumented. Sri Lanka nonetheless anticipates around 7–9 per cent contribution to GDP from the construction sector during the next decade, deemed possible with the new economic measures to boost the sluggish housing sector in the country.

Construction investment in the region saw uninterrupted growth until the Asian financial crisis hit the region. The building and construction industries in the Asia-Pacific region seem to be recovering and (with the exception of Indonesia) appear poised for growth from 2000. Real growth in investment remained negative even during the recovery phase (1999–2000) in Singapore, Indonesia and India. The Philippines, South Korea and Vietnam were the countries expected to produce the highest building and construction industry growth rates in the period 2000–2002. The industries in Hong Kong, Malaysia and Thailand were also recovering but no upturn was apparent. In China, construction will be boosted by the increasing investment in civil engineering, averaging 720 billion yuan a year.

The construction sector slipped into a severe and protracted downturn in Indonesia as a consequence of the 1997 economic crisis. Building activity remained subdued and was likely to remain so in the short term, with very few new projects. The value of total completions fell to decade-low levels in 1998 and 1999, and fell again in 2000. The expected value of total completions in 2002 was 34 per cent below the pre-crisis levels, and new floor space was only about a third of the average level of the mid-1990s.

Table 1.5 Construction industry in 1999

Countries	*GDP current prices (US$)*	*Construction share in GDP*	*Construction volume in 1999*	*Population (million)*	*Construction spending per capita (US$)*
Hong Kong	166	8.8	8.9	6.6	1,348
India	352.4	6.7	50.4	1,016.0	49
Indonesia	95.5		12.6	204.4	61.7
Japan	4,335.2	10.5	622.1	126.7	4,910
Korea	422.4	11.5	70.6	46.9	1,505
Malaysia	79.0	3.6	11.4	22.7	502
Philippines	76.7	5.3	2.5	76.8	32
Singapore	84.9	7.6	6.9	3.2	2,156
Sri Lanka	12.2	1.6	2.0	19	105
Vietnam	28.7	7.5	n.a.	76.9	n.a.

In the Philippines the construction sector grew by an average of 10 per cent per year over the past 2 years of 1999–2001 since building activity plunged to the mid-1990s levels. The Singapore construction sector likewise reached a seven-year low in 1999 at only 43 per cent of its 1997 peak but took a turnaround of 27 per cent driven by a sharp increase in residential demand and return of consumer confidence and robust economic growth. However, even if this growth persisted, total construction value was expected to be 18 per cent below its 1997 peak overall and residential building activity was expected to be 35 per cent below the 1997 peak. Civil engineering construction contracts rose to S$4.9 billion with new projects that included several multi-million dollar rail transit and expressway projects, two desalination plants and the third Changi Airport Terminal.

In South Korea, the growth of the construction sector that continued to climb until 1997 plunged by 10.4 per cent in 1998 (2.3 per cent in 1997). The amazing turnaround in Korea's economy in 1999, coupled with deregulation of the property market, had resulted in an estimated 64 per cent rebound in the value of construction orders received by the residential building sector. This rebound had offset continued steep declines in the non-residential sector. The civil engineering sector grew by over 16 per cent in 2000 boosted mainly by a major high-speed rail project along the Seoul–Pusan corridor and several port projects. Again, the total value of construction was far below the 1997 peak.

Table 1.6 indicates the region-wide fall in construction investment during the Asian financial crisis. In general, Hong Kong, India, Indonesia, Japan and Singapore had experienced a slow return to the investments of 1997 levels. Malaysia, the Philippines and Vietnam have shown strong resilience to the crisis and the recovery had been quicker. However, the results should be viewed with caution as some countries like Malaysia saw their investments fall again in 2000 despite growth in exports and other optimism over economic recovery.

Table 1.6 Construction investment in the period 1995–2000

	Construction investment (current price) in billions US$ in 1997			*Construction Investment change YoY*					
	Total	*Private sector*	*Public sector*	*1995*	*1996*	*1997*	*1998*	*1999*	*2000*
Australia	38.4	27.4	11.0						
China	183.2	60.6	122.6	16.5	10.0				
Hong Kong	16.9	7.3	5.4	0.2	9.1	3.2	8.6	4.8	13.7
India	–	–	–	–	–	–23.0	–40.0	–4.8	
Indonesia	25.1	11.2	13.9	58.0	2.0			–25	
Japan	574.4	322.7	251.6	0.2	3.8	–10.9	–0.6		
Korea	72.3	44.0	28.3	19.3	19.3	8.1	34.0		
Malaysia	27.2	17.8	9.4	17.3	14.2	11.0	9.0	44.1	–8.8
Philippines	3.9	2.5	1.5	7.9	15.9	15.8	–	2.5	
Singapore	10.9	5.5	5.5	7.0	26.2	21.2	–	–13.1	–8.3
Sri Lanka	1.5	1.21	0.31	18.3	–	41.6	8	10	–1
Vietnam	2.8	0.8	2.0	23.0	55.3	16.6	18.4		

1.3.4 Construction quality, buildability and private sector involvement

In Asia, as globally, the impetus for adopting ISO standards will affect the industry. The standards relating to construction are ISO 9000 (Quality assurance and Business standards) and ISO 14000 (environmental management). They have, perhaps accidentally but indeed increasingly, become trade barriers in different countries. Companies seeking opportunities must therefore adhere to such regulations if they wish to compete successfully in international markets. There were more than 250 organisations with ISO 14001 accreditation in Malaysia. Of that figure, however, not one was from the construction sector where the ISO 14001 had been obtained on a voluntary basis.

In Vietnam quality factors in construction had received considerable attention over the years. The Ministry of Construction issued a guidance note 'Implementing Rules for Enterprises to fulfil ISO 9000 Serial Standards' to boost enterprises in applying GB/TI9000–ISO 9000, and over 200 enterprises were issued ISO 9000 certificates by 1998. The China National Accreditation Council (abbreviated as 'CNACR') approves and authorizes third-party quality system registration bodies. The number of large firms adopting the ISO 9000 quality management system has been increasing. About 836 construction companies in Hong Kong and 30 in Indonesia have been issued with ISO 9000 QA certificates. Several certification bodies are fully operational there. In India local standards are fairly elaborate and are gradually being modified to conform to ISO Standards. Apart from this, several scientific and Research and Development Institutions, CSIR, CBIR, IITs and material testing laboratories also provide necessary input support.

The level of quality awareness in the construction industry is not very high in the countries studied. However people are gradually becoming aware of the necessity of observing the statutes and improving work practices to meet the norms. Malaysia and Singapore with QLASSIC and CONQUAS systems of quality assurance have been successfully running an indigenous quality assurance mechanism. The industry average CONQUAS score in Singapore improved steadily from 67.9 in 1989 to 77.8 in 2000. A target median CONQUAS score was set, beginning with 79 by 2005 and reaching 82 by 2010. QLASSIC is a method for measuring construction quality and for benchmarking the level of quality achieved in a construction project. Singapore also monitors construction quality with such objective standards as the buildability factor, which had been enhanced by the slow but wider recognition of the design and build approach. The average buildability score in Singapore had been 78.75 between the years 1997 and 2000.

It is observed that the construction industry in Asia is changing the way it executes construction projects in response to customer demands. Traditional design *then* build projects are losing favour to the design *and* build approach which is essentially a one-stop shop approach to design and construction services. Although this poses challenges to firms incapable of delivering turnkey projects, Asian countries have shown a great deal of enthusiasm for developing projects on a Build-Operate-Transfer (BOT) basis. There is some growing interest particularly in the power and transportation sectors. A BOT project is usually a contract between the State and a consortium for the construction of an infrastructure project. Potential subjects are roads, bridges and tunnels, power-generation stations, waste-treatment facilities, seaports, railway systems, airports and telecommunications systems, etc. Vietnam's BOT contracts are 'regulated' (it is difficult to regulate non-existent enterprises) in greater detail pursuant to Decree 87/CP dated 23, November 1993, and in Circular 333/UB/LXT dated 28, February 1994. As of January 2001, less than ten BOT projects had been licensed, whereas only one was in full operation indicating lack of consistency in pursuing privatization of projects.

Private sector involvement

As liberalization and privatization policies had swept infrastructure activities in Asian countries throughout the 1990s, the private sector became an important financier and long-term operator. India, Vietnam and several other countries continue to see BOT projects as a form of private sector involvement. These countries are hoping that private sector involvement in domestic infrastructure projects will prime the pump for direct investments from abroad. Japan, too, is beginning to achieve efficient infrastructure development by involving the private sector. One case in point is the fact that it was working to adopt a Private Finance Initiative system based on the British model. In addition to government funds, infrastructure projects in China also benefit from private sector funds (i.e. BOT projects, etc.).

Government spending is not allowed to increase faster than the rate of economic growth. This crimps the flow of government funds when the economy is sluggish, for which reason the government actively seeks to tap into private sector funds. Recent financial turmoil, however, had had an adverse impact upon such efforts. India with a major shift in investment and trade policies, is succeeding in attracting investment from foreign firms. For example, private sector firms are involved in the development of most power plants. They had invested a combined total of approximately \$33 billion, which had been used to construct power plants of some 32 giga watts of capacity. US firms were involved as partners in many of these projects, and had invested approximately US\$5.7 billion. Likewise, in Indonesia the private sector seems considerably attracted to the range of development projects such as power plants, telecommunications infrastructure, transport and water systems including development of tourism facilities.

By adopting a Private Finance Initiative (PFI) system the Japanese government aimed to achieve efficient infrastructure development and lighten the government's financial burden in this area. The PFI system was also expected to play a role in revitalizing the Japanese economy. Likewise in Korea, the private sector had been involved in a number of infrastructure projects, including the construction of a freight terminal and the highway for Inchon new international airport, the Busan New Harbor, the Kyoung-In Canal, and an LNG power plant. To successfully involve the private sector in infrastructure development in Malaysia, close cooperation between the public and private sectors was needed. The Malaysian government provided loans to private sector firms engaged in infrastructure projects, and was expected to continue to promote privatization in the future. The private sector in the Philippines was also being called upon to get involved (through BOT projects, etc.) in infrastructure development.

At present, Singapore does not practise BOT, Build-Operate-Own (BOO) nor BLT for infrastructure projects. However, private sector involvement in infrastructure projects had been increasing due to the corporatization of several major public agencies. In Sri Lanka, most new investments in power generation and telecommunications projects were expected to be offered to the private sector on a BOO/BOT basis. Investment in the Colombo port by the government with the assistance of OECF and private sector investment on a BOT basis was expected to expand the cargo handling capacity of the Colombo port.

From this overview the following broad trends of privatization are identified and summarized as:

- A large proportion of the Asian market involves infrastructure privatization – supplementing public funds with private capital as a way to develop infrastructure cheaper and faster.
- By contrast foreign contractors have made few inroads to the Asian market.
- Asia provides a leading construction industry in the areas of project design, management, environmental controls, safety and construction quality.
- Construction is a capital-intensive industry. Large up-front investment is required to begin and execute projects, and success or failure rests on the ability to arrange a suitable finance mechanism. JV, PP partnering and other creative funding approaches are common in order to share the risks involved.

1.3.5 Research and development

Countries in Asia undertake a negligible amount of research and development (R&D) activities when compared with their international competitors, such as Sweden, which invests up to 2.8 per cent of turnover in business research at the time of writing. Although high-tech industries have been a priority for the governments in recent years in Hong Kong, according to data from the OECD, Hong Kong's expenditure on scientific and technological research and development amounted to only 0.25 per cent of its GDP in 1998, and ranked fortieth among forty-seven major countries and regions. Serious competitors of Hong Kong, such as Taiwan and Singapore, ranked eleventh and fourteenth respectively, while China ranked thirtieth and India, thirty-second.

Asian countries have however shown a considerable interest in fostering competitiveness, accelerating innovation and encouraging appropriate technology transfer in the construction industry. The transfer of innovative technologies, construction services, business practices and government policies has increasingly become a two-way flow between Asian countries. The characteristic fragmentation, diversity and fierce competition in the construction industry combined have made R&D important, with the government sector playing the major role accounting for almost three-quarters of total R&D expenditure in Asia. Major contractors in Japan are also active in R&D work principally in such areas as structural safety, fire, building materials, intelligent transport and energy systems.

The role of academia in different countries in Asia, despite their huge contribution in the R&D over the years, has been largely neglected and undermined. Little interaction has been observed between industry and academia in terms of exchange of findings and their incorporation in practice and the academic research is deemed largely 'curiosity driven'. Attempts to bridge the gap, however, between industry and academia were evident in Hong Kong and Singapore.

Construction-related R&D suffers from various problems that minimize its impact on the industry. In a fiercely competitive environment, with thin profit margin, individual firms especially the smaller ones cannot afford to conduct R&D or pay added regulatory costs of introducing new technologies. Consequently R&D costs to the industry have been consistently low. In 1999, the industry in Asia spent an average of only 0.6 per cent of GDP on research. The Asian construction industry faces the risk of losing its competitive edge if it cannot make its R&D efforts more effective as compared to European and North American construction industries.

Information-technology enhancing productivity, competitiveness

Information technology is increasingly transforming the way business is done and has been an important tool for improving competitiveness and efficiency (in areas such as design, project management, documentation, procurement, electronic commerce and building operation). A new trend of smaller firms embracing information technology for diverse activities was observed in many countries in Asia and not just a few larger players or those in consulting professions such as architects, engineers and quantity surveyors. This was a marked departure from the trend that hitherto prevailed in the industry. While technology transfer had been instrumental in reducing project overheads and communication systems between distant locations (Moavenzadeh, 1978), the wider diffusion of IT within the industry remained a challenge even though countries had been seen gearing up to the IT revolution. Japan for example, established the Construction CALS/EC Action Program in June 1997 aiming at the use of electronic procurement for all projects, contracting through EDI (Electronic Data Interchange), online submission of applications and bid notification for all public projects, creation of a comprehensive database of project information, and information linking and integration using GIS. Closely following suit and almost at par with Japan in this sector were Hong Kong and Singapore.

1.3.6 Safety

The construction safety record in the region had been mixed over the years with accident frequencies ranging between 0.075 per million man-hours worked in India and 2.7 in Singapore (based on the available data). Hong Kong had one of the worst safety records in the developed world. In 2000, reported accidents ran at a rate of 143 per thousand workers per year, of which 29 were fatal. In Singapore from January to end of August 2001, there were 17 fatal accidents. For the same period in 2000, there were 40 fatal accidents, and in 1999, 35 cases. While there appears to be a marked recent reduction in the number of accidents, it may be questioned by attributing the improved safety to the slowdown in the industry. During the economic crisis in 1997, 45 fatal cases were recorded in Singapore for the same period, 62 per cent higher than the figure for 2001. This reduction in the number of fatal cases was probably the result of a genuine improvement in the standard of safety performance in the industry. It can also be attributed to the presence of increasing numbers of skilled workers in the industry. The legislative framework within which safety management operates and the legal processes stemming from unsafe incidents are other important dimensions that have largely been overlooked in different countries.

Safety performance as a competitive advantage

One of the benefits of achieving a good safety performance record is the greater likelihood of securing new contracts. Developers and contractors today, being more aware of the importance of developing a safe project, scrutinize the safety records of contractors who bid for their jobs. Contractors have come to realise that a good track record of safety performance can be a competitive advantage for securing new business.

New legislation

To further enhance safety at work, governments in Asia have introduced new legislation. Two new items of legislation were introduced in Singapore in 2001. They were the Factories (Singapore Standards and Codes of Practice) Order 2001 and the Factories (Safety Training Courses) Order 2001. The first Order made it mandatory for the industry to comply with 15 Codes of Practice (CP) and Standards, including CP for Scaffolds, Formwork, the Safe Use of Tower Cranes, and Mobile Cranes. The Factories (Safety Training Courses) Order 2001 made the attendance on relevant safety courses mandatory for various categories of personnel in the industry, such as, project managers, formwork supervisors and forklift drivers. It was believed that safety training and the competence of the persons working on site are critical to the industry's bid to improve safety at the workplace. Aiming to make safety in the work place a matter of legal policy, the Chinese government was drafting a national work safety law.

Skilled labour

Until a few years ago, the issue of safety in the construction industry was regarded as not important. Concern began, however, due to multinational companies entering the market and their uncompromising adherence to the existing norms in countries such as India, Sri Lanka and Vietnam. The construction safety issue in Asia is complex however due to the level of disparity in the economy among countries. In 2001, about 90 per cent of foreign construction workers were certified skilled in Singapore, whereas in India and Indonesia, a large proportion of construction workers came from a rural base and were thus not well trained. To add to this problem, a large proportion of construction workers lacked technical knowledge. In China for example, the millions of small contractors who were previously associated with traditional work units lack knowledge on safety regulations adopted in the increasingly privatized construction industry.

1.4 Regional cooperation and business opportunity

East Asia had long been considered an area lacking in regional cooperation. Since the end of the Cold War, however, a sense of regional partnership began to emerge in Asia notably in East Asia (Fukushima, 2000) and in the broader Asia-Pacific region. Asian integration had been stimulated by economic globalization, and the establishment and expansion of regional organizations such as Asia-Pacific Economic Cooperation

(APEC), ASEAN's Regional Forum (ARF) and ASEAN+3, (which includes Japan, Korea and China besides the ASEAN countries).

Whether this trend towards 'new regionalism' in East Asia grows stronger to enhance regional cooperation and integration and eventually perhaps contribute to the emergence of a common construction market remains to be seen. It is clear that in East Asia, frameworks of regional cooperation had contributed to stronger connections between countries of the region. These frameworks, however, had not yet succeeded in providing the type of dynamism that has led Europe towards closer integration leading to the promotion of further integration. First and foremost, among the reasons that had contributed to preventing the development of this dynamism is the religious, cultural and historical diversity of East Asia, factors reflected in the social, political and economic systems of each country. Although some countries of the region had achieved a broad level of social and economic maturity, others were still in the position of prioritizing catch-up with the industrialized nations as their major national goals. As such, there was a gap in the level of economic development among these countries, which means they responded in different ways, among other things, to global environmental problems. In other words, some diversity existed in the attitudes of these countries when faced with such common problems.

Asia's regional cooperation in construction broadly rests on:

- encouragement for institutes and enterprises in the promotion of international corporations;
- encouragement for institutes and enterprises in mutual participation in professional associations, in the Asian region and overseas; and
- diversification of cooperation through science research, technology transfer, training, conferences, professional and informational transfer.

By the late 1990s, penetration into domestic services sectors by foreign companies became commonplace for large construction projects and financial services in Asia. The Asian economic crisis that began in fiscal 1997 had a largely negative impact on this however. Japan, the leading construction service exporter, was especially affected, as orders placed with Japanese construction companies in the Asian region fell sharply as a result of the crisis. The volume of orders received in fiscal 1999 fell to less than half of that for fiscal 1996. However, general recovery in the Asian economies after that time helped restore those order volumes to just under ¥700 billion for fiscal 2000.

There are four types of service trades according to the General Agreement on Trade in Services (GATS) which came into being as part of the Uruguay Round agreement which brought GATS, as well as General Agreement on Tariffs and Trade (GATT), into the newly founded World Trade Organization (WTO). These are:

1. cross-border supply where services can be produced in one country and delivered to another;
2. consumption abroad by domestic residents, such as tourism;
3. commercial presence – activities provided to foreign nationals by foreign branches and subsidiaries of domestic firms; and
4. relocation of natural persons.

Most countries under study however as members of the WTO, had adopted and continued to pursue an import liberalization policy under which import taxes and duties for goods and products covered by the GATT have been either reduced or removed. Commitments under the WTO GATS cover several sectors and subsectors. In the Philippines for example, at the time of writing, the commitments for 45 sectors/subsectors had been firmed up under specific schedules particularly for major sectors like communications, financial, transportation and tourism/travel-related services. Foreign direct investment and technical collaborations approved during August 1991 to March 2001 in consultancy services accounted for about 25 per cent in India indicating greater fluidity of the market. In Hong Kong and Singapore, there were no limitations on market access and national treatment under cross-border supply, consumption abroad and commercial presence at the time of writing. There was no restriction on ownership. Foreign construction firms, like local firms, needed only to register with the Registry of Companies and Businesses. There was no specific further requirement for construction firms to meet for private sector projects. In Australia among the major barriers, government required notification of (but normally raises no objections to) investment proposals by foreign interests above certain notification thresholds, including acquisitions of substantial interests in existing Australian businesses with assets of Aus$5 million or more (Aus$3 million for rural properties) and new businesses involving an investment of Aus$10 million or more.

Major differences persisted in existing laws, rules and regulations prescribing the limits of foreign equity participation in construction contracting as well as the type of constructor's license which may be issued by respective governments. In the Philippines locally funded public construction would have to limit foreign equity participation to 25 per cent as opposed to BOT projects that enjoy a 100 per cent equity limit. In Sri Lanka although foreign companies were allowed to bid for projects in local competitive bidding, the construction volume concerned was deliberately limited in order to protect the domestic construction services. Therefore foreign competition is minimized. The government however is actively encouraging private sector investment in infrastructure on the basis of BOO and BOT particularly in power, ports, water sanitation, transportation (highways and rail) and IT sectors. Likewise introduction of the new Enterprise Law in Vietnam had led to the removal of restrictions on foreign lenders to accept mortgages of land use rights in return for loans including introduction of new tax incentives for foreign investors. These include reduction in the profit remittance tax rates, allowing parties to business cooperation contracts (BCCs) to carry forward tax losses for 5 successive years, extension of the five-year import duty exemption for raw materials used for investment projects in especially encouraged activities, a 20 per cent reduction in corporate income tax and a 3 per cent profit remittance tax rate for overseas, etc. Malaysia implemented a liberal policy on equity participation which allowed up to 100 per cent foreign ownership in projects which export 80 per cent or more of their production. For projects which export less than 80 per cent of their production (down to a minimum of 50 per cent), up to 100 per cent foreign equity ownership was allowed when:

- foreign investment in fixed assets (excluding land) is at least RM 50 million or when the project has at least 50 per cent value-added; and
- the company's products do not compete with products presently being manufactured for the domestic market.

1.4.1 Outlook for construction sector

In 2001, the outlook for the construction sector in Asia looked positive. In Hong Kong, massive infrastructure and building projects had been planned and construction tenders were to be invited. Examples include works arising from the Railway Development Strategy, the redevelopment of the Kai Tak Airport and its adjacent areas, the construction of the Lantau cable car, and the construction of the Disneyland Theme Park development. In the private sector, the property market was expected to recover in tandem with the general economy.

In India approximately 45 per cent of the investment outlay in the country's ninth five-year plan was expected to be in the physical infrastructure sector. Over the next decade an average of estimated INR 1800 billion (US$39 billion) investment per annum, is needed in the infrastructure sector to support an average annual GDP growth rate of 7.4 per cent over the next 15 years targeted by the Planning Commission. Projects in sectors like power, ports, civil aviation, roads, telecommunications, mining, chemical and fertilizer, etc. offered investment opportunities of around US$350 billion for 2001–2010. By estimates of 2001, inclusive of the investment backlog in areas such as urban services, the investment needed in the infrastructure sector until 2010 to support an annual GDP growth rate in the range of 7 to 8 per cent is around INR 15,000 billion (US$429 billion). The long-term dimension of this massive investment programme deserves special mention as does also the robust growth potential demonstrated by several infrastructure sectors. Power generation, for instance, had increased annually at the rate of 7–8 per cent both over the very long term and the first 4 years of the reform phase. Electricity consumption per capita was however less than one-tenth the global average. Basic telephone connections had grown annually at 10–22 per cent over the 1990s; but penetration ratio was still low at 1.5 per 100 population as compared to the global average of 10 per 100. The need to accelerate infrastructure development therefore presents enormous opportunities for the investor.

In Indonesia since 1997 the construction industry sector generally experienced tight conditions. It was estimated that the contraction will gradually stop in 2001, provided that political stability was achieved and maintained. The government had introduced Law no. 18 for the construction industry to include the involvement of the construction industry community in the management of the construction industry. This was expected to strengthen the capability of local consultants and contractors as well as to provide a comprehensive system to back up this policy. The government of Indonesia was also taking steps towards decentralization with the implementation of Law no. 22/99 on regional autonomy and no. 25/99 on central and local government budget/revenues. With this, most works will be managed by the regional manager. Although the government has no clear and agreed strategy and implementation plan for decentralization, many of the regions appear ready for more autonomy, and show a fairly practical attitude towards issues such as local reorganization and the absorption of central government personnel . The transition period for decentralization, which will take about 3–5 years to complete, is expected to strengthen the local construction industry further.

In Japan, with the beginning of the regional economic crisis, the Japanese economy experienced a tremendous negative impact due to its close ties with Asia, and this was one of the contributing factors to Japan's descent into a severe recession in

1997. To stimulate the economy, the government had increased investment in public construction. This is because public works projects were actively promoted in the successive economic policies enacted to achieve economic recovery. Construction investment was in a phase of short-term contraction. Private investment was expected to be stimulated as economic recovery progresses, but massive improvements could be anticipated even over the medium to long term due to the negative effects of fiscal budget-tightening, the ageing of the population and population decrease. The recovery of the Asian economies is to result in a gradual recovery of construction orders received by Japanese construction companies in Asia.

The Korean economy was heading for a full recovery with a sharp upturn in the years following the financial crisis. Activity in the construction industry had been highly cyclical. The construction activities have begun to emerge from a cyclical downturn, which started at the beginning of 1999. It was believed that the construction economy had already been through the worst. Infrastructure became the top priority policy issue in mid-1990s, which was reflected in the over 20 per cent growth rate of infrastructure-related expenditure. In 1997, almost 40 trillion won (approximately 9 per cent of GDP) was invested in the infrastructure, the burden of which was shared by the central government (25 per cent), local governments (30 per cent), public corporations (15 per cent), and the private sector (30 per cent). Even under the economic crisis, consensus had emerged for a cut in infrastructure expenditure to only a minimum amount. It was nonetheless planned to invest approximately 100 trillion won between the years 1999 and 2003.

Government policy initiatives in Malaysia regarding the financial system and banking structure would help make the country and its economy more resilient to outside shocks. While the construction industry may not achieve pre-crisis boom conditions in the near future, a steady improvement throughout the sector can be expected as demand picks up. With the onset of globalization, the national construction sector has to prepare to become more resilient and competitive in the face of the greater challenges expected from world-class competitors because of the more open market.

In the Philippines the attainment of positive growth in the country's macroeconomic indicators in 1999 and for the first six months of 2000 (less than 2 years after the Asian financial crisis) affirms the resilience and stability of the Philippine economy. The achievement had also brought renewed optimism that the targets set for the economy will raise overall growth to pre-crisis levels. Although investments and the construction market are still down, the outlook in the near future is bright. Private sector demand for construction is perking up and government infrastructure development plans were expected to push through especially in the countryside. The bulk of public sector demand will come from transportation and energy projects under the government's Public Infrastructure Program; low-cost housing under the National Shelter Program; social infrastructure; and local government projects. Private sector participation in these projects was expected to continue to be encouraged under the BOT schemes. Private demand for residential and non-residential construction was expected to rise in the growth areas outside of Metro Manila.

In the longer term, the outlook for construction in Singapore remains bright. With the proposal to plan for a larger population of 5.5 million, the Singapore construction industry can look forward to a total of S$200 billion worth of projects over the next 20–30 years. Future buildings will be more environment friendly, fitted with intelligent and energy efficient features and electronically connected. To meet the need of housing a larger population, residential buildings are going to be taller. Major civil engineering projects include expansion of the surface road networks and another 450-kilometre long MRT/LRT lines were to continue to put in place an ambitious rail system network to meet the rising transport needs in the future. There will be more process engineering buildings such as waste-water treatment plants and bio-engineering plants. Underground construction to free up land for other uses is to be promoted. Projects in the pipeline included the Common Services Tunnel and the South Tunnel of the Deep Tunnel Sewerage System.

Sri Lanka anticipated a contribution of 7–8 per cent to GDP from the construction sector during the next decade and believes that this can be achieved with its strengths being coupled with effective leadership and continuous dialogue with policy makers, planners and other stakeholders of the construction industry. Even though the economic crisis in East Asia had little impact on the Sri Lankan economy, the terrorist domestic crisis has taken a heavy toll on the construction industry. In the long run the construction industry is expected to grow steadily and systematically while being a significant and integral part of the economy.

Despite economic difficulties, the national reconstruction programme in Vietnam was still underway to overcome the backward infrastructure, and as a part of the construction demand-stimulus policy. To overcome the economic slowdown, major strategies in the construction sector are to improve the funding and operations management, and to diversify activities. While Vietnam was aware of the need for self-support, on the one hand, it was also keen on the development of international cooperation for mutual interests. National development required the construction industry to develop faster, providing opportunities for more foreign participation. Official plans estimated that transport and communication required at least US$23 billion during 2000–2010. The power industry was expected to have a need for about US$1.2 billion over the next 5 years, since the time of writing, while development of the oil and gas industry will require US$5–7 billion. In addition, the Vietnamese government had taken an official stand encouraging the import of equipment and machinery. The government had decided on an investment of 51,500 billion Vietnamese dong (about 35 million dollars) and to recruit 180,000 workers to develop the construction materials industry to enable it to meet domestic demand and provide a surplus for export in 2010.

The overall construction market in China was around US$340 billion a year. About 15 per cent required some form of foreign expertise, providing more business opportunities for overseas companies suggesting some US$50 billion worth of construction open to foreign participation every year. China's WTO accession, the Beijing 2008 Olympics, the western region development strategy and the housing reform in the mainland was expected to be crucial in boosting the construction market in the coming years. The whole road construction and improvement programme planned for the next 10 years in China was expected to be around 700 billion yuan (US$85 billion) to 800 billion yuan (about US$100 billion). The report also says the housing reform in 1998 has triggered a boom in China's real estate sector. In the period 2002–2006, China will build a total of 5.7 billion square metres of housing, up 15 per cent from that of 1997–2001.

Construction activity in Australia was likely to soar again (in real terms) in the near term and exceed Aus$10.8 billion, before peaking in 2006 at a level well above the previous peak of the late 1990s with engineering construction activity at Aus$20.5 billion. Road, rail and harbour construction is set to boom over the next 5 years despite the fact that the Aus$4.5 billion speed rail is unlikely to proceed. The combined investment in several infrastructure projects such as Melbourne Deocklands, SW Metro Railway, VicRoads, and several other urban redevelopment projects spanning 2001–2010 will probably exceed Aus$7 billion. In turn these projects will give the economy a huge boost.

1.5 Construction industry in asia: challenges ahead

1.5.1 Open access to markets

In wrestling with the reality that globalization has created, the Asian governments have moved, though unevenly, through privatization and the introduction of market-based incentives to foster trade liberalization. Governments have continued to offer open access to overseas markets. These policies include reducing trade barriers, standard procurement policies, currency restriction, procurement, barriers to investment, tax exemption, etc. Currency restrictions have ramifications on the transfer of receipts or payments of foreign exchange manifested through advance deposit requirements, capital transfer and profit repatriation limits, procedural and processing delays, and the application of multiple exchange rates to foreign and domestic producers. Foreign capital invested in India for example, for industrial projects with the approval of the Government of India is now allowed to be repatriated from India, along with capital appreciation, if any, provided approval is first obtained from the Reserve Bank of India. Dividend and profits are now allowed to be remitted with much lesser controls. Under the rules and regulations governing licensing and accreditation of contractors in the Philippines, for example, foreign entities intending to engage in any construction activity in the country were required to obtain a special license from PCAB even before bidding. This requirement is waived under any of the following conditions: (i) the project is foreign-financed or internationally funded and international bidding is required; (ii) the project is procured under the terms and conditions of the bilateral agreement between the Philippine government and foreign or international financing institution; or (iii) the project is to be undertaken under the BOT Law. In addition, there were several other conditions for a foreign company to qualify. Even under the adverse conditions resulting from stringent policy barriers, as of 25 April 2001, the government of the Philippines issued special licenses to 54 foreign contractors (representing 22 nationalities composed of Americans, Australians, British, Chinese, Dutch, Japanese, Koreans, among others) planning to bid for government projects. Barriers to investment, including performance requirements, limits on equity participation, and restrictions on repatriation of profits, caused foreign investors to shy away. While there are no rules on investment in the GATT, the challenge is for the countries to revise policies so as to attract more foreign capital for the obvious impact it exerts on lifting a country's economy and its construction industry.

1.5.2 Research and development strategy

Devising a coherent strategy that coordinates the disparate efforts of academia, industry and government departments is a daunting task. The countries should lower barriers that inhibit wider incorporation of new technologies. Unfocused and uncoordinated efforts in R&D would only make matters worse. The challenge is to integrate where possible, through exchange of technical know-how and research among countries such that a network evolves. To train and equip smaller firms is another challenge.

1.5.3 Infrastructure investment and privatization

Construction firms in Asia were hesitant to respond to privatization initiatives because of the significant economic and political risk that comes with it. Many infrastructure projects are highly politicized. Many firms lacked internal, design-build-operate-control capabilities while others fall out of the core competencies desired by the client thus inhibiting a wider participation and response. The Asian financial crisis had also prevented firms from increasing their stakes in infrastructure spending. The challenge for the governments was to devise mechanisms for investment protection and insurance in order to boost private sector confidence.

1.5.4 Standards

The construction industry faces a myriad of complex regulations, codes and standards. Countries should coordinate where possible to consolidate, simplify or clarify these requirements in a bid to achieve increased harmonization. With the standardization of work practices occurring throughout the world with the acceptance and implementation of ISO standards, the framing and drawing-up of these standards has become crucial. That is, a country that does not provide any input into the process of ISO standard-making, or communicating its wants and needs, could eventually face standards that are very burdensome or ill-suited for its work environment. The construction industry in Singapore is one of the most well regulated when compared to other countries in terms of quality and skill enhancement, research and development. The country's open economy is reflected in the construction industry which has steadily improved its performance over the years. Singapore's construction industry seems to be far more mature than that of many other Asian countries including that of the similar 'city state' Hong Kong and techno hub Taiwan for that matter. The introduction of bench marking and buildability scores is seen only in Singapore while other Asian countries such as Korea and Malaysia have been far slower to move in this direction. Harmonized standards will remain the key to a harmonized construction market in the region.

1.6 A preview of the book

In this book we present a detailed account of the construction industry of the 11 countries in Asia plus Hong Kong, a Special Administrative Region of China, over the past 7 years, 1993–2000, focusing on the context of regional cooperation and the new challenges posed by greater integration of the construction market. Each chapter gives an overview of the construction industry for one particular country, giving information broadly in three categories: macroeconomic perspectives under which the broader economic condition and its impact on the construction industry is discussed along with the impact of the financial crisis on the industry and the recovery measures introduced; a section under which different features internal to the construction industry such as productivity, cost, quality safety and R&D issues are discussed; and the 'Construction Outlook' section, under which the scope for investments, opportunities and cooperation is discussed. This book does not intend to exactly aim at a comparative study between the different countries under study. It does, however, provide information about the scope and the experiences of the different countries and provides a useful regional perspective as it currently exists of the commonalities and diversities.

As a prelude to the forthcoming overview, seven broad trends were identified in the construction industry in Asia:

1 Globalization: The impact of globalization is manifested by construction being increasingly a global business, characterized by a trend towards firms gaining increased access to overseas markets.
2 Privatization: A renewed enthusiasm for private enterprises and the desire to limit government spending have led to enhanced claims for the provision of public services by the private sector.
3 Deregulation: Restructuring institutional frameworks concurrent with privatization.
4 Information technology and telecommunications: Development in information technology and communications technologies allows firms to more easily operate internationally.
5 Environment protection: The increasing community concerns about the prospects for a sustainable environment is affecting many aspects of building, construction and operation.
6 New project delivery methods: New project delivery approaches have been emerging, such as BOO, BOOT, partnering and project alliances.
7 International market change: Economic and population growth, and increasing per capita income, as well as urbanization in Asia have created huge demand for construction, including both buildings and infrastructure, a demand that can not be satisfied by local companies alone in terms of technology and management skills as well as capital.

Despite difficult challenges confronting the construction industry in Asia it is heading for considerable growth in the short and medium term along with the economy. The long-term health of the industry nevertheless is dependent on the state of that economy. While the construction sector expanded considerably during the unprecedented growth economic expansion of the 1990s, any economic downturn will inevitably adversely affect its ability to maintain itself. The industry must shed its image as an unglamorous profession and compete with other sectors for skilled workers and technology.

Note

1 In terms of volume Asia is the largest regional market with Europe closely following at near $1 billion.

References

ADB (1999) 'Asian Development Bank', http://www.adb.org/Documents/News/1999/nr1999027.asp.

Ball, M. (1988) *Rethinking Construction* (London: Routledge).

Bjornsson, H. (1998) 'IT revolutionises construction', *Building Innovation and Construction Technology*, 2.

Bon, R. and Crosthwaite, D. (2000) *The Future of International Construction* (London: Thomas Telford Publishing).

Bon, R. and Pietroforte, R. (1990) 'Historical comparison of construction sectors in the United States, Japan, Italy and Finland using input–output tables', *Construction Management and Economics*, 8:3, 233–247.

Carrillo, P. and Heavey, I. (2000) 'UK contractors' acquisitions strategy for Central and Eastern Europe', *Engineering, Construction and Architectural Management*, 7:3, 322–328.

Carrillo, P.M. (1998) 'Mergers and acquisitions: the impact of information systems and information technology', *Engineering, Construction and Architectural Management*, 5, 276–284.

Chiang, Y.H., Tang, B.S. and Leung, W.Y. (2001) 'Market structure of the construction industry in Hong Kong', *Construction Management and Economics*, 19:7, 675–687.

Drewer, S. (1990) 'The international construction system', *Habitat International*, 14:2/3, 29–35.

The Economist (2002a) 'Getting through the roof', 28 March.

The Economist (2002b) 'The lost (half) decade', 6 July.

ENR, *Engineering News Record* (1995), 23 October, 26–29.

Fukushima, A. (2000) http://www.nira.go.jp/publ/review/2000summer/fukushima.pdf.

Hillebrandt, P. and Cannon, J. (1995) *The Construction Company in and out of Recession* (Basingstoke: Macmillan).

Hillebrandt, P. and Cannon, J. (1996) *The Management of Construction Firms: Aspects of Theory* (Basingstoke: Macmillan).

Kerr, A. (2001) *Dogs and Demons: Tales from the Dark Side of Japan* (NY: Hill and Wang).

Low, S.P. (1991a) 'World markets in construction: I. A regional analysis', *Construction Management and Economics*, 9:1, 63–71.

Low, S.P. (1991b) 'World markets in construction: II. A country-by-country analysis', *Construction Management and Economics*, 9:1, 73–78.

Male, S. and Mitrovic, D. (1999) 'Trends in world markets and the LSE industry', *Engineering, Construction and Architectural Management*, 6:1, 7–20.

Mera, K. and Heikkila, E.J. (2000) 'Housing as a tool for reviving the crisis-hit economies in southeast Asia', in Mera, K. and Renaud, B. (eds) *Asia's Financial Crisis and the Role of Real Estate* (NY: M.E. Sharpe), 243–254.

Mera, K. and Renaud, B. (eds) (2000a) 'Introduction' in *Asia's Financial Crisis and the Role of Real Estate* (NY: M.E. Sharpe), 3–26.

Mera, K. and Renaud, B. (eds) (2000b) 'Conclusion: toward efficiency and stability', *Asia's Financial Crisis and the Role of Real Estate* (NY: M.E. Sharpe), 277–290.

Mitrovic, D. (1999) 'Winning alliances for large scale construction projects on the world market', in Ogunlana, S. (ed.) *Profitable Partnering in Construction Procurement: CIB W92 (Procurement Systems) and CIB TG 23 (Culture in Construction) Joint Symposium* (London: E&FN Spon), 189–199.

Moavenzadeh, F. (1978) 'Construction in developing countries', *World Development*, 6:1, 97–116.

Najjir, G.R., Love, P.C. and Runeson, G. (2001) 'Issues for the global construction market', *The Australian Journal of Construction Economics and Building*, 1:2, 73–81.

Ofori, G. and Chan, S.L. (2000) 'Growth paths of construction enterprises in Singapore, 1980–98', *Engineering, Construction and Architectural Management*, 7:3, 307–321.

Porter, M. (1990) *The Competitive Advantage of Nations* (NY: Free Press).

Raftery, J., Pasadilla, B., Chiang, Y.H., Hui, Eddie C.M. and Tang, B.S. (1998) 'Globalization and construction industry development: implications of recent developments in the construction sector in Asia', *Construction Management and Economics*, 16:6, 729–737.

Reina, P. and Tim, G. (2000) Global Costs Cooling, Fourth Quarterly Cost Report, 245:24, 112, The McGraw-Hill Companies, Inc.

Seymour, H. (1987) *The Multinational Construction Industry* (London: Croom Helm).

Stumpf, I. (2000) 'Competitive pressures on middle-market contractors in the UK', *Engineering, Construction and Architectural Management*, 7:2, 159–168.

Weddikkara, C. and Devapriya, K. (2001) 'Demand and supply trends and construction industry development', *The Australian Journal of Construction Economics & Building*, 1:1, 91–105.

World Bank (1993) *The East Asian Miracle: Economic Growth and Public Policy* (NY: Oxford University Press).

Chapter 2

Australia

Construction Industry Institute, Australia and S. Sharma

2.1 Background

Australia became a Commonwealth of the British Empire in 1901. "It was able to take advantage of its natural resources to rapidly develop its agricultural and manufacturing industries and to make a major contribution to the British effort in the First and Second World Wars. A referendum to change Australia's status, from a commonwealth headed by the British monarch to an independent republic, was defeated in 1999" (CIA, 2001).

Australia is located in Oceania, the continent between the Indian Ocean and the South Pacific Ocean at geographic coordinates of 27 00 S Latitude and 133 00 E Longitude. The country has a total area of 7,686,850 square kilometres consisting of land (7,617,930 sq. km) and 68,920 square kilometres water area. This nation has a coastline of 25,760 kilometres with no land boundaries. The Australian terrain is diverse ranging from mostly low plateau with deserts to fertile plains in the southeast with elevation extremes. Lowest point is at Lake Eyre (15 metres) and the highest point at Mount Kosciuszko (2,229 metres).

According to CIA (2001) Australia has a population of 19,357,594 2 (July 2001 est.) with the population growth rate currently standing at 0.99 per cent (2001 est.). Migrants comprise a sizeable portion of Australia's population with 4.19 migrant(s) per 1,000 population (2001 est.). The population in the country has constantly increased because of immigrants from the all over the world, with the migration policy aimed at assisting the economic development. The estimates in the same year indicate proportion of females slightly higher than that of males (0.99 males/female) in the country. Australia is a diverse country by virtue of the great degree of cultural variation among the nation's different ethnic groups. The indigenous aborigines constitute 7 per cent of the total population while the country has a sizeable population of Asian (7 per cent), Caucasian (92 per cent) and others (1 per cent). Religious diversity in the country shows Anglicans (26.1 per cent), Roman Catholics (26 per cent), other Christians (24.3 per cent) and non-Christians (11 per cent). English is the official language of Australia.

2.1.1 Geography

Australian climatic variation is characterized by generally arid to semiarid, temperate in the south and east, and tropical in the north. The country's staple natural resources include bauxite, coal, iron ore, copper, tin, silver, uranium, nickel, tungsten, mineral sands, lead, zinc, diamonds, natural gas and petroleum. It also includes arable land (6 per cent of total lands), permanent pastures (54 per cent), forests and woodland (19 per cent) and others (21 per cent). According to the 1993 estimates, a total of 21,070 square kilometres area is irrigated land. There are no permanent crops in Australia. Cyclones along the coast and occasional severe droughts characterize natural hazards in the country (CIA, 2001).

2.1.2 Political structure

The Commonwealth of Australia is a democratic federal state within the Commonwealth of Nations with the legal system based on English common Law. Australia has a democratic, federal-state system of government that recognizes the British monarch as sovereign. Executive power is vested in the governor-general, who represents Queen Elizabeth II. The legislature consists of a federal parliament comprising a 76-member Senate and a 148-member House of Representatives. Elections are held at maximum intervals of 3 years, when half the senators and all the representatives are elected by compulsory ballot. Administrative divisions in the country include 6 states and 2 territories – New South Wales, Queensland, South Australia, Tasmania, Victoria, Western Australia with dependent areas: Ashmore and Cartier Islands, Christmas Island, Cocos (Keeling) Islands, Coral Sea Islands, Heard Island and McDonald Islands, Norfolk Island, and Australian Capital Territory and Northern Territory. Each of the six states in the federation has its own government, with a governor and a bicameral legislature, apart from Queensland, which has a unicameral legislature.

2.1.3 Development process in Australia: through the years

As a part of the long boom in the international economy particularly during the period between 1953 and 1974, Australia laid its economic foundation to become the fifteenth largest economy with output – as measured by gross domestic product (GDP) by the turn of the century. The same period was characterized by economic growth, low inflation and rising living standards. The Australian GDP per capita increased by 3.2 per cent a year on average during the 1960s. In the 1970s and 1980s GDP per capita growth was lower than in the 1960s. On average, across all sectors of the Australian economy, the 1960s were years of high GDP growth – 5.3 per cent a year in real terms (Government of Australia, 2002).

2.1.4 Industry structure

Since the beginning of the twentieth century there have been significant changes in the industry shares of GDP for the mining, agricultural, manufacturing and service sectors of the economy which is represented mainly by (Government of Australia, 2002):

- the rise in the services-sector share of GDP after the middle of last century with the services sector also representing the largest portion of the Australian economy;

- a peak in manufacturing after the Second World War, and subsequent decline as a share of GDP since that time (consistent with the experience of many industrialised nations) to a level similar to that of a century ago; and
- the gradual decline of agriculture as a share of GDP after the first four decades following Federation as the economy became less reliant on the land.

Australia has been regarded predominantly as a producer of agricultural commodities, minerals and energy. These products have played an enormously important historical role in the country's economic development, particularly in the first 100 or so years after European settlement began. However, their relative importance declined throughout the twentieth century. As a result, despite the importance of agriculture, mining and manufacturing, even during the 1960s, more than 60 per cent of GDP was produced mainly by the services sector. There has been some change in the industry composition of the economy since then with most structural changes in industry shares occurring in the 1970s and, to a lesser extent, in the 1980s. Following the joining of OECD in 1970, Australian manufacturing in particular saw some significant structural change since the mid-1970s, as previously high tariffs and other forms of protection against manufactured imports had been gradually removed leading to a gradual decline from the peak (27.6 per cent) in 1963 (Government of Western Australia, 2002). By the end of the twentieth century, agriculture and mining accounted for just 7.5 per cent of Australia's total output, 6 per cent of its workforce and 42 per cent of its exports whereas manufacturing accounted for about 12.5 per cent of both output and employment. "Reflecting Australia's relatively high rate of population growth, the preference of most Australians for detached dwellings, and the dispersed pattern of settlement, the construction sector is somewhat larger than in most other industrial economies, representing around 7.5 per cent of overall output and providing 6 per cent of total employment" (OECD, 2000).

2.1.5 Microeconomic reforms

At the end of the 1960s, exports and imports, together, made up about 28 per cent of GDP, as against 42 per cent at the end of the 1990s. The landmark decision of the Whitlam Government to reduce the across-the-board tariff by as much as 25 per cent marked the beginning of microeconomic reform, with far-reaching implications on the agricultural, mining and manufacturing sectors of the economy. The reduction in tariff however was not significantly supported by widespread reforms to boost market productivity as well as flexibility (Government of Australia, 2002). It was not until the early 1980s that progress began to be made in these areas. As that decade began, few could have anticipated the scale of economic policy reform that would occur over the next 20 years (Government of Australia, 2002).

Other significant aspects of the reforms included floating of the exchange rate at the end of 1983, accompanied by a relaxation in capital controls. Since then, most of the areas of market operation have been subject to new reforms resulting in across-the-board reduction on tariff. The wave of new reforms continued through to the mid-1980s and 1990s particularly in the non-traded goods sectors of the economy. These reforms had wide implications on the potential deregulation of sectoral markets of transport, utilities and communications provision in the country.

Table 2.1 National accounts in Australia between 1987 and 1993

	1987	*1988*	*1989*	*1990*	*1991*	*1992*	*1993*
Exports of Goods and Service (a)	46.99	52.79	57.49	53.89	67.34	72.03	78.04
Imports of Goods and Service (b)	−49.72	−55.95	−66.62	−66.67	−65.78	−73.08	−80.09
(a) + (b)	−2.73	−3.16	−9.13	−12.78	1.56	−1.05	−2.05
Government Consumption	50.38	54.76	59.2	65.08	69.49	72.46	74.96
Private Consumption	166.15	184.89	207.92	226.26	236.33	248.8	259.55
GDP (real terms, in 1990)	343.06	358.07	373.86	378.51	374.43	384.18	399.85

Source: IMF (1994).

Note
(Unit: billions of Australian dollars).

Towards the end of 1980s and in the beginning of the 1990s new reforms were introduced that were set to encompass other areas of the national economy. Labour market and industrial activity in particular were subjected to considerable reforms followed by education and training sector. Sectoral approach to reform was complemented by National Competition Policy that made provision for the Commonwealth and State governments to review existing legislation to promote competition between the public and private sector without overriding the broad public interest. This resulted in a situation where both public and private sectors could equally vie for infrastructure projects.

2.1.6 *Labour productivity and unemployment*

The 1970s were difficult years for the Australian economy with regard to employment fuelled by policy complacency due to high growth, low inflation and low unemployment of the 1960s. In the 1960s the unemployment rate averaged less than 2 per cent which increased sharply in the 1970s eventually reaching 10 per cent in the aftermath of the early 1980s' economic recession. As the economy bounced back in the second half of the 1980s the rate of unemployment fell back below 6 per cent, but then increased sharply once more to peak in excess of 10 per cent in the recession of the early 1990s (Government of Australia, 2002).

Over the last 40 years, labour productivity growth in Australia has averaged 2 per cent a year, and GDP per capita growth averaged 2.2 per cent. In order to repeat the GDP per capita performance of the yesteryears, Australia aims to achieve average labour productivity growth of 2.3 per cent a year. Table 2.1 shows the national accounts in Australia between 1987 and 1993.

2.2 The national economy

Australia has a Western-style capitalist economy, which uniquely differs from its counterparts in terms of its structure. The country is a major exporter of agricultural products, minerals, metals and fossil fuels such that the share of commodities in the total exports figure exceeds 57 per cent. The implication of this is that commodities market has been sensitive to vagaries of the international market. Despite the government's best intention to maintain the levels of export market, competitions from international market, particularly from those in the Far East and China have stymied the Australian exports over the years. Ironically, defying most international trends, the major boost for Australian economy was seen during the Asian financial crisis. Australia registered a robust GDP growth rate of 4 per cent in 1997 and continued to maintain that level of growth rate in the following years. On perspective, the Australian national economy saw the troughs in the 1990s characterized by low growth and high unemployment, and peaks in the late 1990s.

In the first half of the 1990s, the share of Australian exports entering Asian countries rose consistently from the levels of a decade ago. In 1996, the total export volume to Asia reached 60 per cent of the total export up from around 50 per cent a decade earlier. This was symptomatic of the increasing reliance of the national economy to economic conditions in the Asian economies. Significantly, the structure of exports also changed to pave way for greater share of services and manufacturing as opposed to primary products. By 1996, primary products saw a drop of 20 per cent accompanied by 10 per cent and 7 per cent rise in manufacturing products and services respectively.

2.2.1 Macroeconomic review and outlook

Macroeconomic developments in Australia in the 1990s have turned out to be more favourable in many ways than would have been expected at the beginning of the decade. "Australia recorded an average annual economic growth rate over the 1990s of nearly four percent a year since the trough of the recession in mid-1991, the second-fastest rate in the developed world (next to Ireland)" (OECD, 2000). Living standards, as measured by per capita GDP, improved at a rate not seen since the 1960s – a result shared only with Ireland among industrial countries. The strong growth in the latter half of the 1990s was more marked in comparison to previous periods when expressed in terms of GDP growth per capita. This provided a better indication of growth in living standards. GDP growth per capita in the 1990s was 2.2 per cent per annum, second only to the 1960s and above the century average of 1.7 per cent per annum. In the second half of the 1990s growth was significantly higher at 3.2 per cent per annum (OECD, 2000).

The average inflation in the last 10 years was 3.9 per cent, considerably below the 8 to 9 per cent inflation experienced in 1987 and 1990. Inflation remains low and had risen only marginally from the trough it reached at the end of 1997. It stood at 0.7 per cent for the year 1998.

The ten-year average unemployment rate also impressively stands approximately at 1.8 per cent. Much employment growth in the past years had been in private sector service industries, particularly property and business. In the near-past decades the shift from manufacturing to services had been rapid, with most new jobs being created in the services sector. Strong employment growth experienced in the early part of 1990s declined in 1996 and 1997, then recovered again in 1998. The annual growth in employment was the strongest in 1999 since it began to decline in 1995. Non-service industries recorded a 4 per cent growth rate in employment between 1993–1994 and

1998–1999, while service industries grew by 18.9 per cent. By 1998, according to the OECD, "service industry employees accounted for almost 77 per cent of total employment". Between 1996 and 1999 the number of computer service businesses grew by just over 50 per cent – from 9,679 to 14,731. Employment and income in the sector grew by 10 and 9 per cent per annum respectively. Strong economic growth and moderate wage and price pressures had underpinned sustained employment growth. In the last 2 years until 2002, employment had grown by more than 2 per cent per annum. In March 2001 the unemployment rate was 6.5 per cent, a 4.2 per cent drop from the previous decade's peak unemployment level of 10.7 per cent. "In 2000 Australia was ranked eleventh in the world for overall productivity, as measured by GDP (PPP) per person employed" (Australian Department of Foreign Affairs and Trade, 2000).

"The current account deficit (CAD), traditionally a watchpoint for foreign investors averaged at around 4.5 per cent of GDP over the past two decades. The CAD in 1999–2000 averaged 5.4 per cent of GDP. With the government's budget in surplus, the CAD was largely the outcome of private sector transactions" (Government of Australia, 2002). These included private investment and borrowings, which in turn were more soundly based following extensive reform initiatives.

Australian manufacturing exports grew strongly over the 1990s expanding from 25 per cent in 1990 to 32 per cent of total exports in 2000. Particularly noteworthy was the growth in elaborately transformed manufactures (ETMs). "At the beginning of the 1990s, the majority of Australian manufactured exports were in 'Manufactures classified by material', the most significant of these exports were aluminium, copper, iron and steel, zinc, pearls/precious stones, nickel, lead (mostly little-processed output of the mining industry) plus textile yarn, leather, paper/paperboard and miscellaneous manufactures of base metal" (Australian Department of Foreign Affairs and Trade, 2000).

While manufacturing exports, and particularly more ETMs, grew strongly over the decade, the growth was stronger in the first half of the 1990s. In this period manufactures increased their share of total exports by some 13 percentage points to 35 per cent. Growth in the second half of the 1990s was less spectacular. In particular, manufactured exports under-performed, compared with the pace of total exports. Manufactured export growth had begun to slow in 1994–1995. Over the 3 years from 1994–1995 to 1996–1997 (just prior to the Asian economic crisis), growth in manufacturing exports grew more slowly than total exports –16.5 and 17.7 per cent respectively. Manufacturing exports under-performed relative to total exports by a wider margin during the "crisis" period. "In the 2 years to June 1998–1999 manufactured exports rose by 12 per cent, while total exports grew by 15 per cent" according to Australian Department of Foreign Affairs and Trade (2000).

Table 2.2 shows a relatively strong growth in the major sectors of the Australian economy since the substantial investment of 1995. Agriculture and construction had recorded a modest leap in growth figures from –4.3 to 6.8 per cent and from 1.1 to 3.8 per cent respectively. Table 2.2 shows the trends in major economic variables over the last few years. Growth in GDP had been relatively strong over the years, although it declined in annual terms in 1997 to pick up relatively strongly again in the June quarter, 1998 (Government of Australia, 2002). Australian economy grew strongly in 1998–1999, despite the continuing downturn in Asia. Growth in the

Table 2.2 Industry structure in Australia between 1989 and 2001

	Annual average growth 1989–1990 to 1994–1995 (%)	*Annual average growth*
Agriculture	−4.3	6.8
Mining	3.9	7.6
Manufacturing	3.2	4.4
Construction	1.1	3.8
Wholesale/retail	3.6	4.1
Tourism related	6.6	8.7
Government	4.7	4.0
Transport	4.1	3.5
Finance/property	5.9	8.8
Education	5.4	5.3
Health	5.8	5.9
Other	4.7	6.1
Total	5.9	5.9

Source: Government of Western Australia (2002).

volume measure of GDP of 4.4 per cent in 1998–1999 was achieved in a domestic environment of low interest rates and low inflation. Growth was broadly based with most services and manufacturing industries contributing to a growth of 4.0 per cent in the volume of industry value-added. The services sector, which accounts for almost 82 per cent of total employment and 69 per cent of production grew by 4.8 per cent, with the strongest growth in communication services, property and business services, and accommodation, cafes and restaurants. Manufacturing grew by 2.4 per cent with the strongest growth in non-metallic mineral products, machinery and equipment, and textiles, clothing and footwear. Australia experienced a slower growth in 2000 provoked in part by tighter monetary policy, a tempering of domestic demand and significant structural shifts following wholesale reform of the taxation system. The country recorded economic growth of 2.2 per cent in 2000 (down from 4.1 per cent in 1999). From a broader perspective, a report by Industry Science Resources (2000) has termed the Australian economy, to be in its best shape in 30 years, enjoying a sustained convergence of steady growth, low inflation, low interest rates and a falling external deficit.

One of the major reforms undertaken in 2000 was the re-structuring of Australia's taxation system. On 1 July 2000 The New Tax System (TNTS) was introduced as one of the important components of ongoing reforms aimed at re-structuring the country's taxation system. A whole range of taxation systems was targeted by this reform including income tax, indirect tax, family assistance, business tax and Commonwealth–State financial relations. Further, abolition of a multi-rate wholesale sales tax (WST) and the introduction of a broad-based 10 per cent goods and services tax (GST) were one of the principal elements of TNTS. This in effect resulted in the raised levels of inflation rate from the levels that were between 2 and 3 per cent per annum to 6 per cent by the end of 2000 (Industry Science Resource, 2000).

Table 2.3 Annual trend growth in major economic and labour market variables

	1995	*1996*	*1997*	*1998*	*1999*	*2000*
GDP at real prices (Aus$ million)	363.7	407.4	407.1	364.2	395.3	381.0
Real GDP growth (%)	4.4	3.7	3.8	5.6	4.7	3.8
Consumption	4.6	3.1	3.4	4.5	5.1	3.8
Private consumption	5.0	3.3	3.9	4.6	5.2	3.4
Government consumption	3.6	2.6	1.6	4.0	4.8	4.9
GDP deflator	1.8	2.1	1.3	0.3	0.9	3.2
CPI	4.6	2.6	0.3	0.9	1.5	4.5
Real effective exchange rate (level, 1997=100)	92.3	101.0	100.0	91.5	91.8	87.3
Employment (%)	4.6	1.1	0.9	2.0	–	3.0
Unemployment rate (%)	8.2	8.2	8.2	7.7	7.0	6.3
Population (millions)	18.1	18.3	18.6	18.8	19.0	19.2

Source: Compiled from APEC Economic Outlook Report (2001).

On the construction sector, the introduction of GST led to a prolonged slump in the construction activity resulting in contraction of the construction activity by over 24 per cent although the construction sector gradually emerged out of this in early 2001. This fall was large enough to drag the overall GDP figure down with it, disguising what was otherwise a far more moderate slowdown in domestic economic activity over the course of 2000; while other sectors generally recorded strong growth over 2000 (Table 2.3) (APEC, 2001).

Economic impact of the Olympics

According to one estimate, the Olympics was expected to tack on an additional Aus$6.5 billion – about 1.6 per cent of the GDP – to Australia's GDP over the 1994–2006 period. Over 35,000 people have worked on Olympic-related construction sites (Economics, 2000). "It was suggested that prior to the Olympics and during the Olympic year GDP growth remained higher than an average – maxing out at nearly 1.5 per cent above average GDP in the third year before the Olympics, while, at the same time, growth rates were lower in the post-Olympics years" (Economics, 2000).

2.2.2 *Economy in the first half of 2001*

The Australian economy continued to improve with GDP registering 1.1 per cent in the September quarter in volume terms over the previous quarter and by an annual growth of 2.5 per cent. The annual GST-added inflation stood at 2.5 per cent in September, well within the target set by RBA. Similarly, the CPI at 2.5 per cent inflation stood comfortably within the 2–3 per cent inflation target band. On the employment front, the unemployment rate decreased to 6.7 per cent in November principally due to a reduction in the participation rate by 0.3 percentage points to 63.5 per cent.

Exports fell 5.9 per cent overall in November to Aus$12.9 billion, with rural goods declined by 8 per cent, other goods by 7 per cent and service credits by 1 per cent. This decline was no doubt due in part to the weak global economy, but also particular factors such as decline in beef exports to Japan due to BSE fears and a strengthening of the Australian dollar during the month (+3.1 per cent in TWI terms). Imports of goods and services declined by 2.5 per cent in November to Aus$12.8 billion, but excluding fuel it would have been down by just 0.6 per cent. Imports of consumer goods rose by 1 per cent, while capital goods declined by 0.6 per cent (Asia Pacific Economic Cooperation (APEC, 2001)).

As in many other industrial countries domestic consumption played a vital role in lifting the Australian economy, in particular on the back of consumption-led surge in building sector and household expenditure. Significantly, the weakening Australian dollar also helped maintain a competitive export environment despite weaknesses seen in the international economy, with net exports adding to growth. With the overall building approvals (7/1) for dwellings falling by 9.1 per cent (seasonally adjusted) in November 2001, despite a rise of 5.9 per cent in private sector houses, Australian building sector, the powerhouse of Australian economy began to show signs of slowdown (Master Builder, 2001). However compared to a year earlier approvals were over 40 per cent higher for all dwellings and 60 per cent for private sector houses (Construction Industry Authority, 2001).

2.2.3 *Role of construction/property market in national economy*

In 1998–1999 the work undertaken by the Australian building and construction industry was valued at $64.3 billion. Average employment during the year was 634,000 with approximately 169,000 firms operating in this sector. The three main segments were housing, engineering construction and non-residential building. Contribution to GDP in 1998–1999 was $34.5 billion, or 6.3 per cent. And these figures did not take into account the extensive flow-on effects such as those associated with the manufacturing of building products and architectural and design work. The industry, by 1996–1997 contributed 14.4 per cent of Australia's GDP and employed a total of 729,400 people (Table 2.4) (ABS, 1997).

Understanding the various components that influence the Australian property and constructions industry can be the foundation for long-term improvement in stability and performance of the industry. With the globalization of the world's property markets, certainty from uniformity of returns will increase the attraction of the Australian property and construction industry to overseas investors. Accordingly, to comprehend the diversity in the Australian property and construction industry, it can be separated into three distinct sectors: residential building, non-residential building and engineering construction.

1 Residential building is defined as a building, which is predominantly used for residential purposes and can contain one dwelling unit (i.e. house) or several dwelling units (i.e. flats);
2 Non-residential building is defined as a building other than a residential building and includes, hotels, shops, factories, offices, etc; and
3 Engineering construction is defined as infrastructure construction. It includes construction other than buildings, e.g. roads, bridges, railways, telecommunication, water and sewerage, electricity generation and distribution facilities.

Table 2.4 Key statistics: building and construction

Constant prices, base year = 1999–2000	*1996–1997*	*1997–1998*	*1998–1999*	*1999–2000*	*2000–2001*	*% of Australian total in 1999–2000*
Turnover ($ m)	67,281.1	75,394.2	75,893.3	83,166.4	n.a.	n.a.
Industry gross product ($ m)*	16,102.0	18,288.0	–	–	n.a.	n.a.
Industry value added ($ m)*	–	19,617.7	20,491.8	21,807.8	n.a.	3.47
Employment (number)	576,100	618,900	648,900	711,500	n.a.	5.46
Wages and salaries ($ m)	10,922.6	12,743.6	12,125.9	12,888.9	n.a.	4.26
R&D expenditure ($ m)	60.3	45.6	38.6	64.3	n.a.	1.59
Exports ($ m)	74.3	30.7	18.4	23.0	52.1	0.02
Imports ($ m)	n.a.	n.a.	n.a.	n.a.	n.a.	n.a.

Source: ABS (1997).

Note

* From 1998 to 1999 the ABS replaced industry gross product with industry value-added as the measure of the economic contribution of industry (see "Glossary"). The share of industry value-added in the Australian total equals this industry's contribution to GDP.

Based on ANZSIC Division E.

The property and construction industry forms an integral part of the Australian economy currently contributing to approximately 7 per cent of the labour force in the region and 6 per cent of the national product output. In the 1990s, both labour and output relative to the wider economy had reduced by approximately one percentage point, although fluctuating erratically during this period. The instability can, in part, be linked to industry size and its close links with the overall economy to generate demand. Accordingly, government can use the association to manage the wider economy, these can be by either direct policies (changes in public expenditure and property taxes, etc.) or by indirect means (interest rates, etc.). The private sector accounts for approximately half of the demand across the various property and construction sectors with the remaining public sector concentrated mainly in the engineering construction category. Under government contracts, the private sector produces about 75 per cent of the public sector accounts.

2.2.4 *Construction investment (or market) volume*

Activity in the building and construction industry had grown steadily in recent years until 2002, with growth in turnover and employment averaging 7 per cent each year between 1996–1997 and 1999–2000. This performance was driven by strong demand in both the residential and non-residential construction sectors. Construction activity by value completed in the 1990s had varied across the individual sectors with residential, non-residential buildings and engineering construction recording an annual nominal growth of 8.7 per cent, 4.1 per cent and 6.0 per cent respectively (Australian Bureau of Statistics (ABS, 2001)).

An overview of the value of construction activity in the 1990s (Construction Industry Authority, 1999) indicates that the value of building works completed (residential and non-residential property) and engineering construction activity stood at $31.7 billion by 2000 a rise 34.6 per cent compared to the figure of $23.66 billion in 1994. The trend indicates that, the value, measured in terms of investment per annum itself had grown by 11 per cent in 1995 in real terms, while it began to fall thereafter. Annual average inflation for most years in the second half of 1990s standing at 3.9 per cent indicates a relatively subdued performance of the construction industry. In part, this can be attributed to the construction boom of the late 1980s and the subsequent oversupply in many property sectors. The restructuring of the labour force towards the financial and services sector alongside major infrastructure projects had provided the improvement in the volume of construction activity. Table 2.5 shows between the same period that the value of building activities rose by nearly 38 per cent and that of engineering construction activity by over 31 per cent.

"Results from the Australian Industry Group Construction Outlook Survey conducted in October 2001 reveal that the value of total construction turnover in dollars declined by 1.1 per cent in 2000–2001 following a 5.3 per cent decline in 1999–2000" (Australian Industry Group, 2001). As such the industry expanded at an average of 10 per cent per annum over the period of last 6 years, which makes this decline even more detrimental to the state of the industry. The same report suggests that the falling levels of non-residential building construction, the completion of a number of key private sector infrastructure projects and the lack of any significant growth stimulus from heavy industrial and resource-based projects are considered responsible for the such a drastic downturn.

A nation-wide survey covering 100 companies with employment capacity of 33,500 and combined turnover of Aus$9.2 billion or 25 per cent of total industry activity to identify the state of non-residential construction. The survey showed the average level of industrial/construction capacity used in September/October was 77.5 per cent, which saw a reduction of 4 per cent year on year. The survey results showed that whilst industry activity declined to a 9 year low in the second half of 2001, the value of private sector work fell only by 1 per cent in 2000–2001 in a sharp contrast

Table 2.5 Value of construction activity

Type of activity	*1994–1995*		*1995–1996*		*1996–1997*		*1997–1998*		*1998–1999*		*1999–2000*	
	$ bn	%	$ bn	%	$ bn	%	$ bn	%	$ bn	%	$ bn	%
Building (a)	10.11	12	11.49	14	12.8	11	13	2	13.7	5	14	2
Engineering construction (b)	13.44	3	14.72	10	15.2	3	16.5	9	17.3	5	17.7	3
Total	23.55	6	26.2	11	27.9	7	29.5	6	31	5	31.7	2

Source: Construction Industry Authority (1999).

to 9.6 per cent decline in 1999–2000. This implied that the private sector, which accounts for over 60 per cent of total industry turnover, was beginning to be sluggish reflecting "the continued weakness in private building activity with positive offsets in 2000–2001 coming from growth in private infrastructure works, mining related activity and manufacturing construction" (Australian Industry Group, 2001). On the other hand, turnover derived from public sector work also experienced a slight drop of 1.0 per cent in 2000–2001, which is however expected to turn around to achieve a rise of 2.5 per cent in light of the anticipated rise in Federal Government-funded projects in the next year. The significant achievement was in the overseas business of construction firms which grew modestly in 2000–2001, although the size of the export market for construction companies (6.6 per cent of total industry turnover) remained below the 10 per cent plus level of 2 years ago (Australian Industry Group, 2001). The infrastructure market continues to account for an increasing share of total industry activity. Approximately 38 per cent of all industry turnover was derived from infrastructure projects in 2000–2001 compared with less than 20 per cent a decade earlier (The Royal Commission into the Building and Construction Industry, 2002).

Based on the reports of Australian Procurement and Construction Council (1998–2002) as indicated in Table 2.6, it is seen that residential construction activity steadily rose over the period from 1997 to 2001 to register a 55.9 per cent overall growth in volume, while the non-residential sector saw a contraction of 5.1 per cent over the same period. Total engineering construction also registered a growth of 30.2 per cent over the period, and overall construction activity on the strength of residential and engineering activities rose over 30 per cent. It is interesting to note that in 2000, residential activities comprised about 50 per cent of the total construction activity while the non-residential sector had 20 per cent of the share. An overview of 1997–2001 indicates activities in all three sectors were brisk until 1999 despite financial crisis in East Asia, began to show signs of slowing down in 2000 and had strong falls in 2001. The substantial drop in activities of all three sectors can be partly

Table 2.6 Australian construction activity

	Value of the work done between 1997 and 2001 ($ million chain volume measures)					*% change 1997–2001*
	*1997**	*1998*	*1999*	*2000*	*2001*	
Total residential building	18,877	28,825 (52.7)	32,567 (12.9)	36,944 (13.4)	29,415 (−20.3)	55.9
Total non-residential building	12,980	13,329 (3)	14,583 (9.4)	14,718 (0.1)	12,259 (−16.7)	−5.1
Total engineering construction	13,658	17,106 (25.2)	19,383 (13.3)	20,118 (3.8)	17,792 (−11.6)	30.2
Total construction activity	45,455	59,270 (30.3)	66,534 (12.2)	71,780 (7.8)	59,466 (−17.1)	30.8

Source: Compiled from the Australian Procurement and Construction Council.

Note
* Constant price 1989–1990.

attributed to the post-Olympics slump experienced by the country in all sectors of its economy.

Residential building sector

The housing sector had experienced a strong period of growth following the downturn in activity after the introduction of the GST in July 2000. This growth had been largely due to increasing housing affordability associated with the current low interest rates and the doubling of the First Home Owners Scheme grant for buyers of new homes. Residential investment fell by 2.2 per cent in September 1998 after rising consecutively for previous seven quarters. Activity in the residential construction industry is highly cyclical. The Australian market had emerged from a cyclical downturn which started at the beginning of 1995. After falling to just over 9,600 dwelling approvals in January 1996, the industry recovered with dwelling approvals increasing by 49 per cent to 14,300 in June 1998. The underlying housing demand prepared by the Indicative Planning Council for the Housing Industry (IPC) in 1997 suggested an underlying demand of 138,000 dwelling units nationally.

Non-residential building sector

Non-residential construction activity in 1997–1998 rose by 6 per cent in real terms to \$30.4 billion (Industry Science Resources, 2000). This places the level of activity above the previous high, recorded during the office-building boom of 1989–1990. After declining since mid-1999, Australia's non-residential building construction industry was forecasted to recover modestly in 2001–2002 before a stronger upturn in 2002–2003. Nevertheless, this recovery was expected to be dependent on the extent to which high-valued projects under consideration move into construction while the weak and uncertain global outlook constitutes an added risk to future growth.

Non-residential public sector construction in the past had been constant. At the time of writing it accounted for approximately 35 per cent of total construction work, although much of the work was contracted out to the private sector. The volatility in the private sector is primarily driven by economic activity. Non-residential building activity rose by 1 per cent to 13.3 billion in the same year. Private activity decreased slightly to \$9.7 billion. Public sector work rose by 7 per cent to 3.6 billion due mainly to the announcement of the Olympics.

Engineering construction sector

The engineering construction industry is an important, if overlooked, part of the Australian economy. Engineering construction activity has been steady since 2000, with analysts predicting a slow recovery through the next few years. Its contribution is recognised in major published projects often associated with a high political profile, such as major new transport links. The annual growth for the last 10 years has been 7.2 per cent, although cyclical growth is evident from an annual growth of −13.6 per cent to 19.2 per cent. Engineering construction activity rose by 11 per cent to 17.1 billion in 1997–1998 (Industry Science Resources, 2000). Activity was driven by a 26 per cent rise to \$7.2 billion in private activity, mining and resources. The private sector, at the time of writing, undertook approximately 65 per cent, a gradual increase from 44 per cent of the total 10 years ago.

2.2.5 *Property market overview*

In a recent report released by Colliers Jardine, *Asia Pacific Office Market Overview* (2001), it was found that compared to many other parts of Asia-Pacific, the outlook for a number of CBD office markets in Australia seems promising. Low vacancy rates and slow but steady growth has attracted developers to many of Australia's CBDs. Development in Adelaide is limited to buildings that are fully committed to, while tight vacancy rates at the top end of the Brisbane market suggest the fundamentals for a development cycle. A lack of available supply for prime contiguous floor space greater than 2,000 square metres has fuelled pressure in Melbourne for refurbishment and new construction, resulting in almost 190,000 square metres of new and refurbished development. Rents are increasing in premium grade buildings in Perth resulting in a new, albeit cautious development cycle. The construction cycle is coming to an end in Sydney, however rental rates are expected to continue to rise even in the face of slowing demand. In broad terms the property market in Australia is improving after going through a period of strong growth in the late 1980s, which was subsequently followed by a slump in the early 1990s Colliers Jardine (2001).

Office properties

Although the Australian office markets operate independently of each other in supply and demand, there had been broad similarities in cyclical movements over the past 10 years until 2002. In the late 1980s high construction levels occurred in Melbourne, Sydney, Brisbane, Perth and, to a lesser extent, Adelaide, as a result of rising rents amid an anticipation of boom times. The high construction levels, coupled with stagnant demand by the early 1990s, led to oversupply of varying degrees in each of these cities, and rental levels began to fall. During this period construction activity remained high as a result of the overhang from the boom, and the time lag in realizing that supply was well ahead of demand. Following the high construction levels in the late 1980s, vacancies increased and rents fell. However, the recovery rate of the markets had varied considerably, with Sydney averaging a net absorption level of around 114,000 square metres per annum over the past 5 years until 2002. Melbourne and Brisbane have had net absorption rates of 37,500 and 14,500 square metres respectively over the same period. The national capital, Canberra, had over the past 10 years until 2002 experienced a quite different office cycle. During this period, the Canberra office market had relatively stable vacancies – generally below the national average – and rising rents. In Canberra, rents were falling and vacancy rates rising, partly as a result of the rationalization of government occupancy requirements and, in the case of the market in Civic – the capital's CBD – decentralization.

It is difficult to generalize about the current and future state of Australia's office markets. Each is at a different point in the property cycle and each is influenced by different future supply/demand relationships. While Sydney led the way in almost every regard with net take-up substantially above any other market in Australia, the imminent imbalance in demand and supply put pressure on the fundamentals of the market that saw rental growth easing and a small increase in vacancy rate towards the run-up to the Olympics. Brisbane was the only other CBD market in Australia to move into a new supply mode in this cycle. The Canberra office market was seen rebounding strongly in 1999. Melbourne saw a double digit (12–13 per cent) vacancy rate in 1999 with the rents, paradoxically, at the premium end of the market pushed to some 26 per cent between 1998 and 1999 according to the reports by Colliers Jardine (2001) and C.H. Williams Talhar and Wong (1998–2001).

By 2000, all Australian office markets were heading in the same direction, characterized by low and decreasing vacancies, upward pressure on rents and the prospect of new office construction in the short term. The Sydney market had settled in for at least 3 years of very tight vacancy. Brisbane's vacancy rate improved to 2.1 per cent in 2000 concurrent with Queensland economy, which was growing at a rate well above the national level. This growth must inevitably manifest itself in demand for offices. The Canberra office market (Civic) enjoyed a bumper 2000, with prime net effective rents moving up 30 per cent in the last 12 months. This is no surprise, as vacancy rates had followed the national trend by almost halving in 12 months, to now be around 6.1 per cent. Melbourne experienced prime rents growing a staggering 48.2 per cent over the last 12 months.

It can be seen from Table 2.7 that all the Australian CBD office markets with the exception of Canberra were in a phase of increasing rents and falling vacancies. Although Sydney was the only market with significant speculative office construction currently underway, the remainder of the capital cities is expected to generally improve over the period 2002–2007.

Residential properties

Historically low mortgage rates had contributed to a steadily improving housing market in most of Australia's capital cities. Low and stable interest rates fuelled activity in the residential market in the past several years. However, with interest rates rising and changes in the taxation structure in 2000, the residential market, Australia-wide, experienced substantial shocks during 2000. The new housing market had slowed substantially in the second half of the year, following a splurge in activity prior to the introduction of the GST (C.H. Williams Talhar and Wong, 1998–2001).

Over the years a new trend emerged in Australia of the building of residential apartments in the CBD and inner-city areas as shown in Table 2.8. This has been particularly common in Sydney and Melbourne and to a lesser extent in Brisbane inner-city medium density development, especially in Sydney and Melbourne continued apace briskly through to 2000.

Industrial properties

Although the retail market has been the strongest performing sector in the Australian property market over the long term, the industrial sector has been performing best over the short term. The industrial market was the highest performing property sector in Australia for the year ending June 1998 according to a survey conducted by the Property Council of Australia. The total returns (income and capital returns) for the year were: Australian industrial – 15.3 per cent, Australian retail – 10.3 per cent, Australian CBD office – 9.4 per cent. Construction levels for new industrial property have reached the long-term average. Favourable conditions in the broader economy, a desire by occupiers to upgrade to more efficient premises and a competitive development industry drove the industrial market in 1999. Total returns (income and

Table 2.7 Office market indicators in different cities in Australia

	Sydney			*Melbourne*			*Brisbane*			*Perth*			*Adelaide*			*Canberra civic*		
	1999	*2000*	*2001*	*1999*	*2000*	*2001*	*1999*	*2000*	*2001*	*1999*	*2000*	*2001*	*1999*	*2000*	*2001*	*1999*	*2000*	*2001*
Prime net effective rents ($ m^2)	465	475	518	153	166	246	169	225	242	178	193	209	130	148	171	211	182	238
Vacancy rate (%)	5.6	6.4	4.5	13.4	13.7	8.1	7.0	8.3	7.9	11.8	12.3	11.0	17.6	18.1	12.9	8.9	8.1	6.1
Prime capital value ($ m^2)	9,133	8,600	8,974	4,190	2,083	4,292	4,255	4,759	4,480	3,724	3,835	3,769	2,130	2,205	2,205	2,537	2,070	2,574
Prime yields (%)	5.75–6.25	5.57–6.25	6.42–	7.0–7.75	7.5–9.25	7.75–9.25	7.0–7.5	7.00–7.50	7.00–8.00	5.50–5.75	5.50–6.00	6.25–6.75	9.0–10.25	9.00–10.75	8.75–11.00	10.0–10.5	9.00–11.0	9.00–11.00
New construction underway (m^2)	306,200	39,377	243,666	–	25,826	47,405	18,000	10,608	40,697	3,900	10,338	13,686	20,200	22,343	7,851	–	0	0

Source: C.H. Williams Talhar and Wong (1998–2001).

Table 2.8 Residential market indicators in different cities in Australia

	Sydney			*Melbourne*			*Brisbane*			*Perth*			*Adelaide*			*Canberra civic*		
	1999	*2000*	*2001*	*1999*	*2000*	*2001*	*1999*	*2000*	*2001*	*1999*	*2000*	*2001*	*1999*	*2000*	*2001*	*1999*	*2000*	*2001*
Median detached house price	262,000	285,000	315,000	189,000	230,500	253,000	142,000	142,000	142,000	140,000	149,300	157,800	118,000	134,000	135,000	156,000	160,000	184,000
Median unit and townhouse price	227,000	238,000	250,200	141,000	179,000	190,000	125,000	135,000	130,000	96,300	105,600	114,900	93,000	95,000	95,000	123,600	133,500	136,000

Source: C.H. Williams Talhar and Wong (1998–2001).

capital) from industrial property remained the highest in the country with total returns of 14.19 per cent, down slightly from 15.3 per cent in the corresponding period to June 1998. Construction levels for new industrial property reached a cyclical high in 1998/99 but declined thereafter plagued with fierce competition amongst builders and developers as can be seen from Table 2.9. While industrial property enjoyed the highest short-term ranking in total returns in 1998–1999, that mantle was lost to the greatly improved office markets and to a lesser extent, the perennial retail sector (C.H. Williams Talhar and Wong, 1998–2001).

2.2.6 Changes of economic policies due to the financial turmoil

In the aftermath of the Asian financial crisis Australia saw one of the important elements of the Australian economy suffer. Export market to the southeast Asia, the star performer during the pre-crisis era, began to slow down largely due to the reduced growth in demand for its products and the much lower real (inflation-adjusted) exchange rates of southeast Asian countries. As the sluggish consumption began to characterize the weakening economy at home, the country's deteriorating trade balance with the ASEAN countries, in particular neighbouring Indonesia, Malaysia and Thailand and to a certain extent Singapore, marked the first six months of 1998. As the effects of the financial crisis spread to major trading partners such as Japan and South Korea, Australia received a repeat ripple on its economy. Traditionally, Australia exports huge quantities of coal, iron ore and meat to ASEAN, making ASEAN countries, collectively, the second most important export destination for Australian products after Japan, according to one report (*Business Thailand Magazine*, 2001). According to one report (*Business Thailand Magazine*, 2001), Australia's trade balance with Indonesia plummeted from a surplus of US$380 million (Aus$644 million) in the first 6 months of 1997 to a deficit of US$357 million (Aus$605 million) in the same period of 1998. At the same time, its trade with Malaysia, which was US$75 million (Aus$127 million) sank to US$262 million (Aus$444 million). Australia's trade with Thailand suffered a similar fate, falling from a US$165 million (Aus$280 million) surplus to a US$76 million (Aus$128 million) deficit. The number of tourists coming to Australia from Japan and other Northeast Asian countries was 9 per cent lower in the June quarter of 1999 than in the same quarter of 1998.

"Estimates of reductions in Australian economic growth resulting from the negative external shock ranged from 0.2 to 1.0 percentage point in the year or two immediately after the financial crisis" (Hanratty, 1997). Difficult dilemmas for the current setting of Australian monetary and fiscal policy were thus created. However the key aspects of Australia's reform process according to the government sources (Hanratty, 1997), which supported the economy through the crisis, include:

- "A clearly articulated and demonstrated medium-term framework was adopted to achieve underlying balance, on average, over the economic cycle in addition to the legislated Charter of Budget Honesty to enhance the credibility of Australia's fiscal policy.
- The landmark tax reform, including income tax cuts, and a range of high priority spending initiatives were adopted to help lift the economy.

Table 2.9 Industrial market indicators in different cities in Australia

	Sydney			*Melbourne*			*Brisbane*			*Perth*			*Adelaide*			*Canberra civic*		
	1999	*2000*	*2001*	*1999*	*2000*	*2001*	*1999*	*2000*	*2001*	*1999*	*2000*	*2001*	*1999*	*2000*	*2001*	*1999*	*2000*	*2001*
Prime net effective rents ($ m^2)	137	103	110	65	67	70	94	65	69	55	57	59	60	60	60	58	57	58
Prime yields (%)	9.00–9.25	9.00–9.25	9.25–10.25	8.50–11.0	8.25–9.50	8.50–10.00	8.50–9.50	8.50–9.25	8.50–9.50	9.00–10.25	9.00–9.75	9.00–9.25	10.00–10.75	9.00–10.75	9.00–10.25	11.0–12.50	10.50–11.50	10.50–11.50
Prime capital value	1,501	1,129	1,201	667	749	744	1,044	743	827	571	640	692	578	615	634	489	573	686

Source: C.H. Williams Talhar and Wong (1998–2001).

- Strong support to the Reserve Bank's low inflation target and its independence in setting monetary policy by the government had enhanced the monetary policy credibility. This together with a more competitive Australian economy, had kept inflation low despite the significant decline in the Australian dollar that accompanied the crisis in Asia.
- Ongoing structural reforms, including those in labour and financial markets, had produced a more flexible and productive economy capable of responding to the Asian crisis. This improved flexibility was demonstrated by the capacity of Australian exporters to diversify their sales outside of the major export markets in Asia to other more strongly growing economies during the Asian crisis".

2.2.7 Economic outlook in the next decade

Following the fairly impressive economic performance during the 1990s and noted as by far one of the most resilient economies during the Asian financial crisis, outlook for the Australian economy is more robust than ever. The survey of 2001 predicted that Australia's real GDP will grow by up to 4.5 per cent in 2002 and by 4.4 per cent in 2003. This outlook is higher than the forecasts for all other major economies including the US (3.2 per cent and 4.3 per cent), Britain (2.3 per cent and 3.5 per cent), the Euro area (1.5 per cent and 3.7 per cent) and Japan (−0.2 and 1.8 per cent). The three reasons highlighted by *The Economist* (2002) for the buoyancy of Australia's economy are the competitive Australian dollar, surging prices for homes and the fact that Australia was less affected by the IT crash than many other developed countries. The magazine pointed out that while most of the world's stock markets fell in 2001, the Australian Stock Market touched new highs.

The outlook for the Australian economy remains bright as it expects to experience a continued solid growth and relatively low inflation. The 10-year growth outlook suggests a real GDP growth of 3.9 per cent between the year 2002 to 2006 and 4 per cent thereafter until 2011 (Table 2.10). Consumer price inflation is also likely to remain stable between 2.5 and 2.8 per cent in the coming decade.

In view of absence of any imminent political or policy changes over the next decade, the Australian government is likely to continue its privatization drive as an ongoing reform in the economic structure of the country. In the next decade or so, industry policy is likely to be the focus of the economic drive. Provisions made by the government to attract large-scale projects will be the engine for growth in the medium term. In particular, the stringent criteria applied in selecting these projects, will ensure that wider economic objective goal is met. Some of the criteria include the

Table 2.10 Ten-year growth outlook for Australia

% annual average	*2002–2006*	*2007–2011*
Real GDP growth	3.9	4.0
Consumer price inflation	2.8	2.5

Source: International Finance Centre (2001).

Table 2.11 Economic growth forecast for the decade comparative to the region

Country/region	*1995*	*1996*	*1997*	*1998*	*1999*	*2000*	*2001*	*2000–2010*
Australia	4.4	4.0	3.9	5.1	4.4	4.0	0.5	3.6
ASEAN*	8.2	7.3	3.7	−9.6	2.2	4.2	4.8	5.1

Source: Ernst & Young International (2000).

Note
* Indonesia, Malaysia, Philippines, Thailand, Singapore, Vietnam.

capacity of the project to generate employment and deliver investment opportunities that otherwise would have been lost reflect both maturity as well as forward thinking while making decisions.

In an alternative forecast, it is expected that Australia is likely to experience boom conditions (similar to the boom conditions in the 1980s with a major bust to follow at the end of the decade) through the middle of the decade, due to simultaneous upturns in a number of sectors (Table 2.11).

2.3 The construction sector

2.3.1 *Labour productivity*

The productivity performance of the Australian economy had been extraordinary. For example, in terms of labour productivity in the market sector, over the period 1988–1989 to 1997–1998, labour productivity growth had increased at an average rate of 2.5 per cent per year, compared with the long-term labour productivity growth over the 35-year period since the mid-1960s to the late 1990s of around 1 per cent per annum. “The average increase in labour productivity across all industries in the market sector between 1998–1999 and 1999–2000 was 1 per cent, while over the period from 1991–1992 to 1999–2000 the average annual increase was 2 per cent” (ABS, 2002). The growth over the period 1988–1989 to 1997–1998 was particularly strong in mining, electricity, gas and water, and communications, each in excess of 6 per cent per annum (Table 2.12). Finance and insurance, manufacturing, and wholesale trade also showed strong labour productivity growth over this period, in excess of the average of 2.5 per cent across all market sectors.

In 1999–2000 construction showed the largest decrease in labour productivity of 8 per cent. However over the period from 1991–1992 to 1999–2000 labour productivity increased at an annual rate of 1 per cent. In 1999–2000 labour productivity decreased for this industry because the growth in the chain volume estimates of gross value-added (2 per cent) was less than the growth in hours worked (11 per cent) (ABS, 2002). In contrast, labour productivity in the communication services industry increased by an average of 8 per cent per year over the period 1991–1992 to 1999–2000, because gross value-added grew faster (11 per cent per year) than hours worked (3 per cent per year). These figures however may not necessarily represent the full picture owing to the fact that changes in the composition of labour, which are not captured in the hours-worked measure, can affect output. This can also be affected by changes in inputs other than labour (e.g. capital).

Table 2.12 Indexes of gross product* per hour worked, by selected industry**

Industry	*1999–2000 Index number*	*Change from 1998–1999 (%)*	*Average annual rate of growth 1991–1992 to 1999–2000 (%)*
Mining	116.9	16.9	4.8
Agriculture, forestry and fishing	100.3	0.3	3.3
Supply manufacturing	101.3	1.3	2.7
Electricity, gas and water	104.2	4.2	7.2
Construction	92.2	−7.8	1.2
Communication services	103.1	3.1	7.5
All industries	101.4	1.4	2.3

Source: Australian System of National Accounts (1999–2000).

Notes

* Reference year for chain volume measures is 1998–1999 = 100.

** Estimates presented in this table relate only to industries in the market sector.

Table 2.13 Labour productivity in the 1990s

	1991–1995	*1996–1999*
Labour productivity growth	1.71	2.75
"Multi-factor" productivity	0.75	1.64

Sources: Oliner and Sichel (2000); Toohey (2000).

Labour productivity in the Australian construction industry ranked near the top of OECD nations, with access economics putting only the UK in front. An international comparison by quantity surveyors Page Kirkland, found the base cost of construction in Australia was less expensive than in Germany, the US and the UK. Australia has the lowest construction output prices of any country. Australian construction workers work very long hours – ranking second only to American workers. Table 2.13 shows the labour productivity in the 1990s.

2.3.2 *Construction cost*

Unit construction cost

Table 2.14 shows the unit construction cost for different building activities in Australia for the year 2002. The application of GST will have a direct impact on the unit construction cost of buildings (Table 2.15). According to a study by Davis Langdon and Shea International (2002), "there will be a net increase in cost due to a GST (net in this instance indicates that, GST has been refunded, assuming the client

Table 2.14 Unit construction cost for different sectors of building activities

Building	*Building*	*Services*	*Total*
Residential			
Multi-unit – low rise	770–795	130–405	900–1200
Medium quality – high rise	705–1030	345–470	1050–1500
High quality – low rise	735–1100	315–450	1050–1550
High quality – high rise	925–1335	495–615	1420–1950
Commercial			
Average standard offices			
Low	655–860	345–490	1000–1350
Medium	625–1110	625–740	1250–1850
High	915–1400	585–800	1500–2200
High standard offices	1330–1760	670–1040	2000–2800
Industrial			
Light industrial	325–445	85–105	410–550
Heavy industrial	400–560	100–140	500–700
Hotel			
Resort	1450–2140	550–1360	2000–3500
3 Star budget	1280–2125	520–775	1800–2900
5 Star luxury	1370–2375	930–1025	2300–3400
Suburban motel	940–1445	310–515	1250–1960

Source: Davis Langdon and Shea International (2002).

Table 2.15 Change in cost anticipated due to application of GST (in per cent)

Development components	*Industrial*	*Hospital*	*Hotel*	*Offices*	*Office fit-out*	*Residential*	*Retail*
Land costs	0.5–1	0–0.5	0.5–1	0–0.5	0–0.5	0.5–1	0.5–1
Construction	–1–3	0–0.5	–3.5–4.5	–2–4	–6–8.5	–1–2.5	–2–3
Finance	1.5–3	8–9	3.5–4.5	1–3	–5.5–1	0.5–2	2.5–3.5
Development charges	0–1	0–1	0–1	0–1	0–1	0–1	0–1
Total development cost	–1–2	0.5–1.5	–1–2	–1–3	–7–8	–1–2	–0.5–1.5

Source: Davis Langdon and Shea International (2002).

is entitled to a refund, and compared against pre-GST costs)". It is interesting to see that the majority of building types reduce in cost over the whole development process. This is primarily due to the savings achievable in WST outweighing any increase in financial restraints. On the other hand, the differences in cost impact are because different buildings have a higher or lower proportion of materials that are subject to WST. "A hotel fit-out will not increase as much in price post GST as an industrial building, which is built from materials not subject to WST" (Davis Langdon and Shea International, 2002).

2.3.3 Construction quality

During the 1970s, the standard Association of Australia published Australian Quality Systems Standards AS 1821, AS 1822 and AS 1823 that were based on the military quality assurance to provide guidance for the manufacturing industry. Recognizing a need for an Australian Quality Systems Standard to reflect more closely the workings of the engineering project industry, the Standard Association of Australia commissioned the preparation of an Australian Quality Systems Standards AS 2990 "Quality System for Engineering and Construction Projects". Modelled on the Canadian CAN3 Z2999 series of Quality Systems Standards. This standard first appeared in 1987.

In order to drive up the performance of construction the government in 1986 commissioned a "Review of Standards and Accreditation and Quality Control and Assurance". This report recommended measures for the implementation of a national strategy to improve quality and to adopt international standards as far as practicable. Since the adoption of the recommendations of the report by the government, the following developments have marked the initiatives for quality assurance and enhancement:

- Adoption of ISO 9000 series of Quality Systems Standards and the withdrawal of AS 2990 in 1987.
- Provision of government funding for the development of quality practices.
- Establishment of the joint Accreditation System of Australia and New Zealand to maintain surveillance and control of Certificate Organisations, the Auditor Registration Schemes and Auditor Training Courses.

Since 1990 the public sector construction industry in Australia began to use quality assurance as the springboard for reform of the industry. The ISO 9000 series of quality system standards began to play an important quality assurance part in the construction industry. This international standard consists of three standards dealing with quality systems that can be used for external quality assurance purposes.

1 ISO 9001 – model for quality assurance in design/development, production, installation and servicing.
2 ISO 9002 – model for quality assurance in production and installation.
3 ISO 9003 – model for quality assurance in final inspection and test.

As a result, approximately 33 per cent of current Australian Standards are fully or substantially aligned with International Standards. (It is important to understand that there are some areas of industry where no significant International Standards exist, such as building, construction and occupational health and safety. Around one-third of Australian Standards simply have no international equivalent.)

A number of key industry associations and government bodies have been working in promoting quality in the industry. Some of them are:

- The Institutions of Australia
- National Public Works Council

- Master Builders Australia
- Royal Australian Institute of Architects
- Metal Trades Industry Association
- Association of Consulting Engineers
- Standards Australia
- Quality Society of Australian
- Department of Industry, Science and Technology
- Joint Accreditation System of Australia and New Zealand.

Australian Building Codes Board (ABCB)

An agreement between the Commonwealth of Australia, the States and the Territories on 1 March 1994 saw inception of Australian Building Codes Board (ABCB) responsible for "developing and managing a nationally uniform approach to technical building requirements, embodied in the Building Code of Australia (BCA); developing a simpler and more efficient building regulatory system and enabling the building industry to adopt new and innovative construction technology and practices" as embodied in the Building Code of Australia (BCA) (ABCB, 2001). The role of ABCB has been significant in promoting industry's best practices, enabling the industry to adopt innovative technology and practices and raising wider awareness in the community. Implicit in the agenda of the ABCB is also to monitor and standardize the vast number of private sector work across the country. To complement ABCB activities, the Government has launched the Building and Construction Industries Action Agenda with a committed funding of Aus$3.6 million for activities for its initiatives including an international benchmarking study. This is widely considered as "a blueprint for generating sustainable growth in the industry to address issues such as information technology, regulations and standards, project delivery and business improvement, environment, workplace relations, training and skills development and exports" (ABCB, 2001).

The ABCB is assisted at both strategic and operational levels by the Australian Building Codes Committee, which is the apex advisory body on technical matters and provides major technical input into the development and maintenance of the BCA.

The DIST prepared a report on Australian Building Regulations, Standards and Conformance Systems to assist the Construction Regulations and Standards Internationalisation Group in achieving its objective of reducing non-tariff barriers to trade in housing and construction goods and services (Australian Procurement and Construction Council (APCC, 2001)). The report is targeted particularly for use overseas by domestic practitioners and for government-to-government discussions when explaining Australian building and construction regulatory systems and how they operate. Performance-based designs have led to a number of developments which might not otherwise have been possible under a prescriptive building code. The Star City Casino in Darling Harbour is an example of industry using innovative solutions to meet the regulator's goals as they apply to fire protection, while increasing functionality and reducing the cost of construction. Similarly, the Great Southern Stand at the Melbourne Cricket Ground was built using the performance requirements of the Building Code, enabling an innovative approach to structural design.

2.3.4 Construction safety record

The construction industry in Australia as internationally, is one of the major industry occupation groups with significant injury risk. Construction safety therefore had been singled out for improvement measures for safety performances introduced in the past two decades. The national tables produced in late 1980s indicated that there were considerable differences in terms of occupational injury experience between industry sectors. The industry divisions experiencing the highest frequency rates were: mining (2.7 times the average), construction (2.0 times the average), and manufacturing (1.6 times the average). The three industries accounted for nearly half of all occupational injuries but just over one quarter of all wage and salary earners.

A total of 234 people employed in the construction industry died in work-related incidents between 1989 and 1992 in Australia. This means there were 10 deaths per 100,000 workers in the construction industry (compared to 14 deaths per 100,000 workers during 1982–1984). Australia's construction injury and disease rate during the same period stood at 44.7 per 1,000 persons which is nearly twice the all-industry rate. While there has been a decline in the rate of death for workers in the construction industry, this figure is double the all-industry average of 5.5 deaths per 100,000 workers according to the government statistics (NOHSC, 1993). The Australian National Occupational Health and Safety Commission estimated that there were 2,900 work-related fatalities in Australia. The total employment in Australia was 8.235 million in 1995. The frequency rate for fatal work-related injuries (work-related accidents and diseases) is 35.2 per 100,000 workers. In 1996, the proportion of fatalities that occurred in each industry division in Australia excluding the ACT revealed that the transport and storage industry recorded the highest number of fatalities (64), followed by the manufacturing (63) and construction (46) industries.

Table 2.16 shows that the fatality rate in the Australian construction industry decreased substantially in the decade spanning 1984–1994, while it picked up again to a figure of 13.1 in 1995. The construction industry therefore has been singled out for safety performances improvement measures in the past two decades. OHS legislation in Australia is a specific regime of protective legislation which dates back to the British Factory Acts of the early nineteenth century. During the 1970s and 1980s following the 1972 Robens Committee report that recommended that an "effectively self-regulating system" be promoted, Australia adopted the new-style legislative approach with the enactment of an array of laws such as the *Health and Safety at Work Act 1974*, and *Industrial Safety, Health and Welfare Act 1981*. These begin with the statutory framework of the *Occupational Health and Safety Act 1985* (OHSA), which is buttressed by a number of regulations made pursuant to that Act and supplemented by a range of measures that provide operational support, assistance and guidance. The latter measures include codes of practice and guidance notes. A number of approved codes of practice are specifically related to construction activities, including those directed to tilt up construction, the safe use of cranes in the building and construction industry, and the code on demolition. Again, much of the guidance material issued, including alerts, guidance notes and safety check lists, is focused on the construction industry.

As part of the Occupational Safety Research programme, research has been carried out in the construction industry. A key part of the work has included a literature

Table 2.16 An overview of fatality rate in Australian construction industry

Period	*Reported fatality rate, Fr/100,000*
1982–1984	14
1989–1992 average	10
1995	13.1
Total fatalities in all industries in 1995	35.2

Source: compiled from NOHSC various reports.

review, and field studies by the research team of some 75 selected work-sites, with interviews with over 550 construction personnel at all levels. The study has identified that apart from the need for improved equipment design, upgraded technology and use of lighter-weight materials, the other key factors adversely affecting safety performance include:

- safety is not factored into the tendering process
- minimal site governance with a large amount of subcontracting
- self regulation with inadequate levels of compliance control.

There is also a lack of accountability for injury risk by system designers such as architects and engineers, who do not typically consider aspects of construction or operation and maintenance in terms of risks to health and safety. There is a clear need to consider the consequence of any design, ranging from exposure to injury risk during construction through to operation, maintenance and refurbishment/demolition.

Initiatives of the New South Wales government

In 1994, the NSW government released the Occupational Health Safety & Rehabilitation (OHS&R) Management Systems Guidelines for managing OHS&R on government construction projects. The Guidelines were initially applied to major projects having a value of $20 million or more, or projects of a lesser value that were of high safety risk. Between 1994 and 1998 government construction agencies progressively lowered the threshold value of $20 million to facilitate improved OHS&R performance across their projects overall. The Guidelines were well received and by mid-1998, 243 service providers' OHS&R corporate management systems had been accredited by CPSC (NOHSC, 2000).

In July 1998, the government's white paper "Construct New South Wales" was released. Construct NSW outlined a vision for the construction industry of 2005 and an ongoing development programme for industry. Subsequently the new edition of guidelines was released in November 1998, which extends the application of systems elements by including design and internal OHS&R reviews as mandatory elements. They also outlined clearer requirements for subcontractors. A guide to assist government agencies and service providers in auditing OHS&R management systems was also released to facilitate further improvements in implementation.

2.3.5 *Skill level of workforce and supervision level*

A skills enhancement programme in the construction industry was introduced in 1996 whose key objectives mere to improve the quality of work, and therefore the efficiency, safety and productivity of construction workers on the project through continual skills improvement and through the gaining of new skills and knowledge; identify current skills held and review skills by random audit; register and obtain where possible certification of skills with Construction Training Australia; and identify and implement training for construction and office staff.

In Australia, more than $200 million has been allocated for a package of reforms which build on the strengths of existing arrangements and address the complexities and shortcomings of the training system. This contributes to the Commonwealth's overall expenditure of $1.7 billion in vocational education and training. The new measures are designed to ensure that the vocational education and training infrastructure delivers skills efficiently and effectively.

2.3.6 *Research and development*

By 1997, the construction industry in Australia had a turnover of Aus$101 billion and a total of Aus$13.6 billion was expected to be spent on construction in the next 3 years by the private sector alone. This made the construction industry greater than any other industry in the country thereby raising some concerns about the need for the industry to be responsible, sensitive to the urban requirements of cities, and most importantly increasing uses of new technologies (APCC, 2000). Paradoxically, the net amount of research and development conducted by Australian building and construction industry is minimal. Research and development by private companies in particular is especially minimal and service companies, especially in those offices which depend upon design and innovative aspects of their work, research lags due to both the complexity of claimable expenses as well as inability of the wider movement to penetrate the operation level of the industry. As an example, the BHP spends 1.6 per cent of its sales in research and development. This represents $200 million a year (APCC, 2000). Likewise, Boral, James Hardie and ICI, some of Australia's most important building companies, spend only 0.2, 1.2 and 1.1 per cent of their sales on research and development. In general, less than 2 per cent of Australia's enterprises invest in any sort of research and development and there are many corporate icons spending at internationally low levels in their industry sector.

Table 2.17 shows that total R&D expenditure grew in the first half of the 1990s, peaking in 1994–1995 and decreasing thereafter. As a significant development, although a minimal contribution in terms of actual expenditure, contractors' share of R&D expenditure continued growing throughout the 1990s indicated awareness among the private contractors of the seriousness of the business. Based on 1996 data, the total expenditure on R&D relevant to the built environment was 9.4 per cent of total R&D expenditure or US$13,546 million. Seventy-six per cent or US$10,327 million of this research funding was provided by private industry although built environment R&D accounted for 13.6 per cent of federal funding as opposed to 8.5 per cent of total industry federal input (ABS, 2000).

Table 2.17 Business enterprise R&D (ANZIC) in built environment sector, by segment

Industry segment	*R&D expenditure ($ m)*		
	1992–1993	*1994–1995*	*1996–1997*
Building products manufacturers	39.47	52.96	73.92
Contractors	15.17	12.66	16.26
Property operators	0.40	0.46	0.94
Building design professions	26.78	54.21	41.59
Construction	81.82	120.29	132.71
Electricity distribution	19.09	66.78	27.94
Gas supply	6.53	20.18	2.91
Energy utilities	25.62	86.96	30.85
Water supply, sewerage, drainage	32.08	13.38	6.52
Transport	17.64	12.40	17.56
Public order and safety services	0.88	0.71	0.75
	158.04	233.74	188.39

Source: Compiled from ABS statistics on business enterprise R&D.

2.3.7 *Main category of research projects*

Market research undertaken for AusIndustry in April 2000 found that there was a low uptake, and awareness, of government R&D programmes in the building and construction (B&C) industry in Australia. Only 12 per cent of the B&C industry had applied for some form of government support. Around 7.3 per cent said they knew of the government's taxation concessions and R&D grants and loans. However, a host of research and development work by Australia's leading science institution has received a new industry focus by close interaction with the Cooperative Research Centre (CRC). All of the research programmes involve collaboration with many of the CRC industry partners and provide an initial foundation for development of further research. Some of them are shown in Table 2.18.

2.3.8 *Policy and initiative on the use of IT in construction*

There is a growing use of IT in the construction industry. An estimate by the Australian Bureau of Statistics (1998) reveals that 35 per cent of businesses in the construction industry use computers in their business operations, with larger businesses far more likely to use computers than smaller businesses. The potential applications identified in a study in 1998 "Information Technology in the Building and Construction Industry: Current Status and Future Directions", by Fujitsu Centre, Australian Graduate School of Management and, Building Research Centre and Faculty of the Built Environment at The University of New South Wales are:

- electronic tendering;
- efficient and more accurate estimation and tendering;
- better project planning using 3-D (and 4-D) modelling for simulation, as well as better resource management using project planning and scheduling software;
- better inventory management and reduced costs;

Table 2.18 Research involving partners of Cooperative Research Centre (CRC)

1 The National Museum of Australia project involves an independent analysis by QUT/CSIRO and RMIT of the benefits of project alliancing and use of IT by the contractors led by Bovis Lend Lease. As well as providing valuable research collaboration experience between the CRC partners, the research is also providing practical feedback to the Australian construction industry through a series of national public workshops and integration into university coursework. Many new research opportunities have been identified and these are being integrated into the CRC programme.

2 The CSIRO/QUT Online Remote Construction Management research project is developing new technology in collaboration with the Queensland Government and industry partners. This will provide a foundation for further research in this area with the CRC partners, including the development and use of virtual construction environments.

3 The CSIRO State of Environment Study for the Built Environment is identifying major research opportunities for future building and other construction, that will help shape the detailed Environmental Sustainability research programme. This will build on the ESD and IT research being carried out by QUT/CSIRO with Queensland DPW and DISR, in collaboration with international research partners.

4 The Australian Greenhouse Office project by CSIRO to determine the issues to be addressed in new building regulations is now being taken up by the CRC partner, Australian Building Codes Board, and developed into new performance regulations. New research is required to increase the scope and accuracy of heating, ventilation and cooling models for building construction, as well as integrating these into building design and development of more energy-efficient materials and construction systems. The Lend Lease North Lakes development provides a ready application site for these new technologies and fits with existing partner collaboration with CSIRO.

5 The International Alliance for Interoperability (IAI) project led by CSIRO, involves close existing collaboration between many of the core partners including Queensland Government Departments, Lend Lease, John Holland, AESmith, Woods Bagot, Jones Lang LaSalle, Rider Hunt and QUT. This will form a substantial part of the ongoing research to develop and implement the common IT language for the building and construction industry across all segments and supply chains. The CRC consortium provides a substantial supply chain base in Australia to be both a leading edge developer and early adopter of this technology. The Australian developments will provide significant inputs into IAI international development and open up wider international project opportunities.

6 The University of Sydney's projects on Virtual Worlds and the Online Graduate School will provide the foundation for the Virtual Environments programme. These have shown how such environments can operate and the potential benefits that can be obtained from them. The projects have indicated what the fundamental research issues are that need to be addressed in order to bring these environments to the commercial development stage. The Lend Lease North Lakes development project will be used to trial the resulting technologies and include research collaboration from QUT's Virtual Environment Lab.

7 The University of Sydney's AI in Design and related projects bring the agent technology into both the Virtual Environments and the Integrated Design and Construction Support Systems programs. Agents are a foundational technology and these projects have delineated the research needed to allow for distributed collaborating agents and for embedded agents. The resulting technologies will be tried across projects involving virtual design, virtual construction and virtual supply-chain management with this CRC's industry partners.

8 The University of Sydney's projects on soft computing, optimisation based on evolutionary systems and constraints, and expert systems provide some of the bases for the research to support the Integrated Design and Construction Support Systems program. The resulting performance improvement tools will initially be used to develop demonstration systems that link to current design tools.

Source: Compiled and inserted from various issues of Built Environment, and Innovation and Construction Technology.

- computer-based work-flow management;
- better project/contract management through data sharing, and document management;
- elimination of duplicate effort through improved access to information;
- better project control and co-ordination because of ready access to latest information;
- better accounting/budgetary control and cash flow management;
- use of Global Positioning Systems for setting out is a standard construction activity;
- improved safety management by using database systems;
- cross-sectoral integration to provide single source solutions;
- sourcing of components and materials using IFCs and from the Web.

The current use of IT in the Australian construction industry by contractors and specialist contractors, which is generally an automation of existing processes, is summarized as follows:

- Project planning and scheduling
- Estimation and tendering
- Project management and cost control
- General administration and accounting
- Asset management
- Electronic funds transfer
- Email for exchange of documents/information
- QA and OHS management
- Drafting.

In addition to the generic benefits noted earlier, other significant benefits that could be achieved are summarized below:

- A significant increase in turnover, in which IT has played a critical role;
- Faster and cost-effective communication to the extent that significant investment in IT is expected to pay for itself within 2 years;
- The ability to manage projects despite large distances between the head and site offices, including real time audio-visual linkage to review and discuss design/construction issues;
- The ability to overcome shortages of skilled workers by integrating processes through intensive use of IT in off-site pre-manufacturing processes;
- Early detection and resolution of problems related to project management;
- Significant reduction in re-work and duplication.

Impediments to successful adoption

Similarly, a large number of factors, often called risk factors, are known to be associated with unsuccessful adoption and implementation of IT in the construction industry in Australia.

- High cost of innovating/learning a new technology;
- Tight margins which make it difficult to fund innovating and learning;
- Fear of over-investing in IT – we were told several stories of companies actually or nearly going bankrupt as a result of spending too much on IT;
- Industry disinclination to invest sufficiently in capital;
- Belief that IT alone, without re-engineering/organisational change, can deliver promised benefits;
- Resistance to re-engineering/organisational change necessary to gain a pay-off from the technology;
- Lack of computer skills;
- Lack of awareness of the opportunities;
- Belief that the industry is doing sufficiently well that IT innovation is unnecessary;
- Client resistance;
- Senior partner and senior manager resistance.

Impediments to cross-sector IT

Cross-sector application of IT is a prime goal of IT use in Australia, however a number of impediments to cross-sectoral use of IT prevail:

- Lack of people with an overarching vision for the industry
- Fragmented structure of the industry
- Adversarial nature of the industry
- Lack of trust among firms
- Lack of a shared language by which to understand the supply-chain processes
- Lack of a shared technology.

Initiatives by NSW government

The NSW Department of Public Works and Services (DPWS) alone spends about $6 billion annually on capital works and maintenance, thus taking the lead in forcing the construction industry to adopt IT in order to improve productivity and reach international benchmarks. Launching its electronic contracting strategy, DPWS estimates that communication errors cost the construction industry about 1–2 per cent of total expenditure, which on $6 billion per annum amounts to $60 million – the price of one hospital, or seven high schools, or 30 Tangarra trains (Built Environment and Innovation, 1998).

The NSW government has a three-tier strategy for implementing electronic contracting in construction. In the first stage starting January 1999, selected projects exceeding $6 million and involving consultants and contractors well versed in electronic communication will be monitored, problems identified and benchmarks established. The process will apply to selected projects exceeding $3 million in July 1999 and exceeding $500,000 from January 2000. In the second stage, electronic project databases will be established by clients and designers, and product databases set up by suppliers. Their gradual uptake will be required on selected projects. With construction activity in NSW expected to continue at the current high levels in excess of $15 billion annually through to 2005, DPWS says people will have to invest in technology to stay in business (Built Environment and Innovation, 1998).

2.3.9 *Buildability*

A 1992 Commonwealth Government research report estimated that a 10 per cent lift in efficiency in the construction sector could boost GDP by 2.5 per cent, so that the CRC outcomes will significantly contribute to wider economic benefits for Australia (Built Environment and Innovation, 2001).

Australia's infrastructure is being expanded and rejuvenated, with a trend away from public sector provision to innovative private sector construction and/or provision, often through Build-Own-Operate (BOO), Build-Operate-Transfer (BOT) and Build-Own-Operate-Transfer (BOOT) projects. This is reflected in the decline of direct public sector construction activity from 24 per cent of construction activity in the mid-1970s to 14.6 per cent in 1996–1997. Nonetheless the public sector remains an important client.

2.4 Construction outlook 2002–2010

2.4.1 *Business opportunities and major projects in pipeline*

Some 80 per cent of all engineering construction work in Australia is classified as infrastructure and includes roads, railways, dams, bridges, major pipelines, electricity, telecommunications, mining and heavy industry. Business opportunities in Australia in the construction sector have seen some decline during the past years but several sectors are likely to remain strong and even strengthen in the years ahead. Among them mining, heavy industry construction, rail, pipeline and harbour sectors have been forecast to perform strongly (The US Commercial Services, 2001).

A survey by a leading Australian project developer and economic forecaster estimated (BIS Shrapnel, 2001) the value of engineering construction work for the year ended 30 June 2001 at US$5.7 billion (down 11 per cent on 1999–2000), with US$5.8 billion forecast for 2001–2002. According to a leading industry forecaster BIS Shrapnel (2001), this signifies a temporary but a noticeable halt in the decade long boom in engineering construction. The reason for the decline is, however, the oversubscribed projects in the 1990s decade. This meant, the biggest and most attractive projects for private investment launched in the late 1990s have largely been done and new government projects have not been brought forward fast enough to replace them. The period immediately following the Olympics saw a lull in the construction activity but only to pick up once again. It is estimated that construction activity will regain its past growth (in real terms) in June 2004–June 2005 and exceed US$10.8 billion, before peaking in 2006 at a level well above the previous peak of the late 1990s says a new study, "Engineering Construction in Australia 2000–2015". Engineering construction activity is expected to exceed $20.5 billion in the year to June 2005. According to the study, railway construction will see an unprecedented level of growth between 2000–2002 and 2004–2005. In the next 5 years the railway construction will increase by 135 per cent primarily with the new funds invested in passenger services. This is despite the fact that the $4.5 billion Speed rail is unlikely to proceed. Concomitantly, the road construction will also have an impressive growth, although modest compared to railway construction growth, of over 30 per cent in the next 5 years. Harbour construction will be reasonably strong over the next 5 years with most activity in Western Australia, NSW and the Northern Territory. In WA harbours will need to be expanded to cope with increasing exports of minerals, oil and gas, metals and agricultural products. In NSW, activity will be driven by four separate port projects in Newcastle. Pipeline construction was expected to surge to very high levels from 2001–2002 driven by deregulation of the gas market, including restrictions on inter-state trade.

Likewise value-added was forecast to grow by 3 per cent per year between 2001–2002 and 2003–2004, driven mainly by residential housing demand. With the industry entering a new cycle of growth in the latter years, growth in value-added is expected to increase to 4 per cent per year in the following 3 years. Increased activity in the industry will boost employment, which is expected to grow at an annual rate of 3–4 per cent per year to 2006–2007 (The US Commercial Services Australia, 2001).

2.4.2 *Government initiatives*

Australia is a signatory to the WTO, but is not a member of the plurilateral WTO Agreement on Government Procurement. The year 1992 marks an important shift in deregulation of banking and broadcasting sectors allowing for foreign banks to open branches in the country and allowing up to a quarter of the time of paid advertisement to be filled with foreign-sourced material (The US Commercial Services, 2001).

The government stipulations require foreign interests to be made explicit above a certain value thresholds, including: "acquisitions of substantial interests in existing Australian businesses with assets of Aus$5 million or more (Aus$3 Million for rural properties); new businesses involving an investment of Aus$10 million or more; portfolio investments in the media sector of 5 per cent or more; all non-portfolio investments irrespective of size; takeovers of Australian companies valued at either Aus$20 million or more, or for more than 50 per cent of the target company's total assets; and direct investment of foreign governments irrespective of size" (Bureau of Economic and Business Affairs, 2000). Investment proposals for entities involving more than Aus$50 million in total assets are approved unless found contrary to the national interest. Special regulations apply to investments in the banking sector, the media sector, urban real estate and civil aviation.

As part of the wider micro-level reforms in the country's investment policy, it aimed at reducing notification obligations on businesses, decentralizing purchasing policies and streamlining administration of foreign investment policy without compromising the national interest. The introduction of Endorsed Supplier Arrangements (ESA) has also been seen to have significant impact on the majority suppliers. Concurrently, the investment policy involving the purchase of real estate aims to channel foreign funds into the real-estate activity targeted to real benefits to homebuyers and suppliers rather than for speculative purposes.

2.4.3 *Major projects in the pipeline*

The predicted decline in Australia's engineering construction industry over the next 2 years since the time of writing was expected to be moderated by strong performances in several sectors, including mining, water storage and supply, heavy industry and pipelines. Some other major development and road construction projects already started or committed are shown in Table 2.19.

Table 2.19 Ongoing development projects in Australia and those in the pipeline

Project	*Value* (US$)	*Estimated completion date*
Melbourne Docklands, Victoria	2.4 billion	2010
Urban Redevelopment of South Sydney	1 billion	2005
Linking Victoria Transport Programme	795 million	2003
Pacific Highway Upgrade, New South Wales	636 million	2006
Rebuilding New South Wales Country Roads	530 million	2010
Rapid Bus Transitways, Sydney	424 million	2010
Darling Harbour Wharves 9-10, Sydney	397.5 million	2005
Honeysuckle Redevelopment Project, Newcastle, New South Wales	371 million	2014
Newcastle High Speed Rail	344.5 million	2003
South West Metropolitan Rail, Perth	265 million	2003
VicRoads, Victoria	243.2 million	2003

Source: The US Commercial Services Australia (2001).

2.4.4 *Areas where foreign investments are needed*

Australia's GDP is more than the combined GDP of Malaysia, Singapore and Hong Kong and is ranked fifth in the world for growth competitiveness, based on the country's technological strength, quality of public institutions and economic conditions. Australia's foreign exchange market is the ninth largest in the world and the Australian dollar is the seventh most traded currency in the world and its share market is the thirteenth largest in the world with total market capitalization of $900 billion. With a reformed labour market delivering higher productivity and a more flexible, productive workforce, Australia has become an attractive destination for foreign investors particularly due to its strategic position providing access to the growing Asian market. Australia has continued to be successful in attracting FDI. "The government's current approach to macroeconomic policy and the impact of microeconomic reform have reinforced Australia's competitive standing as an investment destination" (FIRB, 2001).

Australia's developed market economy is dominated by its services sector (65 per cent of GDP), yet it is the agricultural and mining sectors (7 per cent of GDP combined) that account for the bulk (58 per cent) of Australia's goods and services exports. Australia's comparative advantage in primary products is a reflection of the natural wealth of the Australian continent and its small domestic market. The relative size of the manufacturing sector however has been declining for several decades, and now accounts for just under 12 per cent of GDP (Foreign Investment Review Board (FIRB) (2001).

The most recent ABS figures show that, in the period immediately prior to the Asian financial crisis, total FDI stocks were increasingly contributed by the US and the UK. From 1995 to 1997, both the US and the UK increased their total investment,

Table 2.20 Extent of US investment in selected industries, US direct investment position in Australia on an historical cost basis, 1999 (US$ m)

Category	*Amount*
Petroleum	3,344
Total manufacturing	7,646
Food and kindred products	862
Chemicals and allied products	2,924
Primary and fabricated metals	471
Industrial machinery and equipment	693
Electric and electronic equipment	166
Transportation equipment	957
Other manufacturing	1,573
Wholesale trade	2,146
Banking	2,705
Finance/insurance/real estate	8,465
Services	2,190
Other Industries	7,167

Source: Bureau of Economic and Business Affairs (2001).

and the US increased its percentage of total FDI in Australia. Strong interest from UK and US investors and other foreign investors has been stimulated by continuing privatization of utilities, including airports and power utilities. Table 2.20 outlines the US investment in Australia, the biggest foreign investor in the country which indicates the major sectors for investment in Australia (FIRB, 2001).

References

Australian Building Codes Board (2001), http://www.abcb.gov.au.

Asia Pacific Economic Cooperation (2001) "Economic outlook", APEC, http://www.apecsec.org.sg/loadall.htm? http://www.apecsec.org.sg/member/memberecreport/aus.html.

Australian Bureau of Statistics (1997) "ABS information consultancies: 5360.0 international trade in goods and services", ITR calculations.

Australian Bureau of Statistics (2001) *Engineering Construction Activity*, January 2002, http://www.abs.gov.au/auststats/.

ABS (2002) Year Book Australia 2002, "Industry Overview: changes in labour productivity", www.abs.gov.au/Ausstats/abs@.nsf/Lookup/4382E7ECDEDCD, F58CA256B35001C9568.

Australian Department of Foreign Affairs and Trade (2000) http://203.6.171.3/media/speeches/department/020910_sec_sydinst.html.

Australian Industry Group, *Construction Outlook Report*, November 2001, http://www.constructors.com.au/pages/splash_shortcut_02.htm.

Australian Procurement and Construction Council (1998, 1999, 2000, 2001, 2002), http://www.apcc.gov.au/apcc/publications/.

Australian System of National Accounts (1999–2000) (5204.0) "Industry overview: changes in labour productivity", http://www.abs.gov.au/ausstats/abs@.nsf/94713ad445ff1425ca25682000192af2/4382e7ec0edcdf58ca256b35001c9568!OpenDocument.

BIS Shrapnel (2001) "Building and construction in the Asia-Pacific region 2001–2004", http://www.bis.internetrix.com.au/cgi-bin/public_view.cgi?action = view&param = 15& process = category.

Building & Construction *Digest* Bulletins, various issues.
Built Environment Innovation and Construction Technology, various issues, http://www.dbce.csiro.au/apcc innovation/past.cfm.
Bureau of Economic and Business Affairs (2001) "2000 Country reports on economic policy and trade practices", Released by the US Department of State, March 2001.
Business Thailand Magazine, 2000, http://www.businessthailandmagazine.com.
C.H. Williams Talhar and Wong, International Property Report (1998–2001) http://www.com.my/resources/pmr/international/aus2001.html.
CIA (2001) The World Factbook, http://www.odci.gov/cia/publications/factbook.
Colliers Jardine (2001) Local Markets Review, Various issues, http://www.colliers.com/China/fr_market_local.html.
Construction Industry Authority (1999) "Country report: Australia", *Proceedings for the 5th Asiaconstruct*, October, Malaysia.
Construction Industry Authority (2001) "Country report: Australia", *Proceedings for the 7th Asiaconstruct*, October, New Delhi.
Davis Langdon and Shea International (2002) http://www.davislangdon.com.au/dlaustralasia/publications/../ index.html?publications/side_publications.html&publications/indextext.html.
Economics (2000) "The Economic Impact of the Olympics", 17 September 2000.
Ernst & Young International (2000), http://www.ey.com/global/vault.nsf/International/Asia_Pacific_Economic_Outlook_2000_2010/$file/AP_Economic.pdf.
Foreign Investment Review Board (FIRB) (2001) Report 2000–2001, http://www.firb.gov.au/policy_pubs/publications/AnnualReports/2000-2001/index.htm.
Government of Australia (2002) "Australia's century since Federation at a glance", Department of Treasury and Finance.
Government of Western Australia (2002) "Structure of the western Australian economy", Department of Treasury and Finance, http://www.treasury.wa.gov.au/TreasuryPublications/waeco02.pdf.
Hanratty, Phil (1997) "Economic and financial turmoil in south-east Asia: origins and consequences", Current Issues Brief 8, 1997–98, Economics, Commerce and Industrial Relations Group, 2 December 1997, http://www.aph.gov.au/library/pubs/cib/1997–98/98cib08.htm#ON.
http://www.dbce.csiro.au/innovation/past.cfm.
Master Builder, 2001, http://www.spec-net.com.au/bulletins/bulletin14.htm.
IMF (1994) "International Financial Statistics", December 1994.
Industry Science Resources (2000) *Building for Growth: An Analysis of Australian Building Construction Industries*, Australia.
International Finance Centre (2001) November.
NOHSC (1993) *Construction Industry*, NOHSC: Sydney, http://www.nohsc.gov.au/PDF/Statistics/case_construction.pdf.
NOHSC (2000) Monitoring the Implementation of Occupational Health Safety & Rehabilitation Management Systems on NSW Government Construction Projects, www.construction.nsw.gov.au/case_studies.
NSW, 2001, http://www.dpws.nsw.gov.au/Home.htm.
OECD (2000) OECD Economic Surveys1999–2000, *Australia*, OECD Publications.
Oliner, S. and Sichel, D. (2000) *The Resurgence of Growth in the late 1990s: Is Information Technology the Story*, FEDS Working Paper 2000–20 March.
The Royal Commission into the Building and Construction Industry (2002) Various Reports, http://www.royalcombci.gov.au.
The US Commercial Services Australia (2001), Various Reports, http://www.usatrade.gov/website/ForOffices.nsf/WebCCG/Australia.
Toohey, Tim (2000) "Information technology and productivity", *ANZ Economic Outlook*, October 2000, pp. 17–19.

Chapter 3

China mainland

Ministry of Works, China and S. Sharma

3.1 Introduction

3.1.1 Background

For centuries China was one of the leading lights of civilization. The country prospered in art, culture and economy to unprecedented heights. In the turn of the nineteenth century it began to change. Natural calamities, civil unrest and foreign occupation crippled the country. In the post-Second World War, the Communist uprising and eventual control brought the economic development to standstill. Since 1979 the country has been pursuing open-door policy in economic development. Significant achievements have been made up as China now has the second largest GDP in the world.

According to CIA The World Factbook (2001) China is the most populous country in the world, its population based on the 2001 census was 1,273,111,290. The proportion of males was slightly higher than that of females (1.06 males per female). In the past decades, China's population was increasing in a phenomenal rate, though the growth rate had been reduced to 0.88 per cent (2001 est.). China is an ethnically diverse country. The majority Chinese belong to the Han ethnic community (91.9 per cent) while there are other ethnic communities such as Zhuang, Uygur, Hui, Yi, Tibetan, Miao, Manchu, Mongol and Buyi; Korean constituting the remaining 8.1 per cent of the population. The country is officially atheist, but scattered Daoist (Taoist), Buddhist, Muslim (2–3 per cent) and Christian (1 per cent) religious practices can be found in China.

The standard Chinese or Mandarin (Putonghua, based on the Beijing dialect) is the official language in China with other recognized dialects such as Yue (Cantonese), Wu (Shanghainese), Minbei (Fuzhou), Minnan (Hokkien-Taiwanese), Xiang, Gan and Hakka. Over 80 per cent of China's population is literate.

3.1.2 Geography

China is located in Eastern Asia, bordering the East China Sea, Korea Bay, Yellow Sea and South China Sea, between North Korea and Vietnam at coordinates 35 00 N and 105 00 E. China has a total area of 9,596,960 square kilometres divided into land (9,326,410 sq. km) and water (270,550 sq. km) with land boundaries totalling 122,147.24 kilometres and a coastline of 14,500 kilometres. Border countries include Afghanistan (76 km), Bhutan (470 km), Burma (2,185 km), Hong Kong (30 km),

India (3,380 km), Kazakhstan (1,533 km), North Korea (1,416 km), Kyrgyzstan (858 km), Laos (423 km), Macau (0.34 km), Mongolia (4,676.9 km), Nepal (1,236 km), Pakistan (523 km), Russia (northeast) (3,605 km), Russia (northwest) (40 km), Tajikistan (414 km) and Vietnam (1,281 km) (CIA 2001).

China, by its sheer size, is one of the most diverse countries in terms of climate, terrain and people. The climate ranges from tropical in the south to subarctic in the north. The elevation extremes range from the lowest at Turpan Pendi (154 m) to highest points in the Himalayas at over 8,000 metres while the topography is largely varied consisting mostly of mountains, high plateaus, deserts in the west, plains, deltas, and hills in the east. China's staple natural resources include coal, iron ore, petroleum, natural gas, mercury, tin, tungsten, antimony, manganese, molybdenum, vanadium, magnetite, aluminium, lead, zinc, uranium, hydropower potential as well as arable land (10 per cent of the total land area), permanent pastures (43 per cent) forests and woodland (14 per cent) and others (33 per cent) and has no permanent crops. A total of 498,720 square kilometres (1993 est.) consists of irrigated lands (CIA 2001).

3.1.3 Political structure

The Western perspective on China is that of a "Communist state with a constitution and a legal system that is a complex amalgam of custom and statute, largely criminal law and a rudimentary civil code" (CIA 2001). State Council is the apex body of the National People's Council (NPC) served by National Legislature with over 2,300 representatives from across the country. China has Regional Assemblies in 23 provinces, 5 autonomous regions and 4 municipalities. China considers Taiwan to be its twenty-third province while Hong Kong and Macau operate as Special Administrative Regions.

3.1.4 Development process in China: through the years

For the most of ancient history, China remained a global super power with a share of global GDP hovering around 25 per cent until the late 1700s. In 1820, China accounted for 36 per cent of global GDP as compared to 29 per cent of the US in 2001 (World Bank, 2002). Chinese per capita income was also higher than most of the rest of the world until eighteenth century. However, internal strife and foreign exploitation quickly dragged the country to unprecedented poverty and misery. Between 1820 and 1950 the country saw its GDP collapse to mere 5 per cent. While other countries (especially industrialized nations) moved ahead with spectacular economic growth, China lagged behind.

With a population of 1.2 billion people and a land area of 9.6 million square kilometres, China is the largest country in the world in terms of population and the third largest in area. The sheer size of the country and its rapidly transforming economy present many developmental challenges. The country has successfully pursued a strategy of reforming and modernizing its economy for 18 years. In general, the reform measures were first introduced on an experimental basis in some parts of the country and were adopted on a national scale only when they had proved successful at the local level. This pragmatic approach to reform has helped transform the economy from a predominantly centrally planned system to one in which markets play an increasingly important role in allocating the factors of production and in increasing economic efficiency (World Bank, 2000).

Since the time the country took the liberalization route, the size of Chinese economy has nearly quadrupled with real output increasing by 9 per cent a year. With the gains in productivity manifesting in considerable increase in real incomes and living standards, China moved successfully towards eliminating poverty. World Bank (2000) estimates, before the end of 1980s, the positive effects of the reforms had already translated into the increased levels of incomes of some 170 million people who were below poverty line in the pre-reform era. The official poverty line was however set in the mid-1980s and is substantially lower than the international standard of a dollar a day. Concurrently, many of distortions and rigidities of the sluggish Soviet-style economy were also eliminated or reduced, and economic agents were allowed to make their own decisions based on market signals (World Bank, 2002). This facilitated the growth of an active non-state sector, which now accounts for more than half of industrial output and two-thirds of the country's GDP.

"Sustained economic growth and targeted-poverty reduction has brought down the number of rural poor from 80 million in 1994 to below 30 million, or about 3 per cent of the rural population, in 2001" (ADB, 2002). The per capita income of urban residents, however, in 1998 was at RMB 5425 (US$656), about 2.5 times that of their rural counterparts with some areas having incomes below half of that average. To put it in real estate perspective, however, a total of 400 million square meters of new residential buildings were completed in urban areas, as compared to 810 million square meters of new houses built in rural areas in the same year (NBS, 1998), signifying the importance the rural population plays in Chinese economy. ADB (2001) estimates about 213 million people, or 23 per cent of the rural population, still live below the poverty line.

3.1.5 Structure of the economy

The structure of the Chinese economy has been transformed by the rapid growth. The economy has been characterized by a large industry share – standing at 61.2 per cent of total GDP in 1990, with a smaller share of 24.4 per cent devoted to agriculture and a much smaller service sector constituting only 14.4 per cent of GDP. This reflects the legacy of Soviet-styled planned economy since the 1950s that had a far greater emphasis on select industrial activities and discouraged small- and medium-scale entrepreneurial growth. The current dominance of industrial sector in China's GDP has stemmed from the high degree of governmental intervention in the country's economy.

China's first transformation is from a rural agriculture-based economy to an urban-industrial economy (World Bank, 2000). While the GDP growth rate was high in the 1990s, there were significant differences among the sectors of the economy. Agriculture, which has been a dominant industry and employs more than half of the labour force in rural areas, grew at 3.8 per cent during 1991–1995. Agriculture grew more slowly. Consequently, the share of the agriculture sector decreased from 28.4 per cent of GDP in 1990 to 19.7 per cent in 1995. The share of the primary sector in employment declined from more than 70 per cent in 1978 to about 50 per cent in the late 1990s. The share of the industrial sector in GDP increased from 37 per cent in 1990 to 48 in 1995. The rise in the industry sector is also evident from its share of employment that shot up from 17 per cent of the labour force in 1978, to almost 25 in 2000 (World Bank, 2000).

3.2 The national economy

3.2.1 Macroeconomic review and outlook

China is the fastest growing economy in East Asia, at nearly 10 per cent p.a. since the beginning of economic reform in 1979. The country has also survived the East Asian currency crisis relatively unscathed. "The country is one of the very few socialist countries that have made a successful transition from a centrally planned to a market economy – the rate of interest (the price of money) and the exchange rate are the only prices that are still administratively determined on the margin" (Lau, 2000). Since the beginning of the reforms, in a 20-year period, the real GDP in China shot up to 1.08 trillion in 2000 from a humble 176 billion in 1979. Accompanied with this rise is the real per capita GDP that soared by over 300 per cent to US$860 in 2000 from a meagre US$182 in 1979 (Lau, 2000) (Table 3.1).

Throughout the 1990s, Chinese economy sustained a good trend of "high growth and low inflation". Between 1991 and 1996 the GDP growth averaged 11.6 per cent per year, making it the fastest growing economy in the world. The economic strength of the country significantly increased during the past decade that did not show signs of slowing down even through the financial crisis. The economy remained highly resilient with growth rates of 8.8 and 7.8 per cent in 1997 and 1998 respectively. During the same decade, the country made efforts to eliminate the shortage in commodities and streamline bottleneck industries such as transportation and power. Meanwhile, the new framework for macro-control aimed at reforming the establishment of the socialist market economy system helped re-structure the economy by enhancing the basic role of the market in resources allocation considerably. Unsurprisingly, as a strategy for a gradual introduction of liberalization policies, China sought to develop and integrate the parallel processes of promotion of public ownership (as the backbone of economy), and diversified forms of ownership, to form a unique economic pattern (UNIDO, 1992). The market-oriented shares in the national economy, in this new economic pattern, noticeably increased such that the private (non-state) sector accounted for more than 60 per cent of GDP by 2000. In the same period China further opened up to the world, in all directions, at all levels and in broad fields. Especially since May 1997, although the economy has experienced a great pressure from the economic crisis of Asia, it has kept its currency intact, providing a powerful support to its neighbouring countries. This stabilized the situation to an extent and continued contributing to the development of the world economy.

The 1990s also gave China a vital opportunity for retrospection and introspection in order to speed up its reforms. In the recent history of Chinese economic reform the key point of government's macro-control has been shifted from excessively

Table 3.1 Chinese GDP change since the reform

	1979 US$ (2000 prices)	*2000 US$ (2000 prices)*
Real GDP	176 billion	1.08 trillion
Real GDP per capita	182	860

Source: Lau (2000).

restraining rapid increase in demand to giving an active stimulus to effective demand. A profound change has been made in the environment for China's economic growth since the national economy successfully completed its soft landing in 1996 (refer Table 3.2). A buyer's market began to appear, and the demand became rather a factor for restraining economic growth than a factor for igniting overheated economic development. At the same time, a new mechanism for restraining demand was being gradually formed in the course of the profound reform. So the shift made by the government is evident.

The Chinese government over the years has consistently sought that the GDP growth stays above 8 per cent and the rate of inflation at less than 3 per cent and applied various measures to ensure this. The major driving force of the high growth was non-state investment, including huge inflows of foreign direct investment. The trade and current account deficits, which have been a concern in other rapidly growing economies in Asia, have not been a serious problem. During 1991–1995, the trade and current accounts remained in balance despite the year-to-year change in their positions. As a result, the country's economic performance has been astonishing throughout the 1990s. The average annual real growth rate of gross domestic product (GDP) for 1991–1995 was 11.7 per cent, one of the highest in the world.

In 1997, the Chinese GDP reached RMB 7477.2 billion (US$911 billion) indicating a threefold increase from the figures of US$363 in 1990. Whilst favourable revenue and expenditure has remained steady during much of the 1990s with an overall favourable balance of payments since 1993, China's balance of payments has been well positioned and the foreign exchange reserves continued to pile up throughout the period. The 1998 result was already the lowest return of the past 7 years and would have been much lower without the government's special infrastructure spending package, which helped stimulate the economy in the third and fourth quarters. With the GDP growth rate still the highest in the world, by 2000, the

Table 3.2 Average annual rates of growth of selected economic indicators (per cent)

	1952–1979 pre-reform	*1979–1998 reform*
Real GDP	6.20	9.82
Real GDP per capita	4.14	8.39
Real gross value of:		
Agricultural production	4.33	8.05
Light industry	7.83	11.30
Heavy industry	11.37	11.34
Real personal consumption	4.99	8.91
Real consumption per capita	2.96	7.51
Real gross fixed capital formation	11.43	11.10
Capital stock	5.93	9.77
Employment	2.52	2.91
GDP deflator	0.59	6.51
Retail price index	0.80	7.03
Exports (in current US$)	10.98	14.66
Imports (in current US$)	10.27	

Source: Lau (2000).

Chinese GDP rose to US$1076.9 billion (World Bank, 2002). The rates of growth of red GDP and inflation are shown in Table 3.3.

By early 1990s the foreign trade stood at $200 billion or roughly 40 per cent of GNP, which continued to increase in the subsequent years along with the country's foreign exchange reserves. At the end of 1999, China had amassed a massive US$139.9 billion as foreign reserve that increased by over US$25 billion by the end of 2000, signifying an overall increase of US$148 billion within a decade. By 2000, total exports reached US$249.2 billion, registering a growth of over 300 per cent when compared to US$62,091 in 1990. Likewise, total imports also rose by over 300 per cent to US$225.1 billion during the same period with a trade surplus of US$24.1 billion, making China the seventh largest trading country in the world. This confirms a slowdown in the growth of exports and a narrowing of the trade surplus. Export growth was expected to be zero or even negative in the near term (Lau, 2000).

By the early 1990s, foreign trade had grown to an unprecedented $200 billion or roughly 40 per cent of GNP. China's foreign exchange reserves reached US$139.9 billion at the end of 1997. Foreign exchange reserves continued to increase. By the end of 2000, foreign exchange reserves reached US$165.6 billion, an increase of US$148 billion compared with that at the beginning of the decade. At the same time, direct investment remained the main form of inflowing foreign investment. The sum of direct investment made by foreign businessmen was US$44.24 billion. In 2000, the actually utilized foreign direct investment (FDI) for the year was US$40.7 billion. "While FDI was negligible before 1978, by early 1999, FDI in joint ventures and wholly foreign-owned companies in China exceeded one-quarter of a trillion US dollars, several times larger than the cumulative FDI since the Second World War in Japan, South Korea and Taiwan combined" (Sharma, 1999).

China's economic reform however has always faced crucial moments of breakthrough at its domestic front. In the mid-1990, the new Chinese government session made series of decisions to extricate most big- or medium-sized state enterprises from a difficult position and help them establish modern enterprise systems, intensifying supervision by the central bank, giving commercial banks the power of independent decision through re-structuring China's financial systems, and streamlining the administrative structure of central and local governments (Tse, 2000). All these tasks

Table 3.3 Rates of growth of real GDP and inflation (per cent p.a.)

Actual	*GDP*	*RPI*	*CPI*
1997	8.8	0.8	2.8
1998	7.8	−2.6	−0.8
1999	7.1	−2.9	1.3
2000	8.0	−1.5	0.4
2001	7.3	–	–

Source: National Bureau of Statistics (NBS), 2001.

Note

The NBS projected that the award of the 2008 Summer Olympic Games to Beijing should add 0.3–0.4 per cent to the average annual growth rate.

are a matter of overall importance to the reform and have been ongoing for some years. They are both arduous and full of challenge, but their completion is bound to greatly facilitate the country's economic and social development (refer Table 3.4 for main macroeconomic indicators).

Table 3.4 Main macroeconomic indicators

	1995	*1996*	*1997*	*1998*	*1999*	*2000*	*2001*
GDP and its components (US$ billion)							
GDP at real prices (HK$ million)	704	772	840	906	–		
GDP at current market prices	704	818	901	958	436	81,910.9 (100 m yuan)	89,403.5
GDP growth (%)	10.5	9.6	8.8	7.8	7.6	8	7.3
Primary sector	144	142	157	163	–		
% growth	5	5.1	3.5	3.5	3		
Manufacturing sector	344	385	427	466	–		
% growth	13.9	12.1	10.8	9.2	9.1		
Services sector	216	233	252	272	–		
% growth	8.4	7.9	8.2	7.6	6.8		
Construction sector	46	50	54	60	–		
% growth	12.4	8.5	7.9	12	–		
Demographic Indicators							
Population	1,211	1,224	1,236	1,248			
Population growth rate (%)	1.08	1.07	0.98	0.97			0.88
Total labour force	679	688	690	699	–		
Labour force growth rate 9%	1.19	1.33	0.29	1.30	–		
Unemployment rate (seasonally adjusted)	2.9	3	3.1	3.1	–		
Financial Indicators							
Changes in consumer price index (%)	17.1	8.3	2.8	–0.8			
Changes in GDP deflator (%)	14.5	6.4	1.3	–1.2			
Short-term interest rate (%)	–	9.18	7.65	6.12			
Long-term interest rate (%) (over 5 years)	–	12.42	10.53	7.56			
Annual average exchange rate with US$	8.35	8.31	8.29	8.28			

Source: Compiled from ADB (2001), World Bank (2002), MOC various reports.

3.2.2 *Economy in 2001*

In 2001, China posted a GDP growth rate of 7.3 per cent to an estimated RMB 9.6 trillion (US$1.2 trillion), above the government's 7.0 per cent target (ADB, 2002). Industry registered a gain of 8.7 per cent, led by the robust performance of foreign-owned enterprises. The services sector grew by 7.4 per cent, supported by high growth in transport and telecommunications. Agriculture expanded by only 2.8 per cent due to a severe drought. Retail sales jumped 10.1 per cent in 2001 due to a hike in incomes. Purchasing power stayed largely in the hands of urban dwellers as the government increased salaries for civil servants and government retirees during 2001. While a spurt in automobile credit and housing mortgage lending stimulated consumer spending, inflation edged up by 0.7 per cent in 2001, compared with 0.4 per cent in 2000. "Yet some deflationary pressure built up as economic growth declined in the latter part of the year and prices fell slightly. Urban unemployment was pegged at 3.6 per cent in 2001, slightly higher than in 2000" according to the ADB (2001).

In 2001, domestic growth continued to be the key driver, with a combination of strong consumer demand (10 per cent year-on-year increase in retail sales) and government investment, including a 150 billion yuan (US$18 billion) bond issuance for infrastructure spending. Strong domestic demand helped offset sharp declines in external growth which declined from a remarkable 28 per cent year-on-year in 2000 to 7 in 2001, as the global high-tech downturn broadened out into a wider slowdown and reduced demand for Chinese manufacturing output. Imports grew 8.2 per cent year-on-year in 2001, narrowing China's trade surplus with the rest of the world to US$22.5 billion, compared to US$24.1 billion in 2000 (China–Britain Business Council, 2001). External factors have previously had a limited impact on Chinese growth due to the small share of trade in GDP, and its relatively closed economy. But trade as a share of GDP has grown from 18 per cent in 1980 to 43 per cent in 2001. The slowing of quarterly GDP growth rates in 2001, from 8.1 per cent in the first quarter to 6.6 in the fourth quarter, appears to reflect growing linkages with the external market, and the slowdown of the global economy in 2001 (ADB, 2001).

Economic outlook for 2002

The growth target for 2002 was to maintain GDP growth at around 7.0 per cent, as indicated in the annual average target the government has set itself for the 2001–2005 Tenth Five-Year Plan. "This equates to an almost doubling of GDP within a decade, but is seen by the leadership as the minimum growth rate for sustaining China's social and economic stability" (UK Parliament, 2001). The business sector has however remained wary of the risks of the downturn particularly after the quarterly decline seen in 2001. Unsurprisingly, at the end of 2001 the IMF revised its forecast of 2002 GDP growth from 7.2 to 6.8 per cent. The 2002 budget deficit as a percentage of GDP was expected to exceed 3 per cent following deficits of 2.9 and 3.0 per cent in 2000 and 2001 respectively (UK Parliament, 2001).

"The government has fiscal and monetary tools at its disposal, and had indicated that expansionary fiscal spending will continue and perhaps be expanded" (NBS, 2001). Around RMB 150 billion (US$18.1 billion) of further bond issues will be available for construction projects. In terms of revenues, 20 per cent year-on-year increases in taxes are helping to improve the government balance sheet. The revenue to GDP ratio, while on a strong upward curve, was only at around 17 per cent. On monetary policy, in February 2002 the government announced a quarter per cent cut in the interest rate on one-year deposits to 2 per cent, and a half percent cut in the one-year lending rate to 3.25 per cent. This was the first change since 1999. With official figures for annual price increases hovering around 0–1 per cent the central bank was more concerned about deflation than inflation according to the NBS (2001).

3.2.3 Role of construction/property market in national economy

China's construction market took off in 1992 as its economy heated up. By 1996, activity was five times higher than in the 1980s. In 1996, the government designated the construction industry a "pillar industry", with significant growth potential, especially since the contribution of the industry to the GDP (about 6 per cent) was still low compared to those of developed countries. At the time of writing, China's investment in housing construction had accounted for 20–30 per cent of the total fixed asset investment and 6–8 per cent of the GDP, according to *People's Daily* (2001). During the recession that followed the Asian financial crisis of 1997, the growth rate of construction began to slow down. But given the continued need for infrastructure construction to facilitate modernization efforts and stimulate the economy, the MOC anticipated an average annual growth rate of 10–13 per cent through 2000. In February 2000, the MOC announced a target growth rate of 15 per cent for housing construction for the first part of the new century. By January 1999, construction ranked third among all production sectors (behind manufacturing and agriculture) in terms of annual value-added output. The MOC expects the industry's annual domestic value-added output to hit RMB 1 trillion (US$12 billion) by 2010, then accounting for more than 6 per cent of the GDP. From 2000 to 2010, the MOC's focus will be on residential buildings, new cities, the burgeoning rural market and infrastructure construction. The building industry is playing a so important role in the economic development of China that the government has always considered it as a support industry for bringing along or promoting continuous, rapid and healthy development of the national economy. It has become the third largest support industry, next only to manufacturing and agriculture industries. Along with a new periodic cycle of China's economic growth in 1998 and in the next few years, the building industry has been provided with more opportunities for its development.

The building market of China is the largest in the world. A steady increase in fixed assets investment has created a sustained brisk demand for building construction. During the Eighth Five-Year Plan (1991–1995) the total sum of society-wide investment in fixed assets had already reached RMB 6163.7 billion (US$745.2 billion by the exchange rate of yuan to US$8.27:1). In particular, the construction and installation works directly completed by the building industry totalled up to RMB 3926.3 billion, constituting 63.7 per cent of the total investment in fixed assets. The total number of projects completed was 240,000 including 845 big or medium-sized ones. Four and half a billion square metres of housing in the urban and rural areas had been completed by 1997 (China–Britain Business Council, 2001).

Fixed assets investment had kept increasing by a big margin in China throughout the 1990s. In the light of the trend in economic growth and the objective of

Table 3.5 Volume of investment in fixed assets (unit: one hundred million yuan)

Year	*Society-wide investment in fixed assets*	*Percentage increases by comparison with last year*	*Total output of building industry*
1996	22,974	14.8	13,784
1997	25,300	10.1	15,100

Source: Ministry of Construction (1999).

Note
The total output of the building industry is calculated at 60 per cent out of the total value of society-wide investment in fixed assets.

macro-control of prices, an increase in fixed assets investment by 15 per cent was considered acceptable by the Chinese People's Bank. In the next 3 years since 2002, fixed assets investment continued to surge with the completed investment in fixed assets for the country reaching 3,261.9 billion RMB in 2000 (NBS, 2001). In terms of different economic entities, the investment in state-owned units and other type of units was 2,328.4 billion RMB, up by 9.2 per cent. That of collective units was 473.9 billion RMB, up by 9.2 per cent over the year 2001. The investment of urban and rural residents was 459.6 billion RMB, up 9.5 per cent according to the China trade statistics (source: NBS, 2001). Table 3.5 shows the volume of investment in fixed assets (NBS, 2001).

3.2.4 Construction investment (or market) volume

During the years from the early 1980s to 1997 the increase in fixed assets investment, vital to the development of the building industry had experienced several cycles of fluctuation. Most notably, fluctuations were seen in the period of the 1980s and 1990s decade and to a smaller extent at the turn of the century. By 1993, the growth in investment had reached its peak resulting in an overheated economy. The government then , in order to regain control over the capital construction, was forced to introduce a short-term macroeconomic re-adjustment for the next three years. As a result, in 1996 the investment growth rate dropped to 14.8 per cent and in 1997 further to 10.1 per cent. In the later half of 1990s, the growth rate slowed down progressively. The annual average growth rate of investment in fixed assets decreased to 12.6 in 1996–1998 and the growth of construction industry also slowed down along with it such that, the number of employees in construction industry was 34.49 million in 1997. Just before the financial crisis, China invested US$750 billion in infrastructure construction earmarked for the next 3 years enabling the building industry to enter a new periodic cycle of steady growth. Changes were made particularly in infrastructure works rather than industries that were in government ownership. The fluctuation at the turn of the century was more of a natural cycle of investment highs and lows rather that did not reflect structural problems.

China's construction industry continued to grow steadily through the years of Asian financial crisis and is today one of the biggest in the world, employing some 34 million workers. This is a fourfold increase since 1979 when the new reforms were

Table 3.6 Annual construction volume (1 million sq. m)

Year changes	*1995*	*1996*	*1997*	*1998*
Private buildings under construction	781.9 (8.18)	961.2 (22.9)	893.1 (−7.1)	–
Private buildings completed	887.1 (11.6)	1,038.4 (17.1)	1034.6 (−0.4)	
Total private construction	1,669 (9.97)	1,999.6 (19.8)	1,927.7 (−3.6)	
Public buildings under construction	826.4	790.8 (−4.3)	762.2 (−3.6)	515 (−32.4)
Public buildings completed	372.9	365.1 (−2.1)	366.7 (−0.4)	179.8 (−51)
Public residential construction	286.8	273.1 (−4.4)	274.1 (−0.4)	
Total public construction	1,199.3	1,155.9 (−3.6)	1,128.9 (−2.3)	694.8 (−38.5)
Total construction output	2,868.3	3,155.5 (10.0)	3,056.6 (−3.1)	

Source: Ministry of Construction (1999).

introduced. In recent years, the annual housing investment reached a volume of RMB 500 billion, accounting for about 20 per cent of the Investment of Social Fixed Assets.

The floor space of completed housing was 1200 million square metres annually. In the urban areas, the annual housing investment has been RMB 300 billion and the floor space of completed housing about 400 million square metres annually. By 2000, the value-added of construction enterprises reached RMB 591.8 billion, total numbers of construction projects peaked at 680,716, and the total floor space of buildings under construction stood at 1551.75 million square meters (NBS, 2001). By 2000, floor space of buildings completed during the year reached 738.35 million square metres based on the statistics from the China Council for Promotion of International Trade (2000) as shown in Table 3.6.

Housing development generally takes up more than 50 per cent of total construction in China. "Many urban landscaping, road paving and highway projects are not included in this total because they fall under areas of responsibility outside of the Ministry of Construction, such as the Ministry of Communication (highways) or municipal engineering bureaus (many local infrastructure projects)" (NBS, 2000). According to statistics by China's Ministry of Construction, from 1979 to 1998, urban housing construction alone totalled 3.8 billion square metres across the country, and the projection for the period from 1999 to 2010 is 3.35 billion square metres. Table 3.7 shows that total construction output in China has been rising over the years. The output in 2000 was estimated at one and half times its output in 1996 (NBS, 2000).

3.2.5 *Overview of property market*

With excess supply in virtually all sectors of the property market except the mass housing sector, the Chinese property market in major cities has considerably slumped when compared to the buoyant conditions of the early and mid-1990s. As a result of

Table 3.7 Total construction output in China (10,000 yuan)

Region	*1995*	*1996*	*1997*	*1998*	*1999*	*2000*
National total	–	82,822,497	91,264,777	100,619,922	111,528,640	124,975,961

Source: NBS (2001).

the Asian crisis, the growth of China's economy slowed down a little, and the volume of overseas investment decreased. While the Chinese economy, to a certain extent was, affected by the Asian financial crisis, the GDP growth rate remained at impressive levels – 8.8 per cent in 1997 and 7.8 in 1998. The continued influx of foreign companies was the evidence of continued confidence in China's economic reforms, and which also ensured that good-quality office buildings in prime locations remained an attractive investment. Total investments in real estate development amounted to 311 billion RMB in 1997, nearly 13 times the total amount in 1990. The accumulated foreign capital increased from 28.5 billion RMB in 1990 to 228 billion RMB in 1995 (NBS 1998). FDI in real estate increased from 22.3 per cent in 1996 to 27.6 in 1997, suggesting that real estate continued to be an attractive arena for foreign investors. Nationwide, FDI use reached US$42.3 billion in 1997, US$5.17 billion of it in the real estate sector. In 2000, RMB 490.173 billion was invested in China's real estate sector, increased by 19.5 per cent from 1999. "The capital source of the investment in real estate development mainly came from the earnest money and prepayments, self-collected funds and domestic loans. Money coming from these three sources amounted to 75.4 per cent of the total capital in real estate" (Tse, 2000).

The impact of the Asian financial crisis on the property market in Chinese cities was minimal. China's land prices in prime areas increased between November 1997 and May 1998, while all the other Asian countries saw a substantial drop in land prices. In China's office markets though, construction virtually ceased; as funding dried up, half-completed buildings were a common sight in the city as a painful consequence of the pre-crisis building boom. In the residential sector, as a result of the excessive drive for construction, not less than 66 million square metres of commodity housing were still unoccupied nationwide by 1997. Likewise over 80 million square metres of vacant residential space was in the market by 1998 whereas over 3.2 million households live in overcrowded housing units with floor areas of less than 4 square kilometres; in the new quasi-capitalist economy of China (Lai, 1998) typically characterizing the mismatch in the units supplied and the type of units required.

China's property prices gained slightly early in 2002 according to the survey by Shanghai Bureau of Statistics sampling 35 large and medium sized cities. In Beijing the general property index rose 1 point to 491. The office index was down 2 points to 818, but the housing index was up 4 points to 291 in 2001. Rental housing rose 5.7 per cent over the same period in 2000. Sale prices of houses rallied at 1.9 per cent. Transfer of lands went up by 1.4 per cent owing to active transaction (*People's Daily*, 2001).

Commercial property

The office market in China has really only established itself since the early 1990s and as such, is still developing and maturing. In Beijing for example with a limited supply of grade A office space and luxury residential premises, new supply in the market has a marked effect on the supply/demand relationship resulting in a dynamic market condition. There was an office stock of over 2 million square metres of varying quality premises and luxury residential property stock at around 10,000 units.

An overview of the period immediately after the Asian financial crisis indicates the office market falling by over 30 per cent in 1998 (Colliers Jardine, 2001). The net, exclusive asking rental for prime office facilities in Beijing declined to US$37.6 per square metre per month in the third quarter of 1998 to regain stability in the fourth quarter of 1999. The Beijing office market witnessed resurgence in demand accompanied by a slowdown in growth in supply, largely due to delay in completions of office developments on the part of developers under the weakened office market conditions in the late 1990s. Following different bilateral agreements with the United States and the European Union on the terms for China's WTO accession, the rapid expansion of IT, hi-tech and telecom companies in Beijing stimulated a sharp spike in demand for quality office facilities and resulted in the continued rebound of the Beijing office-leasing market. The average asking net rental for prime Beijing office accommodation was recorded as rising sharply by 29.3 per cent to US$33.8 per square metre per month, from US$26.1 per square metre per month in 1999. As of year-end 2000, the average vacancy was recorded at 17.6 per cent, a decline of 11.1 per cent points over the end of 1999 according to the market research by Colliers Jardine (2001).

Beijing saw a huge slump in demand for office space in 2001, up to 52 per cent YoY decline in net absorption. Accordingly, average vacancy rates for the city reached 15.83 per cent at the end of the same year. The market therefore offered a wide assortment of options for large-space users in the primary and secondary business districts, with rentals and concessions on average more favourable than in the previous year. At the end of 2001, total stock of office space in Beijing reached 4.63 million square metres.

Net absorption figures for 2001 clearly indicated the weakening of the office market as total net absorption fell by 52 per cent YoY in 2001, the equivalent of nearly 500,000 square metres. As a result, the average vacancy rate increased by 4.40–15.83 per cent by the fourth quarter of 2001. In 2002, total net absorption was forecast to reach just over 490,000 square metres. Net take-up was expected to significantly increase to 730,000 square metres for 2003 (refer Table 3.8).

Table 3.8 Beijing property market: key market indicators

By Grade	*Rents (US$/sq. m/month)*			*Office stock*			*Vacancy rate (%)*		
All Districts	*March 2001*	*March 2002*	*March 2003 (f)*	*March 2001*	*March 2002*	*March 2003 (f)*	*March 2001*	*March 2002*	*March 2003 (f)*
Prime	46.55	41.30	41.30	144,000	144,000	144,000	3.75	5.33	3.75
Grade A	30.39	28.20	27.58	1,714,713	2,154,713	2,588,713	14.10	18.48	23.79
Grade B	21.55	21.43	20.61	2,133,846	2,471,816	3,152,816	10.09	14.45	20.57

Source: Colliers Jardine (2001).

Residential Property

The residential market in China had been a thriving one during most of the 1990s except mixed results in the higher-end apartment buildings, which saw an actual rate of decline closer to 10 per cent on average in the major cities. The reason for this increased volatility in quality apartment rentals is the dramatic increase in supply of dwellings within this sub-sector since 1996. Second, higher-end residential properties that had come on stream since year-end 1997 have been of outstanding quality, and generally provided a good array of support facilities, resulting in a highly competitive market. In Beijing, unlike the office market, conditions in the city's residential sector remain mixed, as the residential market did not respond immediately to the more positive sentiment in the investment market.

Since 1999, annual completions of luxury condominium apartments dropped significantly, attributable to the 1995 government-imposed clampdown on granting sites for high-end or non-essential development projects according to Colliers Jardine (2001). As a result in 2000, only three overseas sale apartment properties came on stream, providing 836 apartment units. At the same time, the downward adjustment in luxury apartment rentals helped to stimulate demand and resulted in a rise in take-up for such properties. As of year-end 2000, the average vacancy level of luxury apartments was recorded at 18.9 per cent, the equivalent of a drop of 4.8 per cent compared with 1999, according to Talhar and Wong's (1998–2001) property reports. At the same time by 2000, Chinese commercial housing under construction reached 635.3 million square metres. The commercial housing sold reached 169.8 million square metres, and the total sales 357.2 billion, increased by 26.9 per cent and 30 per cent respectively over 1999 figures. The average market price of Chinese commercial housing was 2,103 yuan per square metre, 2.5 per cent higher than 1999.

At the beginning of 2002, Beijing's residential stock of foreign-sales apartments reached 28,154 units, of which nearly 25 per cent were luxury apartments, while new supply totalled 1,298 units, all of which were classified as mid-range apartments. Total new supply was projected to reach 4,988 units in 2002, representing a stock generation rate of 15.66 per cent for the year, about 3 per cent lower than for 2001 (refer Table 3.9).

Table 3.9 Residential market in Beijing – key market indicators

By District	*Rents (US$/sq. m/month)*			*Vacancy (%)*		
All Grades	*March 2001*	*March 2002*	*March 2003(f)*	*March 2001*	*March 2002*	*March 2003(f)*
All Districts	17.04	16.87	17.47	23.02	32.26	27.51
Chaoyang	17.10	16.77	17.60	21.38	30.34	27.45
Chongwen	14.20	14.80	14.20	43.78	23.83	17.12
Dongcheng	19.05	20.31	20.38	22.02	28.47	27.02
Haidian	16.57	14.67	15.29	17.20	48.55	35.10
Xicheng	15.00	14.18	14.25	15.28	14.60	11.56
By Grade, All Districts						
Luxury	22.75	23.30	24.27	17.65	20.71	20.21
Mid-range	14.63	14.00	14.33	25.02	35.98	29.77

Source: Colliers Jardine (2001).

3.2.6 Changes of economic policies due to the financial turmoil

When the stories of financial crisis swept through the tiger economies of South east Asia in mid-1997, speculation was rife as to how soon China would fall. There were good reasons to speculate given China's trade and investment relationship with its Asian partners along with its inherent structural problems arising from overheated economy and property bubble, akin to those countries already affected by the crisis. It was expected that China's growing manufacturing industry would be directly hit by the falling demand as well as the new competition coming from Thailand and Indonesia which saw their currencies devalue to very low levels. China, however, proved speculations wrong and escaped the financial crisis without much damage to the economy. The country's ability to sustain a GDP growth performance through the crisis years (8.5 in 1997 and 7.8 in 1998 and 7.0 per cent in 1999), and continual confidence in attracting foreign direct investment (FDI) put the economy back in track. China, steadfastly maintained the stability of its currency, the RMB, in the face of plummeting currency devaluations and steep property price deflation elsewhere in the region (Sharma, 1999), which proved to be biggest gamble for the country in which it emerged victorious. In fact, the financial crisis worked to an extent the opposite way, as, increasingly wary investors found China a more reliable investment destination.

3.2.7 Impact of the Asian crisis on construction

As a consequence of the Asian crisis, certain impacts were evident in the construction sector that required the government to prepare immediate corrective measures.

Floor space of unsold housing, office and commercial building increased

By the end of 1998, the total floor space of unsold housing, office and commercial building increased to 80 billion square metres, including 60 billion square metres of housing. Buildings that had not been sold for over one year formed 50 per cent of the total. The main reasons were:

1 Since the economy was over-heated during 1992 and 1993, the development of real estate lacked effective control and a large quantity of land was supplied.
2 The drawbacks in the projects completed were that:

- they were not aligned with market needs in terms of location, type and standard;
- the locations of some new towns are remote from cities and the municipal facilities had not been extended to these areas;
- service utilities, layout of housing, construction quality, environment standards could not meet market requirements; and
- the prices were not easily affordable.

Real estate prices varied

Because of this oversupply, real estate prices varied. According to the China real estate index system, for Beijing the housing price index was 875 in 1995 and 879 in 1998.

The office index decreased from 1,641 in 1995 to 1,479 in 1998. The comprehensive index also decreased. The Shanghai index indicated that the price indices of residences and offices decreased even more drastically.

Impact on construction industry

The scale of fixed assets investment is the dominant factor affecting the growth of the industry. Approximately 65 per cent of investment in fixed investment are construction related. During the first 5 years of the 1990s, the volume of investment in fixed assets in China had increased at an exceptional speed, compared to the rest of the world. The annual average growth rate was 36.9 per cent. During that period, the construction industry also attained unprecedented growth and the number of employees in the construction industry increased rapidly from 24 million in 1990 to 33 million in 1995 in China. In the later 1990s, the economic growth rate of China slowed down progressively. The annual average growth rate of investment in fixed assets decreased to 12.6 per cent during the period 1996–1998. The growth of the construction industry also slowed down with it.

3.2.8 Policies and strategies adopted by public and private sectors to overcome slowdown in construction sector

A number of policy measures were taken by the government to overcome the slowdown experienced by the construction sector in the aftermath of the financial crisis.

Issuance of supplementary treasure bonds

In order to cope with the economic crisis in Asia, the Chinese government adopted active financial policies to enlarge investment and domestic demand. The government announced in 1998, about US$750 billion was to be raised by the government and invested in agriculture, water conservancy, traffic, transportation, environment protection, public utilities and other infrastructure works for 1999–2001. The Chinese government issued supplementary treasury bonds of 100 billion RMB in 1998. The money raised was targeted mainly towards the construction of 601 infrastructure projects. The total investment, expected to be over 200 billion RMB, was to be resourced from different government funds, bank loans and public investment.

Policies for clearing up unsellable housing

The state council stipulated three policies in order to clear up the unsold housing and office space in August 1999. First, the local government aimed at eliminating various unreasonable fees in order to decrease housing prices. Unsellable housing was diverted to the category Economic-Suitable Housing, with low-rent housing. All such unsaleable housing sold before 31 December 2000, would be exempted from business tax, contract tax and various fees under the restricted price scheme. Second, the land supply and new development would be strictly limited to cities where there were large quantities of unsellable housing. Third, state-owned banks would clear up the loans of real estate through recourse of debts and auction of mortgaged real estate.

Control of the numbers of construction companies

Since the beginning of 1999, the Chinese government postponed approvals of applications for funding new construction and design enterprises for one year throughout the country, in a bid to reduce the intense competition in the construction market and to control the numbers of construction enterprises.

3.2.9 Economic outlook in the next decade

Driven largely by domestic factors and supported by WTO accession, the Chinese economy is likely to continue to grow robustly over the medium and long terms. Economic growth was forecast at 7 per cent in 2002 and 7.4 in 2003, led by strong domestic consumption and investment. The industry and services sectors are expected to grow by 7–9 per cent, and the agriculture sector by 2–3 per cent. The Tenth Five-Year Plan (2001–2005) aims to achieve (NBS, 2000):

- an indicative plan rather than a mandatory plan;
- doubling of real GDP between 2001 and 2010, with an implied rate of growth of 7.2 per cent p.a;
- an inflation target of less than 3 per cent p.a;
- an increase in the share of central government revenue in GDP (the introduction of a comprehensive individual income tax) – tax revenue rose 22.8 per cent to 1.266 trillion RMB, or 14.2 per cent of GDP, in 2000;
- indirect (macroeconomic) control of the economy using instruments such as money supply, interest rate and exchange rate rather than direct (microeconomic) control through administrative directives, commands and central planning with mandatory targets.

One estimate for the period 2000–2010, is that the Chinese economy will have a more likely growth rate in the range of 6–7 per cent based on positive factors such as high national savings, abundant and low-cost labour supply, and the improvement of human capital and the dynamics of the non-state sector, influx of FDI, low level of capital stock per capita and its large domestic market (NBS, 2000). It is expected that China will continue to maintain a better than 30 per cent gross national savings rate and a similarly high investment rate in 2000–2010 (refer Table 3.10). According to a World Bank estimate, China's physical capital per worker in industry was US$1,600 in 1995 and will be US$13,200 in 2020 (Wu, 1999).

Accession of China to the WTO will undoubtedly boost the growth potential and attract a flood of FDI in China. "Long-term growth of 6.5 per cent is forecast for the coming decade by the market watchers in and outside China with internal demand being a more important driver than exports" (People's Daily, 2001).

Domestic consumption and investment was expected to remain strong in 2002–2003 with a growth rate of about 9–10 per cent. Increased domestic consumption and continued solid economic growth were expected to exert some upward pressure on prices. While the fiscal deficit was forecast remain at around 2.8–3 per cent of GDP in 2002–2003, inflation was expected to be moderate at 1–1.5 per cent due to significant excess capacity in many industry subsectors, large unutilized supply potential of the agriculture sector, and cheap imports after WTO-related tariff liberalization according

Table 3.10 Economic forecast for China in comparison to total Asia-Pacific region and ASIAN

Country/Region	*2001*	*2000–2010*
China	7.3	5.0
Total Asia-Pacific	5.1	5.3
ASIAN	4.8	5.1

Source: Ernst & Young International (2000).

to the ADB (2001). ADB (2001) assumed that the global economy would begin to recover in 2002 and gather further momentum in 2003, and exports were forecast to grow at the rate of 6 and 10 per cent respectively. With WTO accession, imports were expected to continue to grow faster than exports, and the trade and current account surpluses to gradually decline. However, the decline in the current account surplus was expected to be partly offset by foreign investment inflows as the government further opens up the economy to fulfil WTO commitments. It was expected that the economy will post a current account surplus of US$12.6 billion, or 1 per cent of GDP, in 2002, which was likely to fall to about US$5 billion, or 0.4 per cent of GDP, in 2003. The official foreign exchange reserves were expected to be close to $230 billion by 2003, while the debt–service ratio to fall below 10 per cent. (ADB, 2001).

3.3 The construction sector

3.3.1 *Labour productivity*

Labour productivity has increased rapidly since China adopted the open-door policy. In the 1980s, the growth rate of national labour productivity (NLP) was 6.3 per cent per year and it increased to 10.6 per cent in the 1990s. The labour productivity (LP) of construction (PLC) is slightly higher than NLP. There are many factors which caused this increase. There are many factors which caused this increase. First, there was an increase in capital investment on equipment. Second, the reform's privatization drive has had positive impacts on construction enterprises. Last, the rapid growth of the economy has had positive impact on productivity. However the assessment of productivity in construction during recession periods had been inaccurate because during this period, the construction industry also took up unemployed workers from other sectors. This balancing system has been seen as the principal reason for the low construction productivity in China.

Table 3.11 shows that value-added per employee rose by over 8 per cent between 1995 and 1997. This is attributed to overall improvement in efficiency, higher level of technology and the privatization drive. Valued-added per employee in construction sector is still much lower than for other sectors.

Table 3.12 shows the LP of construction enterprises in various years calculated in terms of value-added construction. According to this, construction value-added had experienced an over 40 per cent growth between 1996 and 2000. This can also be attributed to increased productivity, private sector participation and greater competition.

Table 3.11 Value-added per employee (US$ per person per annum)

	1990	*1995*	*1997*
Construction		1342	1457
Manufacturing	681	2200	2686
Services	592	1283	1576
Primary sector	157	407	485

Source: Ministry of Construction (1999).

Notes
Primary sector includes agriculture, mining, etc. Data for 1997 onwards not available.

Table 3.12 Labour productivity of construction enterprises in various years (calculated in terms of value-added of construction)

Region	*1995*	*1996*	*1997*	*1998*	*1999*	*2000*
National Total	–	11,337	12,089	13,350	14,451	15,929

Source: China Council for Promotion of International Trade (2000).

3.3.2 *Equipment for construction enterprises*

Since China adopted its open-door policy and began its reform policies, many highways, energy and water projects have been completed and these have improved overall level of mechanization in the construction industry, especially in city areas. Multi-storey buildings were increasingly common. In 1996, there were more than 100 high-rise buildings over 120 metre tall. Since nearly all multi-storey buildings are reinforced concrete structures, there has been an increase in demand for the following types of machinery.

1 Concrete machines, including concrete mixers and concrete pumps. In the early 1990s, the readymixed concrete volume made up 60 per cent of the total concrete volume, and up to 80 per cent for some cities that are closer to developed countries in nature. Upto 1995, 616 ready-mixed concrete plants were built with a design capacity of 60 million cubic metres and the actual demand was 26 million cubic metres in China.
2 Tower cranes were imported in the early 1980s. Some brand names are POTAIN from France, LIEBHERR from Germany, and SIMMA from Italy. Factories for construction machinery in Beijing, Sichuan, Shenyang, etc. have capability to satisfy the basic needs of the nation, and to export part of their products.
3 Excavation machines, such as machines for diaphragm wall construction, drilling, and various types of piling plant and equipment.

In 1999, the average utilization percentage for construction equipment was 89 per cent for the whole of China; with a high of 95–98 per cent for the best-practice

Table 3.13 Level of mechanization of construction by projects

Comprehensive projects	*Earth work*	*Stone work*	*Pavement*	*Stake base*	*Concrete projects*	*Structural*	*Fitting up*
60%	80%	90%	80%	100%	98%	96%	30%

Source: Ministry of Construction (1999).

enterprises; and a low of 52 per cent for the worst enterprises (refer Table 3.13). The two reasons for low utilization were, first, lack of maintenance and second, management's understanding of the benefits of mechanization. A reduction of maintenance budget will also subsequently lower the performance in construction. There were several reasons for the low utilization of equipment (50–65 per cent). One example is the inadequacy of equipment inventory and management, and the lack of information on available equipment and technology.

3.3.3 Mechanized construction

At the end of the Eighth Five-Year Plan, the Ministry of Construction issued a Summary of Industrial Construction that included:

- using advanced industrial techniques and equipment;
- organizing scientific and logical construction;
- developing professional construction;
- improving the level of mechanization and decreasing labour-intensive wet trades;
- developing construction materials, fittings of structures, and forming suitable scales for business enterprises;
- providing the construction market with various, suitable, widely used and systematized building materials;
- improving the standard of construction;
- using modern management roles and strategies, implementing scientific organization and management;
- educating and developing technical market and information management system.

In China, the future development for the construction machinery will focus on promoting the use of large- and small-sized machines. The emphasis on large-sized machinery development is to substitute manpower in construction where the human capacity is limited, especially in multi-storey construction. Special machines are also used for constructing tunnels, channels, for water conservancy and electric power projects, etc. The emphasis on small-sized machinery is to satisfy special needs, for example small equipment are required for decoration. These small-sized machines can reduce labour dependence and improve the rate of LP. Accordingly, the level of mechanized construction was expected to grow with development of the rental market for construction machinery.

3.3.4 Construction cost

Average construction material price

Table 3.14 shows that construction material had been cheaper over the years. The costs of major materials such as cement and steel had dropped by 15 and 22.5 per cent respectively between 1992 and 1998. Such a drop can be attributed to higher productivity, competition, application of new technology and the introduction of the open market system.

3.3.5 Construction industry salaries and wages

Table 3.15 documents the data available for average sectoral wages in China between 1995 and 1998. Construction sector wages have seen only 29 per cent growth between the period which is far below the wage increase in other sectors such as retail (38 per cent), manufacturing (52 per cent) and transport (41 per cent).

3.3.6 Construction quality

The Chinese economy has moved into a new period of rapid growth since 1978. During this period, the construction scale and the construction teams swelled so fast that the quality of some construction projects could not be fully guaranteed and the

Table 3.14 Indicators of construction cost in the case of state-owned construction companies

Items	*1992*	*1995*	*1996*	*1997*	*1998*
Direct cost	116.6	104.1	104.1	103.2	100.0
Wages	116.5	120.9	116.2	113.6	104.8
Materials	117.2	99.5	101.6	100.8	98.9
Steel	124.6	94.8	99.8	99.1	96.5
Wood	112.6	102.0	101.8	101.3	98.5
Cement	115.8	100.7	102.7	100.6	98.8
Local materials	109.3	99.5	104.0	101.9	100.5
Other materials	113.8	104.7	103.2	102.2	100.8
Other cost	107.0	112.2	101.7	105.3	103.5

Source: Ministry of Construction (1999).

Table 3.15 Average sectoral wage (US$ per person per annum)

	1995	*1996*	*1997*	*1998*
Construction	697	753	802	898
Manufacturing	623	680	715	951
Retail and catering	512	562	584	707
Transport and telecommunications	837	948	1,036	1,182

Source: Ministry of Construction (1999).

original system of administering, construction quality based on their own control and assessment by the planners, designers, constructors and supervisors themselves, gradually revealed some malpractices. The Chinese government by and large has been concerned with improving the quality of construction owing to the large number of defective buildings ever since. A countrywide survey of construction quality by China's Ministry of Construction indicated that the rate of unacceptable, defective housing is 16 per cent. Assuming that there are 0.477 billion square metres of newly built housing annually in China, this means a huge amount of defective housing (Guan *et al.*, 1997). Already in the existing stock, there are 27.07 million square metres of housing in a dangerous state in the whole country of China (1996), which is 8 per cent of total housing.

3.3.7 Governmental systems established to supervise construction project quality

As far back as 1984, the Ministry of Construction together with the State Planning Commission and other relevant departments made a joint proposal to establish governmental systems for the supervision of construction quality with an early anticipation of new problems to come. The proposal was drawn with reference to international experience. For instance, Document No. 123, Interim Provisions on Reformation of the Administrative Systems of Construction Industry and Infrastructure Projects, promulgated in 1984:

> the method of supervising construction quality shall be reformed. Those common local Projects for civilian purposes shall be put under supervision of the authoritative administrators relating to the local governments in accordance with the relevant laws and regulations. It is required that the authoritative organizations for supervising construction quality be established in all various cities.

Thereafter, the three-sided efforts have been made to establish governmental systems to supervise constructional quality.

Accordingly, the governmental legal systems on supervision and administration regarding construction quality were established. Many laws and regulations have been drafted and issued by the Ministry since 1983, such as Supervisory Rules for Construction Project Quality, Interim provisions on Supervisory Stations for Construction Project Quality, Regulations on Examination and Testing for Construction Project Quality, Supervisory and Administrative Rules for Construction Project Quality (revised version in 1990) and lastly, Administrative Rules for Construction Project Quality (revised and issued in 1993). Various local administrative departments in charge of construction also drafted and issued local laws to enable enforcement of rules on construction quality supervision in line with their respective local conditions. At the time of writing 22 provinces, autonomous regions and municipalities have issued such local laws and regulations. With these central and local laws and regulations being implemented, a good basis was built to bring supervisory practice within the legislative framework. The two major initiatives taken by the government which have acted as the foundation for quality control in Chinese construction are:

1 A national supervisory network for construction quality was established and implemented – Since 1984, many supervisory stations for construction quality across the country were established in the capitals of various provinces, autonomous regions and municipalities. The professional departments concerned under the State Council also instituted their own supervisory stations, for the carrying out of supervision on the nuclear, forestry, coal, hydropower, transportation, railroads, telecommunications and petrochemical industries. By the end of 1998, there were 450 supervisory stations under the specific departments of the State Council. At country level, 2,600 other such stations across the country had 90 per cent, with 70 per cent of 40,000 professionals serving in these stations possessing professional titles in engineering and technology, which covered 88 per cent of all construction projects.
2 Institutionalized operating system for examination and testing approval – A three-level operating system necessary for examination and testing of construction quality has been implemented across the country. By 1997, there were 5 centres at the state level for construction, air-conditioners, elevators, construction machinery and gas appliances, and another 5 centres approved by the Ministry of Construction for glass-wall buildings. There were 50 centres at the provincial level, 600 centres at prefecture level and another 600 centres at county level, with a total of 6,000 professionals engaged in examination and testing. These centres are equipped with rather sophisticated testing instruments and facilities to assist them in fulfilling their duties.

Table 3.16 gives us a rough idea on the quality of completed projects in China over the 1990s and 2000. It is observed that the proportion of high-quality

Table 3.16 Construction quality of construction enterprises

Year	*No. of projects completed*	*High quality projects (no.)*	*High quality projects (%)*	*Floor space of buildings completed (10,000 sq. m)*	*High quality*	*% of high quality floor space completed*
1991	222,304	70,024	31.5	19,791.7	6,374.1	32.2
1992	238,909	77,895	32.6	23,476.9	7,828.1	33.3
1993	305,041	118,443	38.8	28,684.8	9,000.1	31.4
1994	286,884	89,346	31.1	32,383.3	10,429.3	32.2
1995	293,698	94,783	32.3	35,666.3	11,906.0	33.4
1996	419,246	124,436	29.7	60,047.9	19,935.3	33.2
1997	442,604	131,337	29.7	62,244.0	22,489.0	36.1
1998	415,501	126,069	30.3	65,682.6	24,353.6	37.1
1999	446,236	142,507	31.9	73,924.9	29,980.5	40.6
2000	534,319	155,172	29.0	80,714.9	35,843.5	44.4

Source: Ministry of Construction (1999).

projects has been about a third of all projects, although in 1993, the proportion went up to 38.8 per cent. While the proportion of high-quality floor space completed in the building projects has been consistently rising. The percentage shot up to 44.4 per cent in 2000, an increase of nearly 12 per cent since 1991.

3.3.8 Governmental supervision and administration effects assessment on construction quality

In the 1990s, the government played a significant role in upholding construction quality by preventing low-quality construction works from flowing into the consumer market:

1 The administrative departments examine and verify the qualifications of all engineering survey and design organizations and construction companies in China with provision for organizations without the stipulated qualified certificates to be penalized. At one time 40,000 such cases were discovered by the supervisory stations across the country.
2 The administrative departments make spot checks and inspections on parts of construction projects strictly according to the relevant standards and regulations. As a result, the previous downward trend of construction quality has been overcome, while construction quality has improved year after year. After two million projects with construction space of two billion square metres were put under surveillance of the supervisory stations, statistics showed that 60,000 cases of hidden danger and 18,000 cases of various accidents resulting from quality problems were uncovered. Eighty thousand rectification notices were dispatched. Projects with governmental supervision are seen to have higher construction quality than those without.
3 Strict supervision has urged construction enterprises to strengthen their managerial control of quality, to improve their internal quality-assurance system and to train their quality managers. To boost enterprises in applying GB/TI 9000–ISO 9000, the Ministry of Construction has issued a guidance note "Implementing Rules for Enterprises to Fulfil ISO 9000 Serial Standards". Over 200 enterprises were issued ISO 9000 certificates by 1998. China National Accreditation Council (abbreviated as "CNACR") approves and authorizes quality system registration bodies.
4 The departments have been paying serious attention to complaints received concerning quality. As a result, enterprises were required to carry out self-examinations and return checks on construction quality of 180,000 projects of 500 million square metres completed during the Eighth Five-Year Plan period. In addition, the administrative departments also re-examined some 160,000 projects with space equivalent to 340 million square metres. Those with substandard quality were required to apply rectification measures.
5 Any defect that is hard to rectify is subject to penalties. The departments at all levels imposed penalties on 823 construction enterprises responsible for quality-related problems during the Eighth Five-Year Plan. Penalties included reducing

the qualification grade, suspension of tendering, fines and circulation of criticisms to newspapers according to the extent of malpractice. The overall quality improvements are reflected as follows:

- Key state projects and large-sized infrastructure works and those which adopted sophisticated technology had achieved a relatively higher quality level almost at par with the international advanced standard throughout the 1990s.
- The qualified rate of common construction works for civilian use is improving annually. The qualified rate for Seventh Year Plan period remained at 62 per cent, but the number for Eighth Year Plan period increased to 82 per cent with the average annual increase of 3 or 4 per cent.
- Some common quality faults have been overcome. For instance, house floor and bathroom leakages recorded during random inspections carried out by the Ministry of Construction in 1991, showed a reduction from the 60 per cent in both in the 1980s to 7.3 and 1.8 per cent respectively.

3.3.9 *Measures to strengthen supervision and administration of construction quality*

Although China has made substantial improvement in construction quality over the years, it was still not at par with the international standard at the time of writing. Accidents, common faults and other quality-related problems are still rampant. The government of China as a way to improve the quality performance introduced new policies to select contractors through open tendering and those of substandard quality performance are excluded from the market. This system was divided into two types. One, a system whereby a contractor has to apply for construction permission before commencement. The other, an employment qualification system requiring a review of eligibility of professionals such as surveyors, designers, constructors and supervisors. China has applied the employment qualification system for many years, and now a registration system for architects, structural engineers, supervision engineers and economists has become mandatory. Other initiatives taken by the government are as follows:

- strict adherence to construction procedures;
- the quality responsibility system for parties engaged in construction should be established;
- to improve content, scope and methods of governmental supervision on construction quality;
- reforms in content, scope and methods of governmental supervision; adopting a utilization approval system as a major supervision measure;
- to adopt a utilization approval system for construction materials. The construction materials and components that do not have admission license would not be allowed on the market. Counterfeit construction materials and components would be banned;
- to promote science and technological progress and improve quality.

Table 3.17 Distribution of fatal events by industries (1991–1997)

Year	*Manufacturing*	*Construction*	*Transport, storage, telecom*	*Service*	*Others*
1991	9 (47)	5 (26)	2	0	3
1992	8 (42)	7 (37)	1	3	0
1993	3 (12)	14 (54)	4	4	1
1994	26 (26)	60 (59)	12	1	2
1995	19 (19)	61 (62)	11	3	4
1996	17 (29)	27 (47)	6	5	3
1997	15 (24)	33 (53)	6	4	4

Source: Zhu *et al.* (2000).

3.3.10 Construction safety record

In 1997, 17,588 fatalities from work accidents, including significant numbers in mining and construction industries, were reported in China. Table 3.17 shows construction is the most hazardous industry in the new development area with fatal accidents accounting for 47–62 per cent of all fatal events from 1993 to 1997 (Lam, 2000). The mortality rate was 9.1 per 100,000 workers over the period of 1991–1997. It is also observed that more than half of the fatal incidents occurred in the construction were falls, collisions and electrocution. These were the major source of fatalities accounting for 81.9 per cent of all fatal events (Zhu *et al.*, 2000).

In a national scale inspection, the Bureau of Work Safety "examined 293,000 enterprises and institutions and detected 325,200 potential risks that could lead to accidents". However, no generic occupational, safety and health (OSH) law exists to prescribe the minimum safety and health standards in China. China has also not ratified the International Labour Organisation's (ILO) Convention Nos 155 and 161 (on OSH) which are the most important international commitments to OSH. National rules and regulations concerning OSH are numerous but are implemented by different government ministries. Furthermore, individual provinces pass their own decrees on OSH which can only be applied at provincial level. In recent years, even local officials are demanding a national OSH law.

The Labour Law of 1994 relates to OSH matters too broadly and is only a guideline. The law's major demand is that employers follow all existing regulations on OSH and provide labour protection facilities for workers. The regulations concerning safe and healthy working environments can be divided into two main categories. AMRC (2000) indicates "the first concerns preventing the sources of hazards or injuries such as fire, dust, poison, sound, extreme temperature, radiation, mining activities, etc. The second category are rules to be implemented industry-wide to ensure good practices and management. Industrial sectors are divided according to old divisions of labour between various ministries".

3.3.11 Skill level of workforce and supervision level

China has a large pool of unskilled labour. At the time of writing there were over 35 million unskilled labourers in the construction industry. Most of them come from the rural areas outside the cities. China has a sufficient labour source of more than 100 million people to call upon. The government has tried to set up a supply and demand market by forming labour-export bases to achieve optimum flexibility in moving labourers from rural to urban areas where they will be most needed. In the meantime, proper training, through a labour channel has been implemented to educate farmers with certain skills prior to placing them on construction sites. According to statistics, 75 per cent of the non-resident manpower in Beijing, came from labour-export offices.

The demand for skilled workers exceeds the market supply. Over 30,000 students graduate from technical schools each year in China but this constitutes only less than 1/1000 of the 35 million workforce in the construction industry. The lack of skilled workers seriously affects the quality of construction, and the problem is getting worse. In order to resolve the problem of training workers, the government is aiming to increase their involvement and expand the public training component.

Despite the rapid development of higher education in construction in China, the total number of graduates from universities and colleges who major in construction was only 33,000 in 1997. An investigation on construction enterprises, carried out by the Ministry of Construction, showed that professionals constitute 15 per cent of the total number of employees in the construction industry. Of these professionals 5–9 per cent graduate from universities and the majority of them are employed in the design arena. In 1997 a study showed that there was a 58 per cent shortage of professional personnel in Beijing. This was lower than computer-skilled workers who rank second in the industries. The shortage rate of managers was 51 per cent, and that of technical personnel in planning and sales was more than 40 per cent.

3.3.12 *Research and development*

As much as it is in the economic growth, the FDI in China has been the driver of technological breakthrough. FDI in the real-estate sector is an effective conduit for technology transfer, in particular, the transfer of specialized technical and management skills during various stages of the project. Technology transfer, as it appears, is the best remedy for China's obsolete construction technology and labour intensive practices as it rapidly transfers essential core 'knowledge' and skills from one country/region/firm to another, in order to facilitate longer-term development. This has been particularly true in the case of Hong Kong, which saw a flood of foreign contractors during its peak construction times. Although real-estate sector in China has a lower R&D investment record (Jiang *et al.*, 1998 quoted in Ganesan, 2000), it has been quick to absorb the results of research in the numerous disciplines from which it secures its inputs (Ganesan, 2000). As part of the investment policy reforms, the Chinese government amended regulations to enable Chinese and foreign partners to contribute to registered capital in the form of intangible assets, such as technology (Ganesan, 2000).

In the past, the Chinese government has consistently sought to maximize the contributions on R&D by urging large construction companies to form joint venture partnership with the foreign partners. This facilitated setting up of R&D centres for building and energy saving technologies and enhancing capacity of the construction companies engaged in large engineering projects. The advanced technology provisions promulgated by the State Science Commission, on October 4, 1997, allowed an investor to contribute up to 35 per cent of the registered capital in the form of technological know-how, provided the technology is complex and novel (Ganesan, 2000). With the accession of WTO, however, China will have to eliminate any conditions related to technology transfer requirements on foreign companies. The terms and conditions of any transfer of technology will be, instead, agreed between the parties. This obviously points towards the greater need for construction and real-estate industries to rely more on domestic research in the future. This will also mean the government will have to introduce measures to ensure that growing deregulation and dispersal of research and the potential risks and uncertainties that come with it do not overwhelm the R&D sector.

3.4 Construction outlook 2002–2010

3.4.1 *Business opportunities and major projects in the pipeline*

China's investment climate has changed dramatically since it opened its door to the outside world in 1979. The additional laws and regulations introduced by the Chinese government to strengthen the 1979 investment law demonstrate how serious the country is on this issue. The government has been quick to recognize the value of foreign companies in bringing both technical and management capabilities apart from the investment."China's leaders and emerging entrepreneurial class understand that foreign direct investment (FDI) and high technology have been critical to China's modernization efforts and the development of its export trade capabilities, served as an engine of job creation and become a crucial earner of foreign exchange" (US Embassy, 2001). According to Chinese statistics, utilized FDI in China since 1979 reached a cumulative total of just over US$267 billion by the end of 1998. Total FDI has consistently stayed at about $45 billion every year since 1997.

Despite efforts in simplifying the legal and regulatory system, investors are concerned with the "arbitrary enforcement of regulations and lack of transparency" in the country. In 1997, the National Planning Committee modified the Investment Guide for Overseas Businessmen. The primary aim of the guide was to re-orient the foreign investment to match the national interest apart from creating a more transparent and simplified process. The guide made explicit requirements that would necessitate foreign investments in certain types of projects such as water supply, heating, gas-supply networks as well as underground projects to be carried out through a holding company controlled by the Chinese. Foreign companies have in the past acted as main contractors in some big projects such as Guangzhou underground railway (partially several projects based on Build-Operate-Transfer (BOT), TOT, etc.) have been adopted successfully in China. Ganesan (2000) argues the "FDI is expected to play a key role in mobilising resources for implementing a number of high profile projects as the Three Gorges Project" thereby boosting the morale of the investors. The government recently released policies to regulate foreign investments that have a twin aim of minimizing financial risks and controlling the foreign debt to an accepted level without stifling long-term investment in the process. Accordingly, short-term investments that are likely to stir the market and generate speculations detrimental to the economy are discouraged. The new policies apparently were also introduced as a result of the discrepancies in implementation. For instance, to steer clear of the supervision from the central government, some local governments issued bonds to foreign investors to raise funds by government guarantee or by false cooperation and/or joint-venture with foreign investors on infrastructure construction.

On the other hand, Chinese construction contractors are looking to expand their share in overseas markets, making it one of their key development priorities. The Foreign Economic Cooperation (FEC) Council, which covers the export of three major services, namely contract engineering, labour service cooperation and design consultation, achieved a contract sum of US$12.32 billion and revenue of US$9.42 billion in 2000. The overseas construction contracting business has witnessed tremendous progress during the past 23 years in China. Nearly all Chinese companies have opened overseas markets – a remarkable increase from the four companies engaged in the business in the late 1970s. According to official statistics, the combined overseas business value of Chinese construction contractors totalled US$111.4 billion in the past two decades, with an annual growth rate of 32 per cent. These projects provided more than 2.25 million jobs for the domestic workforce. In 2000, 34 China-based firms were included into the top 225 international contractors, and five leading Chinese design firms were listed in the top 200 international design firms (ENR, 2001).

2008 Olympics

Budget estimates contained in the Beijing 2008 Olympic bid total US$2.06 billion for the construction of the Olympic village, stadiums and venues. Another US$3.6 billion is earmarked for urban infrastructure and US$8.6 billion for improving the environment. Based on the Preparatory Office for the Beijing Organizing Committee of the Olympic Games (the PO), many of these planned projects will be open to foreign involvement. However, given the nature and magnitude of these proposed projects, it seems likely that new Beijing regulations may be issued to promote foreign participation in more complex design and construction projects.

Numerous construction projects are planned for the Beijing 2008 Olympic Games. These include a total of 32 sporting stadiums and venues in Beijing's Olympic bid. Of these, 19 stadiums and venues will be constructed with another 13 existing facilities to be upgraded or rebuilt. Beijing also plans to develop the city's transport, education and utilities infrastructure. These projects include construction of over 100 kilometres of new rail lines, construction or renovation of over 900 kilometres of roads, building a second natural gas pipeline from Shania to Beijing, increasing the digital capacity of universities and tertiary colleges, improving water quality and increasing electricity supplies.

3.4.2 *Areas where foreign investments were needed*

The overall construction market in China is around US$340 billion a year and about 15 per cent requires some form of foreign expertise, providing more business opportunities for overseas companies, according to a report released by the Trade Development Council in Hong Kong (TDC, 2002). The figure suggests that around US$50 billion worth of construction works is open to foreign participation every year.

China's WTO accession, the Beijing 2008 Olympics, the western region development strategy and the housing reform in the mainland will be up for all international professionals and companies to participate in the infrastructure development of China's mainland. The report says the whole road construction and improvement programme planned for the next 10 years in China is expected to be around 700 billion RMB (US$85 billion) to 800 billion RMB (about US$100 billion). The report also says the housing reform in 1998 has triggered a boom in China's real estate sector. During 2002–2006, China is expected to build a total of 5.7 billion square metres of houses, up 15 per cent from that of 1997–2001 (Trade Development Council (TDC, 2002)).

Infrastructure in China still laggged far behind when compared to other countries in the world. Telecommunications and energy sectors are now catching considerable attention both within and without China. Increasingly infrastructure projects were are executed through international tendering. Advanced international designs have been actively introduced and up-to-date construction equipment and technologies have been applied. Besides, many well-known projects have had the direct involvement of foreign companies. For example, the Day a Bay Nuclear Power Station Project was undertaken by a French Company which won the tender as a general contractor. Other French companies have won tenders for the design of Pudding International Airport and Shanghai Grand Theatre through international tendering. In addition, several foreign companies have been awarded projects through tendering. The use of foreign construction machinery and equipment has been getting more popular recently. For example, the shield tunnelling machines used for the construction of underground railways in Guangzhou and Shanghai and the equipment used for construction of the first 50,000 voltage DC transmission/substation from Guangzhou to Shanghai have all been introduced from abroad.

There are especially significant opportunities for foreign investors in the construction of roads. In 1995, the highway network grew at 3.5 per cent per annum, while passenger traffic increased at 9.1 per cent per annum and freight traffic in ton-kilometres expanded at 4.6 per cent per annum. In 1996, China had only 855 kilometres of highways per million people compared to 25,300 kilometres in the USA and 8,860 kilometres in Japan. JVs have been formed with foreign parties, and these parties have issued bonds in the international market to raise additional money. Conventional bilateral and multilateral funding are also used to finance highway projects (Woodward, 1998). For BOT projects, the foreign parties can have more than a 60 per cent share of the JV, and a concession period of upto 25 years was available at the time of writing, with an estimated return to equity investors of 18 per cent per annum (Woodward, 1998).

Another focus of investment in China is the railways sector. During, the first 2 years of the Ninth Five-Year Plan Period in 1996 and 1997, a total of 75.1 billion RMB had been invested in railroads, new railroads of length 2,944 kilometres had been built; multiple tracks of length 2,142 kilometres had been constructed; and 2,132 kilometres of the existing railway system had been electrified through upgrading. By the end of 1997, the national railroad business mileage reached 66,000 kilometres; multiple track 18,000 kilometres; electrified railroad 12,000 kilometres. In spite of these achievements, there is merely 68.75 kilometres of railroads per square kilometre in China. The railroad density is less than one-third that of the US and India, and one-ninth that of Japan and France. The total length is about 32 per cent that of the US. On top of that, the rate of multiple tracks and electrification is low and the running speed is slow. In developing countries, the running speed is generally 160 kilometres per hour and the highest speed of the passenger train is 300 kilometres per hour and above. In comparison, the speed of passenger trains in China is only 50 kilometres per hour on average; the average total weight of a cargo train is 2,600 tons, which is 50 per cent that of the US. The investment in 1998 for railroad construction amounted to 53 billion RMB and the estimated investment in the last 3 years of the Ninth Five-Year Plan Period (1998–2000) has been 150–170 billion RMB.

The following activities within the overall construction sector have attracted large-scale foreign participation:

1 real estate development projects, such as office buildings, hotels and public buildings;
2 commercial housing for foreigners and residents;
3 manufacturing of building materials and components, building services equipment, construction machinery, transport equipment and interior décor;
4 infrastructure projects and public utilities involving major capital construction outlays, such as tunnels, highways and power generation; and
5 design and construction services.

The scale of foreign investment in construction grew especially rapidly during the period 1993–1997. Approximately over 10 per cent of all foreign-funded projects were in the construction sector according to an MOC official. These included the following:

1 Industrial construction, building materials – 8,100 projects (US$160 billion)
2 Real Estate – 30,000 projects (US$135 billion
3 Public utilities – 330 projects (US$7.8 billion).

It was significant that nearly 16 per cent of all construction sector projects with at least 5 million RMB investment were funded from overseas. The industrial construction sector accounted for 50.4 per cent of these projects, the real estate sector for 32.6 per cent, and others made up 17.0 per cent. Of the projects funded from Hong Kong, Taiwan or Macau, 57.9 per cent were in the real estate sector, according to a survey carried out by the State Planning Commission, the State Economic and Trade Commission and the State Statistical Bureau.

References

ADB (2001) "Annual Report 2001", ADB: Manila.
ADB (2002) "Annual Report 2002", ADB: Manila.
AMRC (2000) http://www.amrc.org.hk/alu/Alu39/013906.html.
China Council for Promotion of International Trade (2000) Various Reports, http://www.ccpit.org/engVersion/cp_tj21/tj2001N.htm.
China–Britain Business Council (2001) "Going abroad: China's multinationals in the making", http://www.cbbc.org/ezine/guides.html..
CIA (2001) The World Factbook, http://www.cia.gov/cia.

Colliers Jardine (2001) "Local Markets Review", various issues, http://www.colliers.com/China/fr_market_local.html.

ENR (2001) *Engineering News Record*, 247:8, 20 August 2001, pp. 66–76.

Ganesan, S. (2000) *Employment, Technology and Construction Development: With Case Studies in Asia and China*, London: Ashgate.

Guan, K., Feng, K. and Zeng, S.X. (1997) "Urban Housing Reform and Development in China", *Building Research and Information*, 29:286–292.

http://www.stanford.edu/~ljlau/Presentations/Presentations/011019.pdf.

Lai, On-Kwok (1998) "Governance and the Housing Question in a transitional economy, the Political Economy of Housing Policy in China Reconsidered", *Habitat International*, 22:3, pp. 231–243, September.

Lam, Tony Fung Kam (Hong Kong Christian Industrial Committee) (2000) "Occupational health and Safety in China", originally presented at the Mekong Region OSH Workshop.

Lau, Lawrence (2000) "The Critical Path for Continuing Chinese Economic Reform", Stanford University: Stanford, October 2001.

Ministry of Construction (1999) "Country report: China", *Proceedings for the 5th AsiaConstruct*, October, Malaysia.

National Bureau of Statistics (1998) "Statistical Communiqué of the People's Republic of China on the 1998", National Economic and Social People's Republic of China Development.

National Bureau of Statistics (2001) "Statistical Communiqué of the People's Republic of China on the 2000", National Economic and Social People's Republic of China Development, http://www.chinatrade.dk/english/communik/main1.htm.

People's Daily (2001) http://english.peopledaily.com.cn/200105/14/eng20010514_69891.html.

Sharma, S. (1999) "Stability amidst Turmoil: China and the Asian Financial Crisis", *Harvard Asia Quarterly*, March 2001.

C.H. Talhar, Williams and Wong, "International Property Report", 1998–2001, http://wtw.com.my/resources/pmr/international/chi2001.html.

Trade Development Council (2002) "Beijing offers various preferential Packages to Olympics Projects", TDC news and speeches, 15 August.

Tse, Raymond Y.C. (2000) "China's Real Estate Market and the Asian Financial Crisis", *Emerging Market Quarterly*, Winter.

UK Parliament (2001) Memorandum submitted by British Trade International, http://www.publications.parliament.uk/pa/cm200203/cmselect/cmtrdind/128i/2061102.htm

UNIDO (1992) *China: Towards Sustainable Industrial Growth*, Oxford: Blackwell.

US Embassy (2001) www.usembassychina.org.cm/econ/991014a.html

Woodward, P. (1998) "On the Open Road", *China Business Review*, March.

World Bank (2000) East Asia: Recovery and Beyond, Washington: World Bank.

World Bank (2002) "Country Brief", Washington: World Bank.

Wu, Renhong (1999) "China's Economic Outlook", Center for Strategic and International Studies: Washington, February.

www.eyi.com.

www.odci.gov/cia/publications/factbook.

Zhu, J., Xia, Z., Christiani, D.C., Sorock, G.S., Ye, T., Courthey, T.K., Zhu, T., Qu, Q. and Fu, H. (2000) "Fatal Occupational Events in a Development Area in East China: 1991 to 1997", *Journal of Occupational Health*, 42, pp. 276–280.

Chapter 4

China Hong Kong

Y.H. Chiang, Edwin H.W. Chan and S. Sharma

4.1 Introduction

Hong Kong, the "pearl of the orient" and the gateway to China, is located at the southeastern tip of China mainland. It became a Special Administrative Region (SAR) of the People's Republic of China on 1 July 1997. Under the Sino-British Joint Declaration, signed between China and Britain on 19 December, 1984, Hong Kong is guaranteed a high degree of autonomy, except in foreign and defence affairs. Under the "one country, two systems" formula, Hong Kong's lifestyle was to continue for at least 50 years after 1997, and China's socialist system and policies would not be practised in the SAR.

According to *Hong Kong Yearbook 2002*, Hong Kong occupies 1102 square kilometres, comprising the Hong Kong Island, Kowloon, New Territories and the Outlying Islands. Since 1887, an additional 66 square kilometres of land has so far been reclaimed. That is more than 81 per cent of the area of the Hong Kong Island and adjacent islands. Hong Kong has a terrain as diverse as hilly to mountainous with steep slopes, and lowlands in the north bordering China. Hong Kong's natural terrain, dominated by mountains and hills with steep slopes that are populated with high-rise buildings, has always posed challenges for geotechnical engineers. The location of her outstanding deep-water harbour has proved to be strategic for Hong Kong's vital link on the international supply chain and global logistics. A 1997 estimate shows that arable land in Hong Kong amounts to only 6 per cent of the total land accounting for only 1 per cent permanent crops and permanent pastures each, whilst forests and woodland amount to about 20 per cent of the total land with the remaining constituting largely barren rocks mostly used for development and partly preserved as country parks (CIA World Fact Book, 2000).

4.1.1 History

Hong Kong, hardly more than the "barren rock" some 160 years ago, has transformed herself into one major world city. Known by the local people, who are mostly Cantonese, as "Fragrant Habour" for the incense factories once built over the island, the enclave had a humble beginning as no more than a fishing village. According to the *Hong Kong Yearbook 2002*, the British presence in Hong Kong began in the 1840s. Before that, they were confined to do business in designated areas in Canton (now Guangzhou), which is about 130 kilometres from Hong Kong. Though there were limitations that the British naturally found unsatisfactory, there was "mutual trust" between the British and Chinese traders. However, ever since the East India Company lost its monopoly of the China trade, foreign free traders had jockeyed for a slice of the lucrative opium trade, which the Chinese had made illegal in 1799. Lin Zexu was sent in by the Ching dynasty in March of 1839 as special Commissioner in Canton to eradicate the opium trade. After a siege of six weeks, he managed to make Captain Charles Elliot, RN, representing the British Government as Superintendent of Trade, to surrender 20,283 chests of opium. Elliot then sailed to Hong Kong in the summer of 1839.

Meanwhile, the British Foreign Secretary, Lord Palmerston, then decided it was time to settle Sino-British commercial relations. He demanded a commercial treaty or the cession of a small island where the British could live under their own flag. The so-called First Opium War (1840–42) then broke out, following an expeditionary force arriving from the UK in June 1840 to back these demands. Eventually, agreement was reached. Under the Convention of Chuenpi signed on January 20, 1984, the Hong Kong Island was ceded to Britain. In June, Elliot began selling parcels of lands. British settlement began.

There were feelings of shame and anger among the Chinese. The Convention did not please the British either. Palmerston was dissatisfied with what he called "a barren island with hardly a house upon it" and rebuked Elliot (*Hong Kong Yearbook 2002*). Later in August of 1841, replacing Elliot, Sir Henry Pottinger arrived and "conducted hostilities with determination". It was only with the Treaty of Nanking, signed on August 29, 1842, that hostilities were finally brought to an end. Hong Kong was secured by the British, and five Chinese ports including Canton were also opened for trade.

Some 150 years later, the British ceded control of Hong Kong back to China in 1997 when their lease expired, though Beijing, ever maintaining that the treaties were signed under duress, had never admitted their legitimacy.

4.1.2 Political structure: one country two systems

With the Joint Declaration initialled by China and Britain on September 26, 1984, came the official reversion of Hong Kong back to her motherland China at midnight of 30 June, 1997, and subsequently the beginning of a new era of public administration under the "one country two systems" principle. The Chief Executive (CE) is the head of the SAR as well as the head of the government. The incumbent was re-elected to his second and last term by an 800-member Election Committee, that was supposed to be representative of the different sectors of the community. However, there have been hundreds of thousands of people marching on the street, not once but twice in the months of July in both 2003 and 2004, asking for election of the CE by universal suffrage in 2008.

The CE is assisted by the Executive Council in policy making. Article 56 of the Basic Law stipulates that the CE shall consult the Executive Council in strict confidence before making important policy decisions, introducing bills into or dissolving the Legislative Council, except for the appointment, removal and disciplining of officials and the adoption of measures in emergencies. In 2002, there were altogether 19 members of the Executive Council. The Legislative Council, as provided for in Article 73 of the Basic Law, is basically the law-making body of the SAR. It also examines and approves government budgets, approves taxation and public expenditure, and raises and debates issues concerning public administration. The Legislative Council is constituted by election under the Basic Law.

Serving their second term are the 60 members of the Council. For its third term after the handover in September 2004, the Council will have exactly half of its 60 members returning from geographical constituencies and the other half functional constituencies. Meetings of the Council are open to the public, and its proceedings are recorded verbatim in *Hansard*, the official record of the Council.

Under the Executive and Legislative Council is the District Administration Scheme. It commenced in 1982 with the establishment of a District Board and a District Management Committee in each district. Its function is to help the Government promote "public participation in district affairs and fosters among the people of Hong Kong a sense of belonging and mutual care" (*Hong Kong Yearbook 2002*).

The execution of policies and routine public administration is done through an intricate web of government bureaus and departments, and quasi-government agencies. In 2002, there were 11 bureaux. Each bureau was headed by a Director of Bureau. Under the bureaux were the 68 government departments and agencies. Their heads were responsible to the Directors of Bureaux, with the exceptions of the Audit Commission, the Independent Commission Against Corruption and the Office of the Ombudsman. Their independence is safeguarded by having their reports submitted either to the Legislative Council in the case of the Audit Commission, and to the CE in the latter two cases *(Hong Kong Yearbook 2002)*. In addition, there are a large sector of quasi-government agencies including, for example, the Mass Transit Railway Corporation and the Hong Kong Monetary Authority. For a full list of these organizations, please refer to the website of the Government of Hong Kong Special Administrative Region (www.info.gov.hk) that contains tons of data and information.

4.1.3 Development process in Hong Kong: through the years

The development process in Hong Kong can be best summed up by a *laissez-faire* policy or "positive non-interventionism". It is a policy that could not be made clearer by the former Financial Secretary, Sir John Cowperthwaite, who once replied to businessmen in the 1960s asking for special treatment for their industries that "I should have thought that a desirable industry was, almost by definition, one which could establish itself and thrive without special assistance in ordinary market conditions" (*Asian Wall Street Journal*, July 1, 1999). The government has limited her role to the provision of basic infrastructure, a stable political environment and a level playing field, that is basically an environment that is conducive to business. There is no legal and institutional barrier to trade and to do business in Hong Kong. There is free flow of information and capital, and a tax regime that is both simple and low. Perhaps much to the credit of the government, Hong Kong has remained the freest economy in the world after the handover, according to two international surveys. Ever since the publication of *The Index of Economic Freedom* in 1995 (O'Driscoll, Jr. *et al.*, 2001), and *The Economic Freedom of the World* in 1970 (Gwartney, J. *et al.*, 2001), Hong Kong has been ranked at the top of both of the scales. Together with the diligence and, perhaps more importantly, the business acumen and entrepreneurship of her people, Hong Kong has become a major economy in the world, with a 2003 GDP per capita that was higher than the UK (www.cia.gov/cia/publications/ factbook/rankorder/2004rank.html). Without the endowment of natural resources (for example, Hong Kong depends on China for water), a group of skilful, talented and hard-working people is the only asset that Hong Kong has had.

However, the government does monopolize the sale of land, and provides subsidized housing to about half of the population. Besides, since the handover in 1997, there have been a few instances where nothing less fundamental than the level playing field of doing business and the rule of law were put into serious doubts. In its defence against the speculative attacks of hedge funds in August of 1998, the government did use public money to shore up equity prices and consequently the Hong Kong dollar. The non-interventionism has thus seemed more like a smokescreen (Jao, 2001).

4.2 The macroeconomy

4.2.1 Macroeconomic review and outlook

Hong Kong is a bustling free market economy highly dependent on international trade. Natural resources are limited and food and raw materials must be imported. The two sectors of agriculture and fishing, and mining and quarrying have become more insignificant throughout the 1980s and 1990s. The contribution of the two sectors combined to GDP had diminished from 1 per cent in 1980 to no more than 0.1 per cent in 1999. Hong Kong had thrived on trade, initially on the trading of goods and later services. It is the world's ninth largest trading economy, and the world's ninth largest exporter of commercial services (TDC, 2001). Indeed, both imports and total exports have exceeded GDP in dollar value since 1988 (*2000 Gross Domestic Product*). In 2000, imports and total exports were 131 per cent and 124 per cent of GDP respectively. The proportions of re-exports in total exports had increased from 25 per cent in 1961 to 88 per cent in 2000, reflecting partly the re-emergence of Hong Kong as an entrepot for China mainland, and partly the increasing dependence on China mainland as the production base for many of the local manufacturers.

Hong Kong has had extensive trade and investment ties with China mainland, before and after its reversion to China. As China mainland evolved as the leading manufacturing base of not only the Asian region but the world, Hong Kong capitalized on her strengths in the provision of financial, legal, marketing, and design and development services to the production, export and re-export of the goods. Many of the goods were indeed manufactured with local capital, as Hong Kong was China mainland's largest source of foreign investment. In turn, China mainland was Hong Kong's second largest source of external investment in stock investment. An unprecedented boom in foreign direct investment (FDI) in Hong Kong had lifted FDI flows to and from Asia to a record level (UNCTAD, 2001). The key role of trade in the local economy was manifested in Hong Kong having the world's busiest airport in terms of international cargoes and the world's busiest container port. Hong Kong was ranked the best city in Asia-Pacific to do business in 2000 (Fortune, 2000).

Hong Kong also had the world's second highest per capita holding of foreign currency, and the world's fourth largest foreign exchange reserves holding. In the year 2000, Hong Kong had Asia's largest gold bullion market or the fourth largest in the world, the second largest stock market in Asia or the ninth largest in the world, and the third largest foreign exchange market in terms of turnover in Asia or the seventh in the world.

The per capita GDP at current market prices was HK$183,219 (or US$23,490) in 1999 (*2000 Gross Domestic Product*). It ranked the fourth among the 21 member economies of Asia-Pacific Economic Cooperation (APEC), after Japan, Singapore and the USA (The Economist, 2000), at more than US$22,507, the average of the 30 OECD member countries in the same year of 1999 (OECD, 2001). The per capita GNI (Gross National Income), at purchasing power parity, of Hong Kong was US$25,670 in 2000, the second highest after Japan among 13 Asian economies (AWSJ, 2001).

GDP growth averaged a strong 5 per cent in 1989–1997. The widespread Asian economic difficulties in 1997 hit this trade-dependent economy quite hard, with GDP down by 5 per cent. The economy staged a sharp rebound in 1999, recovering robustly from the severe setback brought about by the earlier Asian financial turmoil. By the fourth quarter of 1999, the level of economic activity had already surpassed the previous peak in the third quarter of 1997. For 1999 as a whole, the GDP grew by 3.1 per cent in real terms, in contrast to a 5.3 per cent decline in 1998. The recovery in 1999 was strongly export-led. Driven by the resurgence in demand in the East Asian markets as well as by sustained import absorption in the United States and Europe, export of goods accelerated to double-digit growth in the second half of the year to give a 3.7 per cent growth in real terms in 1999. The combined visible and invisible trade account nevertheless remained in surplus, at HK$12 billion (US$ 2 billion) in the first half of 2000. Economic recovery had become more entrenched in 2000, with GDP attaining a double-digit growth of 10.5 per cent in real terms (at constant (1990) market prices) or 3.3 per cent at current market prices in the first half of 2000 over a year earlier.

The economy of Hong Kong however worsened during the first half of 2001. Demand from developed countries, in particular the US, weakened towards the end of 2000 which adversely affected the territory's merchandise exports, of which most were re-exports originating from China mainland. The interest rate cuts that followed the US cuts have so far failed to revive consumer confidence. Private consumption and investment remained subdued. Property prices continued to fall. The unemployment rate climbed to 4.7 per cent in the second quarter of 2001. In mid-August of 2001, the government made further downward adjustments to its economic growth forecast from 4 per cent (as stated in the March Budget) to 3 per cent. There had been much concern that economic recovery in Hong Kong could be further held back by the repercussions of the "9–11" tragedy in New York. The GDP projected growth in 2001 was revised substantially downward to be zero, if not negative.

The situation in Hong Kong was double blown by this cyclical global downturn and the structural changes that were going on in the local economy at the moment. There was an international perception of deterioration in the local operating environment. According to the Economist Intelligence Unit, Hong Kong, which ranked as the third best place to do business in 1996–2000, is expected to drop to tenth place in the forecast period of 2001–2005. Its press release on 13 August 2001 commented that Hong Kong is the only economy where the business environment is expected to deteriorate in absolute as well as relative terms. Apart from this perceived deterioration due to regulatory and political changes, there was the mismatch between the skills of the workforce and the demands of an economy increasingly focusing on high-value-added sectors. Whether there was really a systematic deterioration in the regulatory and political institutions is debatable. There did appear, however, to be a consensus in the community that there was a problem with workforce skills, especially as the economy of Hong Kong has become even more services oriented.

Property appears to have ceased to be an economic growth driver in Hong Kong. A reduction in interest rate used to be accompanied by rise in property prices. The negative real interest rates in the mid-1990s accounted for much of the property boom. However, since the property market collapsed following the outbreak of the Asian crisis in 1997, property owners saw their wealth diminish by more than half and many of them faced financial hardship caused by negative equity. The local economy in general and the property sector in particular have so far not benefited from interest cuts by the Federal Reserve of the US. Consumer confidence had plunged and not recovered since then. The unrelenting high unemployment rate, the burst of the technology bubble, the lacklustre performance of the stock and property markets, the large proportion of property owners being trapped in negative property equity, and the implementation of the Mandatory Provident Fund had been attributed as major causes of this lack of confidence. Prolonged slump in market confidence had led to a growing backlog of unsold residential units. Renewed downturn of the housing market, if prolonged, was expected to hit consumer confidence and put a drag on Hong Kong's economic recovery.

Yet the US dollar peg had remained but not completely without some pessimism within Hong Kong. Perhaps because the recession had been longer lasting and much more serious than any previously experienced there had been an ongoing debate on 'dollarization', some advocating the dollar peg as a fundamental problem.

Though Hong Kong had gone through a structural change from being a manufacturing-based economy to a service-based one, much of the territory's service sector is still largely intertwined with the external trade of merchandise goods. Whilst services contribute to all but some 14 per cent of the GDP now, much of the local economy's performance had been and will continue to be tied to import demands of the US and other overseas countries. The immediate economic prospect for Hong Kong depends very much on the US economy. Performance was expected in turn to depend to a large extent on how effective the Federal Reserve's interest rate cuts and how effective the Bush administration would be in turning around the economic recession. Table 4.1 shows the macroeconomic indicator during the period 1994–2000.

4.2.2 *Role of construction/property market in the economy*

Hong Kong's construction boom of the 1970s and 1980s is attributed to its essentially infrastructure-led development policy. The share of construction in total GDP had been in the range of around 5–7 per cent in Hong Kong generally, but at times soaring up to 12.5 per cent in 1996. The sector is also important as a major employer. The share of the sector in total employment had been almost 10 per cent. In 1999 the size of the construction market in Hong Kong was in the order of US$20 billion accounting for 5.8 per cent of GDP. Much of the construction sector work was

Table 4.1 Macroeconomic indicators

	1994	*1995*	*1996*	*1997*	*1998*	*1999*	*2000*
GDP and components							
GDP at constant (1990) market prices (HK$ million)	727,506	755,832	789,753	831,317	785,338	808,826	893,402
GDP at current market prices (HK$ million)	1,010,885	1,077,145	1,191,890	1,323,862	1,259,306	1,227,658	1,266,840
GDP at factor cost (production-based) (HK$ million)	950,172	1,016,115	1,130,212	1,233,034	1,169,064	1,139,363	n.a.
GDP growth (%)	5.4	3.9	4.5	5.3	−4.7	−3.1	10.5
Primary sector (HK$ million)	24,021	25,349	28,743	30,949	35,376	35,835	n.a.
% growth	2.0	−4.1	−0.8	−1.0	n.a.	n.a.	n.a.
Manufacturing sector (HK$ million)	87,334	84,770	82,769	80,049	70,849	65,767	n.a.
% growth	−5.6	−3	−2.4	−3.3	−11.5	−7.2	n.a.
Services sector (HK$ million)	792,472	851,235	953,642	1,050,386	992,902	980,391	n.a.
% growth	17.4	7.4	12.0	10.1	−5.5	−1.3	n.a.
Construction sector (HK$ million)	46,325	54,761	65,058	71,650	69,937	66,111	n.a.
% growth	7.5	18.2	18.8	10.1	−2.4	−5.5	n.a.
Demographic indicators							
Population	6,119,300	6,270,000	6,421,300	6,617,100	6,689,000	6,761,600	6,865,600
Population growth rate (%)	2.0	2.5	2.4	3.0	1.1	1.1	1.5
Total labour force	2,929,000	3,000,700	3,093,800	3,216,000	3,358,600	3,342,500	3,353,400
Labour force growth rate (%)	2.5	2.4	3.1	3.9	4.4	1.2	3.2
Unemployment rate (seasonally adjusted) (%)	1.9	3.2	2.8	2.2	4.7	6.3	5.6
Financial indicators							
Changes in consumer price index (%)	9.5	7.2	6.6	5.2	−1.6	−4.0	−3.8
Changes in GDP deflator (%)	6.9	2.5	5.9	7.2	0.8	−6.5	−5.4
Short-term interest rate (%)	5.66	5.55	4.45	7.50	5.04	3.77	4.39
Long-term interest rate (%)	8.42	6.31	6.70	9.27	6.17	5.76	5.4
Annual average exchange rate with $US (HK$)	7.728	7.736	7.734	7.742	7.745	7.758	7.780

Sources: Macroeconomic review: *2001: Economic Prospects*, Govt of the HKSAR
2000: Economic Background, Govt of the HKSAR
Economic Report, First Quarter 2001: Govt of the HKSAR
GDP and components: *Estimates of Gross Domestic Product 1961–1999.*
Demographic indicators: Government of the HKSAR web-page: http://www.info.gov.hk/constatd/ home.html
Financial indicators: *Annual Report on the Consumer Price Index 1999* (for "Changes in consumer price index"). Hong Kong Monetary Authority, *Monthly Statistical Bulletin June 2000* (for "Short-term (and long-term) interest rate" and Annual average exchange rate with $US).

Note
n.a.: data not available.

prompted and driven by the property market in Hong Kong. Economists tend to argue that the property market is both a strength and a weakness in Hong Kong's economy. Real estate sales, development and related activities account for about 40 per cent of Hong Kong's economy. Many of the major Hong Kong companies traded on the stock exchange were property developers. The majority population and particularly the middle and upper class had their wealth tied to property business. The government was again the biggest developer in Hong Kong as the land owner who gets large quantities of revenue off big developers from auctioning of leases. Besides, quasi-governmental bodies like the Kowloon Canton Railway Corporation (KCRC) and the Mass Transit Railway Corporation (MTRC) frequently initiated major developments in tandem with the train and metro networks at strategic locations.

Since the early 1990s, soaring property prices have boosted the stock market all through Hong Kong's phenomenal economic growth, which saw its major setback only following the 1997 financial crisis. However the construction sector in Hong Kong continued to benefit from the large volume of construction work initiated by the government, even though the target construction of 85,000 domestic units per year had been considered highly ambitious before it finally became "non-existent", as announced by the Chief Executive in 2001. Public housing construction however was in full swing around the turn of the century. Annual production went as high as 50,000 units as opposed to the average of 30,000 units a year for most of the 1980s and 1990s. The Housing Authority in Hong Kong plans to provide, based on the Public Housing Development Programme of 30 June 2000, approximately 142,000 housing units by the year 2005.

4.2.3 Construction investment (or market) volume

Construction activities in Hong Kong, essentially divided into building and civil engineering works, had been steadily rising since early 1990s. Interesting trends could be observed during this period. While building construction activities were brisk in 1990–1992, civil engineering projects were stable, but the latter took a giant leap with the Airport Core Project in the following years. After 1997 however, civil engineering projects much declined whilst building activities had not lessened as much.

In terms of construction spending per capita Hong Kong ranked ninth in 1998 after Japan and Singapore within Asia, with Japan leading at US$4,975 (Crosthwaite, 2000). Over the years, the total construction market expanded by about half, in real terms, in 1996 when compared with 1990. Since 1996, despite the financial turmoil, the total construction market remained relatively stable for three consecutive years. The adverse effect of the financial turmoil was not felt by the local construction sector until 1998 primarily due to many large-scale projects not being affected so seriously. However, both the government and the private sector considerably delayed announcing new projects. In 1997, the total value of construction works increased by 5 per cent when compared with the year before but since then it had been following a downward trend. One year later in 1998, it fell by 5 per cent. In 1999, it fell further by 11 per cent.

Much of the "stability" of the construction market in terms of contract values could be attributable as mentioned above to the building programmes in both private and public sectors. During the period of 1992–1997 the building component within the construction sector had a share of over 60 per cent. There have been steady increases in residential construction in both private and public sectors in the last decade and noticeably in the 3-year period from 1997 to 1999.

During the same period, on the other hand, the civil engineering sector continued contracting. The Hong Kong Airport Core Programme pushed the volume of public civil engineering construction to the peak in 1996 (accounting for 40.2 per cent of total construction output of US$116.3 billion representing 12.6 per cent of GDP), when the value was more than double that in 1990. Since then, following the completion of the programme, the sector contracted by 15 per cent in 1997, 23 per cent in 1998 and a further 7 per cent in 1999. Civil engineering construction in the private sector also contracted in 1997 and 1998, though the size of the private sector was only about one-sixth of the public sector in the three-year period.

Expenditure on building and construction by the public sector also dropped by 9 per cent in 2000 reversing the 1 per cent rise in 1999. The cut was due to the slowdown on public housing programmes, delays in the construction of Disney Park and the last phase of the KCRC West Rail.

In the year 2000, overall expenditure on building and construction fell by 13 per cent in real terms in the first half of the year, and 3 per cent in the second half amounting to an average decrease of 8 per cent in real terms as opposed to 11 per cent in 1999. Expenditure in the private sector in particular fell by 8 per cent in real terms, considerably smaller than the 19 per cent decrease in 1999. This was attributed to the completion of some of the large-scale residential development projects in Tsing Yi, Sha tin and Fan Ling, and of some commercial development projects in Central Tsim Sha Tsui. By comparison the reduction in private sector civil engineering output was more moderate by virtue of land formation works for the Cyberport and construction commencement at Container Terminal 9. But construction work on the Kowloon Canton Railway (KCR)'s Ma On Shan to Tai Wai Link is expected to gather momentum and the Mass Transit Railway (MTR)'s Tsuen Kwan O extension work is expected to remain intensive. Phase II of the Airport railway Hong Kong Station and enhancement work on the MTR Kowloon Tong station were about to start at the time of writing. Hence public sector building and construction, although still slack, is poised to pick up visibly in the latter part of 2001. For the private and public sector combined an overall expenditure on building and construction was forecast to a further decline of only 1 per cent in the year 2001.

4.2.4 *Land prices, and property prices and rents*

The year 1997 marked the beginning of turbulent financial turmoil for Asian economies including that of Hong Kong. It also marked the end of the upward trends in land prices, property prices and rents ever since the end of the Gulf War. The banking and property sectors were among the hardest hit. There was still optimism for the first half of 1997 in the property sector of Hong Kong. However, this was followed by consolidation in the third quarter and then a sharp downturn towards the end of the year. This downward adjustment process has continued right into and throughout the year of 2001. Land prices, property prices and rents have fallen by 50–60 per cent from their peaks.

Land prices

According to the land sale records of the Lands Department (2001), in a period of 7 years (until the third quarter of 2001) a total of 122 lots had been disposed of through auction or tender, involving a total area of more than 2.01 million square metres of land, or less than 0.2 per cent of the size of Hong Kong (1,098 Sq. km).

The total premium recorded was more than HK$100.8 billion. On an average, one square metre of land was sold for about HK$50,111.

In 1995/1996, more than 0.9 million square metres of land was disposed of. This alone was almost 46 per cent of the total for the 7 years combined. In the following financial year of 1996/1997, the area of land disposed of dropped substantially to less than 0.16 million square metres. However, the average price of land dramatically peaked at about HK$115,000 per square metres, an increase of 354 per cent over the year before. Indeed, unit price of land topped in 1996/1997 for residential use, industrial use and other uses. Unit price of commercial land topped one year later (1997/1998). The unit price of all land dropped by almost 18 per cent in 1997/1998. The full devastating aftermath of the Asian financial crisis did not arrive until 1998/1999 when there was an even bigger drop of 45 per cent, only to be followed with a further drop of 23 per cent in 1999/2000. The land market began to recover in 2000/2001, when the average premium paid for the seven lots rose by 44 per cent compared with a year ago. Yet, the average unit price of land dropped again by more than 42 per cent in the first five months of the year 2001/2002. However, these figures must not be made too much of. They are only a rough indication as other factors like location play a crucial role in determining prices (refer Table 4.2).

During the seven-year period, residential land disposed of amounted to about 50 per cent of the total but the premium it brought was over two-thirds of the total. Residential land had been an important source of public revenue in recent years. It also appeared however to be the most severely hit when the property prices fell in 1998/1999, as the unit land premium dropped by more than 51 per cent from the peak, a magnitude roughly comparable with the corresponding decline in residential property prices. In the commercial land sector in 1998/1999, the unit land premium dropped by almost 34 per cent from its peak, less than that of the residential land. For the industrial land and land marked for other uses, the lots disposed of are too few to make any quantitative analysis meaningful. In conclusion, the effect of economic crisis was quickly reflected in land sales, as in the property market. However, as discussed in Section 4.2.3, the building and construction market had not been as adversely affected as the property market.

Property prices and rents

It appeared that prices and rents had fallen more in the office and industrial sectors than in the residential sector. Office prices fell by an average of 63 per cent in the year 2000 when compared to the peak in 1997. The average office rents in the first quarter of 2000 were about 40–50 per cent down when compared with that of 1996. In the industrial sector, the sale prices of flatted factories plummeted by an average of 62 per cent from the peak levels of 1994, whilst their average rents dropped from 30–40 per cent for the period 1996–2000.

Residential prices steadily rose to peak in the year 1997 only to begin to fall afterwards. In comparison with the peak levels of 1997, flat prices were down by an average of 51 per cent in the last quarter of 2000. In the residential rental market, the overall decrease in average rents was between 20 and 30 per cent for most of its sub-sectors. The drop in residential rents was relatively moderate when compared with office rents. The population influx as well as existing demand for housing were

Table 4.2 Land sales summary

	1995/1996	*1996/1997*	*1997/1998*	*1998/1999*	*1999/2000*	*2000/2001*	*2001/2002**	*1995–2002*
Residential land								
No. of lots	10	10	17	5	21	6	3	72
Area (m^2)	180,669	125,230	242,227	133,843	226,141	87,489	20,588	1,016,187
Premium ($ m)	8,363	15,642	28,646	7,722	8,909	4,989	266.4	74,536
Premium/m^2 ($)	46,291	124,905	118,259	57,697	39,394	57,022	12,940	73,349
Commercial land								
No. of lots	9	7	2	3	2	1	2	26
Area (m^2)	63,299	15,133	10,886	10,574	21,513	14,900	20,754	157,059
Premium ($ m)	13,383	1,624	2,998	1,931	988	850	1,095	22,869
Premium/m^2 ($)	211,418	107,343	275,400	182,618	45,921	57,047	52,766	145,609
Industrial land								
No. of lots	4	4	3	2	0	0	0	13
Area (m^2)	13,869	13,899	13,317	14,483	–	–	–	55,568
Premium ($ m)	150	421	269	286	–	–	–	1,126
Premium/m^2 ($)	10,823	30,255	20,231	19,720	–	–	–	20,263
Other uses								
No. of lots	2	2	3	3	1	0	0	11
Area (m^2)	667,202	2,386	74,782	36,069	2,578	–	–	783,017
Premium ($ m)	1,502	290	337	146	8	–	–	2,283
Premium/m^2 ($)	2,251	121,668	4,502	4,058	3,103	–	–	2,916
Yearly total								
No. of lots	25	23	25	13	24	7	5	122
Area (m^2)	925,039	156,648	341,212	194,969	250,232	102,389	41,342	2,011,831
Premium ($ m)	23,398	17,977	32,250	10,085	9,904	5,839	1,362	100,815
Premium/m^2 ($)	25,294	114,761	94,515	51,728	39,581	57,026	32,933	50,111

Source: Compiled from Land Sales Records of the Lands Department (http://www.info.gov.hk/landsd/lsr/lsr.htm).

Notes

* up to and including land disposed of (by auction) in August, 2001.

Residential land includes land uses designated as PSPS (Private Sector Participation Scheme), R1, R2, R3, R4 (Residential R1, R2, R3, R4) and V (village type development).

Commercial land includes land uses designated as C (commercial), C/MCP (commercial/multi-story car park), C/OU (commercial/other uses) and C/R (commercial/residential).

Industrial land includes land uses designated as DGG (dangerous goods godown), G (godown), I (industrial), I/G (industrial/godown), IO (industrial office).

Other uses include land uses designated as HTL (hotel), MCP (multiple-storey car park), OU (other uses) and PFS (petrol filling station).

attributed to the slightly better performance of the residential rental market relative to the other sectors.

It seems that yields in the residential and industrial sectors have increased more than for the office sector. This was because the prices dropped more than rents, implying an increase in yields and thus the risk premiums of investing in these properties.

4.2.5 *Changes of economic policies due to the financial turmoil*

The impact of the 1997 financial crisis was readily felt in Hong Kong. On the one hand, wages, rental costs and prices became too high for Hong Kong to remain competitive. On the other hand, given the dominance of the real estate sector in the stock market and that 40 per cent of bank loans were exposed to the property market, the ability of the financial sector to absorb asset deflation went into question. The existence of fears that the dollar peg may break down and the currency devalue, in turn, attracted the speculative hedge funds that had already assaulted Hong Kong twice in years before the Asian financial crisis. The Hong Kong government was then forced to immediately intervene in the stock market. This was a serious and controversial departure from the policy of positive non-interventionism long held by the Hong Kong government. The government used foreign reserves to buy Hong Kong dollars dumped by the hedge funds and purchased stocks, thus lifting the index and protecting the peg. In the 10 trading days beginning 13 August 1998, some HK$118 billion of reserve money was used to purchase 33 different blue chip stocks. Soon after the August intervention, a number of small measures were introduced such as revision of interbank trading rules to boost liquidity in the banking system and the monetary base was substantially enlarged to ensure that any effort to create an arbitrarily high interest rate would require tens of billions of dollars. The previously loose rule that stock transactions must be settled within two days was to be be strictly enforced from then onwards. Stockbrokers were ordered to report short-selling trades on promise of stiff penalties for non-compliance. Futures brokers were required to disclose the holders of large positions to regulators and to strictly observe margin limits.

Hong Kong in the end succeeded in defending the peg with the US dollar. In the property and construction sectors, it was observed that Hong Kong took decisive steps somewhat belatedly as the situation worsened. There was initially a denial to the seriousness of the situation as the government claimed shortly after the Asian financial crisis that Hong Kong would be amongst the first to recover. The fact is that Hong Kong was not amongst the first but the last (Jao, 2001). After long debates and much deliberation, the government first scrapped the "85,000 housing units plan" in 2000 and then, in the following year, formulated further short to medium measures aimed at improving property market conditions. The measures were essentially to reduce supply and stimulate demand.

In the property sector, there would be further reduction in land sales and public housing completion. To prevent land auction prices from falling further, large land sites were postponed from being auctioned. Instead, the government had put land available for sale on to a reserve list. This mechanism was aimed at creating buying confidence whilst showing government's flexible attitude towards land supply at the same time. Concerning the supply of housing flats, developers have been calling for a reduced supply of subsidized flats for sale under the Home Ownership Scheme

(HOS) and the Private Sector Participation Scheme (PSPS). The government had since then been considering schemes to assist eligible families to buy their flats from the private sector, rather than provide them with direct housing. Subsequently the Chief Executive had interpreted the government's 50,000 public housing target as "50,000 housing assistance opportunities for eligible families". This hinted at the possibility that instead of continuing the direct housing schemes (HOS and PSPS), the government may provide eligible families with low-interest loans or rental coupons to buy or rent their own flats from the private sector. This policy, if implemented, would leave more of the residential market to the private developers than was the case then. Eventually, in an abrupt change in its stance the government bowed to the pressure of the developers and the ongoing economic slump. It finally froze the sale of HOS flats from August 2001 for 10 months. However, due to the development time lag, the effect of these policies was not expected to be felt in the next couple of years or more. Meanwhile, there would be more low-interest home purchase loans made available for eligible families.

In the construction sector, the government accelerated and expanded public works programmes to stimulate economic growth. Indeed infrastructure works were identified as agents of economic growth in the *1999 Budget*, about one year after the full blow of the Asian financial crisis was felt. In 2001, minor labour-intensive projects were included in the works programme to generate employment opportunities. The effect was yet to be seen at the time of writing as the projects would not be carried out until towards the end of the year. Nonetheless, downside risks remained especially pertinent as asset prices declined further in the wake of the worsening external conditions.

4.2.6 Economic outlook in the next decade

With the new policies to regulate the stock market and stimulate the property market, Hong Kong's economy in 2000 did begin to recover from the severe setback brought about by the Asian financial crisis. The labour market improved with the unemployment rate declining. The cost of doing business came down visibly. Together with the process of corporate restructuring and downsizing seen in previous years, this had enabled the economy to restore competitiveness. However, with the external environment becoming more unstable in the wake of the American war on terrorism, economic activity in Hong Kong was expected to remain subdued at least for the short term.

The prospects for the Hong Kong economy beyond the short term however remain bright, benefiting from the sustained robust growth in China as well as the further economic reform and liberalization in China upon its accession to the World Trade Organization (WTO). However, Hong Kong's future cannot continue to depend on being just a gateway to the Mainland only, but must evolve into an international metropolitan centre. Hong Kong has assumed its role as "integrator and packager" of services. However, there was expected a flooding of firms into the Mainland market once China's WTO accession takes effect. Hong Kong was expected to have to face competition in markets once practically closed to foreign firms. Hong Kong's role as a trading hub was also expected to face challenges from the Mainland as water transportation from and within China continued to improve, and as China completed her port facilities. Domestic financial centres were rapidly developing in Shanghai and Shenzhen. They were expected to compete with Hong Kong as the regional financial centres.

In a bid to revive the "old magic" the Hong Kong government is now promoting Hong Kong to become Asia's world city, contributing to the continual modernization of China through the city's global networks and services-oriented economy. There had been a critical mass of professionals, professional services firms, institutional investors and traders doing business in and helping create wealth for Hong Kong. The city's competitive advantages lay in the creation, maintenance and enhancement of an environment that was conducive to high value-added business. Hong Kong had excelled in free flows of information, people and capital, and in currency convertibility. Together with the established legal framework and the regulatory institutions, these factors had helped Hong Kong to be consistently among the top ranks in various league tables of competitiveness and business-friendliness. It is imperative that the government must strengthen the competitive advantages that Hong Kong has, and rectify the mismatch between the skills of the workforce and the demands of an economy increasingly focusing on high-value-added sectors. There is a consensus in the community that the problem on workforce skills will get more acute as competition for trade of services will become only the more globalized.

Yet, in the short to medium term, Hong Kong's position as the international financial window for the mainland is unrivalled. Hong Kong and Shanghai (or Shenzhen) were seen to perform complementary roles. Hong Kong was expected to cater for the huge financial needs of China's state-owned enterprises (SOEs) as they restructure, helping the SOEs raise finance from investors coming from all over the world. A listing in Hong Kong is seen as having complied with stringent international regulations. This has the effect of lowering the cost of money back home. Shanghai, at the time of writing, was to instead help Chinese firms raise finance from within the domestic market.

In the construction sector, local contractors in Hong Kong had been active in infrastructure and building projects in the mainland. The local Trade Development Council had geared up its marketing effort to help local contractors to enter the Mainland market, especially now that China is turning her attention to the "Big West". The Construction Industry Review Committee (CIRC, 2001) has recommended that, leveraging on the city's expertise in accounting, financing and legal services, Hong Kong has strong potential to develop into an infrastructure service integrator for the Mainland market and elsewhere.

In the 5 years to 2005, the business environment that Hong Kong was to offer is "very good", the best qualitative assessment a country could get in the Economic Intelligence Unit's Global Business Environment Rankings (EIU, 2001). Hong Kong was home to 3,000 overseas companies that have established their regional headquarters or offices here. As long as Hong Kong continues to provide businesses with a strategic location, a level playing field, excellent infrastructure, a dense network of financial and professional service firms and political stability, Hong Kong could truly be and call herself Asia's world city..

4.3 The construction sector

4.3.1 Value-added per employee

Hong Kong's construction sector was marked by low productivity over the years, significantly lower than the manufacturing sector and only about half that of the service sector. A quick overview of the 1990s leads us to believe that, in general,

Table 4.3 Composition of GDP and employment

Sector	*GDP*		*Employment*	
	1980	*1998*	*1980**	*1998**
Primary	0.9	–	1.9	–
Industry	32.2	15.3	49.9	15.4
Services**	66.9	84.7	47.2	84.6

Source: Census and Statistics Department, *Annual Digest of Statistics*, various issues.

Notes
* At 4th quarter of the year.
** Excluding civil services.

the value-added per employee rose tremendously as Hong Kong's industrial structure underwent a fundamental structural change from being predominantly manufacturing to being service oriented with the services' share of GDP risen to 86 per cent in 2000. Until the first half of the 1980s, the industrial sector accounted for about half of total employment (59 per cent in 1975 and 50 per cent in 1980) by virtue of foreign investors establishing industry in Hong Kong due to cheap labour. Since the mid-1980s Hong Kong's industries changed to being service oriented with its output growing at the rate of 16 per cent (in nominal terms) between 1984 and 1996. During the early 1990s the value-added by this sector to GDP rose at an average annual rate of 14.4 per cent which is 2.3 per cent higher than the average growth rate (Table 4.3).

Within the services sector, the sub-sector "FIRE (finance, insurance and real estate) and Business Services" shows the most spectacular productivity gains. The productivity growth of this sector was 22 per cent per annum between 1985 and 1990, slowing down to 15 per cent per annum between 1990 and 1997. This sub-sector alone accounted for 22 per cent of total service employment in Hong Kong. The value-added by this sector to the GDP grew at an average rate of 17.9 per cent between 1987 and 1997, which is 2.2 per cent and 3.8 per cent higher than the average growth rate of all services and the general economy respectively between the period. This sub-sector has also recorded the highest productivity per worker (Tables 4.4–4.6).

4.4 Construction cost

4.4.1 *Unit construction cost*

Unit construction cost in Hong Kong had changed little when compared to property prices (Table 4.7). However, unit construction costs for almost all kinds of buildings rose to their peaks in 1997 and had come down since then to 1995–1996 levels. As the labour and material index remained relatively unchanged, the rise and fall seen in the unit construction costs over the years must be attributed mainly to the changes in profit margins, brought about in turn by the property and thus construction cycles.

4.4.2 *Average construction material price*

Table 4.8 shows the average construction material price.

Table 4.4 Value-added per employee

Activities	*1995*	*1996*	*1997*
Construction ($)	799.1	796.5	878.5
% change	–	−0.3%	10.3%
Manufacturing ($)	225.6	254.6	284.4
% change	–	12.9%	11.7%
Services ($)	464.7	499.3	616.9
% change	–	7.4%	23.6%
Primary sector* ($)	n.a.	n.a.	n.a.
% change	n.a.	n.a.	n.a.

Source: Chiang, Y.H., Hui, C.M., Raftery, J. and Tse, Raymond, Y.C. (2000) *Hong Kong Country Report for the 6th AsiaConstruct Conference,* September, 2000, Malaysia (Construction Industry Development Board Malaysia), p. 46.

Notes
* Primary sector includes agriculture, mining, etc.
Figures in 1998 not available.

Table 4.5 Services sub-sectors performance

Sector	*Product per worker* (HK$)*			*Productivity growth* (%)	
	1985	*1990*	*1997*	*1985–1990*	*1990–1997*
FIRE and business services	218,953	408,960	833,345	22.1	14.9
Wholesale, retail and export import	87,992	169,628	344,808	18.5	14.2
Transport, storage and communication	206,361	393,826	336,852	18.1	−2.1
Community social and personnel services	204,373	324,998	327,470	11.8	0.2

Source: Census and Statistics Department, *Annual Digest of Statistics*, various issues.

Note
* Average over the period at market price.

Table 4.6 Average growth rate of value-added

Sector	*1981–1987** (%)	*1987–1997** (%)
FIRE and business services	13.5	17.9
All services sector	17.3	15.7
Economy	15.7	14.1

Source: Census and Statistics Department, *Monthly Digest of Statistics*, various issues.

Note
* Average over the period.

Table 4.7 Unit construction cost

Year	Residential (High-rise)		Commercial office		Industrial		Hotel 5 Star
	Standard	*Luxurious*	*Standard*	*Prestige*	*Light*	*Heavy*	
1995	8,000–8,500	8,500–10,000	8,500–9,500	12,000 up	4,800–5,000	5,300–6,300	14,000 up
1996	9,000–9,500	9,500–12,500	9,500–10,700	13,000 up	5,400–5,700	6,100–7,200	16,000 up
1997	11,000–12,000	11,500–15,000	11,550–13,500	15,500 up	6,500–7,000	7,100–8,500	19,500 up
1998	10,000–11,000	11,000–13,500	10,500–12,500	14,300 up	5,600–6,200	6,300–7,500	18,000 up
1999	9,800–10,700	10,800–13,200	10,300–12,100	13,800 up	5,500–6,100	6,200–7,100	17,500 up
2000	9,300–10,000	10,500–12,800	9,700–11,500	13,500 up	5,500–6,000	6,000–6,900	17,500 up
2001	9,100–9,800	10,300–12,600	9,700–11,500	13,500 up	5,500–6,000	6,000–6,800	17,500 up

Source: Davis Langdon and Seah International. *Current Building Cost Information Data in Hong Kong*, various issues.

Notes

Unit: $/square metre of gross floor area (unit costs exclude sub-structural works).

Table 4.8 Average construction material price

	Cement in bulk ($ per tonne)	*Steel bars* ($ per tonne)	*20 mm Aggregates* ($ per tonne)	*Concreting sand* ($ per tonne)	*Common bricks* ($ per thousand pieces)	*RMC* Grade 30* ($ per m^3)**
1995	711	2,961	60	60	8,894	600
1996	738	2,921	64	61	8,550	610
1997	734	2,891	59	63	8,722	630
1998	719	2,348	54	60	8,342	620
1999	689	2,239	53	61	8,252	580
2000	658.14	2,272.85	70.43	48.94	7,824.49	–
2001	648.91	2,290.66	85.23	49.10	7,359.60	–

Sources: 2000 and 2001: Consumer Price Index Section, Census and Statistics Department.
1999: Department of Architectural Services.
1995–1999: Since the average wholesale prices of building materials are compiled by the Commissioner for Census and Statistics for internal use, the figures for the period 1995–1998 were estimated by using index numbers of the costs of labour and materials used in public sector construction projects.

Notes

* RMC: Ready-mix concrete.
** Prices measured at April of the year.

4.4.3 *Construction industry salaries and wages*

Table 4.9 shows the construction industry salaries and wages.

4.4.4 *Construction quality and safety*

The construction industry in Hong Kong has been pre-occupied with supply for the last few decades in Hong Kong. Owing to the booming population in the 1960s and 1970s, quantity and speed of construction rather than quality were the major

Table 4.9 Construction industry salaries and wages

	1995	*1996*	*1997*	*1998*	*1999*	*2000*	*2001*
Professionals (monthly salary)	41,726	45,190	49,404	51,824	51,365	51,333	50,782
Technicians (monthly salary)	9,563	10,367	11,134	11,481	11,333	10,961	11,146
Skilled worker (daily wage)	710.9	766.5	882.3	933.7	947.9	985.4	960
Unskilled Worker (daily wage)	460	503.4	549.9	604.2	595.8	607.6	630.3

Source: Census and Statistics Department, HKSAR Government, *Hong Kong Annual Digest of Statistics*, 2000 Edition.

Notes
All figures are nominal.
(unit: $).
The wages of unskilled workers are based on wages of unskilled labourer, excavator, concretor's labourer, bricklayer's labourer and plasterer's labourer.
The wages of skilled workers are based on wages of concretor, bricklayer and drainlayer.
The salary of technicians is in the category: Financing, insurance, real estate and business services.
Only the index (not absolute figure) for salary of professionals in the construction industry is available. Based on the *Manpower Survey Report on the Building and Civil Engineering Industry*, we estimate the weighted average monthly salary of professionals in the construction industry. Then, we use the index to estimate the salary figures in the period 1995–1997.

concern of public housing development. The notions of quality and safety were sidelined. Since 1990, however, the construction industry seems to have woken up to the fact of deteriorating quality of building, both the product and the process. The heavy toll on workers that the industry has taken, aside quality, has taken the centre stage as maintenance, repair and replacement have become increasingly expensive. Maintenance problems associated with thousands of, particularly, public residential buildings have arisen such as water leakage, spalling of concrete, with rusting of steel reinforcement, honeycombing, etc.

Construction accidents

Accidents have been a recurring event on construction sites in Hong Kong. The record shows that about one in every three accidents occurred in the construction industry between 1994 and 2000. Furthermore over 60 per cent of industrial fatalities have occurred on construction sites in the 1990s. Accident rates have been generally extremely high when compared to other local industries, and the construction industry of the other countries as well. This is mainly attributed to three reasons. First and foremost, construction industry in Hong Kong, especially in the building sector other than public housing, has remained very labour-intensive. Second, Hong Kong construction predominantly deals with high-rise buildings. The two major causes of accidents have been falling people and falling objects. Third, owing to limited space in general, most construction sites do not have sufficient storage space or space for auxiliary works. Accidents frequently happen because workers step on, strike against or are struck by objects.

It was observed that the number of accidents have been increasing in tandem with the volume of work over the past decade. However, it peaked in the year 1997 and

1998. Incidentally during this period construction activities had been brisk. However, there appears to have been a significantly lower fatality rate in the same period compared to the early 1990s. The trend of both accidents and fatalities however seemed broadly to be lowering in the construction industry. In 2000, 11,925 accidents including 29 fatalities were recorded for the local construction industry. The accident rate of the construction industry per thousand workers in 2000 was 149.81, lower than the 198.45 in 1999 by 24.5 per cent. Over 2000–2001 the construction industry recorded a significant decrease by 40 per cent in the accident rate (Tables 4.10 and 4.11) (OSHC 2002 at http://info.gov.hk/labour/osh/index.htm).

Fundamentally, safety and health training for employees is perhaps the most significant element for a safety management programme. Not only is it essential for employees to be fully aware of safety and health matters, but the regulations also require all employees to be trained in those aspects related to their employment, as it is essential for them to perform their tasks safely. Since 1997, most government contracts have explicitly indicated that contractors are required to arrange for their workers to enrol in the Enhanced Green Card Course, a safety induction course for the construction industry. Moreover, the Factories and Industrial Undertakings (Training) Regulations also stipulated that construction workers should pass the approved safety training course before being allowed to enter construction sites.

The alarming record of construction accidents aroused concern on site safety. Accordingly the government decided in 1991 to introduce safety management systems in the Airport Core Programme projects. The practice was then extended to all public works projects. A series of initiatives to enhance site safety was implemented in 1996, including the Pay-for-Safety Scheme and the Independent Safety Audit Scheme. The Pay-for-Safety Scheme was aimed to remove site safety consideration from competitive bidding. Payments would be made to contractors in interim valuations after their compliance with safety management-related elements had been certified. The payments would be pre-determined and included as fixed sums in the bills of quantities. The Independent Safety Audit Scheme would be a mechanism whereby safety audits were to be carried out on a quarterly basis by independent safety auditors. The

Table 4.10 Industrial and construction accidents (1994–2000)

Year	*Total number of accident cases in industrial undertakings (in thousand)*	*Accident cases in construction trade (in thousand)*	*Fatal cases in construction trade*	*Accident rate/1000 workers*
1994	44.2	16.4	51	275.0
1995	41.0	15.3	63	222.8
1996	40.4	16.5	51	219.9
1997	43.3	18.6	41	227.4
1998	43.0	19.6	56	250.1
1999	36.0	14.1	47	196.1
2000	33.7	11.9	29	143.0

Source: The data for 1999 and 2000 are extracted from HK Government Labour Department Statistics (http://www.info.gov.hk/labour/osh/index.htm).

Table 4.11 Construction accidents (1989–1998)

	1989	*1990*	*1991*	*1992*	*1993*	*1994*	*1995*	*1996*	*1997*	*1998*	*1999*	*2000*
Number of accidents	26,399	25,138	23,115	18,815	16,573	16,422	15,268	16,469	18,559	19,588	14,078	11,925
Number of fatalities	59	58	54	48	80	51	63	51	41	56	47	29
Employment*	70,505	71,113	63,450	62,232	56,226	59,710	65,611	74,907	81,629	79,007	71,780	83,924
Accident rate/1,000 workers	374.4	353.5	364.3	302.3	294.8	275.0	232.7	219.9	227.4	247.9	196.1	143.0
Fatality rate/1,000 workers	0.837	0.816	0.851	0.771	1.423	0.854	0.960	0.681	0.502	0.709	0.655	0.350

Source: Labour Department (http//:www.info.gov.hk/ld).

Note

* The employment size only covers manual workers on construction sites.

contractors would be paid if the audit scores were 70 per cent or above. The two measures were later incorporated in the PWP Construction Site Safety Manual. The measures were initiated to encourage the public works contractors to set up efficient safety management systems and to enhance the standard of safety performance in order that the accidents remain below 60 per 1,000 workers per year.

In the following years, the safety issue was also addressed by the Buildings Department and the Labour Department. The Buildings Department formulated the Site Safety Supervision Plan System in December 1997 with the issuance of the Technical Memorandum for Supervision Plans and the Draft Code of Practice for Site Safety Supervision. In the following years, the practice had been further codified and incorporated in the Code of Practice for Site Safety Supervision (BD, 2000) in the private sector.

The Occupational Safety and Health Council together with the Labour Department had devised the Safety Charter for wider adoption by the industry in general. It is "a written document highlighting the commitment of both the employers and employees in creating and maintaining a safe and healthy work environment. It spells out the safety goals of the organization and points out the responsibilities of various parties" (OSHC, 2001a). Among the 557 organizations that have subscribed to the Occupational Safety Charter were the Hong Kong Construction Association and a large number of contractors (OSHC, 2001b).

Quality assurance

The construction industry was the first commercial sector to look for ISO 9000 certification, because the Hong Kong Housing Authority (HKHA), as the biggest housing developer in the economy, decided to tackle the problem of sick buildings by enhancing a quality management system in the construction industry. Following the pre-mature demolition of its 26 blocks of residential housing in the early 1980s due to inferior concreting, itself the result of corruption scandals involving government officials and contractors, the HKHA had since then committed itself to producing buildings of good quality. It had been made mandatory that all concrete suppliers and contractors have to be audited and certified against the ISO 9001:1994 or ISO 9002:1994 quality management standard (HA, 2000). This requirement was later extended to all public works (besides public housing) with an estimate of $50 million or more. With effect from 1 November 1999, the Works Bureau only accepts ISO 9001:1994, ISO 9002:1994 or ISO 9001:2000 certificates issued under the rules of the Hong Kong Certification Body Accreditation Scheme operated by the Hong Kong Accreditation Service (HKAS) or equivalent (WB, 2001).

The ISO 9002 (for building contractors) sets out how companies can establish, document and maintain an effective management system, and compliance with it will help to provide the necessary consistency. The main components of which are: tender and contract; planning and documentation; control of measurement and test equipment; quality procurement; sampling, inspection and testing; incoming inspection; in-process inspection; final inspection; inspection and test status; material identification and traceability; handling, storage, packaging and delivery; control of production/construction; quality records; control of non-conformity; corrective actions; use of statistical techniques; and auditing the quality system.

At the time of writing the Hong Kong Accreditation Service (HKAS) was the single government-supported organization that provided accreditation to both laboratories and certification bodies on a voluntary basis. One of the three accreditation schemes that HKAS ran is the Hong Kong Certification Body Accreditation Scheme (HKCAS), launched in November 1998. There were six accreditation bodies under the Scheme, as at 16 February 2001 (HKAS, 2001). They provide certification for: product conformity; supplier and services quality management systems; and personnel involved in quality verification, within various sectors of the construction industry and within specific fields of activity.

4.5 Research and development

4.5.1 Research

Research within the construction industry was very minimal. Expenditures on research and development by local contractors, particularly the indigenous local contractors, had been practically nil with one or two isolated exceptions. Throughout the 1990s the construction industry operation has remained labour-intensive despite the labour shortage and consequently high wages. The private building sector had shown little sign of any significant technological advancement. The government had not, only until the time of writing, made any effort to raise the general technology level in construction.

Belatedly Hong Kong had begun to recognize research as the key to improving overall performance and competitiveness by enhancing productivity and improving quality through innovation (CIRC, 2001). The construction industry will inevitably benefit by development of better designs and materials as well as construction techniques and management practices. A strong R&D culture will not only produce design solutions that are suited to local conditions but will also maintain Hong Kong's competitiveness and keep the local construction industry abreast with the latest technological enhancement in other countries. In 2002, investment in construction research in Hong Kong was largely administered through two funding mechanisms:

1 Research Grant Council (RGC) Fund which includes Built Environment; and
2 Innovation Technology Fund available to the Industrial Community component of all sectors.

The RGC fund is available on an annual competitive basis for individual university academics. The general University Grants Committee (UGC) support for universities, in addition, includes a component for the support of research at the discretion of each university.

The research projects of 1990s were nearly all undertaken by the tertiary institutions and mostly funded by the UGC and the RGC. Quantitatively the investment made in R&D in construction research was significantly low and those efforts that were made were marred by lack of coordination between industry and the academia. Construction research was mostly initiated by individual research teams from within the universities, instead of the industry itself. While the academic sector was recognized as the principal research body, there was little interaction between academia and industry to identify and articulate local research needs. Besides, there was also relatively low awareness of the research outputs in the construction fields, and there was even lower take-up of these research results. In order to avoid duplication of efforts to bring in cost efficiency, research funding bodies, research providers and potential research users need to work in partnership, under the aegis of an all-embracing body. Better coordination will also encourage private sector clients to begin to commit resources for the construction research and development in Hong Kong. As a result of the CIRC report, a Provisional Construction Industry Coordinating Board (PCICB) had been established and intended to address this issue in the future.

In early 2002, the government had allocated more fund to its Innovation Technology Fund for wider enhancement of technology. The construction industry fell within the coverage of this Fund, but the industry had yet to make any significant use of this Fund due to the low awareness of the scope as well as the application procedures of the Fund, and also due to the inherently fragmented nature of the industry. While the RGC fund was focused more on basic research, the Innovation Technology Fund supports applied research that contributed to innovation or technology or the upgrading of local industries.

New technology development and prefabrication technology

The construction industry in Hong Kong had remained prudently conservative when it comes to application and research of new technology. On the whole the local construction is labour intensive and involves extensive wet trades on site. Both quality and cost are dependent on the availability of skilled labour. The labour intensity also puts heavy demand on site logistic management necessitating intensive supervision for quality assurance. Excessive wastage, increased risk factors and environmental damages are other factors that make *in situ* construction costly, in both material as well as non-material sense. Some degree of mechanization could be seen with public housing construction where large panel formworks, tower cranes, concrete pumps and other mechanical equipment and plans are used and early pioneering work was undertaken by the Housing Authority (HA). Otherwise, local contractors' technological ability was generally poor compared to international standards. The lack of competence in technology has restrained them from undertaking large and technologically savvy projects. In the Airport Core Programme, Hong Kong's biggest project in the 1990s, out of the total 145 contracts awarded by the government, Airport Authority, MTR and Western Harbour Tunnel Co., Japan gained the largest share (by value) of 25 per cent followed by Hong Kong (23 per cent), UK (16 per cent), China (6 per cent), Netherlands (6 per cent) and the remaining distributed among France, Belgium, New Zealand, Australia, Spain, Italy, Germany and the USA (Chiang, 1996).

It is believed that wider use of prefabrication can effectively overcome many of these problems. The Hong Kong construction industry is now more aggressively pursuing the possibility of using more prefabrication as a step forward towards efficiency and reduction in cost. The Construction Industry Review Committee (CIRC) has now placed a growing emphasis on the use and development of newer technology in building and construction to suit the changing context of labour, prices, materials and environmental standards. The committee in its report calls for an efficient, innovative and productive industry. The commitment of the government behind this particular "impetus" for upgrading construction methods and techniques stemmed from the experience of nearly a half a century of public housing construction in Hong Kong. The committee considered that "[p]refabrication, coupled with the use of standardized and modular components, will contribute to improved buildability and should be widely promoted, with public sector clients taking the lead" (CIRC, 2001). It had been increasingly recognized in Hong Kong that prefabrication along with the use of standardized and modular components will help enhance buildability.

Public sector clients such as the Housing Authority (HA) in Hong Kong had taken a noticeable lead in promoting them and could play a more instrumental role in facilitating technology change and development. The HA adopted prefabrication in the mid-1980s along with the introduction of a new Trident block series design that attempted to standardize the size and shape of walls. Application of a precast façade system was first experimented with on a 33-storey block by a French company, which unfortunately was stopped due to excessive water leakage from the joint. Metal-formed formwork subsequently became mandatory. Prefabrication technology eventually made its way, as architects began to concentrate on the simplification of the structural design to facilitate mechanized construction as well as standardize building using basic building components. With the advent of the Harmony series design – characterized by large panel formwork, prefabricated façades, prefabricated stairs, semi-prefabricated slabs and fabric reinforcement – prefabrication technology became the hallmark of public housing construction to achieve cost effectiveness, improved durability and enhanced buildability. At the time of writing precast façades were mandatory in standard domestic blocks. A variety of other precast materials such as precast concrete structural elements and panel wall partitions were also widely used. In the infrastructure sector, leading quasi-government clients such as KCRC and MTRC have been increasingly using precast concrete materials as evident in the construction of West Rail viaducts. The KCRC has also tried out components such as an acoustic layer laid on precast parapets that are prepared in the factory rather than on site, despite greater chances of wear and tear in the transport to site.

Overseas experience suggested that the local construction industry could adopt a manufacturing approach with prefabrication to upgrade its technology base and to become competitive in the international market. Many studies (Crowley, 1998; Veeramani *et al.*, 1998; Gann, 2001) had turned to the manufacturing industries for technology transfer in improving construction productivity and quality. Though pre-assembly is no panacea, such technology could definitely be adapted by the construction industry to improve building productivity and even national competitive performance (Gibb, 2001; Matsumura, 2001). A study of innovations in the construction sector of 15 countries (including Japan, the US, the UK) also showed that in general "[i]deas from advanced manufacturing and information technology will migrate more rapidly to the construction sector and that novel approaches for site assembly will be developed" (Manseau and Seadan, 2001). In Singapore, greater use of prefabrication, modularization, standardization and componentization had been promoted to enhance quality, safety, productivity and efficiency. In the UK, the CIPS (1994) concluded that it took 20–30 per cent more man-hours there than in the US for comparable projects, due to lesser extent of standardization and pre-assembly in the UK than in the US. In the US, The CII (1999) anticipated increasing use of prefabrication to minimize site work. In most of the industrialized countries leading contractors have achieved superior market position through use of innovative practices. In Japan, which Kerr (2001) called the "construction state", progress in construction has been accomplished mainly in the public housing sector. Principle drivers for innovation were often created at the firm level, within a stimulating macroeconomic context. The public housing sector was such a context and the "incubator" for the development of advanced technology and management techniques, to be applied, adopted and adapted by the construction industry at large.

There has always existed the need for continued strife for greater efficiency and raising standards. Housing remained one of the most coveted and expensive commodities in Hong Kong while construction itself was a complex task of converging diverse skills and techniques to produce a tempting "object". Probably stemming from the then downturn in the property market, it was observed that a number of developers in the private sector have shown a heightened interest in the adoption of more efficient construction methods. With the drastic reduction in land prices, construction costs have become more significant in the equations of developers' profits. Such interest will inevitably lead to greener building designs and innovative construction methods. At the time of writing, we were concerned that the economy might be trapped in a deflationary spiral. However, it was expected that when the economy picks up again, there would be again, a growing demand for housing, escalating wages as well as rising prices of materials. By virtue of Hong Kong's status as a market-led economy, such efforts will eventually be driven by essentially market forces. Yet it was important for public sector clients as pioneers of new construction technology to continue to lead in this sector. In a bid to streamline construction practices in Hong Kong the government further proposes a dedicated standardization office to facilitate better coordination in development and application of construction standards in Hong Kong. The construction industry can clearly exploit the fast-growing IT sector for these purposes. It is time for the government and the private sector to work towards bringing ample efficiency, innovation and creativity in techniques and materials. It is never too early to prepare.

Policy and initiative on the use of IT in construction

The construction industry needed to harness information and communications technology. Multiple parties were involved during the multi-stages of the multi-disciplinary project delivery process. IT was increasingly recognized to have great potential to enhance access and exchange of project information and transact people and physical borders electronically in a seamless real-time manner. The CIRC in Hong Kong had recognized IT as an important means for better information dissemination among all the parties concerned, improved design capabilities and enhanced project logistics management. By 2002, Hong Kong's construction industry had begun to reap the fruits of IT particularly through the use of software applications for selected tasks such as computer-aided-design and drafting (CADD), structural analysis, budget estimation and contract administration. Yet the full potential for IT in the construction management and logistics planning sector was still far from realization. It was observed that despite its huge potential for bringing all participants together to work efficiently and more productively from the improved information flow along the construction value chain, wider use of IT in the local construction industry had been (for quite some time now) stifled because of a number of reasons including (CIRC, 2001):

1 Fragmented nature of the industry which impedes wider adoption of common IT tools across disciplines;
2 Absence of conducive ambiance due to lack of common standards and data infrastructure;
3 Lack of practical application solutions in the market to suit the local context;
4 High initial development costs and uncertainty over returns; and
5 Low awareness of IT benefits at management levels and low IT literacy.

It has been noted that the major inhibition for IT use in the construction sector in Hong Kong had been the lack of common standards and data infrastructures. The government of Hong Kong, realizing the urgency of the situation, took a number of initiatives that had significantly contributed to the development of a common platform for electronic communication in the industry as well as raising awareness in the potential of IT in construction project management. The government initiatives include: ongoing consultation study on the alignment of planning, lands and works data; study of computer-aided-design (CAD) standard for drawings for works projects; pilot scheme to digitize building plans in the archives of Buildings Department; and feasibility study of the development of an electronic system for building-plan submission and document management. In particular, Web-based Project Management, Electronic Supply Chain and Logistics Management, Mobile Computing for Project Delivery and Electronic Document Management and Electronic Maintenance Management System are other initiatives that are under consideration by the government in a bid to enhance IT usage.

The Buildings Department had been monitoring the development of artificial intelligence technology with a view to testing the feasibility of introducing electronic checking of building plans. The Works Bureau had taken active steps in finding ways to enhance the efficiency of electronic service delivery. The implementation of Electronic Mark Plant Circulation System and the Electronic Tendering Initiatives are a few examples of such efforts.

Apart from the government's initiatives to harness the full potential of IT, it was imperative that the private sector must understand that IT itself is an enabler and not a solution. Return on any IT investment is bound to be long term. For example, in the US, the productivity gains had eventually come only towards the end of the 1990s, long after decades of IT investment in automating office work. Accordingly, to drive more extensive use of IT throughout the construction delivery process, there was a need to build a critical mass of IT users within the industry. Leading clients and construction companies should effectively take the lead in setting up an IT establishment and in training and educating the work force at all levels of the construction industry.

In order to enhance implementation of IT usage in the local construction industry a three-fold approach was recommended (CIRC, 2001):

1 Consultation – Wider consultation with industry participants in the development process was essential with particular attention paid to the compatibility and interoperability of the global standards.

2 Legalization – At the time of writing there were many exemptions made under the Electronic Transaction Ordinance for construction-related legislation which should begin to be withdrawn.
3 Collaboration – A closer collaboration with the software companies and construction industry is essential to map out the demand and supply of software applications in the market. Around 2001 Hong Kong had seen a launch of many IT solutions and integrated industry portals catering for construction industry. Closer ties between the construction and IT industry will thus be a key to developing applications suited to the vernacular context.

4.5.2 Environmental conservation

Perhaps to a large extent construction activities are inherent adversaries to the environment, for their very aim is to physically change the existing landscape. But for their contribution to the economy and our basic need for accommodating and making contact with people, we have to keep on building. However, construction does generate significant environmental nuisances in the form of noise, dust, muddy runoffs and improper disposal of waste which should be minimized as much as possible. Hong Kong's vibrant construction industry saw a 9 per cent rise in the number of complaints for non-compliance with the environmental protection measures in 1999 compared to the figure of 1998. The number of convictions also increased by 40 per cent over the same period. Disposal of construction and demolition materials also presented a major problem in Hong Kong. In 1999, construction industry produced 7.52 million cubic metres of construction and demolition materials, a 14.3 per cent increase over 1998. In 1999 about 21 per cent of construction and demolition materials were disposed off at landfill sites, which accounted for over 40 per cent of the total waste disposed (CIRC, 2001).

The impact of construction activities on the environment in Hong Kong goes beyond simply aggravating air, noise and water pollution (CIRC, 2001). Energy consumption and land contamination also adversely affect the environment in more indirect ways in the long term. Research and development as part of the construction process is intricately linked to the introduction and development of environment-friendly materials and techniques to reduce the impact on the environment. However efforts to simply minimize the symptoms of such pollution will only be a myopic approach to the problem. The solution still depends upon our ability to make the industry more responsible with the application of sustainable and recyclable materials that eliminate or reduce the causes of pollution. The concept of environmental conservation dwells even further in the realm of policy and planning much in line with the principles of sustainability. The government had been aggressively coming forward with multi-tiered regulatory measures to curb these problems.

Corrective measures

The Noise Control Ordinance (NCO): Construction sites in Hong Kong have always been a significant source of noise pollution. Regulations for specific control such as the Noise Control (Air Compressors) and Noise Control (Hand Held Percussive Breakers) were promulgated to exercise specific noise control on particular machinery used in construction site. These regulations required machinery to comply with the statutory noise standard and be issued with a noise emission label. The two most common mechanical equipment that fell within the control of the said regulations were Concrete/Rock Breaking Equipment and Portable Air Compressors.

Water Pollution Control Ordinance (WPC): This ordinance along with its subsidiary regulations (CAP 358A–CAP 358W) allowed designated areas to be "water control zones". Within the zones, licence was required to discharge effluent. The criteria for license and limits were set out in the Technical Memorandum of Effluent Standards. This catered for development in places where connections to public sewerage were not feasible besides construction and detail design of sewage treatment were usually attainable with extra cost and time for the development programme.

Building (Energy Efficiency) Regulations: Promulgated in 1995, it aims to regulate commercial and hotel buildings such that they are designed to achieved energy efficiency. The criteria were measured against the Overall Thermal Transfer Value (OTTV). The external walls and roofs were designed to achieve at least the minimum standard required to avoid waste of energy. The regulation has however been criticized in that the requirements are too rigid in terms of choice of materials, colour scheme and for stifling innovative options in building design.

Air Pollution Control Ordinance (APC): Similar to noise control, the major impacts of the Air Pollution Control Ordinance on property development were experienced during the environmental impact assessment stage and during site layout design. It also aimed to control excessive noise from the percussive machinery.

Building (Demolition Woks) Regulations: Previously the enactment of the Air Pollution Control (Construction Dust) Regulation, this regulation made under the Buildings Ordinance provided for, inter alia, preventing dust nuisance and removal of construction waste caused in the process of demolition. Contractors were required to erect dust screens to protect the works and notify the Environmental Protection Department of the demolition works for monitoring through the Buildings Authority. Under the Air Pollution Control (Construction Dust) Regulation, it was a mandatory requirement of a contractor to notify the Environmental Protection Department prior to commencement of works or else be liable to a penalty.

The Factories and Industrial Undertakings Regulations: This regulation, concerning the safety of people, laid down control measures and requirements for protective equipment, storage and distribution of asbestos materials and conditions for working with asbestos. The (Asbestos) (Administration) Regulation provided for registration of asbestos consultants, supervisors, laboratories and contractors. Since May 1993, the Environmental Protection Department had been operating a "cradle-to-grave" licensing system to control the manufacture/import of asbestos and keep track with their sale and use, removal and disposal.

Environmental impact assessment (EIA)

Environmental impact assessment (EIA) had been an integral part of development and construction projects in Hong Kong following the enactment of EIA ordinance on 4 February 1997. This ordinance covered both public and private projects (does not cover residential projects except for development projects in the vicinity of sensitive zones), enabled the Director of Environmental Protection to regulate the process and designated projects that must apply for Environmental Permit prior to its commencement. In addition, larger urban environmental projects covering 20 hectares or more and involving over 100,000 existing or new population were required to submit an EIA report. Along with EIA, strategic impact assessment for major planning projects had also been enforced.

Hong Kong building environmental assessment method

Modelled on the UK Building Research Establishment's Environmental Assessment Method, The Hong Kong Building Environmental Assessment Method (HK-BEAM) was set up in the early 1990s and operated by the Centre for Environmental Technology Ltd (CET), the executive arm of the Private Sector Committee on the Environment. It was developed and funded by the Real Estate Developers Association with technical assistance of the Department of Building Services Engineering at the Hong Kong Polytechnic University, the Welsh School of Architecture and the ECD Energy and Environmental Ltd. The scheme was a voluntary initiative by property owners. It gave recognition for enhanced environmental performance in the design, operation and maintenance of new and existing commercial buildings. Certification to HK-BEAM was completely voluntary and was undertaken by independent specialists.

Environmental planning

The environmental control framework in Hong Kong was moving towards a more integrated approach to ensure that environmental factors are taken into account at all stages of planning and project development. For territorial and sub-regional planning, the Environmental Protection Department was involved for environmental input in formulating the Territorial Development Strategy carried out by the Planning Department. The strategy will identify further strategic growth areas to meet long-term development needs and comprehensively assess the environmental impact and development potential of these areas. At the local planning level, environmental factors recommended in the Environmental Chapter of Hong Kong Planning Standards and Guidelines had to be considered when planning applications were submitted by planners, architects and engineers. With the effect of EIA Ordinance, environmental permit was a prerequisite for planning approval for designated projects. Table 4.12 shows environmental inputs to different stages of forward planning.

It was observed that efforts for environmental conservation in Hong Kong should ideally be based on life-cycle benefits rather than short-term benefits. The quality of materials and construction techniques are important to attain long term benefits. Ongoing rebuilding and renewal of the city's structures in itself is hazardous to the environment in the long term, hence a more sustainable or durable production should be the aim of the construction activities based on the concept of life-cycle costing, which refers to systematic evaluation of all relevant costs associated with acquisitions and ownership of built form. The Building Safety and Preventive Maintenance Task Force issued a proposal to implement a "defect liability warranty" with the aim of promoting more durable buildings.

As a long-term strategy to preserve the environment the construction industry in Hong Kong needed to adopt greener and more energy-efficient designs. To encourage this a

Table 4.12 Environmental inputs to different stages of forward planning

Levels of planning	*Objectives of environmental inputs*	*Means of inputs*
Strategic planning through territorial development strategy	Address key strategic environmental issues, environmental carrying capacities, environmentally suitable areas for developments	The environment chapter of Hong Kong Planning Standards and Guidelines (HKPSG); strategic environmental assessment
Sub-regional planning and feasibility studies for new towns	Address the environmental acceptability of major development strategies and plans	The environment chapter of HKPSG, environmental planning studies at the regional level, environmental assessment of plans, site search and EIA studies
Outline development plans, outline zoning plans, and layout plan	Address environmental compatibility of land uses, environmental acceptability of plans, and environmental facility requirements	The environment chapter of HKPSG, environmental assessment of plans, environmental advice at planning committees
Project planning and implementation	Address the environmental impacts of project design and implementation, and monitor the actual environmental impacts	Project EIA, environmental monitoring and audit, the environment chapter of HKPSG

Source: Au, E. (2000) "Environmental Planning and Impact Assessment of Major Development Projects in Hong Kong", in Wong, W.S. and Chan, E. (ed.) *Building Hong Kong: Environmental Considerations* (Hong Kong: HKU Press), pp. 257–273.

supportive regulatory framework, market demand as well as access to information on the environmental performance of the building were needed. Construction of green building should be given adequate incentives. The industry also needed sufficient motivation to comply with regulatory requirements by making environmental performance a factor of consideration in tender evaluation and ongoing performance assessment.

Sustainable construction is a global trend and an integral part of the wider dynamics of sustainable development. Hong Kong's construction industry however was still several steps away from making headway towards this direction. The private sector in particular still perceived environmental regulation as an obstruction to economic development and a major setback in the pursuit of profit. In this context, a holistic effort from all the stakeholders became essential to promote sustainable construction.

4.6 Construction outlook 2002–2010

The outlook in the construction sector in Hong Kong was fairly good. At the time of writing massive infrastructure and building projects had been planned and construction tenders were to be invited shortly. Examples include works arising from the Railway Development Strategy, the redevelopment of the Kai Tak Airport and its adjacent areas, the construction of the Cyberport, and the construction of the Disneyland Theme Park development. In addition, the construction of the Shenzhen Western Corridor connecting Hong Kong with Shekou, the Deep Bay Link and the Lok Ma Chau Spur Line was expected to start earlier than planned. The construction of the three projects, that form part of the regional transportation network, was expected to help revive the dampening momentum in the construction sector today. In the private sector, the construction prospect hinged very much on the revival of the property market. It was not certain when the market will bottom out, as the local economy was preparing for the projected global economic downturn in 2002. The construction industry in Hong Kong was likely to look for work more from the public than private sector in the next few years. Civil engineering contractors probably had more opportunities than building contractors.

4.6.1 Business opportunities and major projects in the pipeline

In Hong Kong there existed no institutional barrier for foreign contractors to enter the local construction market. For public contract procurement, foreign and local contractors were treated alike and no distinction was made. Contractors' eligibility to tender depended solely on their financial, technological and managerial capacity; and on their track record, including local experience. In the private sector, firms once registered in the Buildings Department could tender freely as main contractors. Quasi-governmental agencies like the MTRC and the KCRC also entertained contractors in the same fashion once registered. A list of projects coming in the short, medium and long terms can be found in the annual reports presented to the *AsiaConstruct* conference.

4.6.2 Business alliance opportunities

Business alliance opportunities exist both in Hong Kong and China mainland. In Hong Kong, as already mentioned, the construction of major infrastructure works needed the participation of foreign contractors. For example, bridge construction and tunnelling required advanced or proprietary technologies, especially when time and budget were tight. Foreign contractors would be very much sought when the massive railways and highways programmes move to the construction stages. In the past, bridge and tunnel contractors had teamed up with local contractors to form joint ventures. This practice was expected to continue into the future. In China mainland, many large-scale constructions were initiated by Hong Kong developers, contractors and investors. Opportunities exist for foreign contractors and consultants to link up with the Hong Kong side to promote packages of design-and-build and build-operate-transfer projects. According to the Hong Kong Trade Development Council, as China mainland is particularly keen on further developing the country's property development market, foreign investment may find ample opportunities in especially the following sectors: luxury office and residential apartments, hotels and affordable housing.

4.6.3 China's accession to WTO

At the time of writing China was to soon become a member of the WTO and will further open up its market to competition. This will be a milestone in history and will herald a new chapter in world trade. That was to offer Hong Kong vast development potential and business opportunities, though at the same time posing local contractors the threat of direct competition from leading international contractors who would otherwise be barred from the market. Hong Kong's economy was linked very closely to that of the mainland. In the year 2000, some US$44 billion, or 70 per cent of the capital raised on the Hong Kong securities market was for mainland enterprises. In 2000, the number of overseas companies with regional offices in Hong Kong increased by 20 per cent because of Hong Kong's obvious value as an intermediary for developing business on the mainland.

References

AWSJ (1999) *Asian Wall Street Journal*. 1 July 1999.

AWSJ (2001) *Asian Wall Street Journal*. 29 October 2001, pp. S9–S16.

BD (2000) *Code of Practice for Site Safety Supervision*, Buildings Department, November 2000.

Chiang, Y.H. (1996) "Globalization and Overseas Construction", in Raftery, J. and Anson, M. (eds) *Hong Kong Country Report for the 2nd AsiaConstruct Conference*, October 1996, Seoul-Korea (Korea, Korea Research Institute for Human Settlements), pp. 49–58.

CIA World Fact Book (2000) *Hong Kong* (http://www.cia.gov/cia/publications/factbook).

CII (1999) *Vision 2020*, Construction Industry Institute, University of Texas at Austin.

CIPS (1994) *Productivity and Costs*, Chartered Institute of Purchasing and Supply, Stamford.

CIRC (2001) *Construct for Excellence, Report of the Construction Industry Review Committee*, Government of Hong Kong Special Administrative Region.

Crosthwaite, D. (2000) The global construction market: a cross-sectional analysis. *Construction Management and Economics*, 18, 619–627.

Crowley, A. (1998) Construction as a manufacturing process: Lessons from the automotive industry, *Computers and Structures*, 67, 389–400.

EIU (2001) Press release of Economic Intelligent Unit on 13 August 2001.

Fortune (2000) Winners of the World, *Fortune*, 142: 13, November 27, p. 232.

Gann, D. (2001) Putting academic ideas into practice: technological progress and the absorptive capacity of construction organizations, *Construction Management and Economics*, 19, 321–330.

Gibb, A.G.F. (2001) Standardization and pre-assembly – distinguishing myth from reality using case study research, *Construction Management and Economics*, 19, 307–315.

Gwartney, J., Lawson, R., Park, Walter and Skipton, C. (2001) *Economic Freedom of the World 2001 Annual Report* (The Fraser Institute).

HA (2000) *Rules for Administration of the List for the Housing Authority List of Building Contractors, January 2000 Edition* (Counterparty List Management Section, Finance and Accounting Branch, Hong Kong Housing Authority).

HKAS (2001) *List of Accredited Certification Body* (Hong Kong Accreditation Service, http://www.info.gov.hk/itc/eng/quality/hkas/hkcas/acc_cert_body/acc_cert_body.shtml, dated 5 November, 2001).

Hong Kong Yearbook (2002) Government of Hong Kong Special Administrative Region.

Jao, Y.C. (2001) *The Asian Financial Crisis and the Ordeal of Hong Kong* (Conn.: Westport, Quorum Books), p. 227.

Kerr, A. (2001) *Dogs and Demons: Tales from the Dark Side of Japan* (NY: Hill and Wang).

Lands Department (2001) *Land Sale Records* (http://www.info.gov.hk/landsd/lsr/lsr.htm, dated 2 November 2001).

Manseau, A. and Seaden, G. eds (2001) *Innovation in Construction: An International Review of Public Polices* (London and NY: Spon Press).

Matsumura, S. (2001) Perspectives on component-based design in Japanese construction, *Construction Management and Economics*, 19, pp. 317–319.

O'Driscoll, G.P., Jr, Holmes, K.R. and Kirkpatrick, M. (2001) *2001 Index of Economic Freedom* (The Heritage Foundation and Dow Jones & Company, Inc.).

OECD (2001) *National Accounts,* Volume 1: *Main Aggregates* (Paris: Organisation for Economic Co-operation and Development), updated July 2001.

OSHC (2001a) *Occupational Safety Charter* (Occupational Safety and Health Council, http://www.oshc.org.hk/eng/safety/main.htm, dated 5 November 2001).

OSHC (2001b) *Occupational Safety Charter* (Occupational Safety and Health Council, http://www.oshc.org.hk/eng/safety/e_orgsc.htm, dated 5 November 2001).

TDC (2001) *Economic and Trade Information on Hong Kong* (Trade Development Council, http://www.tdctrade.com/main/200010s5.htm, dated 29 October 2001).

The Economist (2000) *Pocket World in Figures: 2000 Edition.*

UNCTAD (2001) *World Investment Report.*

Veeramani, D., Tserng, H.P. and Russell, J.S. (1998) Computer-integrated collaborative design and operation in the construction industry, *Automation in Construction*, 7, pp. 485–492.

WB (2001) *ISO 9000 Certification Bodies* (Works Bureau, http://www.wb.gov.hk/publications/iso.htm, dated 5 November 2001).

Chapter 5

India

Construction Industry Development Council, India and Lennon H.T. Choy

5.1 Introduction

5.1.1 Background

Indian civilization started in about 2500 BC after the inhabitants of the Indus River valley had developed a primitive trading system. In about 1500 BC, Aryan-speaking tribes migrated to the subcontinent from the northwest and created the now renowned classical Indian culture. In the tenth and eleventh centuries India was invaded by Turks and Afghans, and sultanates were established in Delhi. The incursions embarked the spread of Islam in the next 500 years' time. The British started trading activities in India in 1619 and expanded their influence gradually. By 1850s, the British controlled all Indian lands virtually and declared colonial governance. Through non-violent resistance to colonialism, India assumed independence in 1947. Pakistan and Bangladesh were separated from India following the independence and subsequent wars. Massive overpopulation, extensive poverty, ongoing dispute with Pakistan over Kashmir and ethnic strife are the major problems confronted by the Indian government (BSAA, 2002).

According to the 2001 Census, the total population of India is 1.027 billion with the proportion of males slightly higher than females (933 females per 1,000 males). Although India is ranked after China as the second most populous country in the world, its land area is only one-third of China. Hence, a higher population density, 324 persons per square kilometre, is resulted. In the past decades India's population was increasing at a significant rate (of over 2.5 per cent) but since 1995, the growth rate has been well below 2 per cent (1.77 per cent in 1995 and 1.59 per cent in 2001). India's economy encompasses traditional village farming, modern agriculture, handicrafts, a wide range of modern industries, and a multitude of support services.

Geography

India is located in southern Asia and borders the Arabian Sea, the Bay of Bengal, Bangladesh, Nepal, China and Pakistan. Total land boundaries and coastlines amount to 14,103 and 7,000 kilometres respectively. India has a total area of 3,287,590 square kilometres with over 90 per cent of land and less than 10 per cent of water.

India is one of the most diverse countries in the world in terms of climate, terrain, people and government. The climate varies from tropical monsoon in the south to the temperate in the north. The elevation extremes range from 0 metre (Indian Ocean) to 8,598 metres (Kanchenjunga). Topography changes from upland plain (Deccan Plateau) in the south, to rolling plain along the Ganges, to deserts in the west and the Himalayas in the north. India has the fourth-largest reserves of coal in the world. Other staple natural resources include iron ore, manganese, mica, bauxite, titanium ore, chromite, natural gas, diamonds, petroleum and limestone. Land uses include arable land (56 per cent of total lands), permanent crops (1 per cent), permanent pastures (4 per cent), forests and woodland (23 per cent) and others (16 per cent). A total of 535,100 square kilometres (1995/96 estimate) consists of irrigated lands.

5.1.2 Political structure

With one of the most complex bureaucratic and political structures in the world, India is a federal type republican government primarily. There are 36 administrative regions, including 28 States and 7 Union Territories. Each state is run by its own local government and regulations. On top of the local administrations, the central government oversees finance, resource allocations, defence and foreign affairs. The 7 Union Territories have direct rule from the central government.

5.1.3 Development process: through the years

With limited participation in the world market, India's development was based broadly on socialism in the past five decades. The state government was expected to take a leading role in the development process. Under international pressure, India has changed its development approach and followed the trend of liberalization and globalization since 1990s. As a result, development of Indian economy relied more on market mechanism and less on central directives. The state government has changed its role and shifted the concern towards making reforms for 'good governance', which demands responsive, effective, accountable and transparent state machinery.

India's five year Plans

Since independence, India has had a central planning system. The resources allocations have been based on the successive Five-Year Plans. The achievements of the Plans are summarized in Table 5.1. Figure 5.1 shows the GDP growth rates.

To comply with WTO and GATT, the Indian government has regarded decentralization and privatization as the key elements of the development process in the past decade. To facilitate decentralization and privatization, it requires an institutional mechanism that gains the support from the general public, in particular from the grassroots in the form of a wide network of 'civil society' organizations. However, support from the general public seems grossly lacking. Privatization of state-owned institutions has been running with difficulties due to excessive politicization of the issue and hindrance of industry unionism. Despite the fact that its market was yet

Table 5.1 Growth performance in the Five-Year Plans (per cent per annum)

	Target	*Actual*
1 First Plan (1951–1956)	2.1	3.61
2 Second Plan (1956–1961)	4.5	4.27
3 Third Plan (1961–1966)	5.6	2.84
4 Fourth Plan (1969–1974)	5.7	3.30
5 Fifth Plan (1974–1979)	4.4	4.80
6 Sixth Plan (1980–1985)	5.2	5.66
7 Seventh Plan (1985–1990)	5.0	6.01
8 Eighth Plan (1992–1997)	5.6	6.78
9 Ninth Plan (1997–2002)	6.5	6.2 (provisional)

Source: Ninth Plan Document, Government of India.

Note
The growth targets for the first three Plans were set with respect to national income. In the Fourth Plan it was net domestic product. In all the Plans thereafter, gross domestic product has been used.

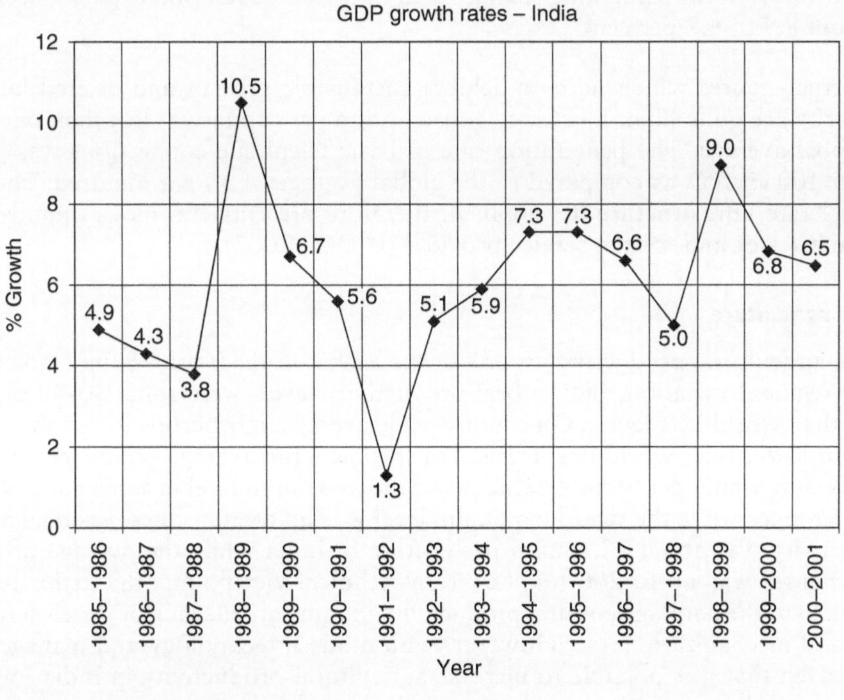

Figure 5.1 GDP growth rates – India.

to open to the world extensively, India was a founding member of GATT in 1947 and its successor organization WTO in 1995. The country is actively participating in her accession to Chinese and Russian markets and has signed bilateral MOUs with China in this context. The following sections discuss the two most significant growth engines in Indian market, namely infrastructure and agriculture.

Indian infrastructure

Approximately 45 per cent of the investment outlay in the country's Ninth Five-year Plan was expected to be in the physical infrastructure sector. Over the next decade an average estimated INR 1,800 billion (US$39 billion) investment per annum, is needed in the infrastructure sector to support the average annual GDP growth rate of 7.4 per cent over the next 15 years targeted by the Planning Commission.

The nature of the growth patterns in the various infrastructure sector components gives some interesting results.

- Whereas the cargo growth rate on the railways had remained stagnant around 4 per cent, the cargo growth rate in aviation had increased from 6 (1981–1991) to 15 per cent (1991–2001).
- In the same time period national highway cargo growth had increased from 0.5 to 0.6 per cent.
- The number of new telephone connections had accelerated from 19 per cent to 27 per cent.
- Port traffic increased by 6.1 and 7.1 per cent respectively.
- However, in the same time span growth in power generation capacity decreased from 8.8 to 6.7 per cent.

The actual figures which need to achieve sustainable growth and desired levels of prosperity are quite high. Electricity consumption per capita was less than one-tenth the global average. The penetration rate of basic telephone connections was low at 1.5 per 100 citizens as compared to the global average of 10 per hundred. The need to accelerate infrastructure development therefore presents enormous opportunities for the investor and for the service provider (CIDCI, 2002).

Indian agriculture

India's agricultural productivity is one of the lowest in the world. Subject to considerable regional variation, India's best productivity levels were some 30–40 per cent lower than world-best levels. On country-wide average, most crops were some 50–80 per cent lower than world-best levels. For instance, the average potato productivity was about 8 tonnes per hectare in India and the maximum level in some parts of India was 35 tonnes while the world maximum level was up to 50 tonnes. Average tomato productivity was about 12 tonnes per hectare in India while the average of Israeli green houses was up to 200 tonnes. From technical viewpoints, the major reasons attributed to the poor agricultural productivity in India are the lack of water, fertilizer, seeds and mechanization, etc. However, with modern technology and management, it is proven that it is possible to increase agricultural productivity in India for most known crops by significant margins.

5.2 The national economy

5.2.1 Macroeconomic review and outlook

With the government playing a leading role to plan the economy, India had chosen an inward looking economic development strategy since the independence in 1947.

The foundations of this strategy, though laid in the pre-independence era, had gained the support of most political parties, no matter it is left or right wing. State control was extensive covering foreign trade, domestic and foreign investments, essential commodities prices and internal trade. The government was also an active player in providing goods and services to the people. Like other similar planned economies, the economic performance of India until 1980 was poor, evidenced by the low per capita income growth rate at around 1.5 per cent. In 1980, the Indian government carried out a few initiatives, which improved the economic development of India significantly. These initiatives include (i) exercising macroeconomic prudence, (ii) abandonment of policy in favour of expansion by borrowing at home and abroad at increasing cost, and (iii) relaxation of state controls on many areas. These efforts were successful, evidenced by a much faster per capita growth rate at 3.5 per cent and a significant reduction of poverty in this year. However, these also rendered an accumulative debt in later years and led to fiscal imbalance in 1991. The government was hence forced to make structural reforms in the 1990s.

Despite the poor economic performance, India's democratic political system is regarded as a success since it is rarely found in the third world. It is conceived that the poor choice of economic policies has a political motivation behind. In Root's (1998) words, "getting elected has required targeting tangible spoils to an increasingly well-organized, but fractured, electorate. Political patronage was the stimulus for interventionist economic management, eventually producing massive fiscal deficits. When the danger of defaulting on foreign debt became a reality in 1991, the country's leadership began to re-evaluate the flawed economic policies without considering the flawed system of governance that accompanied and sustained the policy matrix. Patronage politics spawned corruption; money, muscle, or influence propelled public services and government, making the system of public administration as incompatible with liberalism as the system of economic regulation. Political and administrative imperatives impelled the country to economic policies that failed."

The results of the economic reforms embarked in 1991 can be evaluated by two distinctive periods. During 1991–1996, the reforms have unleashed progressive forces in the past decades. Domestic and foreign investment increased without altering the industrial structure that could damage the benefits of existing firms. During 1996–1999, the slowdown of the economy unveiled the internal weaknesses of Indian industry. It is believed that the liberalization process carried out in 1991–1999 has spurred fundamental changes in technology and innovation in particular at the micro-level (Forbes 2001).

The thrust to open up its economy has induced India to undertake a number of reforms. The successes of the reforms, however, have also exposed India to the crude world. India was influenced by the 1997 financial crisis, but the extent was much less than other counterparts in Asia. Indian economy had picked up quickly after the financial turmoil. In 2000, India's real GDP growth rate was over 6 per cent while all leading economies were still growing slowly. To remove the bottleneck effects in some sectors of the economy, the government decided to increase public investment significantly. For instance, there was a double-digit growth rate for physical infrastructure.

It is believed that the root cause of the economic crisis in the past decades is the fiscal imbalance caused by the over expenditures of the Indian government. The problem was further aggravated by the vast amount of external commercial borrowing. These fiscal deficits led to high levels of borrowing by the government from the Reserve Bank of India (RBI). The gross fiscal deficit surged from 9 per cent in 1980–1981 to 10.4 per cent in 1985–1988 and peaked at 12.7 per cent in 1990–1991. Major causes of the fiscal deficit include the global recession due to the oil crisis in the early 1980s and the high economic growth in the late 1980s. To remedy the situation, the government took a number of measures such as phasing out of monetization, tax reform, placing government securities to the market and devaluation of the currency in the 1990s. As a result, the fiscal deficit was brought down to as low as 5 per cent in 1999. However, one of the fundamental weaknesses that still remains unsolved with some of the state governments is the increases on non-developmental expenditure and interest payments as a proportion of revenue receipts. Table 5.2 shows the overview of national economy.

Economy in first half of 2000

The global slowdown in late 2000–early 2001 has affected the infrastructure sector also. Except for coal mining, petroleum and telecoms, all sub-sectors, in the first

Table 5.2 Overview of national economy

	1997	*1998*	*1999**	*2000**	*2001***
GDP and components (INR in trillions)					
GDP at real prices	9.99	10.5	11.43	12.21	13
GDP at current market prices	12.9	14.3	17.02	19.89	22.57
% GDP growth (real prices)	6.60	5.01	8.96	6.82	6.47
Primary sector (mining, agriculture, etc)	8.43	9.41	11.27	13.23	14.81
(%) Growth	3.95	11.6	19.77	17.39	11.94
Manufacturing sector	0.84	0.93	1.03	1.12	1.23
(%) Growth	11	8.5	10.75	8.74	9.82
Service sector	1.69	1.86	2.45	2.91	3.5
(%) Growth	13	10	31.72	18.78	20.27
Construction sector	1.9	2.1	2.27	2.63	3.00
(%) Growth	12	12	8.10	15.86	15.18
Demographic indicators					
Population (million)	958	975	991	1007	1023
Population growth rate (%)	1.79	1.77	1.64	1.61	1.59
Total labour force in Construction (million)	30	31	31	31.5	32
Construction labour force growth rate (%)	7	3	0	1.61	1.59
Financial indicators					
Short-term interest rate (%)	18–20	18–20	17–18.5	17–18.5	14–16
Long-term interest rate (%)	10–13	10–13	10–12.5	10–12.5	9–11.5
Changes in consumer price index (%)	7.2	6.1	4.2	5.8	7

Source: Central Statistical Organisation and others.

Note
Base year (1993–1994)
* Updated figures.
** Projected.

half of 2001, had shown a decline in the year-on-year growth rates. The projected 6.5 per cent GDP growth rate had come down to 5.4 per cent due to this. However, services-Sector growth had been largely sustained due to the high year-on-year growth in the IT sector, despite the global slowdown (CIDCI, 2002).

Mid-term appraisal of the Ninth Plan (1997–2002)

The growth rate of GDP during the first 3 years of the Plan was estimated to be 6.2 per cent per annum on average as against a target of 6.5 per cent. Significant shortages in growth performance had been recorded in agriculture, mining and quarrying and manufacturing, while construction, communications, public administration and community services had exceeded the targets. Both domestic savings and investment had fallen short of targets by over 5 per cent. The entire shortfalls were in the public sector, where public savings had recorded a shortfall of 70 per cent and public investment a shortfall of 23 per cent. Private investment had exceeded the target by over 3 per cent.

Investments in agriculture and allied services, mining and financial services had fallen short of the targets by over 20 per cent. Registered manufacturing, construction and other services (including software) had exceeded their investment targets. In the final 2 years of the Plan, both public investment and public savings were expected to recover strongly, but were unlikely to attain the targets set in the Plan. The investment targets in agriculture and allied activities, mining and quarrying and electricity were unlikely to be met in the last 2 years of the Plan. All other sectors were likely to meet the targets according to current trends. The balance of payments position was likely to remain comfortable in the last 2 years of the Plan despite the rise in the international price of petroleum. The current account deficit was not expected to exceed 1.4 per cent of GDP for this period. External financing was expected to remain comfortable, with a mild acceleration in foreign direct investment and significantly higher inflow of foreign portfolio investment. The deterioration of the fiscal position was primarily due to serious slippages in the tax revenues, particularly during 1999–2000. The tax/GDP ratio was expected to be 10.4 per cent of GDP in 1999–2000, but the realized ratio was only about 8.7 per cent. Despite an expected revival in the tax/GDP ratio, it was unlikely that the central budget support to the Plan can be maintained at the target level. It was estimated that only about 87 per cent of the Plan target may actually be attained by the end of the Plan period. This compares unfavourably with the 93 per cent realization during the Eighth Plan. Total public investment, as a result, would be about 81 per cent of Plan targets. During the Eighth Plan, the realization in public investment was 85.4 per cent of the target. Due to serious slippages in public investment in the physical and social infrastructure, the pipeline investment for the Tenth Plan was expected to be low. This was expected to weaken the possibility of significant acceleration in the growth rate during the Tenth Plan period (CIDCI and PCGI 2002).

Role of construction/property market in national economy

The construction sector in India has fully utilized the indigenous science and technology, the physical infrastructure as well as the manpower trained from Indian academic and technology-based institutions. Only falling behind the agriculture industry, Indian construction industry generates the second highest number of job opportunities among all sectors in the economy. For every 1 million rupees spent on construction activities, it was estimated that 3,000 man-days of skilled and semi-skilled labour and 1,300 man-days of managerial and technical employment would be generated. Overall, the industry provided employment to around 16 per cent of the working population in the country. For a country like India needing to generate employment opportunities on a massive scale for its large and expanding population, the contribution of the construction industry is indispensable. Furthermore, the construction industry also facilitates the growth of a number of other industries, like the production of construction materials such as cement, steel, glass, wood, paint, pipes and electrical goods, etc. It was estimated that these materials altogether constitute 60–70 per cent of the total cost of construction for a typical project. On the one hand the construction industry generates income to the country, on the other hand it boosts the overall economic growth. According to one estimate, construction adds 0.78 rupee to GDP for every rupee of investment while agriculture adds only 0.2 rupee per rupee invested. Construction output, in total, accounts for 5.7 per cent of the GDP in India.

As an instrument for economic transformation, the role of the construction industry has been well recognized. It constitutes an important component under the Five-Year Plans. The outlay of construction has been enhanced over successive years. The construction industry, with an accelerating tempo, facilitates the creation of infrastructural activities as well as housing, industrial, agricultural and social activities. The total budget for construction over the past five decades exceeds a trillion and a half rupees.

Construction investment (or market) volume

Over the past two decades, construction investment in India has increased tremendously. The value of new construction in India moved up from Rs. 1,365 million in 1980–1981 to Rs. 13,890 million in 1998–1999. The increase in construction value was significant till end of the 1990s. In 1998–1999, the increase in construction investment averaged 27 per cent per annum.

For the year 2001, the infrastructure sector was expected to assume the biggest share (over 50 per cent) of construction volume followed by the industrial sector (35 per cent). Residential sector only took up 5 per cent of the total volume. Table 5.3 shows the total construction investment and Table 5.4 shows the total construction volume in India. For the past years total construction investment has been increasing at a double-digit rate except for 1998 when it grew only by 5 per cent. In the private construction sector, more remarkable growth was observed. In 2000, private construction investment increased by 17.55 per cent compared to the year before. With the ongoing large-scale infrastructure projects, total construction investment by the public sector has also been rising steadily.

Civil engineering construction investment has dominated the construction market for the past years. Investment in this sector almost doubled in 1996–2001. Although the residential and non-residential sectors also experienced a comparable rate of increase in the same period, the increase in total amount could not be compared with

Table 5.3 Construction investment (in billion INR)

	1997		*1998*		*1999*		*2000*		*2001**		*2002**	
	Total	%	*Total*	%	*Total*	%	*Total*	%	*Total*	%**	*Total*	%
Residential construction (A)	100	–	90	(–)11.50	95	+3.38	150	+55.26	175	+14.95	180	+1.42
Public	48	–	43	(–)11.88	46	+4.73	65	+38.94	75	+13.73	77	+1.20
Private	52	–	47	(–)11.15	49	+2.14	85	+70.59	100	+15.90	103	+1.56
Non-residential construction (B)	150	–	160	(+)4.87	165	+1.01	150	–10.60	170	+11.31	185	+7.65
Public	53	–	56	(+)3.77	58	+1.55	55	–6.73	60	+7.47	65	+6.82
Private	97	–	104	(+)5.46	107	+0.73	95	–12.66	110	+14.03	120	+7.57
Civil engineering construction (C)	1,650	–	1,850	(+)10.22	2,010	+4.60	2,330	+13.99	2,655	+12.26	3,035	+12.72
Public	1,190	–	1,350	(+)11.51	1,480	+5.55	1,690	+12.29	1,900	+10.76	2,155	+11.84
Private	460	–	500	(+)6.87	530	+3.82	640	–72.27	755	+16.22	880	+14.93
Total construction investment (A + B + C)	1,900	–	2,100	(+)8.65	2,270	+4.10	2,630	+13.93	3,000	+12.38	3,400	+11.75
Total public	1,291	–	1,449	(+)10.33	1,584	+5.31	1,810	+12.32	2,035	+10.77	2,297	+11.30
Total private	609	–	651	(+)5.09	686	+3.22	820	+17.55	965	–15.94	1,103	+12.71

Notes

* 2001, 2002 are forecast constant prices at 1997 level. GDP deflator applied.

** Repeal of ULCRA, tax benefits and other reforms, initiation of highways and port projects.

Table 5.4 Construction volume (in billion INR)

Type of development	*1996*	*1997*	*1998*	*1999*	*2000*	*2001 (forecast)*
(Local currency at current price)						
Residential	90	100	90	95	150	160
Commercial	100	150	160	165	150	160
Industrial	540	650	750	770	900	1,000
Infrastructure	970	1,000	1,100	1,150	1,500	1,600
Others (included in above categories)						
Total	**1,700**	**1,900**	**2,100**	**2,180**	**2,700**	**2,920**
Gross built-up area (thousand square metres)						
Total	22,727	19,445	29,625	30,000	35,000	40,000
Residential	10,227	10,695	9,625	11,000	16,000	18,000
Commercial	12,500	18,750	20,000	19,000	19,000	22,000
Industrial						
Infrastructure	Not represented in square metres					
Others						

Source: CIDC survey.

the civil engineering sector. As the government was committed to provide homes to millions of its people, the residential sector will certainly increase its share in the construction market in the near future.

The Indian construction industry will play an important role in the Tenth Five-Year Plan by creating millions of jobs. With a capacity to handle diverse works, there are about 90,000 construction companies countrywide. In parallel, there are also a number of companies engaged in the productions of various types of machinery including heavy construction machinery, instruments, tools and appliances as well as materials for civil construction works for small- and-medium-sized contractors. About 83 firms are registered for the manufacturing of supporting and ancillary equipment for the construction sector. About 40 categories of products are produced by these firms, including pumps (4,000 models in 55 types), prime movers, IC engines, lifting equipment, tower-hoists, forks lifts, ropes, chains, pulleys, blocks, tackles, jacks, winches, conveyer belts, trolleys, concrete mixers, vibrators and allied equipment, steel scaffolding, plates, workshop and testing equipment. There are also firms focusing on the production of bigger size machines like engines, motors, cranes, rollers, compressors, lathes and testing machines.

Commercial prices showed a steady growth in the first half of the 1990s and a peak in 1997. Prices dropped as a consequence of the Asian financial crisis till the year 2000. After a year of rebound, prices fell again in 2001. In just a single year, residential land prices plunged about 30 per cent. Compared to the peak in 1997, office land prices decreased by an average of 30 per cent in the year 2000. With the Indian economy further opening up to the whole world, the commercial prices were likely to rise during the period of the Tenth Five-Year Plan. Table 5.5 shows the property prices and rents in New Delhi.

Table 5.5 Property prices and rents in New Delhi

	1997	*1998*	*1999*	*2000*	*2001*
Land prices					
Commercial land	245	235	225	230	210
Residential land	270	260	250	255	185
Industrial land	170	150	150	160	130
Average rents					
Commercial building	12.5	10.5	9	9.5	8.5
Residential building	5.4	3.5	3	3.5	2.5
Industrial building	10	9	9	9.5	7

Residential land prices rose steadily in the first half of 1990s. The prices reached a peak in 1997 and then dropped at about 5 per cent per annum afterwards. Residential prices plunged 30 per cent in early 2001. Likewise, industrial land prices plummeted around 15 per cent from the peak levels of 1997 in the year 2001.

Since the financial crisis of 1997, average rents have been dropping in India. Office rents fell by 30 per cent over the past 5 years while residential rents fell almost by half. Industrial rents also recorded a drop of 30 per cent during the same period of time.

The opening-up of the real estate market to large private sector realtors, the implementation of the Foreclosure Law, development of the Mortgage Act, dropping of the Urban Land Ceiling Act and various tax benefits offered by the State and Central Governments have led to a huge oversupply of opportunities and products in the real estate sector, thereby pushing the prices down. Further impetus has been given by the government-owned financial institutions like the Housing and Urban Development Corporation (HUDCO), with an expanded range of housing finance products. Norms for lending by banks and other financing institutions had been liberalized by the Central Bank. This in turn had led to reduced speculative activity in the parallel economy with the resultant sharp drop of overpriced real estate across the country. New real estate products had been assessed to be of better quality and of reasonable pricing (CIDCI 2002).

5.2.2 *Changes of economic policies due to the financial turmoil*

As part of the ongoing structural reforms to improve the economic performance of India, the government had been taking a number of measures at both macro- and micro-levels. For instance, the government has undertaken fiscal measures like reductions in reserve bank lending rates as well as statutory liquidity ratio of banks. Constructions of infrastructure projects like highways were being speeded up. All pending projects like airports development were being resumed and expedited. The role of infrastructure construction as a tool to revive the economy was clearly recognized. According to the pre-Budget Economic Survey 2001, the government had re-defined its role in the infrastructure sector – from being a "provider" in the past to being a "facilitator" now.

The Indian government had been undertaking structural reforms for more than a decade. The economy is still with a certain degree of protective shield although it has mingled with the world market for years. This has proved useful to protect the country from the adverse effects of the financial crisis. Among the reforms undertaken are a widening of the tax-base with regard to the services sector, recovery of arrears and enforcement of tax compliance, especially in case of direct taxes; strict control over wasteful expenditure through adoption of a zero-base budgeting system for all Plan and non-Plan schemes; optimization of budget provisions through re-prioritisation along with speeding-up of the process of dis-investment in non-strategic PSEs in order to mobilize more resources for Plan investment (PCGI, 2002).

It had however been imperative that a second generation of reforms will have to embrace all sectors of the economy and cover the central and state governments. There was a need to bring down the fiscal deficit and the rate of inflation which will help to reduce the real interest rates. Fiscal deficit should be brought down through resource mobilisation efforts and austerity. Private initiatives in infrastructure were not expected to be forthcoming until the services are appropriately priced to cover the full cost of operations. Regulatory mechanisms need to be introduced which will oversee pricing of services and other issues. The industrial sector will have to adjust and respond to the challenges and opportunities posed by the WTO regime. Competition policy and reforms in labour policy will have to be introduced. Administrative and legal reforms should be an integral part of the reform agenda (PCGI, 2002).

India had it own financial crisis embarked in 1991. There was a setback in the fiscal system due to the unresolved problems accumulated over time. Heavy borrowing from the international market was called upon to avert the crisis as a consequence. The crisis, however, turned out to pave the way for the subsequent economic growth as the government had committed to reform the economic, financial and business system drastically. Since then, there has been a virtual consensus among the political and bureaucratic lines on the directions of reforms.

Compared with other Asian economies, India had suffered less by the impacts of the financial turmoil in 1997. By the time that the Asian financial turmoil hit, India had already been recovered from the domestic crisis that happened in the early 1990s. Because of the robust domestic market, India did not depend too excessively on exports. As a result, the impact of the Asian financial turmoil to India had been kept to the minimal. Furthermore, the currency control also kept the country out of the hands of speculators who played a substantial role in deepening the adverse influence of the turmoil.

Impact of financial turmoil on the construction sector

Growth in construction during the crisis period was fairly steady and in line with the national economic growth. Underlying weaknesses were identified and addressed to streamline the working to meet global work ethics, quality of products and efficiency.

Several institutional frameworks and linkages were absent or grossly underdeveloped. Necessary steps have been taken to build them up to provide comprehensive dynamism to the growth of the construction industry. Specifically, measures taken include:

a declaration of the construction sector as an industrial concern, to enable smooth flow of institutional financing to the industry;
b modifying the standard contract document to avoid cost and time over-runs in construction contracts. This basic document is used in varying modified forms by various government ministries/government bodies/government enterprises/public sector undertakings and other organisations connected with the central and state governments;
c development and getting acceptability of bankers and financial institutions to adopt standardised financing norms for construction entities;
d grading of construction agencies;
e development of a large number of insurance products to mitigate construction risks. Under development are various bonding systems for reduction of financing costs of projects; and
f other measures include establishing construction equipment bank(s), evolution of construction cost indices, human resource development including a nationwide programme for training of construction workers under a harmonized system.

Others measures such as development of a Unified Construction Business Act were in progress.

Policies and strategies adopted by public and private sectors to overcome the economic slowdown in the construction sector

a Restructuring the corporate governance structure.
b Infusion of better and cost-effective technologies and work practices.
c Resorting to technical audits and third-party surveillance systems.
d Hiring professionally qualified managerial and technical manpower.
e Training and skill upgradation of the workmen engaged in the works.
f Adhering to better and more effective dispute resolution mechanisms.
g Better Industry – Academia interaction and closer participation.

5.2.3 *Economic outlook in the next decade*

The Planning Commission had set a target of average 7.4 per cent year-on-year growth target for 2001–2015. Targets for the plans are:

- Ninth Plan (1997–2001) – 6.5 per cent (revised), achieved 6.2 per cent.
- Tenth Plan (2001–2006) – 7.7 per cent.
- Eleventh Plan (2006–2011) – 8 per cent.

To achieve these targets, the demands for infrastructure were expected to be huge. The demands projected for the just concluded Ninth Plan are shown in Tables 5.6 and 5.7.

Airports

The Airports Authority of India (AAI) manages five international airports, 87 domestic airports and 28 civil enclaves. The annual number of passengers handled

Table 5.6 Projected demand for electrical energy

	Energy requirement (MkWh)			*Peak load (MW)*		
	2001–2002 End of 9th Plan	*2006–2007 End of 10th Plan*	*2011–2012 End of 11th Plan*	*2001–2002 End of 9th Plan*	*2006–2007 End of 10th Plan*	*2011–2012 End of 11th Plan*
Northern region	181,649	254,161	350,185	31,735	44,009	60,007
Western region	176,732	239,731	320,956	28,430	38,538	51,562
Southern region	134,671	178,690	234,164	21,975	29,070	37,996
Eastern region	68,243	96,884	135,049	11,846	16,722	23,228
N-E region	8,148	12,062	17,553	1,722	2,527	3,661
A N Island	180	295	475	41	67	108
Lakshadweep	27	40	58	8	11	15
All India	569,650	781,863	1,058,440	95,757	130,944	176,647

Source: Central Electricity Authority, Government of India.

Table 5.7 Projected demand for other forms of energy

	1994–1995	*2001–2002*	*2006–2007*	*2011–2012*
Petroleum products (MMT)*	63.8 (7.3)	104.9 (7.9)	153.0 (8.1)	226.3
Coal (MMT)	79.6 (5.1)	114.0 (4.4)	140.0 (5.1)	179.5
Natural gas (MCM)	12,110 (3.8)	15,730 (3.1)	18,291 (2.7)	20,853

Source: Ninth Plan Report, Government of India.

Notes

* Excluding demand for power generation. Figures in bracket are the CAGR over the period.

is 12.91 million international, and 24.02 million domestic. Cargo handled annually is 474,000 tonnes international plus 224,000 tonnes domestic.

Private sector participation

- The Government of India keeping in line with its policy of liberalisation decided to corporatize Delhi, Mumbai, Chennai and Calcutta airports in order to induct the much needed capital for expansion and modernization of these airports to world-class standards. Accordingly, AAI sought consultants, both financial and legal to advise AAI on the entire process. AAI issued advertisements seeking Expression of Interests (EOI) globally from prospective financial and legal consultants. Action for appointment of financial consultant was on hand.
- Government had decided to set up a parallel cargo terminal at IGI Airport by way of a joint venture between AAI and the private sector, in order to offer an option to users so as to achieve better services and increased efficiency through healthy competition. AAI had invited EOI for appointment of a consultant to advise on the modalities of forming a joint venture.
- AAI signed an MOU in May 1999 with Karnataka State Industrial Investment & Development Corporation Limited (KSIIDC) to establish a new airport of international standards at a site near Devanhalli, Bangalore by bringing in funds from

the private sector to handle both passengers and cargo, domestic and international traffic to meet the growing demand of Bangalore city.

- AAI signed an MOU in May 1999 with Cochin International Airport Limited – the airport constructed at Cochin (in Kerala State) jointly by State Government and the private sector. The MOU provides for limited equity participation of AAI towards the cost of CNS/ATM facilities and also for rendering the services for operation and maintenance of CNS/ATM facilities on chargeable basis.

The AAI has drawn ambitious long-term plans to meet challenges posed by ever-increasing air traffic and advancement in aircraft technology. Some of the major plans for implementation of ICAO CNS/ATM programme are:

- Replacement of ground-based Communication, Navigation and Surveillance (CNS) with satellite-based CNS system.
- Establishment of Differential Global Positioning System (DGPS).
- Automation in Air Traffic Control Services (ATCS).
- Establishment of Automatic Dependent Surveillance (ADS).
- Coverage of the Indian land mass through satellite communication, VHF data links and monopulse secondary surveillance radar with mode "S" capability.

Ports

In its 6,000 kilometres of coastline, India has 11 major ports and 179 minor ones with a total capacity of 215 million tonnes. The available capacity was grossly inadequate and it was planned to expand existing facilities, while building up new ones (Table 5.8).

Urban infrastructure

Urban infrastructure projects were sought to be developed on Build-Own-Operate (BOO), Build-Own-Operate-Transfer (BOOT) and Build-Own-Lease-Transfer (BOLT) systems. The central and state governments welcome private initiatives and public–private participation in sectors like water supply, sanitation, public transport, and township and land development. The Indian government is committed to provide support in the form of equity contribution, a package of concessions, dedicated levies to repay loans and a transparent regulatory framework. Private investors are encouraged to negotiate the concessions required to make their investments safe and paying.

One significant fact that investors should consider is that local agencies in India had shown phenomenal progress in the recovery of costs of services and some have achieved full cost recovery. The city of Visakhapatnam in Andhra Pradesh offers a good example of how cross-subsidisation between consumer groups can make water-supply systems run on commercial principles. The municipal corporation of Ahmedabad in Gujarat has performed a remarkable turnaround from a perennially loss-making body to a highly profitable organisation and has already launched the country's first Municipal Bond.

Table 5.8 Project opportunities in ports

Project particulars	*Estimated costs* (US$ million)
Jawaharlal Nehru Port, Liquid cargo berth	35
Mumbai (JNPT) marine chemical terminal	535
Six-berth terminal at Nhava Creek	335
Kandla container freight station	5
New Mangalore bulk handling terminal	100
Port facilities for expansion of Mangalore Refinery	35
Tuticorin construction of new outer harbour (including modern container terminal)	1,200
Container handling facilities at Berth No. 7	80
Chennai construction of new outer harbour	700
Visakhapatnam construction of outer to outer harbour (Part construction through private sector)	800
Construction of captive and multipurpose berths	140
Kochi construction of container terminal	800
Construction of LPG & LNG terminal at Puthuvypeen	100
Calcutta cargo handling facilities at Budge container terminal	40
Second dock arm at Haldia	250
Mormugao construction of outer harbour	700
Construction of FRH master plan berths	75
Construction of berths west of breakwater	200
Paradip dry dock ship repair facilities at Paradip	470
Upgradation of facilities for container berth	35
Captive fertiliser handling system	35
Estimated total investment	6,670

Source: National Informatic Centre, Government of India.

Water supply

Privatization could be introduced in case of new townships and projects for planning, designing, source development, execution of works, operation and maintenance including billing and collection. In case of metros and mega cities, water supply augmentation schemes for source development, conveyance of raw water, its treatment and bulk supply to the city water supply authorities, maintenance of pumping stations, water treatment plants and city distribution systems can be undertaken by private agencies.

Sewerage

Similarly, in case of sewerage and sewage treatment, works such as maintenance of pumping stations, sewage treatment plants and city sewerage systems could also be taken up. Keeping in view the ever increasing demands for fresh water, the private agencies may also install tertiary treatment plants for reuse and recycling of sewage and industrial effluents for various non-domestic uses.

Solid waste management

Solid waste management was another activity which could be taken over profitably by the private sector provided resources recovery is contemplated to make the system self-sufficient and financially viable. In addition, efforts should be made to manufacture various items of equipment and machinery such as pipes, pumps, quality control equipment and machinery required in the water and wastewater treatment plants, etc. within the country by various foreign manufacturing concerns in collaboration with Indian companies as joint ventures for the Indian market.

Urban public transport

India has 23 metropolitan cities. The number was expected to increase to 40 by the year 2001. All offered attractive investment opportunities in public transport. Citywise studies have been carried out for Delhi, Bangalore, Calcutta, Chennai, Hyderabad, Mumbai, Ahmedabad, Jaipur, Surat, Jammu, Nagpur, Vijayawada, Lucknow, Cuttack and Bhubaneshwar.

India welcomes private investment in Mass Rapid Transit Systems (MRTS) and Light Rail Transit Systems (LRTS). Governmental support for such projects may include rights for development of property, foregoing returns/dividends on any investments made by the government, the availability of budgetary sources for part-repayment of loans and tariff agreements.

The proposed MRTS for Delhi offers good potential for public–private partnerships and the project was already in an advanced stage of planning. Bangalore and Hyderabad were also planning rail-based public transit systems.

Roads, bridges, flyovers

Bypasses to large cities and bridges are investment opportunities. There exists tremendous potential for private investment in construction and maintenance of ring roads, arterial and sub-arterial roads, bridges, flyovers and other facilities in cities.

Housing

The National Housing Policy 1998 was formulated to address the issue of sustainable development of infrastructure and for strong public–private partnership for shelter delivery. Private investment in the sector is brisk and the opportunities are unlimited. The government would provide fiscal concession to carry out legal and regulatory reforms and create an enabling environment.

As for the action plan under the 2 million Housing Programme, the Ministry of Urban Affairs and Employment has embarked upon facilitating construction of 7 lakh additional housing units in urban areas every year. HUDCO was entrusted with financing 4 lakh units and the balance of 3 lakh units per year was to be met by other HFIs recognized by the National Housing Bank, cooperative sector and corporate sector. As of 30 March 1999, HUDCO had sanctioned schemes for the construction of over 4 lakh houses under the 2 million Housing Programme.

The Urban Land (Ceiling and Regulation) Act 1976 was repealed through an ordinance notified on 11 January 1999. This has since been approved by parliament and the Repeal Act notified on 22 March 1999. Government had issued detailed guidelines to all state governments and union territories to protect the interests of people belonging to economically weaker sections and lower income group.

The Ministry of Urban Affairs and Employment offered incentives to non-resident Indians and foreigners of Indian origin as well as overseas corporate bodies that are predominantly owned by them, for investment in the housing and real estate sector.

Land and township infrastructure development

Returns on projects for development of land in extended areas of large cities and new townships can be well above 20 per cent. A package of concessions was being worked out.

A similar surge was expected in all the other sectors too. There was a slippage in the achievements in the Ninth Plan (6.2 per cent vs targeted 6.5 per cent) which were expected to be compensated in the Tenth Plan, thus increasing the pressure on future growth targets.

5.3 The construction sector

5.3.1 Value-added per employee

The value-added per employee for the construction sector had been taking two steps forward and one backwards. The fastest value-added growth rate recorded with regard to construction was in 1996, while it shrank to negative territory the following year. The service sector has recorded only a modest growth in value-added per employee (Table 5.9). Construction sector had been relatively efficient with this regard. Table 5.10 shows the physical measurement of construction productivity.

Table 5.9 Value-added per employee (annual)

	1996	*1997*	*1998*	*1999*	*2000*
Construction	71,429	67,242	67,741	73,225	75.355
% change	+13.45	−5.5	+0.7	+8.1	+2.8
Manufacturing	2,11,538	2,13,418	2,15,000	2,17,150	2,19,015
% change	−2.00	+0.8	+0.75	+1.00	+0.8
Service	2,10,512	2,15,305	2,18,520	2,23,436	2,28,811
% change	+1.50	+2.20	+1.50	+2.25	+2.4
Primary sector*	64,583	64,666	62,500	65,300	64,850
% change	+0.46	+0.59	−2.78	+4.5	−0.8

Source: CIDC database.

Note
* Primary sector includes agriculture, mining, quarrying, etc. (Factored Costs).

Table 5.10 Physical measurement of construction productivity (production)

Square metre per man-day	*1996*	*1997*	*1998*	*1999*	*2000*
Public residential	0.50	0.50	0.50	0.50	0.50
Private residential	0.50	0.50	0.50	0.50	0.50
Commercial	0.60	0.60	0.60	0.60	0.60
Industrial	0.70	0.70	0.70	0.70	0.70

Note
Area (sq. m) refers to total built-up area. (The measurement may not reflect the overall trend since the housing sector constitutes 8–9 per cent of total construction activities.)

5.3.2 *Construction cost*

Unit construction cost

The trend of construction cost in India between the period of 1996–2000 was mixed (Table 5.11). While unit construction cost in the residential sector increased by over 30 per cent during the period, the unit construction cost of higher-end residential, commercial offices and hotels showed a relatively modest increase.

Average construction material price

Table 5.12 shows the average construction material prices.

Construction industry salaries and wages

Table 5.13 shows the construction industry salaries and wages.

Average sectoral wages

Table 5.14 shows the average sectoral wages.

Table 5.11 Unit construction cost (INR/Sft)

	*Residential (terrace)**		*Residential (high rise)*		*Commercial office*		*Industrial*		*Hotel*
	Standard	*Luxurious*	*Standard*	*Luxurious*	*Standard*	*Prestige*	*Light*	*Heavy*	*5star*
1996	750	1,000	625	1,000	600	900	525	700	3,500
1997	825	1,100	650	1,100	625	950	600	800	3,500
1998	870	1,200	650	1,100	625	950	600	850	3,500
1999	890	1,300	650	1,100	625	950	600	850	3,800
2000	865	1,350	640	1,150	610	1,050	600	870	3,950

Notes
Unit price excludes sub-structural works.
* Including boundry walls, gates and other utilities.

Table 5.12 Average construction material prices

	Cement in bulk (INR per tonne)	*Steel bars* (INR per tonne)	*20 mm Aggregates* (INR per M3)	*Concreting sand* (INR per M3)	*Common bricks* (INR per thousand pieces)	*RMC Grade 30* (INR per M3)
1997	2,750	15,000	550	350	1,700	2,700
1998	2,800	15,000	550	370	1,700	2,800
1999	2,800	15,000	575	370	1,800	2,800
2000	2,850	15,500	600	440	1,970	2,850
2001	3,050	15,900	670	460	2,100	3,050

RMC: Ready-mix concrete (very limited availability).

Table 5.13 Construction industry salaries and wages (INR)

	1997	*1998*	*1999*	*2000*	*2001*
Professional (Per month) mid-management Level	13,000	15,000	15,000	15,000	15,000
Technicians (Per month)	5,000	5,000	5,500	6,000	6,000
Skilled worker (Daily wage)	140	140	150	160	160
Unskilled (Daily)	72	72	72	80	84

Table 5.14 Average sectoral wages (INR per month)

	1997	*1998*	*1999*	*2000*	*2001*
Construction	6,500	6,700	6,700	6,800	6,800
Manufacturing	14,000	14,000	14,000	14,000	14,000
Services	12,000	12,000	12,000	12,000	12,000
Primary sector	4,900	4,900	4,900	4,900	4,900

5.3.3 *Construction quality*

Construction quality/quality assurance

The civil construction industry with all its uncertainties and physical challenges to be overcome especially in the infra-structure segment was not considered amenable to the rigours of a systematic quality management approach. Identification and action on quality problems is a seemingly impossible task in circumstances where most such problems are "brushed under the carpet". With the exception of a few elite civil engineering companies in India, quality management has not been practised in India by the broad spectrum of the operators in the construction arena ranging from the small time culvert contractors to the bigger companies.

It is observed that to establish a quality philosophy and a system such as ISO 9001 is an evolutionary process that will take considerable time. Past track records and experience of similar organizations indicated that a period upwards of 2 years is needed to achieve a total change of heart in an organization.

The specifications, norms and the quality of the product is inspected on the basis of codes specified by Bureau of Indian Standards. The Indian standards are fairly elaborate and are gradually being modified to conform to ISO standards. Apart from this, several scientific and research and development institutions, CSIR, CBIR, IITs, material testing labs also provide necessary input support.

Construction Industry Development Council (CIDC) together with the Building Materials Technology Promotion Council, Bureau of Indian Standards and several other institutions had initiated a Performance Appraisal Certification Scheme (PACS) which has authorized test labs as well as R&D institutions in the country to authenticate the performance-related claims, made by several manufacturing organisations.

The number of organizations which have ISO certification was very large. However, due to the relatively small turnover of each, the volume of production represented by these entities was very small. The process of acquisitions and mergers had started in India, which was liable to lead to larger corporate entities with higher levels of standardization.

Training to workers and supervisors

The CIDC had started a nationwide initiative, in collaboration with three leading universities – Indira Gandhi National Open University (IGNOU) New Delhi, Dr Bhimrao Ambedkar University, Hyderabad and Thapar Institute of Engineering & Technology, Patiala, each of which is located in a different part of India.

The initiative was part of the CIDC campaign to develop a harmonised system of training of construction workmen across the nation. There are over 32 million construction workers in India out of which barely 25 per cent have some type of rudimentary training received from their guilds or their parents. Even these workers need some retraining to have a well-rounded knowledge of the construction trade that they follow. Approximately 22 million workers have received no training at all except that which has been learned at the sites. Moreover, the level of literacy is very low. Further, with over 18 languages and 400 plus dialects, there is a great deal of opportunity to address this need for training.

In collaboration with CIDC, the universities develop the course contents and the testing and certification procedures. For the training modules, 38 construction trades had been identified. Dozens (soon to be hundreds) of centres established in various institutions/companies/workshops/guilds across the country have been authorized to impart the training. At the end of each course, the CIDC–University team conducts tests and certifies eligible candidates. Emphasis of the course is on theoretical and practical aspects, and weighting is also given to previous experience. The course contents/tests have been translated into seven languages so far and work is proceeding on other languages.

The main features of this training are:

- National-level standardization of curricula and competency standards, testing and recognised university certification;
- On-the-job learning;

- Distance learning methods with self-instructional materials in vernacular languages;
- Maximum use of audio and visuals;
- Minimum loss of wage earning time;
- Packaged programmes to be taken up at a pace convenient to learner;
- For candidates who may possess or acquire the minimum prescribed educational level, access to gradual upgrade of education to diploma/degree level;
- Successful candidates receive formal Certificates of Competence from the universities.

The result expected from this exercise is a pool of trained workers to be available to the industry, and also to help mobility and raise the living standards of a large neglected pool of productive potential consumers.

5.3.4 Construction safety record

India's construction safety issue is complex. Until a few years ago, the issue of safety in the construction industry was far too unimportant, but concern began due to multinational companies entering the Indian market and their uncompromising adherence to safety norms. Adding to this problem is that, of roughly 21 million construction workers (or about 18 per cent of the working population of India Construction Industry Development Council's estimate), 85 per cent of this workforce was drawn from the rural background. They lack technical perspective and industrially relevant common sense. Even the other 15 per cent of semiskilled and skilled workers have had their training, through the trial-and-error route, making their acquired skills short of the desirable. In addition, the visible lack of a safety rationale in India's social systems and cultural beliefs further makes safety management a difficult proposition.

Unfortunately, more exact statistics on construction accidents are not available for the Indian construction industry because of the absence of a mandatory monitoring agency and statutory reporting system (Table 5.15). Construction being the second largest economic activity in India, next only to agriculture, with the bulk of the

Table 5.15 Safety record

	Accident frequency rate (accidents/million man-hours worked)
1995	0.06
1996	0.11
1997	0.08
1998	0.10
1999	0.10
2000	0.10
Target level	Near zero

Table 5.16 Skill level of workforce and supervision level

	Total construction workforce (Nos)	*No of skilled workers*	*Skill level (%) (Skilled worker)*	*Ratio of supervisor to workers (1:XX)*
1996	28,000,000	8,000,000	60	1:25
1997	30,000,000	9,000,000	65	1:24
1998	31,000,000	9,000,000	65	1:22
1999	31,000,000	9,000,000	68	1:22
2000	31,500,000	9,050,000	68	1:22

Benchmark: CIDC/University parameters; percentages determined on the basis of intake survey statistics by CIDC's partner institutions.

workforce being engaged in this unorganised sector, one does not really exactly know how bad the picture is. Today the workplace is faced with rapid changes in technology.

In order to build and maintain effective accident-prevention programmes various techniques like job safety analysis, site safety audit, site safety plan, safety inspection, safety sampling, hazard and operability studies, work permit system, method statement, etc. were being adopted with mixed results. Some of the techniques like job safety audit and safety plan, besides safety inspection, have been successfully tried and have proved effective. Raising general awareness of the construction workers will remain a priority task for all concerned within the industry.

5.3.5 *Skill level of workforce and supervision level*

The HRD programme also extends to the supervisory level and the training, testing and certification for supervisory level has been commenced simultaneously under the same HRD scheme (Table 5.16).

5.3.6 *Research and Development*

In previous years, information on the R&D initiatives taken by the Indian Industry had been amplified, some are recapitulated below:

a Centres for development – India possesses a large pool of engineering and technological institutions and research laboratories. With over 416 engineering institutions, and industry-supported research institutions, work on the areas of environmental protection, re-engineering of work techniques and development of new technologies were being done on a continuous basis. However, comparing the outputs with peer groups, specially those from the European, American and advanced Asian countries, substantial ground needs to be covered.

b New materials and practices – As an outcome of globalization and relaxed entry norms for businesses from the world over, the Indian construction industry too is adopting different and better work techniques and practices.

Manufacturing industries were continually upgrading the quality and performance parameters of their products, to compete with the products of foreign origin or being manufactured in India, based on foreign technology/processes.

c Thrust areas – Specific to the housing sector, the major area of thrust and concern is to make the structures withstand seismic loading. The work in this area includes:

   i monitoring and forecasting of the events including the development and design of structures and also that of the work practices;
   ii development of a regulatory framework to ensure implementation of the practices thus developed; and
   iii to work in the field of disaster management. It is noteworthy to mention that other natural calamities, such as cyclones, typhoons, fire hazards, etc. are also being brought under this ambit.

Main category of research projects

Tables 5.17 and 5.18 show new technology in automation, mechanization and prefabrication, and category of new technology in the underground structure.

Policy and initiative on the use of IT in construction

Project design, development, implementation and monitoring are gradually getting transferred to the computers by consultants, project owners and contractors. IIIT, a leading software development centre established by the Government of Andhra Pradesh had set up an educational institute at Hyderabad, which is dedicated to the comprehensive incorporation of IT into every aspect of the construction industry. Similar initiatives were being planned by some leading corporate houses like Unitech Ltd for the Web-enabled design, control and monitoring of construction projects.

The CIDC has also developed the concept of a Wide Area Network (WAN) based system, for the monitoring and control of projects across dispersed centres using different IT platforms.

Table 5.17 New technology in automation, mechanization and prefabrication

Technology and brief description	*Developer/Facilitator*
1 Trenchless pipe-laying technology: INDSTT is working for developing, advancing and promoting the technology in India	INDSTT (Indian Society for Trenchless Technology c/o NBCC Ltd, NBCC Bhavan, Lodi Road, New Delhi – 110 003)
2 Directional Drilling	INDSTT (Indian Society for Trenchless Technology c/o NBCC Ltd, NBCC Bhavan, Lodi Road, New Delhi – 110 003)

Table 5.18 Category of new technology in the underground structure

Category A: emerging technologies	These are the technologies which are still in the experimental stage. The experimentation work for the development of technology is carried out in two stages and at times in two different types of institutions. The first stage is fundamental or theoretical in nature and is generally carried out in the universities. The second stage is practice-oriented or field-based and is carried out in such specialised institutes which can provide necessary infra-structure and expertise to carry out experiments on a large scale. Below are details about two such international institutes. Unfortunately in India, importance is not given to the construction industry and as a result institutions exclusively devoted to construction technology and funded by contractors cannot be found in the country.
ISMES	This institute is located in Bergamo, Italy and is funded primarily by the National Electricity Board. The approximate percentage distribution amongst shareholders is as follows: i National Electricity Board 69% ii Major Italian Contractors 20% iii Italcementi SPA 6.5% iv Design and Consulting Companies 3.25% v Independent Power utilities 1% vi Other Industrial Companies 0.25% The nucleus of ISMES was established in the year 1947 as an "Experimental Institute" by the Societa Adriatica di Elettricita, a major power generation and supply company, Italcementi, Italy's major cement producer and Torno, a large civil engineering contractor. These three companies were involved in the construction of the Piave arch-gravity dam which was characterised by exceptionally low thickness to height ratio. Since the mathematical calculation methods available at that time were considered not so dependable for a thorough understanding of the structural performance, it was suggested that the dam could be evaluated by means of a large-scale model tested to failure. The scope of activities at ISMES have been increasing ever since and today a significant percentage of ISMES' projects consist of providing integrated services in the fields of civil, off shore and environmental engineering. Many problems referred to them are from the construction industry. The fact is that, today, we cannot imagine a research institute of this nature in India which is funded by private contractors and carries out research and developmental work for the Industry.
NGO	The other institute worth mentioning is the Norwegian Geotechnical Institute in Oslo. The institute is a private research and development foundation, where 90% of the revenue comes from contracts within Industry. NGI carries out useful research leading to cost-effective, safe and economically responsible solutions for various problems in geo-engineering.
Category B: proven technology	In this category are the technologies which have a proven track record abroad, but for some reasons they have not been used much in India. Some of the technologies and the possible reasons for their lack of popularity in our country are discussed below.
Rock bolting and shotcreting	Many tunnels are constructed in India. These are for water resources development projects e.g. for diversion of water, head race tunnels, tail race tunnels, access tunnels, etc. and transport tunnels for roads and railways.

	Most of these tunnels are supported by heavy steel sections connected with laggings and backfilled with concrete. Supporting by steel girders is the most backward system of providing supports. The steel sets provide a passive support which tries to hold the movement only after it has reached a certain stage. On the other hand, supporting systems such as rock bolts and shotcrete provide an instantaneous strengthening of the surrounding strata. The system is flexible so that its stiffness can be increased as and when required. The system is very efficient and hence very economical. This is particularly useful in squeezing ground conditions, which are very common in the lower Himalayan region, because of poor rocks and high stresses. When combined with adequate instrumentation, these supporting systems can be used to great advantage. However, one of the pre-requisites of successful rock bolting and shotcreting is the use of proper equipment and the installation of the support system on time. The experience in India has been that some of the contractors who have tried it have come to the conclusion that it does not work in India. The reason put forth is that Indian rocks are different. The secret of successful rock bolting lies in drilling immediately after blasting, i.e. before the rock movement starts. Rock bolts are of two types – active and passive. The active rock bolts are anchored at the end and a specified amount of tension is applied by pulling the other end before the plate and nut are fixed. The remaining portion of the bolt is also, sometimes, grouted. As soon as the bolt is tensioned, a compressive stress is generated in the rock mass. Systematic bolting provides a ring of rock under compression which supports the tunnel. On the other hand, the passive system of rock bolting or soil nailing is simply grouted into the soil or rock and therefore depends upon the ground movement to create compression in the surrounding rock. It is therefore essential that installation of passive rock bolts be done before the rock moves. Besides the lack of understanding and improper equipment for drilling, there is one more factor which plays an important part. This factor is the contract document. If the contract document has been framed in such a way that it helps the contractor if he/she indulges in overbreaking and installing steel supports, he/she will do only this. No matter how much we persuade him/her, he/she will never do it unless it is beneficial to him in financial terms.
Fabric reinforced shotcrete	Fibre reinforced shotcrete is much more effective than plain shotcrete due to the random distribution of fibres. The success of shotcreting – whether plain or fibre reinforced – depends partly on the proper equipment and the training of the nozzleman. A properly trained nozzleman will see to it that the shotcreting is done uniformly and that the rebound is a minimum. Again, the experience in India has been that an average contractor does not want to spend on the purchase of proper equipment and the training of the nozzleman. Improperly done work can lead to poor results and bring a bad name to the technology itself. The steel fibre reinforced shotcreting of Nathpa Jhakri Desilting Chamber has not been considered very successful. The reasons are being investigated. Steel fibre reinforced shotcrete has several other applications besides supporting the underground openings, such as for slope stabilization

	providing surfaces of air fields and strengthening repair work, etc. These applications have also not been very popular for the same reasons.
Roller compacted concrete	Roller compacted concrete permits use of a higher percentage of flyash in the concrete mix, than has been the conventional practice. The technology has a tremendous potential and utility for large gravity structures such as dams. It is now being all over the world for the construction of gravity dam except in India. In India, although many researches, Ph.D studies and papers in national and international fora have been conducted, there are no standards established so far. In contrast, other countries such as China have passed ordinances making it obligatory only to construct gravity dams by roller compacted concrete. Several structures constructed with this technique exist.
Trenchless technology	Trenchless technology is used all over the world for installation and repair of utilities below the ground surface. Even though there exists an Indian Society for Trenchless Technology, it has a limited membership and the technology has not become very popular. One of the primary reasons for this is that while working out different alternatives we do not consider the environmental cost or the social cost. Thus it does not cost anything if the road is dug open and left in the same condition for months. The only popular activity in this area appears to be the laying of telephone cables. One notable project involving pipe jacking, however, is under execution in Mumbai. The other reason why the technology has not picked up in India is that there is no expertise available among consultants or contractors who have any worthwhile experience or expertise. The system does not permit consultants/contractors from outside to come and operate in India. Unless these activities are permitted in a controlled manner, it will not be able to develop any expertise indigenously.
Reinforced earth technology	Reinforced earth technology is one of the most exciting inventions used in India. Invented in the early 1960s in France, the technology became very popular in Europe and America in a very short time. It is now being used all over the world. However, its application in India started only a few years ago. Now that use of the technology has picked up in India, one can look back and try to find out the prerequisites before any new technology can be introduced or popularised.

Source: Mr Sharma, V.M.

5.4 Construction outlook 2002–2010

5.4.1 *Business opportunities and major projects in the pipeline*

Government policy on liberalization

There has been a steady progress on the path towards liberalization by the government, despite the compulsions of a mammoth democratic polity. There was an across the board consensus on the need to take the liberalization process towards its next phases. The issues of debate have shifted towards the methods rather than the objectives of liberalization (refer Tables 5.19–5.21).

Table 5.19 Major initiatives taken during the year 2000–2001

Industry

1 Strong thrust to knowledge-based industry by reducing customs duty on several items of IT, telecom and knowledge-based industries.
2 Foreign direct investment permitted through automatic route in all industries except for a small negative list.
3 Non-banking financial companies allowed to hold foreign equity up to 100% if they are the holding companies. Their subsidiaries, which are the operating companies, also allowed to hold foreign equity up to 75%.
4 Dereservation of the garment sector from the purview of SSI reservation.

Infrastructure

5 Securitisation of dues of central sector power and coal utilities for assisting the SEBs in clearing dues.
6 Domestic long-distance service opened up without any restriction on the number of operators.
7 Corporatisation of Department of Telecom Services (DTS) and Department of Telecom Operations (DTO) by creating Bharat Sanchar Nigam Limited (BSNL) w.e.f. 1 October 2000.
8 Revenue sharing regime, in place of existing fixed licence fee, introduced for both basic and cellular service operators.
9 Thrust to accelerated implementation of Prime Minister's National Highways Development Project (NHDP) from petrol and diesel cess and additional fund-raising measures for NHAI.
10 Divestment of Government equity proposed in Indian Airlines and Air India.
11 Extension of tax holiday benefit to solid waste management and water treatment for developing urban infrastructure.

Direct taxes

12 Non-agricultural income of farmhouses made taxable.
13 Venture Capital Funds accorded complete pass through status with the income being taxed only in the hands of investors.
14 Interest from bonds issued by local authorities, as specified by Central Government, made tax free to make funds available for infrastructure.
15 Minimum Alternate Tax (MAT) to be charged at 7.5% of the "book profits" by all companies as determined under the Companies Act instead of the effective rate of 10.5% earlier.
16 Tax holiday benefits liberalised in respect of newly established industrial undertakings in Free Trade Zones, Software Technology Parks, Electronic Hardware Technology Parks and 100% Export Oriented Undertakings.
17 Weighted deduction for expenditure incurred on scientific research on in-house research and development facility enhanced from 125 to 150%.
18 Benefit of exemption of export income by entertainment industry extended to non-corporate assesses.
19 "One-by-six" criteria, introduced in the Union Budget 1998–1999, for identifying potential taxpayers, extended to 79 more cities (from 54 cities) having population of 2 lakh or more.

Indirect taxes

20 Peak protective customs tariff rate reduced from 40 to 35% *ad valorem*.
21 The existing five major *ad valorem* rates of basic customs duty reduced to four *ad valorem* rates.
22 The system of central excise was overhauled with the introduction of a single Central Value Added Tax (CENVAT) of 16% *ad valorem* on all manufactured goods with a few exceptions.

Fiscal management

23 The Fiscal Responsibility and Budget Management Bill, 2000, was introduced in the Lok Sabha in December 2000. The proposed legislation provides for a legal and institutional framework to eliminate revenue deficit, bring down the fiscal deficit and stabilise debt as a proportion of GDP within a time frame.
24 The interest rate on general provident funds reduced by 1–11% with effect from 1 April 2000.
25 Several measures taken for controlling growth in non-plan, non-developmental expenditure.

Financial sector

26 Tightening of entry norms for IPOs through modifications to SEBI (Disclosure and Investor Protection) guidelines.
27 Modified guidelines issued for 100% one-stage book building process.
28 Legislation initiated for reducing minimum Government shareholding in nationalised banks to 33%.
29 Establishment of IRDA.
30 Enlargement of functional area and greater autonomy to NABARD through amendment to the NABARD Act, 1981.
31 Revised norms for entry of new banks in private sector.
32 Permission to banks and NBFCs for undertaking insurance business.

Trade policy

33 Setting up of Special Economic Zones (SEZ) to encourage export production.
34 Evolution of a scheme for granting assistance to states based on their export performance for development of export related infrastructure.
35 Permission to import second hand capital goods, less than 10 years old without obtaining any license on surrender of SIL.

Capital account

36 Foreign Direct Investment (FDI) up to 100% permitted in e-commerce, subject to specific conditions.
37 The dividend balancing condition for FDI in 22 consumer goods industries removed.
38 The existing upper limit of Rs 1,500 crore for FDI in projects involving electricity generation, transmission and distribution (other than atomic reactor plants) dispensed with.
39 FDI under the automatic route permitted up to 100% for all manufacturing activities in Special Economic zones (SEZs), except certain activities.
40 Foreign equity participation up to 26% in insurance sector allowed under the automatic route.
41 Policy liberalisations effected for facilitating the use of ECB as a window for resource mobilisation.
42 Policies pertaining to international offerings through ADR/GDR by Indian companies further liberalised.

Source: Economic Survey, Government of India, 2001.

Horizontal commitments: India

MARKET ACCESS

Unrestricted except for measures affecting the entry and temporary stay of natural persons who fall in any of the following categories:

a *Business visitors* – Persons who visit India for the purposes specified in (i) and (ii) below and who will not receive remuneration from within India

Table 5.20 Commitments in General Agreement on Trade in Services (GATS)

Sector or subsector	*Limitations on market access*	*Limitations on national treatment*
Construction work for Civil Engineering (CPC Ex. 513)	*Unbound* Unbound	*Unbound* Unbound
Roads and bridges only:	Only through incorporation with a foreign equity ceiling of 51%	None
Construction of highways, streets, railways, runways, bridges, tunnels, subways, waterways, harbours, dams, pipelines, communication lines, power lines and construction work of constructions for mining and manufacturing not elsewhere classified, e.g. power plants, iron foundries, blast furnaces and coke ovens. It excludes construction work of warehouses and industrial buildings, residential and non-residential buildings	Unbound except as indicated in the horizontal section	Unbound except as indicated in the horizontal section

i for business negotiations, or
ii for preparatory work for establishing a commercial presence in India

Entry for persons in this category shall be for a period of not more than 90 days.

b *Intra-corporate transferees* – At the level of managers, executives and specialists who have been in the employment of a juridical person of another member for a period of not less than one year prior to the date of application for entry into India and are being transferred to a branch or representative office or a juridical person owned or controlled by the aforesaid juridical person.

Managers Persons who direct a branch office or one or more departments as their head, or supervise or control the work of other supervisory, professional or managerial personnel and have the authority to appoint or remove the personnel and powers to exercise discretionary authority over day-to-day operations.

Executives Persons who are in senior positions within a juridical person including a branch who primarily direct the management, have wide decision-making powers and are either members of the board of directors or receive directions from the board or the general body of share holders.

Specialists Persons who possess high qualifications and knowledge at an advanced level relevant to the organization's activities or the organization's research, equipment, techniques or management and may include persons who are members of accredited professional bodies.

Entry for persons in the above categories shall be for a maximum period of 5 years.

c *Professional* – Natural persons to be engaged by a juridical person in India as part of a services contract for rendering professional services for which he/she possess the necessary academic credentials and professional qualifications with 3 years experience in the field of physical sciences, engineering or other natural sciences.

Entry and stay in this category shall be for a maximum period of 1 year.

Table 5.21 Construction and related engineering services

Mode of supply / Group	*1 Cross-border supply*	*2 Consumption abroad*	*3 Supply through commercial presence*	*4 Supply through presence of natural persons*
	GATS: Specific commitments			
1 Construction work for Civil Engineering (CPC Ex. 513) (M.A.) Roads and bridges only: Construction of highways, streets, railways, runways, bridges, tunnels, subways, waterways, harbours, dams, pipelines, communication lines, power lines and construction work of constructions for mining and manufacturing not elsewhere classified, e.g. power plants, iron foundries, blast furnaces and coke ovens. It excludes construction work of warehouses and industrial buildings, residential and non-residential buildings.				
	Unbound	Unbound	Only through incorporation with a foreign equity ceiling of 51%	Unbound except as indicated in the horizontal section
2 Engineering Services (CPC 8672) M.F.N.	Unbound	Unbound	None	Unbound except as indicated in the horizontal section
3 Engineering Services (CPC 8672) M.A.	None	None	Only through incorporation with a foreign equity ceiling of 51%	Unbound except as indicated in the horizontal section
4 Construction and Related Engg. Services M.F.N.	Unbound	Unbound	None	Unbound except as indicated in the horizontal section
5 Construction and Related Engg. Services M.A.	Unbound	Unbound	Only through incorporation with a foreign equity ceiling of 51%	Unbound except as indicated in the horizontal section

Notes
Abbreviation: M.F.N. = Most favoured nation; M.A. = Market access.

5.4.2 *Areas where foreign investments are needed*

Rules and regulations for participation of foreign contractors and professionals

The rules for foreign contractors are no different from those of any other service suppliers. For incorporation permission is needed from the Reserve Bank of India and Foreign Investment Promotion Board (FIPB).

Foreign participation in the domestic market

With the rapid escalation in the quantum of infrastructure projects over 2001, there has been a large influx of foreign contractors, especially in the highways and power sectors. Most major international contractors and consultants have established permanent bases here and are actively participating in the infrastructure contracts. For example in the national highways projects, 48 foreign contractors, directly or with joint ventures have qualified as eligible to bid. A list is given in Table 5.22.

Impact of liberalization under the WTO

Mandated negotiations under the GATS commenced at the WTO on 1 January 2000, involving a further round of progressive liberalization of specific commitments on market access undertaken by members. This was a comprehensive round covering all services sectors and all the modes of supply of services.

At the time of writing special sessions of the Council for Trade in Services (CTS) were being held for these negotiations and the CTS was in the process of formulating negotiating guidelines and procedures. India, along with 22 other developing countries, had filed a proposal in the WTO containing the essential elements of such negotiating guidelines and procedures.

The main features of this proposal are that the basic architecture of GATS should remain unchanged; the "request–offer" approach should be the principal modality of these negotiations; there should be appropriate flexibility for developing-country members while undertaking commitments; and the starting point of these negotiations should be the bound level of commitments undertaken by the members.

India's primary interest in these negotiations lies in ensuring freer movement for natural persons especially professionals. Accordingly, India has filed a proposal in the WTO suggesting possible strategies and approaches for bringing about effective liberalisation in the movement of professionals.

The main features relate to broader and deeper horizontal and sectoral commitments in sectors of interest to India, commitments to be based on uniform and well-defined categories of personnel, removal of existing practices requiring social

Table 5.22 Foreign direct investment and technical collaborations approved during August 1991 to March 2001 in consultancy services

Areas of consultancy services	*No. of collaborations approved*			*FDI approved* (Rs million)	%
	Total	*Technical*	*Financial*		
Design and engineering services	304	60	244	12306.29	54
Management services	195	22	173	5049.87	22
Marketing	54	7	47	526.68	2.5
Construction	15	2	13	567.58	2.5
Other consultancy services	46	7	39	4277.47	19
Total	614	98	516	22727.59	

Source: Secretariat for Industrial Assistance Newsletter – April 2001.

security contributions for temporary movement, greater facilitation and use of mutual recognition agreements for recognition of qualifications, etc.

References

BSAA (2002) Background Note: India, Bureau of South Asian Affairs, U.S. Department of State, http://www.state.gov/p/sa/ci/in/.

CIDCI India Country Report, Papers for Asiaconstruct Conference, Construction Industry Development Council, India, various issues.

Forbes, Naushad (2001) Doing Business in India: What has liberalization changed?, Conference paper on Indian Economic Prospects: Advancing Policy Reform, Stanford Centre for International Development.

Root, Hilton L. (1998) India: Asia's Next Tiger?, Essays in Public Policy, Hoover Institution, 2/98.

PCGI (2002) Mid-term Appraisal–Highlights, Planning Commission, Government of India.

Chapter 6

Indonesia

Ministry of Public Works, Indonesia and Eddie C.M. Hui

6.1 Introduction

6.1.1 Background

Indonesia is the world's largest archipelago. It proclaimed its independence on 17 August 1945 soon after Japan's surrender in the Second World War. Until the Asian financial crisis in 1997, the country had progressed to become a strong economy in East Asia. At the time of writing, issues in Indonesia include implementing IMF-mandated reforms of the banking sector, and effecting transition to a popularly elected government after four decades of authoritarianism. Strong commitment was implemented for good governance initiatives by combating cronyism and corruption, resolving human rights violation and separatist movements in Aceh and Irian Jaya. On 30 August 1999, East Timor province, a former Portugal colony integrated into Indonesia in 1976, was given a referendum for independence which was then overwhelmingly approved by its people. Concurrence by Indonesia's national legislature followed and after a transition period of UN administration, East Timor's independent status was finally established in 20 May 2002.

With a population of 228,437,870 (July 2001 est.) people, Indonesia is the world's fourth most populous nation behind China, India and the United States. In the past decades, Indonesia's population has been increasing at the rate of 1.6 per cent. The island of Java is one of the most densely populated areas in the world, with more than 107 million people living in an area the size of New York state. Indonesia is a multiethnic country with Javanese constituting 45 per cent of the total population while Sundanese, Madurese, coastal Malays and others constituting 14 per cent, 7.5 per cent, 7.5 per cent and 26 per cent respectively. Indonesia includes numerous related but distinct cultural and linguistic groups, many of which are ethnically Malay. Since independence, Bahasa Indonesia (the national language, a form of Malay) has spread throughout the archipelago and has become the language of most written communication, education, government and business. Many local languages are still important in many areas, however. English is the most widely spoken foreign language.

Geography

Indonesia is located in Southeast Asia, an archipelago between the Indian Ocean and the Pacific Ocean, at geographic coordinates 6 00 S latitude and 141 00 E longitude.

Indonesia has a total area of 1,919,440 square kilometres with land (1,826,440 sq. km) and water (93,000 sq. km). Indonesia has land boundaries totalling 2,602 kilometres (border countries: Malaysia 1,782 kilometres, Papua New Guinea 820 km) and a coastline of 54,716 kilometres.

The Indonesian climate is generally tropical hot and humid in most of the lowlands but more moderate in the highlands. Most areas in Indonesia are coastal lowlands with larger islands having coastal plains and interior mountains. Indonesia has more than 17,000 islands, 6,000 are inhabited and 1,000 of them are permanently settled. The elevation extremes range from the lowest at Indian Ocean (0 metre) to Puncak Jaya at (5,030 metres). Indonesia's staple natural resources include petroleum, tin, natural gas, nickel, timber, bauxite, copper, fertile soils, coal, gold and silver. It also includes arable land (10 per cent of total land), permanent crops (7 per cent) permanent pastures (7 per cent) forests and woodland (62 per cent) and others (14 per cent). A total of 45,970 square kilometres (1993 est.) constitutes irrigated lands.

6.1.2 Political structure

Indonesia is a republic based on the 1945 constitution providing for a limited separation of executive, legislative and judicial powers. The newly elected government has been fashioning political reform legislation that – without changing the 1945 Indonesian Constitution – will formally set up new rules for the electoral system, the House of Representatives (DPR), the People's Consultative Assembly (MPR) and political parties. A separate law redefining the term limits for the presidency is also on the drawing board. Substantial restructuring has been observed since President Suharto's resignation.

The president, elected for a 5-year term, is still the dominant government and political figure. He/She is selected along with the vice president by the MPR. The president has the authority to conduct the administration of the government and is accountable only to the MPR. The president appoints a 41-member cabinet to assist him.

Under the government proposal, the House of Representatives will increase in size (by 50) to 550 members. Totally, 420 members are to be elected on a "first past the post" basis from districts, 210 located in Java and Bali, and 210 more in other islands. Another 75 seats will be allotted proportionally to parties on the basis of nation-wide showings, a mechanism to provide for some minority representation. Fifty-five legislative slots will be reserved for the military whose members will not be permitted to vote in the general election. The MPR will still select the president and vice president of Indonesia. Its members traditionally had been appointees hand picked by President Suharto. Under current proposals, the new MPR will be reduced from 1,000 to 700 members, with no presidential appointees. It will consist of the 550 members of the DPR, plus 81 members elected from provincial parliaments, and 69 appointed by social and community groups. Under the Suharto regime, the ruling "functional group" (not party) GOLKAR dominated, and the United Development Party (PPP), and the Indonesian Democratic Party (PDI) were the main opposition parties. In the new system there is theoretically no limit on competitive political parties. Since May 1998, some 80 parties have emerged. Most new parties, however, are small and local, and probably will fail to satisfy the party registration criteria. These criteria include a requirement to have party branch offices in at least 14 of Indonesia's 27 provinces or to demonstrate support by collecting a million signatures.

The armed forces have shaped and staffed Suharto's New Order since it came to power in the wake of the abortive 1965 uprising. Military officers, especially from the army, have been key advisers to Suharto and have great influence on policy. Under the dual function concept, military officers serve in the civilian bureaucracy at all government levels, although there has been a recent tendency to somewhat reduce the military's direct involvement in the civilian bureaucracies. And the public calls for an end to the military's dual role have increased since Suharto's resignation.

6.1.3 Development process in Indonesia: through the years

According to the World Bank (2001), from 1967 to 1997, Indonesia saw its economic performance soar, with growth at 7 per cent a year; its poverty rates fall from 60 per cent of the population to 11 per cent; and standards of living improve. Adult literacy went from 56 per cent of the adult population to over 90 per cent at the time of writing. Fertility rates declined from 5.4 in early 1970s to 2.6; life expectancy increased by nearly 20 years – from 46 years in 1967 to today's age of 65. Production of food doubled; trade volume of exports grew at 8 per cent a year; Indonesia was considered a "model" of development success. Then the Asian Financial Crisis struck in mid-1997 when everything changed. Indonesia's public sector debt had risen to over 110 per cent of GDP. Sovereign debt was expected to rise with the infusion of international funds into the economy. The World Bank, the Asian Development Bank, the International Monetary Fund, and other bilateral and multilateral development agencies had also contributed their technical expertise and approximately $118 billion dollars in the hope of maintaining political and economic stability for Indonesia. However, with an estimated 80 per cent depreciation in the value of the rupiah since July 1997, loan repayment has become even more burdensome (source: http://www.indonesia-ottawa.org/Handbook99/contmpid_R.htm).

Indonesia began with the liberalization of its economy about 35 years ago following a change in government in 1966. Throughout this period there were three distinct periods of "liberalization" each motivated by different events (Soesastro, 2000). The first period was between 1966 and 1972 during which measures were introduced to rehabilitate the economy that was in near collapse. The second wave of liberalization occurred between 1982 and 1991, which was essentially driven by necessity to restructure the economy following the sharp fall in the oil prices worldwide. It began diversification of its economy to avoid over-dependence on the oil and mineral sector and attracted manufacturing to create a more diverse and stable economic climate. In the mid-1980s, the government began eliminating regulatory obstacles to economic activity. These steps primarily were directed at the external and financial sectors and were designed to stimulate growth in non-oil exports and revenues and to strip away import substitution barriers. The third liberalization period from 1994 onwards was inspired by wider liberalization and globalization processes taking place all over the world and was designed to keep the economy competitive. The May 1994 and May 1995 deregulation packages helped level the playing field for competition and reinforced the country's involvement in global liberalization initiatives while the January 1996 package helped cut tariffs. The most important, unfinished deregulation steps were removal of non-tariff barriers, the privatization of key industries, and the removal of domestic subsidies and export restrictions.

Although Indonesia's growth model was technically a free-market economy that was dominated by the private sector, the government still played a significant role in the economy through state-owned firms and the imposition of price controls in selected industries. During the 30 years of President Suharto's rule, Indonesia's economy grew steadily, from a per capita GNP of $70 to a per capita GNP of about $1000. It was recognized as a newly industrializing economy. Annual real GDP growth averaged close to 7 per cent from 1987 to 1997. By employing a restrictive monetary policy and a conservative fiscal stance, the government held inflation in the 5–10 per cent range. With strong export performance and manageable import growth, Indonesia saw a trade surplus of about $3 billion in 1995.

Indonesia was one of the flash points of, and had been hit hard by, the Asian financial crisis. The effects of the financial crisis have been severe. The economy contracted by 13 per cent in 1998 year on year, unemployment reached 35 per cent and was likely to worsen as the real economy feels the financial crisis. In 1998, soon after the crisis hit the country, the rate of inflation was 59 per cent. The destruction of banks and small shops in wake of the May 1998 riots and flight of ethnic Chinese businessmen added to significant capital flight. Petroleum, one of Indonesia's major exports, had been hit especially hard by both the financial crisis and by the recent drop in world oil prices. The oil and mineral sector that had been an integral part of Indonesian economic growth over the past decades with its contribution approximately 8 per cent to GDP and 18 per cent to government revenues by fiscal year 1996/1997, saw the sector's share of export earnings and government revenue drop to about 10 per cent. Although the oil prices tumbled by 30 per cent after the crisis it remained an important part of the economy in which many US companies have heavily invested.

Two years after the crisis, the Indonesian government was trying to prevent a total economic and political collapse. Analysts feared that economic deterioration could undermine the usually fragile government. The government asserted its commitment to work with the IMF in moving forward with economic reforms, including increased transparency and open markets. In late 1997, Indonesia agreed to a three-year stabilization/structural adjustment programme with the international financial institutions. This programme was designed to stabilize the rupiah in foreign exchange markets by the government's adoption of tight fiscal and monetary policies. Non-viable banks have been closed, and a broad range of structural reforms have been implemented including liberalization of foreign trade and investment; dismantling of domestic monopolies; allowing greater private sector participation in provision of infrastructure; and expanding the privatization programme.

6.2 The national economy

6.2.1 Macroeconomic review and outlook

The Indonesian economy in the fifth five-year development plan (1989–1994) grew by 6.8 per cent while in the sixth plan (1995–2000) the growth was 7.1 per cent. Economic growth during the years after the completion of the first 25-year development plan ending March 1994 remained high, exceeding the annual average growth rate during the first 25-year development plan which was at 7 per cent.

In 1994 the growth rate was 7.5 per cent while it was 8.1 per cent in 1995 and 7.5 per cent in 1996. GDP growth slowed to 4.9 per cent in 1997, reversing the trend of sustained high growth in the last decade. As the crisis deepened in 1998, GDP forecasts for the year were continually revised downward. GDP contracted by 4.3 per cent in the first quarter, 12.3 per cent in the second quarter, 18.4 per cent in the third quarter, and 19.5 per cent in the fourth quarter. The final estimate of GDP contraction of 13.7 per cent in 1998 was far in excess of the initial forecasts of the Country Economic Review 2000 by the ADB (ADB, 2001). However since 1999, Indonesian economic stabilization had been taking place arising from the 1997 multi-dimensional crisis that very badly hampered the construction industry among other sectors. The recovery had shown optimistic signs. This was manifested in many ways: better exchange rate, lower inflation and lower commercial bank interest as shown in Table 6.1.

Other indicators have also shown that national economic stabilization had been taking place, one of them is economic growth, which contracted as much as 13.60 per cent in 1998 but grew to 0.8 per cent in 1999. The GDP further grew to 4.8 per cent in 2000. Indonesia's severe economic contraction seems to have bottomed out and clear signs of economic recovery have emerged. Although the fiscal stimulus was smaller than envisaged, it helped to stabilize the contraction. Private consumption then picked up and fuelled the recovery. Investment contracted further in 1999, but the decline slowed markedly in the fourth quarter. Exports also declined further in 1999, but more slowly. On the output side, although the recovery was initially underpinned by a strong rebound in agriculture, it spread to virtually all non-primary sectors by the end of the year. Contraction of the services sector was marginal compared to the sharp decline in 1998, and reflected mainly the continued decline of financial services.

Macroeconomic stability achieved towards the end of 1998 set the stage for recovery, and was greatly consolidated in 1999. Year-on-year inflation at the end of 1999 was only 1.9 per cent compared to 78 per cent at the end of 1998. Declining food prices, rupiah appreciation and the tight monetary stance together accounted for the sharp drop in inflation. High rupiah volatility experienced in 1998 was contained, and the rupiah traded within a narrow band of Rp. 6,800–7,500/US$ in 1999. Low inflation and rupiah stability permitted the key one-month SBI interest rate to fall from over 35 per cent in early 1999 to 12.5 per cent at the end of the year. The consolidation of macroeconomic stability not only supported the economic recovery

Table 6.1 Macroeconomic indicators: 1998 and 1999

	1998	*1999*
Exchange rate (1 US$/Rp.)	10.088	7.850
Inflation	77.63	2.01
Annual interest rate (commercial bank)	3.7	20.7
Working capital investment	26.2	17.8

Source: Bank Indonesia, Annual Report 2000.

Table 6.2 Major demographic indicators (1995–1999)

	1994	*1995*	*1996*	*1997*	*1998*	*1999*	*2000*
Population (millions)	192.3	195.3	198.3	201.3	204.4	207.4	210
Households (millions)	38.44	46.9	47.2	48.3	49.4	50.5	n.a
Unemployed (millions)	3.7	6.2	4.4	4.2	5.06	6.03	5.8
Unemployed (%)	4.4	7.2	4.9	4.7	5.5	6.4	6.1
GDP (trillion rupiah)[2]	382,219	452,380.9	532,568.0	625,505.9	989,573.1	379,558	397,666
GDP (billion US$)	176,943	201,303.6	223,486.0	134,517.0	93,042.2	n.a	n.a
GDP – constant price (trillion rupiah)	354,640	383,792.2	414,418.9	434,095.3	374,718.8	n.a	n.a
GDP (% change at constant price)	7.54	8.29	8.00	4.75	−13.6	0.8	4.8
Consumer price (% change)	9.24	8.6	6.5	11.1	77.6	n.a	n.a
Wholesale price (% change)	5.39	11.6	7.5	9.3	101.8	n.a	n.a
Construction price (% change)	5.16	9.8	6.1	4.6	68.1	n.a	n.a
GDP per capita ($ current)	842.3	927.7	1,043.3	1,155.4	1,073.5	468.3	
Short-term interest rate (%)	17.75	18.8	19.2	22.0	25.1	29.0	9
Long-term interest rate (%)	14.96	15.8	16.5	19.2	19.4	22.3	16

and reduced transient poverty, but also played a crucial part in improving market sentiment towards Indonesia (Table 6.2).

Economy in first quarter of 2001

The economic recovery process in the year 1999 had by and large triggered off impetus for the years to come and provided ground conducive for further growth. While the optimistic indicators in 1999 had been dominated by incremental consumption expenditures, in the first quarter of 2000, the national economy had achieved 4.8 per cent with a more balanced composition between consumption expenditure and other transactions. This recovery was even faster than any of the economists' predictions. In the whole year of 2000, the national economic indicators tended to be stronger regardless that some controversial issues had not been overcome yet, such as bank debt restructuring, financial intermediation of commercial banks, and the limited national fiscal stimulus. These issues had been prime obstacles for the Indonesian economy to achieve a more stable economy recovery level. Tables 6.3 and 6.4 map out the Indonesian economy recovery since 1998.

In the first half of 2001, GDP growth slowed to 3.4 per cent from 4.8 in the same period in 2000, mainly because of continued household and public sector spending and only a modest recovery in private investment. Growth was most striking in

Table 6.3 Major macroeconomic indicators (A)

Description	*Annual growth (%)*		
	1998	*1999*	*2000*
Real gross domestic product (The base year is 1993)	−13.1	0.8	4.8
By expenditures			
Consumption	−7.1	4.3	3.9
Fixed gross capital gain	−33.0	−19.4	17.9
Export of goods and services	11.2	−31.6	16.1
Import of goods and services	−5.3	−40.7	18.2
By sector			
Agriculture	−1.3	2.7	1.7
Mining and quarrying	−2.8	−2.4	2.3
Manufacturing industry	−11.4	3.8	6.2
Gas, electricity and water supply	3.0	8.3	8.8
Building construction	−36.4	−0.8	6.7
Hotel, restaurant and trade	−18.2	0.1	5.7
Transportation and communications	−15.1	−0.8	9.4
Finance	−26.6	−7.5	4.7
Services	−3.8	1.9	2.2
Inflation	77.6	2.01	9.35
Exchange rate (Rp./$)	10.088	7.850	8.400

Source: Bank Indonesia, Annual Report 2000.

manufacturing – averaging 4.2 per cent in the first half – as well as in some services, particularly utilities and transport. Financial services showed only modest expansion, reflecting the still weak financial capacity of most banks. Inflation grew at double-digit levels after February, running as high as an annualized 13 per cent in September (ADB, 2001). A weaker domestic currency, accommodative monetary policy, and increases for price-administered goods and services, such as oil and electricity, all contributed to firmer inflation.

The inflation for the first quarter was 10.62 per cent which was higher than the fourth quarter of 2000 (9.35 per cent). Surprisingly big contributors to the inflation were incremental price for housing (2.49 per cent), agricultural product (2.30 per cent) foods (2.87 per cent). Exports fell by 0.3 per cent in the first six months of 2001 while imports grew by 28.4 per cent, reflecting some pickup in household demand and investment in late 2000–early 2001. As a result, the merchandise trade surplus fell by 25.6 per cent in the first half of 2001, relative to the same period in 2000, while the current account surplus fell by approximately 25 per cent year-on-year. The current account surplus ending the year 2001 was at around $3.5 billion, down from $8 billion in 2000, while the balance of payments was likely to post a deficit of approximately $1.0 billion. In the forex market after the inauguration of the new President in August 2001, the rupiah became stronger from Rp. 11,300/US$ in July 2001 to Rp. 8,850/US$ was seen as a sign of economic growth associated with political stability.

Table 6.4 Major macroeconomic indicators (B)

No.	*Description*	*1995/1996*	*1996/1997*	*1997/1998*	*1998/1999*
1	Inflation	6.47	11.05	77.63	–
2	Balance of payment				
3	Goods and services				
	Export, f.o.b.	47.754	52.038	54.605	59.484
	• Oil and gas	10.616	12.771	11.019	10.232
	• Other than oil and gas	37.138	39.267	43.586	49.252
	Import, f.o.b.	–41.502	–45.819	–45.957	–48.067
	• Oil and gas	–3.905	–4.693	–3.804	–3.385
	• Other than oil and gas	–37.597	–41.126	–42.153	–44.682
	Services	–13.239	–14.288	–14.990	–16.813
	• Oil and gas	–3.238	–3.541	–4.403	–4.401
	• Other than oil and gas	–10.001	–10.747	–10.587	–12.412
4	Running transaction	–6.987	–8.069	–6.342	–5.396
	• Oil and gas	3.473	4.537	2.812	2.446
	• Other than oil and gas	–10.460	–12.606	–9.154	–7.842
5	Inflow	5.730	5.298	13.147	9.051
	• CGI which current transfer	0	0	7.572	3.800
	• Non-CGI	5.730	5.298	5.575	5.251
6	Private capital, net	11.672	13.488	3.169	2.210
7	Debt repayment	–5.939	–6.118	4.884	–4.789
8	Total A to D	1.262	4.599	5.090	1.076
9	Monetary movement	–616	3.898	4.615	–1.076
10	Errors and omissions, net	–646	701	475	0

6.2.2 Role of construction/property market in national economy

The construction industry in Indonesia is a relatively young industry. However, it has grown significantly since the early 1970s. Its contribution to GDP increased from 3.86 per cent in 1973 to 7.94 per cent in 1996. It constitutes about 60 per cent of gross fixed capital formation. And the manpower employed has increased significantly, from about 413,000 in 1978 to about 3,796 million in 1996. In 1994 GDP growth was led by construction and domestically oriented manufacturing industries (Arditi and Mochtar, 1996).

Indonesia is an archipelago with more than one thousand islands. More than 60 per cent of the construction works are in the island of Java, and more than half of this is in the capital city, Jakarta. Most public works projects, including any construction projects under government authority or under state-owned companies, were awarded through a competitive bid. Professional designers and constructors were engaged in separate contracts. The contractors usually would not be involved until the designs have been completed. The private sector usually followed the same approach in executing its construction projects.

The construction industry of Indonesia had been included in the national development plans by the People's Deliberative Assembly, and constituted an important sector of economic activity in terms of its contribution to the national economy. The national development plan is executed by the government. The GDP originating from construction between 1985 and 1990 had been 5–6 per cent of the GDP. In 1990 the construction share of the GDP amounted to Rp. 10,827 billion ($5,414 m) equivalent to 5.48 per cent of the GDP (Indonesia Country Report, 1997). Investment in the same year amounted to Rp. 116,634 billion ($8,317 m) equivalent to 8.41 per cent of the GDP.

Construction investment (or market) volume

Construction was the sector hit hardest by the 1997 financial crisis, resulting in a contraction of nearly 40 per cent in 1998 (ADB, 2000). This sharp contraction in manufacturing and construction attributed to the outcome of the corporate debt overhang and the banking crisis. The services sector also suffered a large contraction of 16.6 per cent; a significant fact in this was the over 20 per cent decline in banking and financial services (source: http://www.adb.org/Documents/CERs/CER_INO_1999.pdf) which also seemed to affect the construction sector due to its backward linkages. The chronological development in the decline in construction investment showed the construction sector investment was already in negative growth −10 per cent) in the year 1996 dropping to −23 per cent in 1997. Investments further contracted, immediately after the 1997 financial crisis, by 40 per cent in 1998. In 1999 it was running a negative growth of −25 per cent.

With the GDP growth in positive territory, Indonesia's construction industry began to show signs of recovery from the prolonged economic crisis. Investment in the construction industry was expected to reach US$10 billion in 2000, an increase from US$5.5 billion in 1999. However, the sharp increase in investment was still below the pre-crisis level of over US$13.75 billion, according to construction analysts. Infrastructure projects were leading the recovery of the construction industry. The most feasible projects amid the economic crisis were infrastructure projects funded by foreign loans. It was estimated that 28 per cent of the construction projects are for roads, irrigation and telecommunications. Road projects were predominantly located on Java island, while irrigation projects were mostly located outside Java. In general, construction projects in 1999 and 2000 were mostly of a small and medium size, ranging from US$250,000 to 1.75 million. Analysts said that projects such as roads, traditional markets, residential estates and industry processing plants were in demand. This contrasts with the situation in 1998 when the property sector, including office towers, five-star hotels, as well as toll roads, dominated the industry. However, the property sector, especially housing, was predicted to recover to its pre-crisis level in 2001, as sales double from 1999.

Construction private sector

In retrospect, in the first half of 1990 strong growth was observed in all sub-sectors of the construction industry while the growth had fallen substantially towards the end of the decade particularly since the financial crisis in 1997. The growth in private sector participation in housing or residential sector in the year 1995, increased sharply by 123 per cent (2.063 trillion rupiah), compared to 43 per cent (0.92 trillion rupiah) in 1994. Comparatively, government investments in the same sub-sector increased by only 62 per cent. However in 1996, the growth rate reduced to 40 per cent, from 2.063 trillion rupiah to 2.890 trillion rupiah. This reduction in growth indicated oversupply and market saturation in the housing market. In 1997 the growth turned negative (−54 per cent), from 2.890 trillion rupiah to 1.336 trillion rupiah.

In the private non-residential sub-sector, the growth in 1995 was 46 per cent (6.76 trillion rupiah) against 29 per cent in 1994 (4.62 trillion rupiah). In 1996 however, the trend was not different from that of the private residential sector. The growth rate remained only 14 per cent, from 6.76 trillion rupiah to 7.74 trillion rupiah. In 1997, there was negative growth of −30 per cent.

The private civil works sub-sector comprises a mixture of works such as electric installation, plumbing, foundation works, sanitation, gas piping, etc. There was no significant growth in 1995. The growth in 1994 was 53 per cent, 32 per cent in 1995 and 12 per cent in 1996. In 1997, the growth rate remained positive although there was only an increase of 1 per cent against 1996.

From the year 1997 onwards private sector investment in the residential sector nose-dived to a negative figure of −54 per cent from 2.89 trillion rupiah to 1.336 trillion rupiah which again plummeted by −76 per cent in 1998 (Table 6.5).

It is unfortunate that for the years of 1999, 2000 and 2001 the data of the whole construction private sector was not available at the time of writing, but partially for

Table 6.5 Construction sector investments (constant prices) by construction type

Type of construction	*Absolute amount in 1996 (billion rupiah)*	*1995 (%)*	*1996 (%)*	*1997 (%)*	*1998 (%)*	*1999 (%)*
Private residential construction	2,890.0	123	40	−54	−76	n.a.
Private non-residential construction	7,738.0	46	14	−30	−48	n.a.
Private civil construction	12,055.0	32	−12	1	−23	n.a.
Total private construction (A)	22,683.0	41	1	−16	−25	n.a.
Public residential construction	2,180.0	62	10	−62	−73	n.a.
Public non-residential construction	5,838.0	6	−10	−44	−34	n.a.
Public civil engineering	9,094.0	−5	−31	−18	−48	n.a.
Total public construction (B)	17,112.0	2	−21	−32	−56	n.a.
Total construction output	39,796.0	19	−10	−23	−47	n.a.
Repair and maintenance (C)	n.a.	n.a.	n.a.	n.a.	n.a.	n.a.
Total construction output (A+B+C)	39,796.0	19	−10	−23	−40	−25

housing the Bank Tabungan Rakyat (People Saving Bank) had released credits for Non-Perumnas housing as follows:

1998	98,985 houses
1999	44,465 houses
2000	15,531 houses (first quarter)

With reference to the construction sector growth of as much as 6.7 per cent in 2000, the whole construction private sector would not increase less than that especially when the most attracted toll project Jakarta Outer Ring Road (JORR) and toll road Cikampek–Padalarang were to be implemented smoothly by the private investor(s). However the spirit of the investors' coming back could materialize if the proposed investments that had been approved by the Investment Coordination National Board (BKPM) could be implemented without interference.

Table 6.6 gives us the hope that the construction private sector will recover soon within the coming years.

Construction government sector

The year 1995 was the turning point for the shift from public to private sector. In the residential sector, however, the government continued to provide housing development for the economically weak community. In 1995 the government raised the budget for residential development to 1.98 trillion rupiah, an increase of 62 per cent compared to 1994 (6 per cent). The activities and budget for residential development by the government decreased in 1996 and the growth rate was only 10 per cent. Negative growth (−62 per cent) occurred in 1997. The private residential sector had been observed to have the largest drop in investment compared with other sectors with investment in public residential construction falling by −73 per cent in 1998. Although the government had increasingly delegated much of construction activities to the private sector, it is still responsible for directly providing public sector housing for low income families. In the public residential sector, the growth of 10 per cent achieved immediately prior to the 1997 financial crisis dropped by −62 per cent in 1997.

Housing development by the government is detailed as follows:

Perum Perumnas (National Housing Development Corp.) The housing development by Perum Perumnas since 1995 is:

1995	313,526	house units
1996	351,856	house units
1997	390,487	house units

Table 6.6 The approved investment proposal by BKPM during the period of 1999–2000

Sector	*1999*		*2000*	
Construction	National (Rp. 10^9)	International (\$$10^6$)	National (Rp. 10^9)	International (\$$10^6$)
Investment	395.1	153.4	449.1	87.8

Sources: Investment Coordination National Board (Badan Koordinasi Penanaman Modal/BKPM).

In almost all provinces, the house-types built in 1997 by Perum Perumnas comprise 198,273 units of core-house, 179,148 units of primitive house and 13,066 units of storied house.

Ministry of Transmigration

The Ministry of Transmigration has built houses in transmigration areas to provide housing for transmigrants as detailed below:

1995	1,032,949	house units
1996	1,096,233	house units
1997	1,151,154	house units

The construction of non-residential buildings by the government annually had shown a declining trend. This decline occurred because the government started to reduce the building of offices for its own usage. This policy was tightly implemented by the government and was the main reason for the decline. This policy was a preventive step taken by the government before the monetary crisis in Indonesia.

Other civil engineering works executed by the government included infrastructure developments required to generate economic growth either regionally or nationally as a whole. Those works include construction of roads, sea ports and airports, hospitals, school buildings, water supply networks, irrigation systems and other public facilities.

These activities represent the government's efforts to stimulate economic activity. It is also a government responsibility to improve public utilities for the sake of the community. The level of these government investments however was declining, due to a deliberate government effort to reduce its intervention in the economy.

There was a 6 per cent growth of government investment in non-residential construction in 1995, from 6.12 trillion rupiah in 1994 to 6.49 trillion rupiah. Government investment in civil construction started decreasing from 1996 onwards, registering a negative growth of −31 per cent in 1996. This negative growth continued in 1997 (−18 per cent), although the decrease was not as much as in 1996.

Since the Law No. 22/1999 for Autonomous Regional Government had been declared, the construction of public infrastructure had been implemented by the city (Kota) and district (Kabupaten) government. The achievement of public infrastructure construction carried out by the regional government in the 3 years of crisis was mainly in low-cost housing.

These were done through Perumnas:

1998	11,863 units
1999	7,109 units
2000	996 units

The National Annual Budget of 2001 was designed mainly to fulfil government debts and international as well as national liabilities, specifically for government official remunerations. The budget was not expansive therefore. However the government still gave high priority to low-cost housing for low-income people and public infrastructure.

6.2.3 Changes of economic policies due to the financial turmoil

The most severe macroeconomic impact of the crisis had been felt in the fiscal sector. Indonesia's public debt was estimated at 95 per cent of GDP at the time of writing, up from 24 per cent in 1997 (*Economist*, 2001). Of the total public debt of $152 billion, almost 59 per cent was domestic – bonds issued to recapitalize insolvent banks and liquidity injection to support failing banks when the crisis hit. Prior to the crisis only external debt was a source of budget financing. The implication of such a large public debt burden was that scarce public resources will have to be allocated to support current expenditures at the expense of public investment, thus undermining the country's growth potential for several years. Moreover, it was expected to keep the macroeconomy more vulnerable to shocks compared to the pre-crisis period. The expansionary fiscal stance of recent years cannot continue indefinitely given the weakness of the government's accounts. As the recovery gathers pace, the expansionary stance may shift to fiscal consolidation and then to fiscal sustainability. In more operational terms the fiscal strategy that had been taken by the government aims

- to optimize the national income from tax and non-tax income;
- to control and improve the efficiency of state government expenditure;
- to reduce subsidy;
- to apply balancing national budget distribution;
- to reduce the dependency of international and foreign financial support.

The quarterly trends in macroeconomic indicators clearly indicated that economic recovery was underway in Indonesia. However, evidence also suggested that the recovery was still fragile due to important vulnerabilities. First, aggregate demand was still weak, driven by private consumption. Investment and exports contracted for a second consecutive year, although their decline slowed. Second, the massive build-up in public debt had imposed a heavy burden on public finances. Servicing the debt burden and preventing any further fiscal deterioration will require extremely prudent public financial management. Third, while there had been progress in structural reform especially in the banking sector, corporate debt restructuring had been very slow. Without rapid progress in this area, investment demand will remain weak and export supply bottlenecks could constrain export growth even if external demand picks up. Surveys indicated that transient poverty induced by the crisis peaked in February 1999 plunging an additional 17 million people below the poverty line in less than 2 years (World Bank, 2001). Although transient poverty may have abated by the end of 1999, its adverse impacts were expected to linger for many years. Survey results also showed that the impacts were not uniform – the poorest in certain regions had suffered the worst deterioration in nutritional status. Moreover, the proportion of children most seriously at risk of malnutrition increased significantly. It was expected that the serious impacts on education, health and nutrition will not be automatically reversed even if transient poverty drops to pre-crisis levels. And likewise, many whose savings and assets had been depleted will take a long time to recover from the crisis. Revised government estimates suggested that even with a return to pre-crisis poverty incidence, about 39 million people would be below the poverty line. In addition, many million people living just above the poverty line were also poor. Combating poverty in Indonesia must therefore be a foremost development priority, and the focus of public policy must accordingly shifted to addressing problems of structural poverty in the post-crisis period.

On structural reforms, there had been substantial progress in bank restructuring in the past year. Major private banks and state banks have been recapitalized and restructured. However, some banks including Bank Mandiri, the largest (state-owned) bank was still experiencing a negative intermediation spread at the time of writing. Such "bleeding" of the capital base of state-owned banks in particular, could require further capital injections straining an already difficult fiscal situation. Resolving the status of the remaining state banks and other banks under Indonesian Bank Restructuring Agency (IBRA) control, and strengthening the legal framework and settling interbank claims transparently were key issues in the unfinished agenda of banking reforms. Another major financial sector issue was the need to accelerate capital market reforms, especially to develop a domestic bond market. This was expected to permit more effective channelling of domestic savings to investment, and promote hedging instruments for foreign exchange risks. Corporate debt restructuring was intertwined with the issue of financial sector reforms and holds the key to a rebound in investment, and to a lesser extent exports. Despite substantial improvement in the climate for resolution of such debt – appreciation of the exchange rate, low interest rates, strong rebound of the equity market, and reforms in the institutional mechanisms for corporate debt restructuring – the process had been painfully slow. A major impediment had been weak enforcement of the bankruptcy law, as reflected in a series of controversial court rulings. Unless credibility of the judicial process was restored quickly, the overhang of corporate debt could stall the recovery. Table 6.7 shows Indonesian near-term economic restructuring and target.

Recovery of construction sector

The construction sector recovery depended on the national economic recovery since construction investment relates closely to economic growth specifically in Indonesia. In 1999 when the national economy began recovering, the construction sector increased its growth to −0.8 per cent compared to a negative growth of 36.46 per cent in the year before. In 2000, the growth rate of construction sector was in line (at 6.7 per cent) with the national economic growth that stayed at 4.8 per cent.

This growth in the construction sector was emulated by the then newly revived property businesses, particularly commercial property, which saw a surge in demand for higher-end malls and supermarkets. Investors (mostly from France) took a bold step ahead to occupy the "empty market" of commercial offices and trading space especially in major areas such as Jakarta, Bogor, Tanggerang and Bekasi (JABOTABEK).

A demand for low-cost housing was expected to increase in 2001, along with the government policy to increase construction credit for small- and medium-sized contractors in order to stimulate the trading of small- and medium-sized businesses. The government's emphasis on low-cost housing had been opportune and had also been supported by the improvement of new credits for the construction sector from the commercial banks.

Table 6.7 Indonesian near-term economic restructuring and targets at a glance

Macroeconomic framework and policies	• Maintain a growth target in 2001 of 3–3.5%, inflation 9–11%. • Reduce base money growth to 12% by March 2002. • Review implementation of 2001 budget and framework for 2002 budget in mid-October. • Draft 2002 budget to include a deficit of 2–3% of GDP. • Submit to Parliament the draft Sovereign Debt Securities Law.
Fiscal decentralization	• Transfer to the regions a total of 2.1 million civil servants by end-2002. • Calculate revenue sharing and General Allocation Fund transfers based on original budget estimate. • Use contingency funds only up to the Rp. 3 trillion allocated in the revised budget. • Finalize modalities to issue bonds in last quarter 2001 to resource surplus regions. • Complete audits of the allocation of the contingency funds by end-2001. • Refine the formula for the General Allocation Fund for use in 2002.
Banking system reforms	• Launch Bank Mandiri IPO (up to 30% of shares) by end-2001. • Publish key financial data for individual banks by end-2001. • Adopt action plan to improve supervision, regulations, and governance of NBFIs. • Complete by end-2001 all outstanding issues from BI's 1999–2000 audit. • Resolve BLBI credits issue between BI, GOI, and Parliament by end-2001. • Finalize replenishment of Government Guarantee Scheme by first week of September. • Remove BII's impaired assets and replace with government bonds by mid-September.
IBRA asset recovery and restructuring	• Respond to the OC review of first four large restructurings by mid-September. • Publish next round of ten OC reviews by mid-October. • Launch sale of unrestructured loans by end-2001 using competitive bidding mechanisms. • Discuss with Parliament sale of 51% of BCA to a strategic partner. • Sell majority stake in Bank Niaga by end-2001.
Corporate restructuring and legal reforms	• Restructure a cumulative total of $14–15 billion by end-2001. • Refer from FSPC by end-December 2001 cases with a total debt of $10–11 billion. • Submit an amended version of the Bankruptcy Law to Parliament by end-2001. • The Anti-Corruption Commission to become fully effective in coming months.
Public sector and other structural reforms	• BPKP audit of the Reforestation Fund will be completed by end-2001. • Publish audits and announce corrective actions of key SOEs by end-September. • Publish audit of the tax office by end-September and announce corrective actions by mid-October.

Source: World Bank, 2001.

The construction sector recovery seemed promising as the demand for low-cost housing was relatively high. However, the recovery of the construction sector was also strongly related to growth in other sectors. Table 6.8 shows the economic growth projection of other sectors.

Table 6.8 The GDP projection for sectors other than construction year-on-year

Sector	*Second quarter – 2001*
Agriculture	0.8–1.8
Mining	1.3–2.3
Manufacturing industry	4.5–5.5
Electricity, gas and water supply	6.5–7.5
Construction (mainly housing)	5.5–6.5
Trading	5.8–6.8
Transportation and communication	6.0–7.0
Finance	3.3–4.3
Service	2.8–3.8
Composite	3.8–4.8

Source: Central Bank of RI (Bank Indonesia).

6.2.4 Economic outlook in the next decade

Looking ahead, the economic recovery that began in 1999 should gather pace with continued consumption growth, a modest increase in investment and moderate non-primary exports growth. Manufacturing growth will underpin recovery supported by a return to trend growth in agricultural production. Accordingly an overall GDP growth rate of around 4 per cent in 2000 was expected to rise to 5 per cent in 2001. The budget deficit may remain unchanged over the medium term, but was expected to decline steadily to 3 per cent in 2001 as private sector activity picks up and outlays on subsidies decline. Imports were expected to increase in 2001 as domestic demand increased, lowering the current account surplus ratio to GDP 0.5 per cent in 2001 from 2.2 per cent in 2000. This should, however, be offset by official financial flows and private capital inflows and the import coverage of reserves would remain at about 6 months. Inflation was likely to exceed the government forecast and reach 6 per cent because of increases in wages and salaries, upward adjustments in fuel and electricity prices, and tariffs on rice and sugar. Inflation was expected to decline to below 5 per cent in 2001. With greater macroeconomic stability and adequate reserve cover, the exchange rate was expected to remain stable at around Rp. 7,000–7,500/US$ in the coming years. However, with a strong increase in reserve growth and exports in 2001, it would drop below Rp. 7,000 per dollar. The public debt/GDP ratio was expected to decline somewhat with GDP growth, decline in interest rates and larger asset sales by the IBRA.

The construction sector outlook for 2001–2010 is upbeat. Since the middle of 1997–1998, the construction industry sector in Indonesia experienced a rather tight contraction as a whole. It was then estimated that this contraction would continue in 1999 because it was predicted that significant economic recovery will not occur in 1999. Although the situation at the time in the construction industry was gloomy, the prospects of this industry in Indonesia present a large opportunity when the economy recovers. This opportunity was supported by estimates prepared by ADB based upon some scenarios; a more conservative scenario in 1999/2000 and 2000/2001 in which the growth of construction industry sector remained negative but smaller than that in 1998/1999. The growth of construction industry on a more optimistic scenario for 2000/2001 was 5 per cent. In the conservative scenario, the growth of construction industry sector in 2000/2001 was −2 per cent. The growth rate estimates of the construction sector in the scenarios prepared by ADB showed that the sector still had potential to develop. These estimates can be supported by the fact that Indonesia had extensive areas which still require many infrastructure developments.

The prerequisite to fulfilling those scenarios was political stabilization. This will lead to economic stabilization and strengthen Indonesia's economic bases upon firm foundations and on the right platform. The GDP for the medium term was projected to grow by 4 per cent according to World Bank figures. While still well above recession levels, 3–4 per cent GDP growth is only about half of the 7.2 per cent average GDP growth Indonesia experienced in 1990–1996. More important from the viewpoint of social and political stability, GDP growth of 3–4 per cent does not create enough job opportunities to absorb significant numbers of Indonesia's millions of unemployed and under-employed workers and new labour market entrants. Indonesia's labour force was increasing by 2.2–2.7 per cent a year, a growth rate equivalent to 2–2.5 million new job seekers each year. A substantial proportion of this is the construction-related labour force. The National Development Planning Agency (BAPPENAS) in turn estimated that 4 per cent GDP growth translates into an increase in the demand for labour of 2.4 per cent, or 2.2 million new job opportunities per year. These figures make it clear that in order to re-employ large numbers of workers who lost their jobs during the 1997–1998 crisis and absorb new labour market entrants, Indonesia needed a sustained period of GDP growth of well above 3–4 per cent. Many circles of society in Indonesia began to realise this. There are appeals and policies appearing, aiming to return to the main economic basis (i.e. the agriculture sector), with a broader implication that Indonesia should not to be trapped in the camouflage of a modern economy. If this is realised, the existing potential market for the construction sector will be wider because much infrastructure and structures which are required to support economic mobility will be in greater demand. Table 6.9 shows Indonesian real GDP for Q1 2001.

Table 6.9 Indonesian real GDP for Q1 2001

	Q1	*Share of GDP*
Production category		
Manufacturing	2.9	26.1
Agriculture	2.3	17.2
Retail, hotel, rest	5.0	14.6
Mining	2.9	13.4
Services	1.2	8.7
Finance and leasing	2.9	6.2
Construction	1.2	7.1
Transport and communication	6.9	5.2
Electricity, gas and water	7.0	1.1
Total (categories weighted)	4.01	100.00
Expenditure category		
Household consumption	4.8	66.8
Government expenditure	6.0	6.7
Investment	10.2	24.4
Export	11.6	43.3
Import	34.1	34.5

Source: Central Bureau of Statistics (BPS).

6.3 The construction sector

6.3.1 *Value-added per employee*

Productivity

Published records of construction productivity are not available. However based on the data compiled by the Central Bureau of Statistics on the volume of works completed and made ready for use in the building sector (comprising residential buildings, office buildings, shops and business centres, hotels and inns, cinema and other commercial buildings including factories) from 1995 to 1997 the productivity of field workers was observed to follow a decreasing trend for both ICA (Indonesian Association of Contractors) and non-ICA employees (Table 6.10). Similarly, the value-added per employee during the same period had also decreased by over a quarter of a percentage (Table 6.11).

From the limited observation of a productivity study in 2001, the data for construction workers are as in Table 6.12. However, the data was taken from a fairly

Table 6.10 Productivity of field workers

Square metre per man-day	*ICA employee*			*Non-ICA employee*		
	1995	1996	1997	1995	1996	1997
Residential building	0.510	0.11	0.12	0.27	0.21	0.12
Non-residential building	0.14	0.05	0.26	0.66	0.03	0.26

Note
Figures for 1998 onwards not available.

Table 6.11 Value-added per employee in the construction sector

	1995 (US$)	*1996* (US$)	*1997* (US$)	*1998* (US$)
Value-added per employee	16,869	15,726	11,580	n.a.
% change	–	6.78%	26.36%	n.a.

Table 6.12 The productivity data of construction workers

No.	*Trades*	*Qualification*	*Productivity*		
			Unit	*per day*	*per hour*
1	Masonry man	Bricks and stones layers	m^2	5.24	0.66
2	Masonry finisher	Brick wall finisher	m^2	7.62	0.95
3	Carpenter	• Door, window wood	m^2	0.07	0.01
		frame	m^2	26.0	3.25
		• Instalment of ceiling	m^2	6.0	0.75
		• Installation of asbestos ceiling	m^2	0.77	0.10
		• Wooden roof truss			
4	Steel binder	Bar binding, stirrup hook, etc.	kg	243	30.38

Source: Indonesia Country Report (2001).

limited area and from among a limited set of construction traders. The government of Indonesia, with technical assistance from the World Bank, planned to conduct a comprehensive study in 2002 and which was expected to be completed within one year. The main objectives of the study were:

- developing a comprehensive "map" of job types in the construction industry, at all levels of management, its competency requirements, relationship among jobs and comparison with similar "maps" of neighbouring countries; and
- developing methods, instruments and programmes for improving construction worker productivity with special reference to the level of ASEAN countries.

The physical measurement of construction productivity (production) is included in Tables 6.A.1 and 6.A.2 of Appendix each representing volume of works finished and ready for use during 1995–1997 and man-days of part-time field workers respectively.

6.3.2 Construction cost

Unit construction cost

During the second half of 1990s, the unit construction cost in Indonesia had shown no significant changes as is evident from Table 6.13 for all types of development. Any increased costs seemed to have been offset by the general slowdown of the economy and the final figure is balanced out. However, it is important to be aware that the huge depreciation of the rupiah during the same period and particularly in the after-years of the financial crisis understandably made a lot of difference especially for the luxury and higher-end buildings depending on the amount of materials imported from overseas. However, falling labour costs may have balanced the increased material costs. The impact of this had however not been widely studied.

Average construction material price

The average prices of building materials fluctuated following the prices of other sector products and wider changes in the economic structure in the country. It was seen that building and construction material prices that rose steadily until 1996 drastically fell by over 50 per cent in 1997 in all sectors. The prices had since gradually recovered

Table 6.13 Unit construction cost (US$/Sq. m of gross floor area)

Year	*Residential (high rise)*		*Commercial office*		*Industrial*		*Hotel*
	Standard	*Luxuries*	*Standard*	*Prestige*	*Light*	*Heavy*	*Star*
1995	450	550	400	600–550	400	500	500–550
1996	450	550	400	600–550	400	500	500–550
1997	450	550	400	600–550	400	500	500–550
1998	450	550	400	600–550	400	500	500–550
1999–2000	450	550	400	600–550	400	500	500–550

Source: BPS, Indikator Ekonomi.

Table 6.14 Price index of building and construction materials

Type of structures	*1995*	*1996*	*1997*	*1998*	*1999*	*2000*	*2001**
Residential and non-residential buildings	248	265	128	219	245	269	295
Agricultural structure/irrigation	264	285	131	193	222	250	284
Roads, bridge and harbour	256	275	132	196	223	252	288
Electrical, gas, water supply and communication	219	226	118	215	240	256	281
Others	244	258	131	208	236	258	284

Source: BPS, Indikator Ekonomi.

Note
* May 2001.

and by 2001 had surpassed the peak levels of 1996. The statistic of building materials are as shown in Table 6.14.

Construction industry salaries and wages

The statistical data on salaries and wages in the construction industry are not available specifically for 1998–2000 but minimum wages that had been established by the Ministry of Labour were as follows:

- Helpers Rp. 15,000/day
- Specific workers Rp. 30,000/day
- Foreman Rp. 50,000/day

The salaries for a manager set up by the labour market in Jakarta and other big cities in Java Island are as follows:

- Site Manager Rp. 1,000,000/month
- Middle Manager Rp. 2,000,000–3,000,000/month
- Executive Director Rp. 6,000,000–10,000,000/month

The variation of salaries and wages were significantly influenced by location, contractor-company size and also the productivity or performance of the workers.

6.3.3 *Construction quality*

The government of Indonesia was concerned about the quality of the works carried out by some 65,000 local construction firms. For these, the government was considering the introduction of a certification mechanism, both individual and institutional for the construction industry. Puslajakon (Education Centre for Construction Services) under the human resources and regional development department was responsible to see this through. The government also hoped to shift the training and certification responsibility to the appointed independent institutions as part of the build-up in construction industry systems in the near future.

The number of large firms adopting the ISO 9000 quality management system had been increasing. By 2000, at least 30 construction firms had obtained ISO 9000 certification from various certification bodies such as: Lloyd's Register, Sucofindo, SGS, etc. The government was trying to introduce the ISO 9000 quality management system for smaller enterprises at affordable cost by providing special concessions.

6.3.4 Construction safety record

Safety management in construction has a legal basis by virtue of Law No. 1, Year 1970 on safety management. The construction sector is also regulated by the Ministry of Manpower No. PER. 01/MEN/1980. In general, safety had been increasingly assuming a larger importance particularly when there are foreign firms involved in construction. Accurate records on accident frequency rates, however, are not available. Based on the only record maintained by the Ministry of Manpower it is observed that both accident frequency and fatality rates increased between 1998 and 1999. In 1998, there were 648 recorded accidents with 641 injuries and 20 deaths. In 1999 there were 723 recorded accidents with 699 injuries and 24 deaths.

6.3.5 Skill level of workforce and supervision level

The labour force in the construction industry had not been classified by skill level but by formal education instead. Table 6.15 gives a general overview of the labour force for the year of 2000.

In 1995, almost 70 per cent of the total workforce were skilled and unskilled labourers, and around 15–25 per cent were supervisors. However, there had been a steady decrease in the quantity of unskilled labour. In 1998, the number fell by 19 per cent although this group constituted 80 per cent of the country's total work force (Table 6.16).

Table 6.15 Construction labour force by formal education for the year of 2000

No.	*Occupancy*	*Formal education*	*Amount*
1	Workers	Not educated	94,481
		Not graduated	372,184
		Primary school	1,497,252
		Junior H.Sc.	765,357
2	Foreman	High school	421,460
		Technical H.Sc.	217,749
3	Supervisor/superintendent	Academy/D3	43,599
		University	85,150
	Total		3,497,232

Source: BPS, Labour Force in Indonesia, August 2000.

Table 6.16 Number of unskilled labourers

Rank	*Education*	*1996*	*1997*	*1998*	*1999*	*2000*
Manager and officials	Bachelor's degree	29,418	44,115	27,450	21,367	22,669
	University	68,383	72,798	74,361	66,523	69,385
	Subtotal	99,797	118,910	103,809	89,889	92,054
Sales or clerical	Senior high school	318,162	344,585	334,086	306,905	314,974
	Senior high school +	7,400	7,458	11,168	3,571	4,473
	Subtotal	325,562	352,043	345,254	310,476	319,447
Professional and technical	Senior high school	262,607	265,005	242,443	225,555	227,615
	Junior high school	77,551	96,503	96,025	103,626	107,194
	Subtotal	340,158	361,508	338,468	329,181	334,809
Others		3,032,707	3,369,736	2,736,149	2,687,600	3,175,807
Total		3,798,224	4,203,197	3,523,680	3,417,146	3,922,117

Source: BPS, Labour Force in Indonesia, August 2000.

6.3.6 *Research and development*

Research and development (R&D) in the Construction Sector is partly done through government research offices such as the Research Institute within the Ministry of Public Works in Bandung which consists of R&D for Water Resources, R&D for Highways and Bridges and R&D for Urban Settlement. By Law No. 18/1999 (for Construction Service) the Indonesian Construction Industry Development Board (LPJK) is also responsible to extend the R&D initiated by the government organizations.

The output of research and development in the past years had not been very substantial. A technology innovation called "Hydraulic Non-Friction Rotating Device" or "Sosrobahu A.S.B. & O" which can be used to rotate a heavier load for flyover pier construction has not developed significantly during the past 2 years, although the device had been put to use by some large contractors.

Buildability

The system of contracting in Indonesia consists essentially of four types, which are:

1 Construction only (traditional procurement system)
2 Design–Build
3 Finance–Design–Build
4 Build–Operate–Transfer

The buildability factor had been enhanced by slow but wider recognition of the design and build approach. The concept had hitherto been little applied due to the fact that contractors require a proper design network. The construction of various government offices and private commercial building and mixed buildings is usually done through a Design–Build approach where engineering detail design is carried out by the contractor, and basic design by the designer or planner. For process plant investments, the engineering procurement and construction (EPC) system has been more widely practised.

Environmental planning

According to the Indonesia Law No. 23/1997 concerning environmental management, project activities which have significant impact on the environment should include the Environmental Impact Assessment (EIA) document. The EIP system was directed by the Government Decree No. 29/1999. The EIA study which is conducted at the project feasibility stage, should take into consideration spatial planning, protected areas, historical buildings, sensitive areas, rare and protected biological species, natural resources and potential and socio-economic cultural society around the project site. EIA recommends significant impact mitigation through the EMaP and EMoP (Environmental Management Plan and Environmental Monitoring Plan). In addition, EIA can also recommend changes to project location or cancellation of project activities if significant impacts cannot be mitigated by technological, economical and institutional approaches. Sustainable development requirements, EMaP and EMoP should be specified in working drawings and technical specifications, and be implemented during construction. Due to the complexity of construction works in Indonesia at the moment, the EIA system should ideally be supported by the EMS, which is based on ISO 14000, where environmental documentation, recording and auditing are carried out to assure environmental conservation.

As part of the implementation of the Indonesian Law No. 23/1997 for environmental management, and the Government Decree No. 29/1999, some large contractors were undergoing the process of ISO 14000 certification by the International Registrar. It was expected that with the establishment of AFTA, many other large contractors will have been certified and registered.

6.4 Construction outlook 2002–2010

6.4.1 Business opportunities and major projects in the pipeline

It was expected that with the right policy framework and a strong commitment to reform, Indonesia will be able to capitalize on its fundamental economic strengths to restore investor confidence. The nation offers a large domestic market and a correspondingly large workforce, abundant natural resources, reasonably modern telecommunications and other infrastructure, a strategic location along some of the world's major trade routes, and substantial experience with market-based economics and the international trade and payments system. Nevertheless, the Indonesian investment climate has suffered from the combination of a worldwide economic slowdown and increasing domestic political instability. Investor confidence remains depressed, with existing and potential investors citing a number of concerns: political uncertainty, the unknown impact of political and fiscal decentralization, uneven implementation of economic reform commitments, the unreliable judicial system, security issues and treatment of existing investors. However, the long-term outlook remained positive with the new government policy to encourage private sector-led growth and foreign investment. In 1998 and 1999, the GOI issued several new regulations to ease the entry of foreign firms and capital into Indonesia. However, the Foreign Capital Investment Law of 1967, which provides the basic framework for foreign investment, is still in effect. Representatives of several ministries were working together to draft a revised Foreign Capital Investment Law that may go into effect as early as 2002. The law had been under revision for almost 3 years and its reform comprises one of the objectives of the GOI's IMF-supported economic reform programme.

Investment in Indonesia is categorized as either domestic (PMDN) or foreign (PMA). An investment with any degree of direct foreign ownership is defined as PMA. The Capital Investment Coordinating Board (BKPM) now subsumed under the Board of Investment and State-Owned Enterprises (BPM-PBUMN) plays a key role in promoting foreign investment and approving project proposals. The relevant technical government departments handle investments in the oil and gas, banking and insurance industries. BKPM, or the corresponding provincial board (BKPMD), approves foreign and domestic investment in all other sectors.

Economic growth is the prerequisite for the growth of the construction industry. With the economic outlook expected to improve, the construction industry was likely to be more stable. The enormous need for the country to upgrade its infrastructure will fuel construction-related growth in the next decade. There is no information as such on projects in the pipeline but the construction industry is likely to be predominantly infrastructure and residential led.

6.4.2 Areas where foreign investments were needed

Foreign investment approvals reached almost US$ 15.5 billion, 41 per cent higher than 1999 in the year 2000. According to BKPM statistics covering 1 January 2001 through 1 March 2001, foreign investment approvals rose by almost 276 per cent, rising from US$ 810 million in the same period in 2000 to US$ 3 billion in 2001. The value per project rose from US$ 2.5 million to US$ 10.7 million during the same period.

For the first quarter of 2001, the chemical and pharmaceutical sectors (approvals valued at about US$ 1.5 billion) were the most attractive for foreign investment applications, followed by the basic metal, engineering, and electronics sectors (US$ 304 million), textile sector (US$ 204 million), hotel and restaurant (US$ 171 million), and rubber and plastics (US$ 128 million). There is however no breakdown on construction sector foreign investments. Through the same period, Japan led in approved foreign investment applications (2 very big projects worth about US$ 1.5 billion), followed by Malaysia (19 projects worth US$ 449 million), Japan (22 projects worth US$ 278 million), Singapore (61 projects worth US$ 200 million, the United Kingdom (35 projects worth US$ 184 million), and the United States (7 projects worth 27 million) (source: http://www.usembassyjakarta.org/econ/investment2001-3.html#F). Tables 6.17–6.19 show the net capital inflows for Indonesia, foreign investment approvals and the top-ten sectors by value per year respectively.

Table 6.17 Indonesia: net capital inflows (US$ billions)

	1993	*1994*	*1995*	*1996*	*1997*	*1998*	*1999*	*2000*[2]
Private	5.2	3.7	10.3	11.5	−0.4	−13.8	−9.9	−8.5
FDI[1]	2.0	2.1	4.3	6.2	4.7	−0.4	−2.7	−4.1
Other	3.2	1.6	5.9	5.3	−5.0	−13.5	−7.2	−4.4
Official	12.8	0.3	0.3	−0.5	2.9	10.0	5.4	3.8
Total	18.0	4.0	10.6	11.0	2.5	−3.9	−4.5	−4.7

Source: Bank Indonesia.

Notes

1 Foreign direct investment.

2 Preliminary data.

Table 6.18 Foreign investment approvals, 1996–2000

Year	*No. of projects*	*Total value* (US$ million)	*Average value/ project* (US$ million)
2000	1,508	15,413	10
1999	1,164	10,891	9
1998	1,035	13,563	13
1997	790	33,833	43
1996	959	29,929	31

Table 6.19 Top-ten sectors by value per year, 1996–2000

Sector	*Project*	*Total value* (US$ million)
2000		
Chemicals	89	7,448
Trade	499	1,441
Transportation/telecom	61	1,218
Metal goods	132	1,005
Basic metal	8	824
Food industry	39	701
Textiles	107	401
Food crops	17	311
Real estate	21	302
Hotel/restaurant	48	260
1999		
Chemicals	75	3,266
Electricity/gas/water	2	2,310
Paper	15	1,412
Food industry	48	681

Metal goods	85	593
Basic metals	9	501
Plantations	17	284
Trade	348	279
Textiles	121	240
Food crops	10	224
1998		
Chemicals	73	6,173
Electricity	6	1,795
Real estate	19	1,271
Metal goods	119	891
Plantations	22	725
Hotel/restaurants	56	451
Basic metals	13	394
Food industry	32	342
Non-metal minerals	15	237
Food crops	10	224
1997		
Chemicals	93	12,563
Transportation/telecom	36	5,900
Paper	14	5,353
Metal goods	190	2,332
Electricity/gas/water	8	1,840
Non-metallic minerals	17	1,457
Real estate	20	1,394
Food industry	26	573
Textiles	5	373
Food crops	3	234
1996		
Chemicals	91	7,361
Electricity/gas/water	8	3,809
Metal goods	186	2,939
Paper	16	2,907
Real estate	35	9,628
Hotels/restaurants	34	1,717
Mining	4	1,697
Plantations	18	1,168
Non-metallic minerals	28	790
Food industry	37	691

Appendix: Physical measurement of construction productivity (production)

Table 6.A.1 Volume of works finished and ready for use 1995–1997

Type of construction work	*Unit*	*1995*		*1996*		*1997*	
		ICA member	*Non-ICA member*	*ICA member*	*Non-ICA member*	*ICA member*	*Non-ICA member*
Residential	m^2	1,453,748	5,464,576	7,491,925	3,113,084	1,213,205	3,635,626
Commercial	m^2	8,575,736	3,505,530	4,665,690	1,712,730	3,923,924	19,266,751
Office building	m^2	6,889,212	1,543,927	2,592,590	607,247	1,926,888	1,630,153
Shops and business centres	m^2	846,724	0	638,430	0	948,918	0
Hotel and inn building	m^2	580,960	0	1,380,620	0	743,321	0
Cinema building	m^2	0	0	0	0	0	0
Other building	m^2	258,840	119,735	54,050	89,872	304,797	219,929
Other commercial building	m^2	0	1,841,868	0	1,015,611	0	17,416,669
Industrial (Factory)	m^2	767,364	0	3,106,120	0	14,712,730	0
Total		10,796,848	8,970,106	15,263,735	4,825,814	19,849,859	22,902,337

Source: BPS, Indikator Ekonomi.

Table 6.A.2 Mandays of part-time field workers

Type of construction work	*1995*		*1996*		*1997*	
	ICA member	*Non-ICA member*	*ICA member*	*Non-ICA member*	*ICA member*	*Non-ICA member*
Residential*	14,940,576	20,382,440	67,000,608	15,083,238	10,029,254	11,858,592
Non-residential**	65,658,594	58,624,867	142,286,323	49,830,904	72,278,924	32,186,956
Productivity						
Residential	0.10	0.27	0.11	0.21	0.12	0.31
Non-residential	0.14	0.06	0.05	0.03	0.26	0.60

Source: BPS, Indikator Ekonomi.

Notes
* In terms of mandays.
** In terms of m^2 of construction/manday.

References

ADB (2000) Country Economic Review, available at http://www.adb.org/Documents/CERs/CER_INO_1999.pdf.

ADB (2001) Indonesia GDP Growth to Ease to 3.2 per cent in 2002, http://www.adb.org/Documents/News/2001/nr2001153.asp.

Arditi, D. and Mochtar, K. (1996) Productivity improvement in the Indonesian construction industry, Construction Management and Economics, vol. 14: 1.

Bank Indonesia, Annual Report (2000) http://www.bi.go.id/bank_indonesia_english/

Economist (2001) Indonesian Economy: In the Dream time, 8–14 December 2000, http://www.usembassyjakarta.org/econ/investment2001-3.html#F.

Indonesia Country Report (1997) *Proceedings of AsiaConstruct3*, Hong Kong.

Indonesia Country Report (2001) *Proceedings of AsiaConstruct7*, New Delhi, India.

Soesastro, A. (2000) The Economic Crisis in Indonesia: Lessons and Challenges for Governance and Sustainable Development, available at http://www.pacfic.net.id/pakar/hadisuastro/economic.html.

World Bank (2001) Country Brief: Indonesia, http://wbln0018.worldbank.org/eap/eap.nsf/Attachments/wb&ind/$File/wb&ind.pdf.

Chapter 7

Japan

Research Institute of Construction and Economy, Japan, M. Anson and J. Raftery

7.1 The country

Japan, with its total area of 377,837 square kilometres, is about the same size as Germany, Finland, Vietnam or Malaysia. It is about 1/25 the size of the US and is smaller than the state of California. The country has four major islands: Hokkaido, Honshu, Shikoku and Kyushu. Japan is a land of volcanoes. Almost one-tenth of the world's approximately 840 active volcanoes are in Japan, altogether it has only about 1/400 of the world's land area (Ministry of Foreign Affairs website: http://www.mofa.go.jp).

7.1.1 Government

Japan has a constitutional monarchy with a parliamentary government. The website (http://www.kantei.go.jp) of the Prime Minister of Japan and his Cabinet describes the Emperor as "symbol of Japan and of the unity of her people". Whilst the Emperor retains his throne as a symbol of national unity, he has no administrative powers ever since the end of the Second World War. Instead the Emperor performs largely ceremonial acts including the appointment of the Prime Minister and the Chief Judge of the Supreme Court, respectively designated by the Diet and the Cabinet. The Diet comprises the House of Representatives and the House of Councillors, that are the "highest organ of state power and the sole law-making organ of State" respectively. Both Houses have similar power except for exceptional circumstances when the House of Representatives could override decisions of the House of Councillors. For adults of 20 years of age or more, universal suffrage is guaranteed by Article 15 of the Constitution of Japan.

The Cabinet, vested with the highest executive power, consists of the Prime Minister, who must be a civilian and is designated from the Diet by a resolution of the Diet. The Prime Minister appoints and dismisses the Ministers at his discretion. One of the Ministers is that of Land, Infrastructure and Transport, overseeing inter alia construction matters. The Ministry was established in 2001 when the Ministry of Construction was merged with the Ministry of Transport. On the other hand, highest judicial power is vested in the Supreme Court. Inferior courts include the High Courts, the District Courts, the Family Courts and the Summary Courts. The Chief Judge is appointed by the Emperor, whilst other judges of the Supreme Court are appointed by the Cabinet. The Legal system is modelled after European civil law system with English–American influence (CIA World Factbook).

The Capital is Tokyo, overseeing the administrative divisions in the 47 prefectures of Aichi, Akita, Aomori, Chiba, Ehime, Fukui, Fukuoka, Fukushima, Gifu, Gumma, Hiroshima, Hokkaido, Hyogo, Ibaraki, Ishikawa, Iwate, Kagawa, Kagoshima, Kanagawa, Kochi, Kumamoto, Kyoto, Mie, Miyagi, Miyazaki, Nagano, Nagasaki, Nara, Niigata, Oita, Okayama, Okinawa, Osaka, Saga, Saitama, Shiga, Shimane, Shizuoka, Tochigi, Tokushima, Tokyo, Tottori, Toyama, Wakayama, Yamagata, Yamaguchi and Yamanashi.

7.1.2 Economy

Japan is the second most technologically powerful economy in the world after the US and third largest economy in the world after the US and China, according to the World Factbook published by the Central Intelligence Agency of the US. The country has been characterized by government-industry cooperation, and by the collaboration of business units including manufacturers, suppliers, distributors and bankers in closely knit groups, better known as "keiretsu". Underpinned by the provision of lifetime employment, "keiretsu" has formed the basic feature of successful Japanese business. As reported by the CIA World Factback, overall real economic growth had been spectacular. Average growth was 10 per cent in the 1960s, 5 per cent in the 1970s, and 4 per cent in the 1980s (see Table 7.1).

Yet, there was excessive speculation in both the stock and property markets. Asset prices surged and large amounts of loans were drawn for financing deals which, with the benefit of hindsight, were speculative. Bubble finally burst and ever since the late 1980s, economic growth slowed down and eventually there were years of economic stagnation and recession. Fiscal and monetary measures were implemented to stimulate the economy and to reverse the deflationary vicious circle, but there was not too much success. It was not until 2003 that the economy began to send clear signals of solid recovery. During the 1990s, which many call the "lost decade", there were re-structuring and turnaround of companies and their associated supply chains. Attempts were also made to reform the banking and capital sectors. After years of endeavour, there are fewer but stronger Japanese banks and many Japanese manufacturers have regained their lost positions. However, with the encouraging signs of business recovery come the eroding features of "keiretsu" and its lifetime employment.

In 2001, Japan had a GDP of 3.45 trillion, or GDP per capita of US$27,000, on the basis of purchasing power parity, and a real GDP growth rate of –0.3 per cent. The services sector contributed the most: about 62 per cent of the GDP, whilst the industry some 36 per cent and the agriculture sector a mere 2 per cent (CIA World Factbook).

The exports and imports of Japan were estimated to be US$404.6 billion and US$331.6 billion respectively in 2001 by CIA in its World Factbook. Major exports comprised motor vehicles, semiconductors, office machinery and chemicals. About 30 per cent were exported to the US, with another 26 per cent to the four economies of Taiwan, South Korea, China and Hong Kong. Major imports comprised fuels, foodstuffs, chemicals, textiles and office machinery. About 19 per cent came from the US, 15 per cent from China, and another 18 per cent from the four economies of South Korea, Taiwan, Indonesia and Australia.

Table 7.1 Economic growth, 1983–1988

	1983	*1984*	*1985*	*1986*	*1987*	*1988*
Total (¥ bn)						
GNP at (2002) prices	280,568	298,453	317,441	331,253	345,292	366,566
GNP at (1980) prices	264,704	278,140	291,807	299,024	312,370	330,208
Real change (%)	3.2	5.1	4.9	2.5	4.5	5.7
Per capita (¥ '000)						
GNP at (2002) prices	2,353	2,487	2,627	2,727	2,828	2,982
GNP at (1980) prices	2,220	2,317	2,416	2,461	2,559	2,686
Real change (%)	2.5	4.4	4.3	1.9	4.0	5.0

Source: Economic Planning Agency, *Japanese Economic Indicators Quarterly*; Management and Coordination Agency, *Monthly Statistics of Japan*.

7.1.3 *Agriculture, energy and minerals*

The relatively very tiny agricultural sector is highly subsidized and protected, with crop yields among the highest in the world. Japan usually produces slight surplus of rice but imports many agricultural products substantially from the US. In fishery, Japan maintains one of the world's largest fishing fleets and accounts for nearly 15 per cent of the global catch (CIA World Factbook). Japanese industries are heavily dependent on imported raw materials and fuels. Japan's dependence on imported agricultural products, raw materials and fuels is reflected on the US, Indonesia and Australia being her principal importers.

7.1.4 *Transportation*

In Japan, there was altogether more than 1.1 million kilometres of highways in 2000, of which more than 0.86 million kilometres was paved including some 6,000 kilometres of expressways. Meanwhile, the renowned Japanese railways stretched more than 23,000 kilometres long, of which almost 16,000 kilometres was electrified (CIA World Factbook). Her experience in infrastructure and railway construction back home goes a long way to explain her dominant position in international construction around the globe.

7.1.5 *Labour*

According to the CIA World Factbook, Japan had a population of almost 127 million in July of 2002 with a growth rate of 0.15 Per cent. About 18 per cent of the population was 65 years old and over, whilst almost 68 per cent was between 15 and 64 years of age. Labor force in 1999 was about 67.76 million. Around 65 per cent of the force was engaged in trade and services, 30 per cent in industry with the remaining 5 per cent in agriculture, forestry and fishing. A comparison of the composition of the labour force with that of the GDP suggests that the industry sector had a slightly higher labour productivity than both services and agriculture. The services sector altogether deployed some 30 per cent of the labour force but contributed more than 30 per cent (that is 36 per cent) of the GDP. It probably also reflects the superiority of the Japanese industry sector. As the CIA World Factbook suggests, Japanese producers are among the world's largest and technologically advanced in the production of "motor vehicles, electronic equipment, machine tools, steel and nonferrous metals, ships, chemicals, textiles, processed foods".

Among the Japanese "labour" are robots. Japan has 410,000 of the world's 720,000 "working robots", which collectively constitute a "key long-term economic strength" (CIA World Factbook).

7.1.6 Japanese economy as of 2002

The Japanese economy was in the midst of a prolonged and serious recession. Japan posted a 0.7 per cent growth in FY 1997, the country's first negative economic growth rate in 23 years until 2002. The unemployment rate had hit an all-time high of 4.3 per cent. A major cause had been a tardy response to the weakening balance sheets and declining profitability that financial institutions were suffering as a result of the collapse of the bubble economy.

Banks took a very cautious approach to lending as they restructured their assets to seek improved profitability in preparation for Japan's "Big Bang" financial reforms. The resulting credit crunch had an adverse impact on the real economy, especially on small- and medium-sized business.

Economic growth in fiscal 2000 was a positive 0.9 per cent. Having climbed from negative growth in fiscal 1998 to positive 1.4 per cent growth in fiscal 1999, the Japanese economy seemed to have found a path towards recovery, but dark clouds of difficulty had already began to threaten future progress. Having decided on a quantitative financial deregulation policy in March 2001, the Bank of Japan had diligently hammered out policies for propping up the economy, which was facing decreasing exports, through deregulation measures that provide large volumes of cash to short-term financial markets. But as of 2002, the tremendous volume of non-performing loans left over from the collapse of the bubble economy had hindered the ability of the Japanese economy to resume growth.

On 21 June 2001, the government's Council on Economic and Fiscal Policy formally decided some basic policies for economic and fiscal management aimed at reviving the Japanese economy (Outline of Basic Policies for Macroeconomic Management and Structural Reform of the Japanese Economy). Based on these, the government had revised the fiscal 2002 budget to include a policy for keeping new government bond issues to a maximum of ¥30 trillion as a first step towards regaining financial health. It was decided to grapple with making annual expenditures more efficient by heavily allocating funds to seven sectors which would have a significant effect on stimulating consumer demand, such as the environment, urban redevelopment, and information technologies, and cutting public works investments and support for local governments and special corporations.

On a more fundamental, long-term level, a number of factors were combined to erode the confidence of the Japanese people. The more important factors include the fact that the Japanese economic system had become outmoded, and the governments worsened budget deficit.

In the Monthly Economic Report for March 2002, the Japanese government revised its assessment of the then state of the Japanese economy upwards for the first time in 21 months, saying that "while the economy continues to remain in a difficult situation, signs of bottoming out can be seen in some areas." The term "deterioration" was not used in the assessment of the then state of the economy for the first time in 10 months. Later, in the Monthly Economic Report for May 2002, the government explicitly stated that the projected bottoming out had occurred, saying that "while the economy continues to be in a difficult situation, it has bottomed out." The prolonged stagnation in the Japan's economy seemed finally to have reached its twilight period thanks to increasing exports.

In fiscal 2001, Japan's GDP was ¥497.2 trillion in nominal terms, causing the real economic growth rate to fall to −1.7 per cent, the first negative growth rate in 3 years and the largest negative growth rate since fiscal 1980. This has a result of the downward pressure on private demand caused by the country's bad debt problems, excessive debt, tight employment conditions, and uncertainty over the sustainability of the nation's fiscal and social security systems. Faced with the need to overcome these problems, the Koizumi Cabinet formed in April 2001 had promoted structural reforms spanning every sector – from economic and fiscal policy to government administration and social reform – under the principle that "there can be no growth without reform."

Due to active fiscal investment and large tax reductions implemented as part of the government's economic policy, however, and to reduced tax collections caused by the recession, the national government's outstanding debt as of the end of fiscal 2001 (the total of the government bonds, loans payable, and government short-term securities, or financing bills) was ¥607.3 trillion, exceeding the past record high of ¥600 trillion. This figure far exceeded Japan's GDP (nominal ¥497 trillion in fiscal 2001). Over the mid- to long term, this enormous amount of outstanding debt was expected to cause economic distortions, thus making the task of fiscal policy reform, along with economic recovery, a major challenge for the Japanese economy.

The Koizumi Cabinet had advanced wide-ranging structural reforms, starting with the "Basic Policies for Macroeconomic Management and Structural Reform of the Japanese Economy" (the "Honebuto Policy"), while adopting appropriate and flexible policy measures to deal with the economic and employment situations, including the adoption of the "Front-Loaded Reform Program" and the "Immediate Action Program for Structural Reform." Furthermore, the Cabinet adopted the "Structural Reform and Medium-Term Economic and Fiscal Perspectives" in January 2002, outlining the kind of economy and society Japan was aiming to be, and provided an outlook for achieving this through economic and fiscal management based on steady growth driven by private-sector demand and an efficient and sustainable fiscal structure. Following this, in June 2002, the Cabinet adopted the "Basic Policies for Economic and Fiscal Policy Management and Structural Reform 2002" which includes "economic revitalization strategies", including the establishment of the special structural reform areas, "tax reform" based on the structure of incidence of the income tax, resident's tax, and corporate tax, and "reforms in major government expenditures" expressed as a commitment to "small government". To overcome deflation, the most significant problem facing economic management over the past few years until 2002, the government and Bank of Japan worked together in their efforts to implement these measures as quickly as possible.

7.1.7 Economy in the first half of 2002

The real GDP announced on 7 June 2002 for the January to March period 2002 was 1.4 per cent over the previous period. Exports have increased because of the recovery of the US and Asian economies from the IT recession and the terrorist attacks of September 11 2001 and individual consumption, which accounts for about 60 per cent of GDP, increased for the second consecutive quarter, rising 1.6 per cent over the previous period. Against the backdrop of a declining desire to invest among businesses, however, capital investment fell by 3.2 per cent from the previous period, and residential investment fell by 2.3 per cent, the second consecutive quarter of decline for both types of investment.

Capital investment has continued to fall since 2001 due to falling production and corporate earnings. According to the "Financial Statements of Corporations by Industry", which provide demand-side statistics, the decline had continued since the January to March period of 2001. "Capital Goods Deliveries", the supply side statistics of machinery and capital investments, had also shown a decline since 2001. Machinery orders, a leading economic indicator, were expected to show a decline for the April to June period of 2002, continued declines were expected in both the manufacturing and non-manufacturing industries. Residential construction was weak due to a low inclination to consume caused by a deteriorating income and employment situation. Public works investments were also at low levels overall since the national government's public works investment expenditures fell 10.7 per cent in the fiscal 2002 budget and expenditures on independent projects undertaken by local governments fell 10.0 per cent. Exports, on the other hand, had received a lift by a worldwide recovery in production, and were increasing overall due in part to increases in electrical equipment such as semiconductors and other electronic parts. Individual consumption had remained fairly level, and seemed to have bottomed out in some sectors. While there continues to be weak upward movement in incomes, it does not amount to a general improvement in the situation. However, since there has been some improvement in consumers' desire to spend, there have been increases in some industries and expenditure categories. This data indicated that the Japanese economy in the first half of 2002 had found a path towards an export-driven recovery.

7.1.8 Five-year outlook for the Japanese economy

With the next 2–3 years, since the time of writing, having been designated as a concentrated adjustment period, the Japanese government is making efforts to see bad loans written off and to undertake extensive structural reforms in the fiscal and regulatory structures.

However, the outlook for the Japanese economy even following these reforms is not necessarily rosy. The ageing of society and a shrinking population are expected to put some constraints on the Japanese economy in the near future. According to studies conducted by the National Institute of Population and Social Security Research, the population of Japan, which was 126.9 million in 2000 (National Census), is going to increase moderately until it hits its peak in 2006 of 127.7 million. After that, there will be a long-term population decline. At the same time, the ageing of society will progress at an accelerated pace, such that the population of senior citizens (aged 65 and older) will continue to increase rapidly from the current level of 22 million people to over 30 million by 2013 and 34.2 million by 2018. The ageing of society will inevitably increase the burden on working people in terms of taxes and social security. The savings rate was also expected to decline, thereby reducing the society's overall resources available for investment. In 2010, the pace of population decline will still be moderate, but if no extraordinary technological innovations or new means of increasing productivity are unearthed, the Japanese economy will have a difficult time getting back on a track of positive growth over the mid- to long term.

The ageing of society will result in an increased reliance on fiscal measures, especially social security expenses, and will make fiscal reform even more difficult. Once the economy gets back on the track of autonomous recovery, fiscal reform must urgently be promoted by rationalizing and streamlining government administration.

On the other hand, the globalization of the world economy is also going to have an impact on Japan's future economy. In a post-Cold War environment where most countries have adopted a market economy and the information revolution has intricately affected the global market, Japan has been able to retain its high level of international competitiveness. This has made it essential to create a framework for implementing deregulation, promoting the IT revolution, and building infrastructures for transportation and telecommunications networks. In terms of its role within Asia, Japan needs to promote technology transfers to other countries and the liberalization of the Japanese market, while also playing a more active role in pursuing economic ties with its Asian neighbours.

Another imminent concern is the increased impact that global environmental problems are going to have on the economy. Cross-national efforts in areas such as global warming have already begun, but Japan is pouring its efforts into creating an international framework for dealing with environmental problems and providing government development assistance in the environmental sector. Domestically, the government is promoting a shift to a recycling-oriented social economy that addresses waste management and recycling issues.

As can be seen from the above, the outlook for the Japanese economy over the next 5 years since 2002 is not one of unbridled optimism, but instead one that recognizes the various problems that remain to be addressed. Increased wisdom and effort will be required if the Japanese economy is to continue to thrive now that it has shifted from a growing to a mature economy.

7.1.9 The construction industry as of 2002

Construction investment, at approximately ¥75 trillion, accounted for roughly 15 per cent of the nation's GDP, while the construction industry's 6.85 million employees represent more than one-tenth of the nation's labour force. Construction firms face an extremely adverse business climate. Construction investment was slumping (see Figure 7.1), and the construction market had undergone major structural changes. At the same time, orders were declined and profitability was sagged. Private construction investment continued to remain in the doldrums. The private sector had shown little activity in the way of capital investment, and housing starts were down.

As for public investment, the government's main budget was drastically reduced in an effort to lower the fiscal deficit. Later, however, economic stimulus measures were adopted which will probably push public investment to a level higher than that for the previous fiscal year.

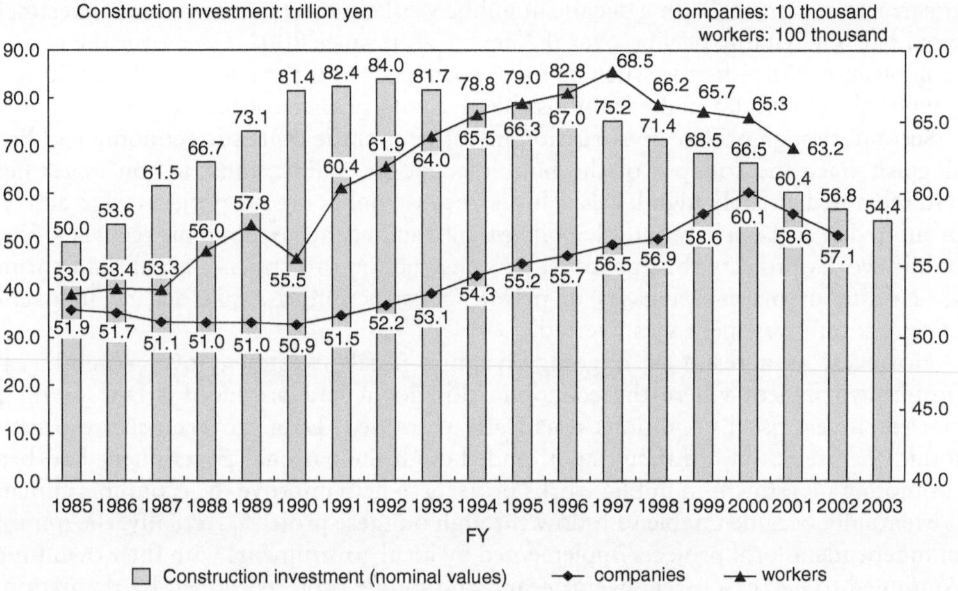

Figure 7.1 Trends in construction investment (nominal values), the number of licensed construction companies, and the number of workers in the construction industry.

Sources: FY 2002 Construction Investment Forecast, Ministry of Land, Infrastructure and Transport; Construction Economy Forecast (May 2002), Research Institute of Construction and Economy; Survey of the Number of Licensed Construction Companies, Ministry of Land, Infrastructure and Transport; *Labour Force Survey*, Ministry of Public Management, Home Affairs, Posts and Telecommunications.

Notes

1 Construction investment values are calculated based on figures published by the Ministry of Land, Infrastructure and Transport up through fiscal 2001. Figures after fiscal 2002 are based on estimates issued by Research Institute of Construction and Economy. Fiscal year basis.

2 Number of licensed construction companies: Figures as of the end of March for each year.

3 Number of workers in the construction industry: annual average.

Since the collapse of the bubble economy, construction firms had suffered capital losses on property assets, and had been burdened with uncollectible accounts receivable. In addition, many firms had not been able to reduce the amount of interest-bearing debts on their books. This had prevented them from putting their businesses on a more stable financial footing.

Private-sector financial institutions were lending with extreme caution and repossessing a great deal of collateral or non-performing loans. This situation had caused severe cash flow problems for small- and medium-sized enterprises. Bankruptcies among construction firms skyrocketed in 1997 to 4,785.

7.1.10 Trends in the construction industry

First construction investment in Japan continued to decline since fiscal 1997. It was estimated to hit ¥56.3 trillion for fiscal 2002, having fallen all the way to only 70 per cent of its 1992 peak. It was expected to be quite some time before economic recovery begins in earnest and before that movement was then reflected in the private construction market. With a decline in public works projects, construction investment would also inevitably decline over the next 5 years since 2002. Even over the mid- to long term, negative factors such as fiscal stringency and the ageing and shrinking of the population would make it impossible to avoid further contraction.

Second, though private construction investment in the domestic economy had been sluggish since the collapse of the bubble economy, public construction investment had remained at fairly high levels. This is because public works projects were actively promoted in successive economic policies enacted to achieve economic recovery. Since public works projects thus propped up construction investment, which had shrunk as a result of major decreases in private construction, a rapid decline in overall construction investment was averted.

However, as a result of ongoing dynamic fiscal investment over several years in an environment where the economic slowdown had produced a lack of financial resources, fiscal conditions drastically worsened. Local governments especially found themselves in a difficult spot, and though the national government had been promoting investment in public works projects to help improve the economic climate, it eventually became unable to follow through on these projects. Actually, the number of independent local projects implemented by local governments with their own funds continued to decrease over several years, and public projects funded by the national government reached their limit. It became difficult to rely on national-government finances to sustain construction investment.

Third, in fiscal 2001, there were serious declines in all three sectors of construction investment. Government construction investment fell for the third consecutive year by a nominal 9.5 per cent, and private residential investment likewise fell by 8.8 per cent, due to a worsening of the employment and income situations. Private non-residential construction investment also fell, dropping 12.2 per cent due to the deteriorating investment environment. Overall construction investment fell by a nominal 9.2 per cent from the previous fiscal year.

In fiscal 2002, overall construction investment fell by a nominal 6.8 per cent over the previous year, sinking under the ¥60 trillion mark for the first time in 16 years to settle at ¥56.3 trillion. In government construction investment, there was a 10.7 per cent decline in public works project expenditures in the initial budget, and without a doubt there was a noticeably large decline in independent local projects. Government construction investment overall fell by 9.4 per cent (nominal terms), about the same level of decline as last year. Private residential investments were supported by tax reductions on new home loans, but because of the time it takes to change the consumer mentality, they still fell for the third consecutive year, by 6.6 per cent (nominal terms). There is little good to add in the areas of general building and civil engineering, as private non-residential construction investment continued to decline, falling 2.0 per cent.

Forecasts indicated that in fiscal 2003, overall construction investment would fall by a nominal 3.9 per cent from the previous fiscal year, hitting ¥54.2 trillion. Government construction investment was expected to fall by a nominal 6.9 per cent from the previous year since, with a stable budget and no additional measures in the revised budget, there was nothing to carry forward from the revised budget. Private residential investment was expected to bottom out in some sectors, such as owned houses and condominiums, but this would not amount to a full-blown recovery, as investment would continue to decline, falling 1.4 per cent. The declining trend in private non-residential construction investment would also continue, as forecasts indicated that it would fall by 1.3 per cent.

Figure 7.1 shows the trends in construction investment (nominal values), the number of licensed construction companies, and the number of workers in the construction industry since fiscal 1985. Construction investment continued to expand after FY 1985 until it reached ¥84 trillion in FY 1992. After the bubble economy collapsed, investments declined but remained at about the ¥70 trillion level until FY 1998. In FY 1999, however, they dipped below ¥70 trillion, and were expected to drop to ¥60 trillion in FY 2002. The numbers of licensed construction companies and workers in the construction industry, however, increased since 1990, with the number of licensed companies rising until the year 2000 and the number of workers rising until 1997. During the time of writing these numbers started declining, but while construction investment was at about the same as its 1987 level, the number of licensed construction companies was about 1.15 times (an increase of 75,000 companies) the 1987 level and the number of workers in the construction industry is 1.19 times (an increase of 990,000 workers) the 1987 level.

Major construction firms: current status and strategies as of 2002

Competition in the construction industry intensified, and major construction firms responded by cutting profit margins, especially for private-sector projects. Healthy firms wrote off extraordinary heavy losses in FY 1997 to put their business on a sound footing, but overall, outstanding interest-bearing liabilities remained high.

As for the latest strategy within the industry, the major firms were organizing project teams to study and gather information on the Private Finance Initiative (PFI), established by the government, and to discuss their participation in the initiative.

In addition, remodelling was expected to become an important part of the industry. For this reason, major construction firms were working hard to develop technologies and organizational units targeted to the needs of remodelling.

The contraction in the construction market since the 1990s had been described above, but Figure 7.2 show that there was a serious supply surplus in the market. Construction investment (nominal value) per licensed construction company dropped by 40 per cent from ¥160 million in 1991 to ¥99 million in 2002 and the construction investment per construction worker dropped by 30 per cent from ¥13.6 million in 1991 to ¥9.6 million in 2002. Also, while future contractions in the construction market would inevitably focus on the government sector, adjustments to the numbers of construction companies and construction workers to suit the size of the market, through bankruptcies and worker lay-offs, for example, were not likely to be avoided.

Local construction companies: management strategies aimed at survival

If the numbers of construction companies and construction workers are inevitably going to be reduced, then companies that focused on public works projects during the favourable business environment of the 1990s will be especially hard hit in the future, and will have to adopt business strategies aimed at survival.

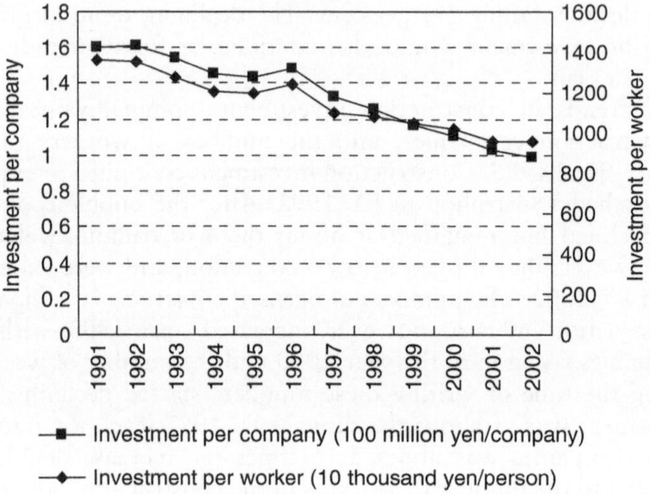

Figure 7.2 Trends in construction investment per construction company and construction worker.

Note

Worker numbers are based on the *Labour Force Survey* (Ministry of Public Management, Home Affairs, Posts and Telecommunications), but figures for FY 2002 are from estimates issued by Research Institute of Construction and Economy. Licensed construction company numbers are from the Survey of the Number of Licensed Construction Companies.

1 If businesses want to survive, they will have to move into construction and construction related fields outside of the public works sector. If they hope to be able to expand, they will have to carve out a path in industry fields with growth potential.
2 A merger may be a potential growth strategy for survival. Specifically, a company could negotiate a merger with a company in a construction field or business area that would provide mutual benefits to both parties. One option would be to negotiate a merger with a company that focuses on private construction projects.
3 A company could also form alliances to facilitate the joint purchase of materials, the joint use of heavy machinery and assets/technologies, or the exchange of engineers.

7.1.11 Top national construction companies: review business focus and range and re-examine organizational strategies

National general contractors need to seek out possibilities for partnerships and alliances that do not generate sentiments of resistance. However, as mentioned above, the industry was starting to see movements towards business consolidation. In April 2002, the businesses of Nitto Daito Kogyo and Mitsui Fudosan, though ranked lower than the top 50 companies, were consolidated (forming the Mirai Group with sales of ¥100 billion) and discussions of business consolidation had began between top-50 companies, for example, Mitsui Construction and Sumitomo Construction.

Amidst the anticipated contraction of the construction market, the major construction companies are naturally reviewing the focus and range of their businesses and considering such options as alliances with other companies. In the long term, however, they may be left with little choice but to develop their businesses through mergers, sales, and acquisitions.

Under these conditions, some companies may select an aggressive merger or sale as part of its strategy, while others may be forced into this course of action by external factors.

First, when the merger or sale is selected as a business strategy a full-scale merger may not be feasible for top-ranking general contractors such as super general contractors. Still, to strengthen certain fields that they consider to have growth potential as they review the organization of their own corporate groups, these contractors will have to give careful consideration to eliminating or selling off unnecessary divisions while strengthening or buying companies in their growing divisions. However, even the five leading companies reported that the sales of their subsidiaries (averaging about 60 companies as of the end of fiscal 2000) account for only about one-fourth of their total sales. This shows that the construction industry has a structure cantered on the parent company. Corporate groups also have the option of downsizing as the construction market contracts, but if they want to try to maintain or expand their corporate group as a whole over the mid- to long term, they may also choose, regardless of what the rest of the market is doing, to take aggressive action in the sectors where they expect to see future growth and to develop into a corporate group that has mixed interests in the construction industry as well as in non-construction industries.

For middle and lower ranking companies (companies in the middle and lower levels of the top-50, lower-ranked listed companies that have faced or are soon likely to face difficulties in sustaining their businesses at the national level due to declining sales until 2002, and that have had problems sustaining their top rankings in the public works sector) may choose to downsize their businesses and transform themselves into local companies, or, if they want to continue to develop on a national scale, may choose to negotiate mergers or purchase other companies. The business consolidation between Nitto Daito Kogyo and Mitsui Fudosan mentioned above was actually chosen as a means for each of the companies, whose sales individually were less than ¥100 billion, to develop their business on a national scale.

At the second case, when merger or sale are necessitated by external factors, the major construction companies covered by this institute's "Financial Analysis" have been reduced by three due to legal measures, leaving 48 companies. Some companies have been replaced, but seven companies reported a consolidated capital ratio (to total capital) of less than 1 per cent (one of which reported a negative ratio) at their fiscal 2001 interim closing. Even when the newly added companies are included, however, there are still six companies reporting less than 1 per cent (with one reporting a negative ratio). As sales continue to decline and construction gross profit rates fall, some companies will find themselves in a difficult position by fiscal 2005 when changes will be introduced in the way companies report the value of tangible fixed assets. Main banks are expected to take the initiative in promoting efforts to revise this system.

Development of construction technologies

A vast array of policies aimed at developing and enhancing the construction industry had been implemented. For example, policies had been implemented to improve product quality assurance through the application of ISO 9000 series standards, to raise the skill levels of on-site workers and managers, and to increase construction safety by preventing human error. Environmental protection measures had also been introduced to promote the use of information technologies in the construction sector, the research and development of new materials and earthquake-resistant construction methods, and the enactment of a construction materials recycling law.

Earthquake resistance reinforcement methods using new materials

These methods provide outstanding earthquake resistance by affixing new materials to or using new materials to encase structures that have sustained earthquake damage, or the existing RC portions of old dilapidated structures. These methods have rapidly gained widespread acceptance in recent years because of their ease-of-use and low cost, their role in eliminating dust and noise pollution, and because they can be implemented while the building is still occupied, thereby eliminating the need to move occupants during construction. New materials include carbon fibre sheets and aramid fibre sheets.

Thermal storage technologies

These are technologies for storing extra heat energy to be used later when needed. Used in the air-conditioning systems in buildings, these technologies are expected to be very useful in facilitating the effective use of heat energy. The most common technique is to create and store cold water, hot water, or ice during the night when electricity costs are low using ice thermal storage technologies that utilize latent heat, and then to use that cold water, hot water or ice to provide air conditioning during the day. In the past, the required thermal storage tanks were installed in building basements, but technologies are now being developed for structural thermal storage using residential floors or walls, and for thermal storage at great depths, which are effective in large regional heating and cooling systems. An example of their use on a regional level can be found in the Minato Mirai 21 (MM21) area of Yokohama.

7.1.12 The use of IT in construction and related policies

The major general construction companies are trying to use information technology in the construction industry for electronically procuring materials over the Internet, electronically collecting bids from project companies, sharing data (CAD, project information), using extranets with cooperating companies, and increasing distribution efficiency.

The Ministry of Land, Infrastructure, and Transport established the Construction CALS/EC Action Program in June 1997, and aims to achieve construction CALS/EC in the projects under its direct jurisdiction by 2004. The major goals of the construction CALS/EC are as follows:

- use of electronic procurement for all projects
- contracting through EDI (Electronic Data Interchange)
- online submission of applications and bid notification for all public projects
- creation of a comprehensive database of project information
- information linking and integration using GIS (Geographic Information System).

In October 2001, electronic bids were used primarily for the large projects under the direct control of the Ministry of Land, Infrastructure, and Transport, and by fiscal 2003 it was expected to used for all projects under that ministry's direct control. Electronic delivery of documents had been enforced in part, for projects since April 2001, and would be expanded for all projects under that ministry's direct control by fiscal 2004. The Ministry of Land, Infrastructure, and Transport has enacted the Regional Development Action Program (National Version) to promote positive efforts toward the implementation of construction CALS/EC by local governments, and plans to have completed the full dissemination of construction CALS/EC to all the institutions that order public works projects by 2010.

7.1.13 Environmental policies pertaining to the construction industry

Environmental impact assessment methods

Japan's environmental impact assessment system used to be left to the discretion of government officials based on the Cabinet's Guidelines Concerning the Performance of Environmental Impact Assessments, but since the process of establishing the Basic Environment Law and the Basic Environment Plan revealed the need for revisions to this system, the Environmental Impact Assessment Law (Environmental Assessment Law) was enacted in 1997.

All businesses over a certain size are automatically required to conduct studies, provide estimates, and perform, at their own expense, assessments of the impact of their business activities on the environment before engaging in those activities. Companies must then use the results of those studies to take the relevant environmental issues into consideration in carrying out their business.

The new law marks a watershed in the history of environmental conservation because it requires that the assessment procedures include mechanisms for ensuring that the views of local residents are reflected in the assessment.

In addition to environmental impact legislation the Japanese government designated 2000 as the "first year of the recycling-oriented society" and decided to enact legislation regarding a basic framework towards that end. The government has thus compiled the Basic Law Promoting the Formation of a Recycling-Oriented Society which sets out the basic principles of recycling policies, such as regulations concerning waste disposal and the collection responsibility of producers.

It has also enacted the Revised Waste Disposal Law which requires waste disposal companies to ensure final waste processing and the Building Construction Materials

Recycling Law (Construction Recycling Law, enforced in May 2002) which systematizes the registration of demolition companies and requires the recycling of concrete, asphalt, and lumber.

Construction waste accounts for about 20 per cent of total industrial waste, about 40 per cent of the volume of final processed waste, and about 90 per cent of illegal dumping, thus making the promotion of construction waste recycling an extremely important issue.

While new laws have been introduced to promote the proper disposal and recycling of waste, such as the Basic Law Promoting the Formation of a Recycling-Oriented Society, Revised Waste Disposal Law, and the Building Construction Materials Recycling Law (Construction Recycling Law), the construction industry is being urged to undertake active environment-oriented efforts and to achieve stringent goals.

It is essential that recycling plans be implemented if the construction industry is to be viewed favourably by the general public as an industry that makes an important contribution to society. Supporting the increased environmental impetus, the number of companies in Japan obtaining certification under the ISO 14000 series of international standards for environmental management systems is rapidly increasing.

Many companies in the construction industry are moving forward with preparations for obtaining certification for such purposes as fulfilling their corporate mission (social demand), preserving the global environment, improving their corporate image, bolstering their company's stature, or establishing an environmental management system. The statistics on ISO 14000s certified companies compiled by the Japan Standards Association as of March 2002 indicate that a total of 8,893 companies have obtained certification. Of those, 628 are general construction companies (7.0 per cent) and 140 are plant and equipment construction companies (1.6 per cent).

In fact, the prospects for eco-business are good. Though the construction market in Japan is now in a period of contraction, steady increases can be expected in some sectors, such as maintenance and repairs, welfare, and information. The environment is another industry where growth can be anticipated, and the construction industry has started to become active in "eco-businesses," the gateway to the environmental sector.

The definition and structure of an eco-business was explained in the OECD manual. The Environment Agency has calculated and released information on the status and future prospects of the eco-business market in Japan as of 2002 based on the OECD manual (see Table 7.2).

According to this data, the eco-business market was valued at ¥24.7 trillion in 1997 and is expected to grow to ¥39.8 trillion by 2010. This corresponds to an expected average annual growth rate of 3.7 per cent. Businesses that will sustain the recycling-oriented society, such as those involved in waste disposal and recycling, accounted for 50 per cent of the market in 1997, and are expected to have about the same 50 per cent share in 2010.

Another environmental action taken by the Government recognizes that carbon dioxide accounts for 90 per cent of global greenhouse gases, but most of that is emitted from necessary energy consumption in everyday living and business activities.

The national government thus developed the Outline for the Promotion of Efforts to Prevent Global Warming in June 1998 as a global warming policy to be urgently

Table 7.2 Status and future prospects of the eco-business market as of 2002

Eco-business	*Market size* (billion yen)		*No. of employees*	
	1997	*2010*	*1997*	*2010*
A. *Environmental pollution prevention*	14,214.0	18,843.0	31,125.8	34,035.0
Manufacturing of devices and materials for preventing pollution	1,347.5	1,786.0	2,234.6	2,189.3
1 For air pollution prevention	305.2	366.0	482.6	428.6
2 For waste water processing	982.4	1,082.8	1,555.0	1,259.3
3 For solid waste processing	8.9	38.7	20.1	61.1
4 For purifying soil and water (including ground water)	1.5	240.8	2.4	296.2
5 For noise and vibration prevention	14.2	10.4	25.4	14.5
6 For environmental measurement,analysis, and assessment	35.2	47.3	149.1	129.5
7 Other	–	–	–	–
Services	8,609.8	10,360.7	24,600.5	25,613.9
8 Air pollution prevention	–	–	–	–
9 Waste water processing	956.9	1,211.1	857.5	799.1
10 Solid waste processing	7,390.4	8,520.2	22,617.4	23,149.6
11 Soil and water purification (includinggroundwater)	35.6	322.5	129.0	522.3
12 Noise and vibration prevention	–	–	–	–
13 Environment-orientedresearch and development	–	–	–	–
14 Environment-oriented engineering	–	–	–	–
15 Analysis, data collection, measurement,assessment	219.7	218.6	951.7	946.9
16 Education,training, information provision	2.1	34.8	13.3	80.6
17 Other	5.1	53.4	31.6	115.4
Installation of buildings and equipment	4,256.7	6,669.4	4,290.6	6,231.8
18 Air pollutionprevention equipment	0.0	5.9	0.0	7.2
19 Waste processing equipment	3,394.2	5,788.4	3,051.5	5,204.0
20 Waste processing facilities	719.6	642.1	1,110.7	786.8
21 Soiland water purification equipment	–	–	–	–
22 Noise andvibration prevention equipment	142.9	259.9	128.5	233.7
23 Environmentalmeasurement, analysis, and assessment equipment	–	–	–	–
24 Other	–	–	–	–
B. *Technologies and products that reduce the environmental burden* (Device manufacturing, technologies, materials, services)	225.6	546.4	351.6	877.4
1 Energy-efficienttechnologies and processes that reduce that the environmental burden	0.0	250.0	0.0	574.7
2 Energy-efficientproducts that reduce that the environmental burden	225.6	296.4	351.6	302.7

Table 7.2 (Continued)

Eco-business	*Market size (billion yen)*		*No. of employees*	
	1997	*2010*	*1997*	*2010*
C. *Effective use of resources* (Device manufacturing, technologies, materials, services, installation of buildings and equipment)	10,303.1	20,704.9	38,037.1	51,788.3
1 Inside air pollution prevention	–	–	–	–
2 Water supply	28.8	105.1	33.7	171.0
3 Recycled materials	3,745.1	8,850.6	8,708.1	16,911.9
4 Facilities that use recyclable energy	169.0	710.9	630.2	1,194.6
5 Energyconservation and energy management	756.0	2,494.9	1,261.9	2,577.7
6 Sustainable agriculture and fishing	–	–	–	–
7 Sustainable forestry	–	–	–	–
8 Natural disaster prevention	–	–	–	–
9 Eco-tourism	–	–	–	–
10 Other (nature conservation, ecology and environment, biodiversity, etc.)	5,604.1	8,543.4	27,403.2	30,933.0
Total	**24,742.6**	**39,844.3**	**69,514.5**	**86,126.0**

Source: Ministry of Environment.

Notes

1 In some cases it is difficult to differentiate between "Manufacturing of devices and materials for preventing pollution" and "Installation of buildings and equipment." Thus items that are ordered only in units of individual devices are categorized under "Installation of buildings and equipment."

2 Data is unavailable for items marked with a dash (–).

3 In some cases some data from 1996 was used to calculate the market size in 1997.

promoted with a target date of 2010. The goal of these principles was to establish a core strategy for combating global warming that called for the reduction of carbon dioxide emissions through the increased efficiency of energy use and the elimination of wasteful energy consumption.

The Law Concerning the Rational Use of Energy (Energy Conservation Law) is a comprehensive law for promoting the efficient use of energy and eliminating wasteful consumption. It has been in effect for about 20 years since it was enacted in 1979, and has contributed much to the efficient use of energy in Japan, said to have the highest levels of energy efficiency in the world.

7.1.14 Administration of the construction industry

Basic structure

The national government body responsible for supervising the construction industry in Japan is the Ministry of Land, Infrastructure, and Transport (MLIT). As part of the central government restructuring that was implemented in January 2001, the MLIT was established through an integration of the Hokkaido Development Agency, National Land Agency, Ministry of Transport, and Ministry of Construction, and was given the responsibility for the comprehensive and systematic use, development, and preservation of Japanese land, the consistent handling of social funds for that purpose, and the promotion of transportation policies. The MLIT is comprised of its own internal departments (the Minister's Secretariat, Policy Bureau, etc.) as well as the Regional Development Bureaus (including the eight Regional Development Bureaus and the Hokkaido Regional Development Bureau) and various other institutions (such as the Policy Research Institute for Land, Infrastructure and Transport).

The Construction Industry Division and Construction Industry Promotion Division of the MLIT's Policy Bureau are responsible for the administration of the construction industry, and positions responsible for the administration of the construction industry have likewise been established within the Regional Development Bureaus. The Policy Bureau, for example, is involved in the following kinds of construction industry supervisory activities: Comprehensive and basic policies for the ministry as a whole, land expropriation, the environment, public property, project coordination, transportation planning, the construction industry, labour resources, construction project implementation, the real estate industry, housing, international construction activities, IT introduction, surveys and statistics, promotion of restructuring in the construction industry, the planning and introduction of bidding systems that are both highly competitive and transparent, the removal of poor or unqualified companies.

Construction industry administration as a buyer of construction services

Public institutions that purchase construction services, including the MLIT (local governments, construction-related organizations) administer construction companies insofar as they are buyers of those companies' services. Like the central government, the 47 prefectural governments (including the prefectures of Tokyo and Osaka), the 675 city governments (including Yokohama, Nagoya, and Kyoto), and the approximately 2,500 town and village governments (as of April 2002) each have personnel in Construction Administration Coordinator positions who administer the nation's approximately 598,000 construction companies (as of April 2002) by evaluating them and certifying whether they are qualified to participate in competitive bidding procedures. Construction-related organizations, including quasi-governmental agencies (such as the Japan Highway Public Corporation) and regional public corporations perform their own administration by evaluating companies and certifying whether they are qualified to participate in competitive bidding, but the supervisory and evaluation practices employed by the MLIT (Business Evaluation and competitive bidding applications) are often used as the basis for the practices employed by the local governments and construction-related organizations (see Annex 5 for information on the Business Evaluations).

These kinds of national and local supervisory institutions penalize construction companies for legal violations in their transactions and construction practices, and handle worker safety issues by suspending the business of or revoking the licenses of construction companies when major damage is inflicted by accidents or other incidents.

Regulating entry into the construction market through construction permits and licenses

In Japan, regulations regarding business licenses for the construction industry are based on the Construction Business Act. Any person or company who wants to operate a construction business, with the exception of contractors who engage only in simple construction work, must obtain a license issued by the prefectural governor (when establishing offices within a single prefecture) or by the Minister of Land, Infrastructure and Transport (when establishing offices in two or more prefectures). Licenses must be renewed every 5 years.

Licenses are divided into Special Construction Licenses and Ordinary Construction Licenses, and are issued for 28 different construction work classifications, such as civil engineering, general building, etc.

Special Construction Licenses are governed in many areas by more stringent regulations than Ordinary Construction Licenses in order to protect subcontractors and to ensure that construction work is done properly. Operators that accept direct contracts must obtain a Special Construction License if the work is valued at ¥45 million or more for general building work or at ¥30 million or more for other types of construction.

Licensed construction companies, regardless of whether they are contractors or subcontractors, must employ a full-time engineer when working on a contracted construction project who can manage all the technical aspects of the project on site. When the total value of the subcontracts that a Special Construction License holder, who accepts direct construction orders from buyers which exceeds ¥30 million (or ¥45 million in the case of general building work), a project manager must be employed, instead of a full-time engineer, who can manage all of the technical aspects of the project on-site.

Management system for public works and government construction projects

In public works projects, the project management organization is the body that commissions the construction work, such as the national government, local government, or quasi-governmental agency (national or local). These organizations evaluate and register construction companies before they can bid on projects those organizations commission and manage. This evaluation process for prequalifying competitive bidding participants is conducted once every 2 years by the MLIT or other major institution. The commissioning organization evaluates the applying construction company's size and performance, technological capabilities, and business status, and only grants permission to those it deems qualified for the work it wants performed to participate in the bidding process. The commissioning organization assigns a ranking to the companies that pass the evaluation process based on their size and performance, and that rank determines the size of projects the company is permitted to bid on. Construction projects commissioned directly by the MLIT (general civil engineering and general building) are divided into five levels, A–E. The Business Evaluation performed by the MLIT or prefectural government is used as the basis for evaluating companies (see Annex 5 for information on the Business Evaluation).

Because companies that accept contracts for public works projects are required by the Construction Business Act to undergo a Business Evaluation, most construction commissioning organizations use the ranking of construction companies based on this evaluation. Consequently, construction companies, even those that have been licensed in accordance with the Construction Business Act, generally have to prequalify to bid on a construction project commissioned by a particular organization. They do this by submitting an application to participate in competitive bidding and undergoing an evaluation.

Bidding system

Contracts for public works projects used to be awarded in most cases by means of a selective bidding process, but in January 1994 the Cabinet adopted the Action Plan on Reforms of Bidding and Contracting Procedures of Public Works. This Action Plan ushered in open bidding, and improved the system of selective bidding.

For public works projects, as a rule, separate contracts are generally awarded for design, construction, and equipment. Also, the client and the general contractor usually enter into a single, comprehensive contract covering an entire construction project. However, a design-build contracting system and a value engineering (VE) system are also being tried out.

Recognizing the trend towards globalization during the period of writing, more construction service buyers were using open and competitive bidding, as well as new bidding methods such as the value engineering (VE) system, in which technical proposals are accepted during the bidding process, are followed by an integrated evaluation system, in which the function and quality of the proposed structure are evaluated together during the bidding process.

In Fiscal Year (FY) 2001, design-build contracting system was used for 14 projects under the direct supervision of the MLIT. As for private-sector construction projects, the design-build contracting system is the most common, while recently an increasing number of projects have involved the CM contracting system, although the number remains small.

Private-sector clients most often select contractors by means of the mission method (a decision is made after price negotiations with a single company) or by getting estimates from several companies. There are two main types of contracting systems, the "blanket construction – only contract system" and the design-build system, and in most cases gross price contracting (lump-sum contracting) is the predominant model used. Examples of Construction Management (CM) and Project Management (PM) contracting have also arisen, but their use remains quite limited.

In addition, electrical power companies have begun to use the VE system, and real estate developers have begun to use direct contracting (in which major real estate developers dispense with general contractors and award contracts directly to specialized contractors).

Market-opening initiatives in the construction sector

The WTO, which was founded in January 1995, established the GATS in order to deal with government measures that have an impact upon trade in the service sector, which includes the construction industry. In accordance with the GATS agreement, Japan has pledged not to set numerical limits upon foreign participation in the construction industry and five other industries, and not to adopt other measures which would have a discriminatory effect against foreign firms in these industries. The Japanese construction market has traditionally not made any distinctions between domestic and foreign companies in terms of requirements for obtaining business

Table 7.3 Public works orders received by foreign firms (unit: million yen)

	FY 1996	*FY 1997*	*FY 1998*	*1999 (April–November)*
Construction work	9,788	13,736	16,213	10,766
Design/consulting	491	256	261	327
Total	10,289	13,991	16,474	11,093

Source: Office for Construction Market Access, Construction Industry Division, Policy Bureau, Ministry of Land, Infrastructure and Transport.

permits or other procedures. An action plan was introduced in 1994 to further increase the transparency, objectivity, and competitiveness of bidding and contracting procedures, and to enhance the rules against discrimination between domestic and foreign companies. Furthermore, Japan became a party to the WTO Agreement on Government Procurement in 1996.

Foreign construction firms were expected to develop a greater presence in Japan's domestic construction market, but the number of such firms licensed to operate in Japan had changed little since 1992. In fiscal 2001, 10 of the foreign construction companies that had obtained business licenses in Japan originated in Asia (9 from South Korea, 1 from Singapore).

The number of public works projects undertaken by foreign firms had steadily increased. The increase in the number of orders received by South Korea had been especially remarkable since it became a party to the WTO Agreement on Government Procurement in 1997. Orders for construction in Japan received by foreign construction firms are shown in Table 7.3.

Policy toward the construction industry

In response to the harsh business climate in the construction industry, the Ministry of Construction in January 1998 adopted the Measures to Improve Business in the Construction Industry. The aim of these measures was to ensure an adequate supply of funds, improve the financial status of construction firms, and provide a better environment for the awarding of construction contracts.

Since June 1996, the Central Council on Construction Contracting Business has been working to formulate measures to promote a highly transparent and competitive construction industry that will foster the success of construction firms with the best technology and management. In February 1998, the Council put forth a recommendation entitled "A Guide for the Construction Industry in Response to Restructuring in the Construction Market" which called for further improvement of bidding systems, and for promotion of further restructuring and reform of the construction industry.

In February 2001, the Act for Promoting Proper Tendering and Contracting for Public Works was enacted to secure the national citizens' trust towards public works and to promote the sound development of the construction industry that performs public works projects commissioned by the national government, quasi-governmental agencies, and local governments. The act is based on

1 ensuring transparency in the tendering and contracting procedures;
2 promoting fair competition between bidders;

3 completely eliminating improper behaviour; and
4 ensuring that public works projects are performed properly – and makes it compulsory for commissioning organizations to

   a announce their projected orders each fiscal year
   b announce information related to tendering and contracting procedures
   c establish measures to eliminate improper behaviour and
   d establish proper construction systems.

The act was applied to tendering and contracting procedures starting in fiscal 2001.

7.1.15 Infrastructure development

Infrastructure development: status and long-term plans as of 2002

Infrastructure development in Japan has proceeded steadily since the end of Second World War. It started with the redevelopment of land that had been abandoned during the war, and then jumped into high gear to meet new demand as the economy grew and the standard of living rose. Nevertheless, in certain areas and for certain types of infrastructure, the level of development still lags.

The long-term planning of infrastructure development in Japan can be classified into (1) a comprehensive plan, and (2) sectoral development plans.

Comprehensive planning addresses the long-term infrastructure needs. There are various comprehensive plans in Japan, including the National Comprehensive Development Plan, the economic plan formulated by the Economic Deliberation Council, and the Basic Plan for Public Investment.

There were 16 sectoral development plans that target such infrastructure as roads, sewerage systems, etc. These plans cover periods of 5–10 years, and they determine the objective and number of projects to be carried out during the planning period.

Under the terms of the June 1998 report entitled "Measures to Promote Fiscal Structural Reform," the Basic Plan for Public Investment has been extended 2 years, and all sectoral development plans have been extended 2 years. The extension of these development plans has effectively reduced the amount of money being invested in infrastructure development.

Though private construction investment in the domestic economy has been sluggish since the collapse of the bubble economy, public construction investment has remained at fairly high levels. This is because public works projects were actively promoted in successive economic policies enacted to achieve economic recovery. Since public works projects thus propped up construction investment, which had shrunk as a result of major decreases in private construction, a rapid decline in overall construction investment was averted.

However, as a result of ongoing dynamic fiscal investment over several years in an environment where the economic slowdown had produced a lack of financial resources, fiscal conditions drastically worsened. Local governments especially found themselves in a difficult position, and though the national government had been promoting investment in public works projects to help improve the economic climate, it eventually became unable to follow through on these projects. Actually, the number of independent local projects implemented by local governments with their own funds continued to decrease over several years, and public projects funded by the national government reached their limit. It has become difficult to rely on national government finances to sustain construction investment.

Funding of infrastructure development

Infrastructure development was funded by the national government, prefectural and local governments, public corporations, local public utilities, local public corporations, public–private partnerships, and private enterprises.

With public works projects, it is unusual for the organization which contracts out a project to bear 100 per cent of the funding burden. Usually the burden is shared by other related organizations. It is also common for a public corporation or other organization to acquire debt to fund a project, and then to service that debt by collecting a fee from those who benefit from the resulting infrastructure.

Different organizations fund infrastructure development in different ways. The national government does so principally by issuing government bonds and by drawing upon revenues earmarked for road construction. Local governments do so mainly by drawing upon general revenues, issuing bonds, drawing upon earmarked funds, and using national treasury disbursements. Public corporations generally do so by taking out loans from the national government under the fiscal loans and investment policy.

Private-sector participation in infrastructure development

Private-sector participation in infrastructure development in Japan has been marked by two especially salient characteristics: one is the role played by the Fiscal Investment and Loan Program in providing funding and the other is the diversity of corporations engaged in infrastructure development (public corporations, public–private partnerships, the Trans Tokyo–Bay Highway Corporation, the Kansai International Airport Co. Ltd, etc.).

Efforts are now afoot to find new ways to tap the resources of the private sector for the sake of infrastructure development. In addition to efforts at all levels of government, vigorous discussion is also taking place in the construction and finance sectors, among others.

To bring actual projects to fruition, the business community and all levels of government must cooperate closely to overcome difficulties and create an environment conducive to private-sector participation in infrastructure development.

7.1.16 International exchange and cooperation

Development aid and technical cooperation by the Japanese government

Japan is the world's second largest economy and the world's top donor through its Official Development Assistance (ODA). Due to depreciation of the yen, Japan's ODA declined in FY 1996 to US$9.608 billion, but this figure made Japan the world's leading aid donor for the sixth year in a row until 2002. This aid consisted of grant assistance, technical cooperation, loans, and funding of international organizations.

While funds thus provided were used for a wide variety of purposes, the highest proportion of this money went for infrastructure development in developing countries. In addition to helping raise the level of trust and respect that other countries have for Japan, the Japanese government grants ODA with the basic understanding[1] that it helps promote the national interests of Japan, a country that is dependent on world peace and stability and that relies on the international community for resources, energy, and food. In the light of this knowledge, technological assistance projects which utilize Japanese firms and take advantage of the experience of specialists at universities, think-tanks, local governments, and private companies, and which aim at transferring useful technologies and managerial skills, are one of the most important means through which Japan provides overseas aid. Various technological cooperation programs are already underway in the construction sector in areas such as construction materials development (Indonesia), new housing technology research (China), and increasing construction productivity (Philippines).

Part of Japan's technical assistance is directed towards the construction sector. The Japan International Cooperation Agency (JICA) dispatches experts, brings persons to Japan for training, and provides equipment and materials under its project-type technical assistance programme.

Countries throughout Asia have high praise for the technology and know-how at the disposal of government agencies and public corporations in Japan. In response to requests from developing countries and international organizations, the Ministry of Construction works through JICA and other channels to dispatch experts in rivers, roads, housing, urban planning, and other specialties. These experts work together with counterparts in other countries to plan and carry out development projects.

Overseas ODA contract amounts for Japanese construction companies were valued at ¥808.3 billion for fiscal 2001. A breakdown of orders reveals that ¥525.4 billion were placed with Japanese companies (branches or sales offices), and ¥282.9 billion were placed with local subsidiaries of Japanese companies. There is a large gap between the value of orders received from Asia, Japan's leading overseas source, whose overseas contract value of ¥583.6 billion accounts for 72.2 per cent of its overseas contracts, and those from the second runner-up, North America, with ¥96.8 billion.[2] By undertaking projects in Asia, Japanese construction companies operating as main contractors or joint venture participants are able to form close cooperative relationships with local subcontracting firms and other joint venture participants. They are also able to engage in technology transfers in such areas of construction as project management, quality control, and construction methods. As a result, they not only help improve the technological capabilities of the local companies with whom they share a cooperative relationship, but also raise the overall level of the construction industry in the countries in which they operate.

Overseas direct investment

Japan's overseas direct investment (all figures converted to dollars) rose for three consecutive years from FY 1992 through FY 1995 before dipping by 5.3 per cent to US$48.019 billion in FY 1996. This decline, however, was primarily due to the depreciation of the Japanese yen versus the US dollar (the yen-based figure rose in FY 1996 by 9.1 per cent to ¥5.4094 trillion).

The Asian construction market was the largest overseas market for Japanese construction companies, and 70 per cent of the sum total of overseas construction orders in fiscal 2001 came from Asia.

Japan's activities in the Asian market include, as part of its ODA efforts, technological cooperation such as the transfer of construction technologies and business know-how. Through the project orders they receive, Japanese construction companies build cooperative relationships with joint venture partners and the local companies that serve as their subcontractors, and transfer technologies in the fields of process management and quality control.

Investment in Asia's manufacturing sector jumped 41.6 per cent in FY 1994 and 55.5 per cent in FY 1995. While this figure fell 17.7 per cent in FY 1996, the absolute total remained a very high US$6.628 billion. These investments were considered a significant spur to construction demand and employment throughout Asia.

Construction materials and equipment

As for construction material (other than lumber), the Japanese construction industry does not depend upon foreign resources very much due to dissatisfaction with quality, delivery time, and selection. However, imports by major construction firms had been on the rise during the time of writing. The top import item, stone, accounts for about 40 per cent of the total. In terms of origin, about half of all construction material imports come from other Asian countries.

Japan exports large quantities of construction equipment, especially to other Asian countries, and the Japanese construction industry, working through various international cooperation schemes, provides a stable supply of high-quality, low-cost construction equipment to countries throughout the world.

Participation in overseas markets

The total value of overseas construction projects awarded to Japan's top-56 construction firms declined in FY 1997 for the first time in 4 years to US$10.403 billion (including the portion of such projects performed by overseas affiliates). Overseas contract amounts for Japanese construction companies were valued at ¥808.3 billion for fiscal 2001. A breakdown of orders reveals that ¥525.4 billion were placed with Japanese companies (branches or sales offices), and ¥282.9 billion were placed with local subsidiaries of Japanese companies. There is a large gap between the value of orders received from Asia, Japan's leading overseas source, whose overseas contract value of ¥583.6 billion accounts for 72.2 per cent of its overseas contracts, and those from the second runner-up, North America, with ¥96.8 billion.[3] Although the post-bubble economic recession and the resulting slowdown in the domestic construction market have spurred Japanese construction firms to make concerted efforts to win overseas contracts, the economic doldrums stemming from the 1997 Asian currency crisis have sharply reduced construction orders from Asia and caused an overall decline in overseas construction contracts.

This cause-and-effect chain shows that strong ties exist between construction markets in Japan and the rest of Asia, and that these markets have a strong influence upon each other. The Asian currency crisis, in fact, had been a contributing factor in the slowdown of the Japanese construction economy since 1997. It not only led to a cooling of the domestic market in the construction sector, but also proved a major blow to the number of orders received by Japanese construction companies located throughout Asia. Nonetheless, the Japanese economy was entering a new phase of its efforts to overcome the blow dealt by Asian currency crisis.

Asia is the largest overseas market for Japanese construction companies. In fiscal 2000, Japanese companies (including those in Asia) received orders for 1,197 projects in that region worth ¥694.7 billion. These accounted for about 70 per cent of the total number of orders for overseas construction projects placed with Japanese companies. The number of construction orders for Asia received by Japanese construction companies increased drastically during the bubble period, exceeding a value of ¥1.2 trillion in fiscal 1996.

However, with the onslaught of the Asian currency crisis in 1997, the number of orders received took a rapid dive downwards (see Figure 7.3). While a large 29.9 per cent increase was recorded for fiscal 1996 over the previous year, there was a huge 27 per cent drop in fiscal 1997, bringing the market to just under ¥892.4 billion. The domestic business slump faced by construction companies in Japan was accompanied by a related paring down of expansionary plans into Asia, resulting in a continuing decline in the number of orders from Asia. This brought the value of orders received in fiscal 1999 to ¥483.5 billion, down by more than half over 3 years.

This decline was halted in fiscal 2000, when the value of orders increased by 43.7 per cent. With a limited domestic market, Japanese construction companies saw overseas expansion as a way to pursue growth. This naturally makes Asia, their largest market with the highest levels of anticipated growth, an extremely important market. The recovery of the Asian economies and the contraction of the domestic market were expected to result in a gradual increase in construction orders received by Japanese construction companies in Asia.

Japanese construction firms active in Asia and Oceania give generally high marks to countries in these regions for ease of market access, but calls have been made for improvement in specific items.

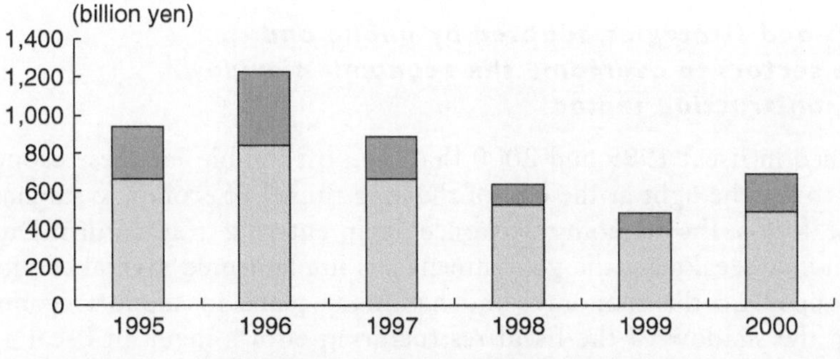

Figure 7.3 Construction orders in Asia.

Source: Overseas Construction Association of Japan.

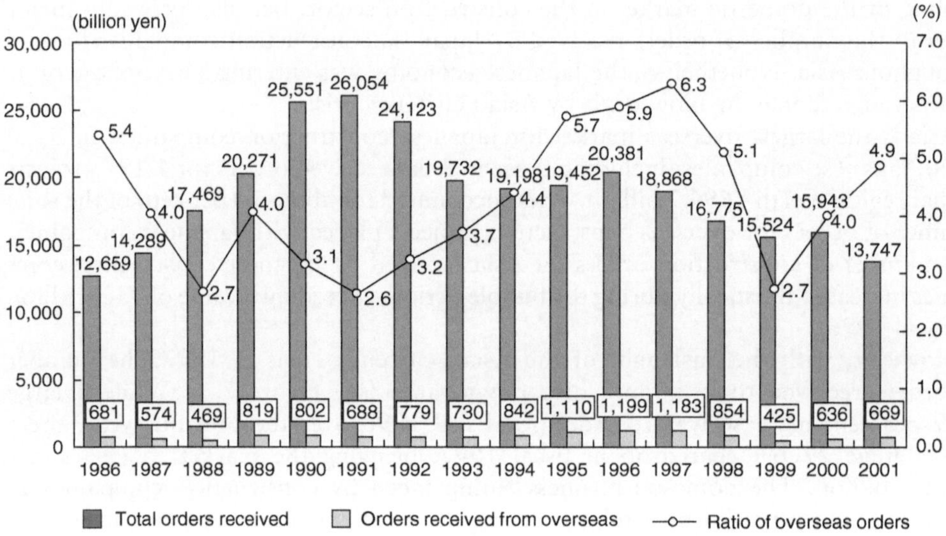

Figure 7.4 Trends in overseas construction orders received by the top-50 general contractors (1986–2001).

Source: Survey A of Orders Received, Construction Research and Statistics Division, Information and Research Department, Policy Bureau, Ministry of Land, Infrastructure and Transport. Survey conducted among the top-50 general contractors.

Notes

1 The number of overseas construction orders does not include those received by local corporations.
2 Figures shown here differ from figures reported by the Overseas Construction Association of Japan due to differences in the companies surveyed.

Figure 7.4 shows the overseas construction orders received by major domestic construction firms in relation to the total orders received and Figure 7.5 shows the geographical locations of overseas orders for 2001.

7.1.17 *Policies and strategies adopted by public and private sectors to overcome the economic slowdown in the construction sector*

Although it seemed in fiscal 1999 and 2000 that the post-bubble Japanese economy was finally able to see the light at the end of the long tunnel of economic stagnation that spanned the 1990s, the economy was once again entering a downturn. During this time of economic weakness, the government has implemented several economic policies, has propped up the economy, and has made plans to support economic recovery. Still in the shadow of the fiscal restructuring efforts taken in fiscal 1997 by the Hashimoto Cabinet, the subsequent Obuchi and Mori Cabinets set economic recovery as their top priority, hammered out two large-scale economic policies in fiscal 1998, and implemented a large-scale economic policy in November 1999 known as the Policy Measures for Economic Rebirth, as well as a policy in October 2000

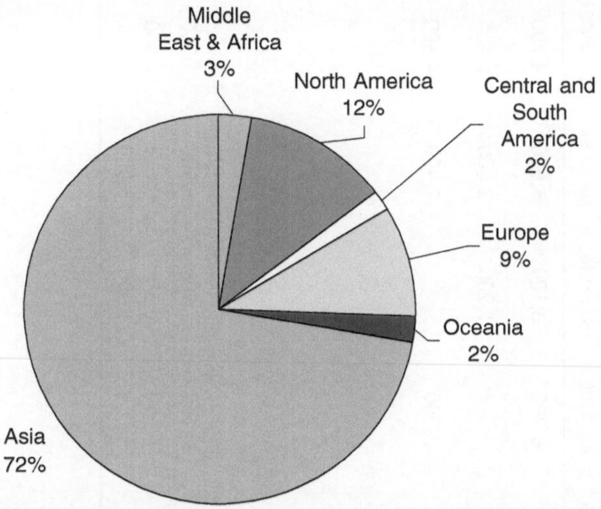

Figure 7.5 Overseas construction contracts in fiscal 2001 (ratio of order values by region).
Source: Overseas Construction Association of Japan. Totals for the 48 member companies.

known as the Policy Package for New Economic Development towards the Rebirth of Japan (Table 7.4).

Restructuring the financial system is essential for solidifying the base for economic recovery. For this reason, the government established a ¥17 trillion framework for protecting depositors. Also, the "Financial Revitalization Law" and the "Early Strengthening Law" were established in the fall of 1998, and the government set aside a government guarantee under each system in the amounts of ¥18 trillion for dealing with failed financial institutions, and ¥25 trillion for capital reinforcements. In fiscal 1999, ¥7.7 trillion was injected into city banks and local financial institutions to strengthen their capital bases. Dealing with failures is also actively promoted in the above framework.

In addition, plans to restore the economy comprised a wide range of efforts including the establishment of a special credit guarantee program (¥30 trillion guaranteed) for making loans available for SMEs, the establishment of the Industrial Competitiveness Council, for researching ways to improve industrial competitiveness, and the introduction of the Economic Strategy Council, for researching economic management.

These policies proved temporarily successful as Japan's GDP showed positive growth in fiscal 1999 (1.4 per cent in real terms) for the first time in 3 years. In fiscal 2000, however, the growth rate again slowed (0.9 per cent in real terms), indicating that Japan has yet to recover from its economic malaise.

Impact on the construction sector

Though private construction investment has been sluggish since the collapse of the bubble economy, public construction investment has been at fairly high levels. This is

Table 7.4 Economic measures since the collapse of the bubble economy in 1992 (billion yen)

	Comprehensive stimulus package	*Statement on implementation of the comprehensive stimulus package*	*Emergency stimulus package*	*Comprehensive stimulus package*	*Emergency measures to counter yen appreciation*	*Economic stimulus measures*	*Comprehensive stimulus package*	*Emergency stimulus package*	*Policy measures for economic rebirth*	*Policy package for new economic development towards the rebirth of Japan*
	Aug. 1992	*Apr. 1993*	*Sep. 1993*	*Feb. 1994*	*Apr. 1995*	*Sep. 1995*	*Apr. 1998*	*Nov. 1998*	*Nov. 1999*	*Oct. 2000*
National public	5,000	5,320	1,450	4,200		4,630	6,200	8,100	6,800	4,700
Local public works, etc.	2,800	3,500	500	1,800		1,000	1,500			
Housing loan corporation, etc.	800	1,800	2,900	1,200		520		1,200	2,000	1,100
Support measures for SMEs	1,200	1,910	770	1,360	1,440	1,290	2,000		7,400	4,500
Promotion of private capital investment	900	500		100						
Tax cuts		150		5,850			4,600	6,000		
Great Hanshin-Awaji Earthquake recovery measures					3,800	1,410				
New infrastructure					330	910				
Other		20	300	740	1,400	4,460	2,350	7,900	1,900	700
Project size	10,700	13,200	About 6,000	15,250	About 7,000	14,220	16,650	About 24,000	About 18,000	11,000

Note
Measures taken in April 1998 incorporate new infrastructure projects in the category of "National public works".

because, as mentioned earlier, public works projects were actively promoted in successive economic policies to achieve economic recovery. Since public works projects thus propped up construction investment, which had shrunk as a result of major decreases in private construction, a rapid decline in overall construction investment was averted.

However, as a result of ongoing dynamic fiscal investment over several years in an environment where the economic slowdown had produced a lack of financial resources, fiscal conditions drastically worsened. Local governments especially found themselves in difficulties, and though the national government had been promoting investment in public works projects to help improve the economic climate, they became unable to follow through on these projects. Actually, the number of independent local projects implemented by local governments with their own funds continued to decrease over several years, and public projects funded by the national government reached their limit. It became difficult to rely on national government finances to sustain construction investment.

Table 7.5 gives investment figures and Figure 7.6, the year-on-year percentage changes. In fiscal 2000, government construction investments fell for the second consecutive year, as did private residential investments, but steady shifts in private non-residential investments made up for these shortfalls, leaving overall construction investments to decline slightly, with 0.2 per cent real growth over the previous year.

In fiscal 2001, overall construction investments fell sharply over the previous year, by 5.7 per cent (real terms). Government construction investments fell 5.8 per cent (real terms) since, with a stable budget and no additional construction investments, there was a decline in public works reserves as well as a decrease in local government projects not subsidized by the national government. Private residential investments were supported by tax reductions on new home loans, but because of the time it takes to change the consumer mentality, they still fell for the second consecutive year, by 3.9 per cent (real terms). Private non-residential construction investment was considerably down, registering a 7.6 per cent (real terms) decline due to the effects of the overall economic slowdown.

Table 7.5 Construction investment (or market) volume

(unit: local currency at current price) (100 million yen)

	1997	*1998*	*1999*	*2000*	*2001*[1]	*2002*[1]
Residential construction (A)	238,568	209,900	218,000	215,400	206,095	202,487
public	13,699	13,300	12,100	11,600	11,333	9,753
private	224,869	196,600	205,800	203,800	194,762	192,735
Non-residential construction (B)	160,298	139,200	128,600	130,400	108,180	108,924
public	40,529	36,500	34,700	33,400	26,016	26,178
private	119,769	102,700	93,900	97,000	82,164	82,746
Civil engineering construction (C)	353,039	358,500	356,300	357,700	340,958	308,134
public	275,414	284,500	271,000	266,900	251,801	220,715
private	77,625	74,000	85,300	90,800	89,158	87,418
Total construction investment (A + B + C)	751,905	707,600	702,900	703,500	655,233	619,545
Total public	329,642	334,300	317,800	311,900	289,150	256,646
Total private	422,263	373,300	385,000	391,600	366,084	362,899

1 2001, 2002 are forecast.

Table 7.5 (Continued)

(unit: change at local currency at constant price)

	1997 (%)	*1998* (%)	*1999* (%)	*2000* (%)	*2001*[1] (%)	*2002*[1] (%)
Residential construction (A)	−14.4	−10.0	4.8	−1.3	−3.8	−1.5
public	−29.7	−1.0	−7.8	−4.2	−2.1	−13.7
private	−13.3	−10.6	5.7	−1.1	−3.9	−0.8
Non-residential construction (B)	2.2	−11.6	−6.5	1.1	−15.5	2.2
public	0.6	−8.4	−3.7	−3.9	−21.8	0.9
private	2.7	−12.7	−7.5	3.0	−13.4	2.6
Civil Engineering construction (C)	−0.3	3.4	0.5	0.0	−3.3	−9.1
public	0.2	5.1	−3.8	−1.9	−3.9	−11.3
private	−2.0	−2.8	16.7	6.0	−1.5	−2.8
Total construction investment (A + B + C)	−4.8	−4.1	0.4	−0.2	−5.7	−4.8
Total public	−1.5	3.2	−3.9	−2.2	−5.8	−10.3
Total private	−7.2	−9.8	4.2	1.5	−5.7	−0.5

[1] 2001, 2002 are forecast.

Source: Construction Economy Forecast (July 2001), Research Institute of Construction and Economy.

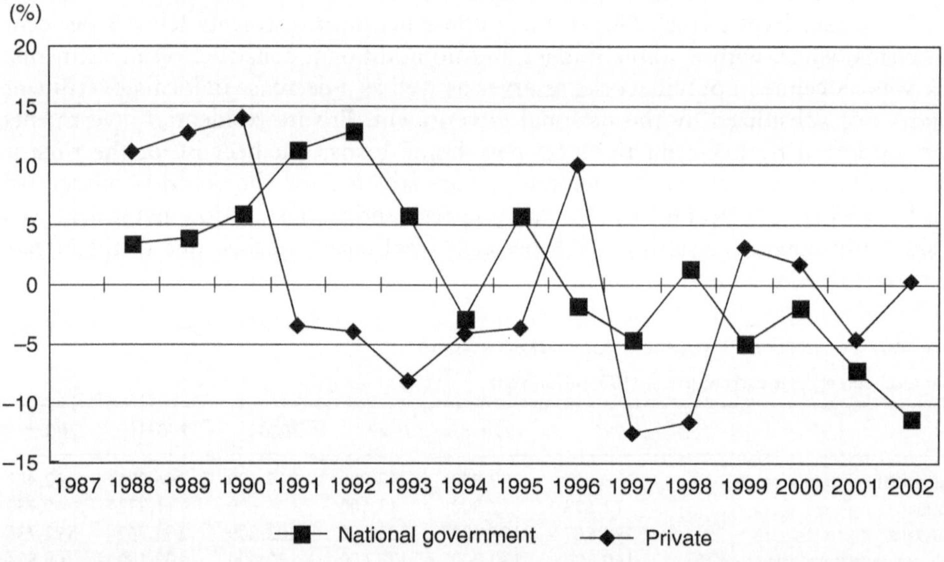

Figure 7.6 Trends in the construction investment growth rate (national government versus private).
Source: Construction Economy Forecast, Research Institute of Construction and Economy.

Note
Assuming no additional national government construction investment in FY 2000 and 2001.

In fiscal 2002, major cuts in the government sector was expected to leave overall construction investments down 4.8 per cent (real terms) from the previous year. With a stable budget and neither any public works reserves nor added measures in a revised budget, government construction investments was expected to drop by a large 10.3 per cent (real terms). Falling 0.8 per cent (real terms), private residential investments would be negative for the third consecutive year, and private non-residential investments would fell again 2 consecutive years, by 0.2 per cent (real terms), due to a slowdown in civil construction field.

The above indicates that construction investment was, as of 2002, in a phase of short-term contraction. Private investment was expected to be stimulated as economic recovery progresses, but massive improvements could not be anticipated even over the medium to long term due to the negative effects of fiscal belt-tightening, the ageing of the population, and population decreases.

7.2 Construction and property markets

After the collapse of the bubble economy, Japan in the latter half of 1993 began to show signs of recovery. Thereafter, however, a rush of consumer demand ahead of the consumption tax increase in April 1997 was followed by a sudden drop-off in consumption once the tax had gone into effect. Then in the fall of 1997, the bankruptcy of several financial institutions undermined confidence in the financial system. The result had been continuing economic recession. GDP for 1997 fell 0.7 per cent from the previous year, marking the first such decline in 23 years. A number of measures had been taken since November to respond to the recession, and in April 1998, Japan adopted the largest level of comprehensive economic measures in the country's history. It appears that this package had successfully put a stop to the economic slide, but the economy has yet to show any signs of an upturn beyond the effect achieved by government measures.

The (former) Economic Planning Agency announced that the economy had hit the bottom of its recessionary valley in April 1999, after which it began to show signs of recovery. After the recession that followed the collapse of the bubble economy in the early 1990s and the further slowdown caused by the 1997 financial crisis, the Japanese economy had finally seemed to be able to see the light at the end of the tunnel, but now the outlook is worsening once again.

Economic growth in fiscal 2000 was a positive 0.9 per cent. Having climbed from negative growth in fiscal 1998 to positive 1.4 per cent growth in fiscal 1999, the Japanese economy seemed to have found a path towards recovery, but dark clouds of difficulty had already begun to threaten future progress. Having decided on a quantitative financial deregulation policy in March 2001, the Bank of Japan had been diligently hammering out policies for propping up the economy, which is facing decreasing exports, through deregulation measures that provide large volumes of cash to short-term financial markets. But today the tremendous volume of non-performing loans left over from the collapse of the bubble economy has hindered the ability of the Japanese economy to get back on track towards growth.

In the Monthly Economic Report for May 2002, the government explicitly stated that the bottoming out projected in March had occurred. The prolonged stagnation in the Japan economy seems finally to have reached its twilight period thanks to increasing exports.

In fiscal 2001, there were serious declines in all three sectors of construction investment. Government construction investment fell for the third consecutive year by a nominal 9.5 per cent, and private residential investment likewise fell by

8.8 per cent, due to a worsening of the employment and income situations. Private non-residential construction investment also fell, dropping 12.2 per cent due to the deteriorating investment environment. Overall construction investment fell by a nominal 9.2 per cent from the previous fiscal year.

In fiscal 2002, overall construction investment fell by a nominal 6.8 per cent over the previous year, sinking under the ¥60 trillion mark for the first time in 16 years to settle at ¥56.3 trillion. In government construction investment, there was a 10.7 per cent decline in public works project expenditures in the initial budget, and without a doubt there was a noticeably large decline in independent local projects. Government construction investment overall fell by 9.4 per cent (nominal terms), about the same level of decline as for 2001. Private residential investments were supported by tax reductions on new home loans, but because of the time it takes to change the consumer mentality, they fell for the third consecutive year, by 6.6 per cent (nominal terms). There was little good to add in the areas of general building and civil engineering, as private non-residential construction investment continued to decline, falling 2.0 per cent.

Forecasts indicated that in fiscal 2003, overall construction investment would fall by a nominal 3.9 per cent from the previous fiscal year, hitting ¥54.2 trillion. Government construction investment would fall by a nominal 6.9 per cent from the previous year since, with a stable budget and no additional measures in the revised budget, there was nothing to carry forward from the revised budget. Private residential investment was expected to bottom out in some sectors, such as owned houses and condominiums, but this will not amount to a full-blown recovery, as investment would continue to decline, falling 1.4 per cent. The declining trend in private non-residential construction investment would also continue, as forecasts indicate that it would fall by 1.3 per cent.

7.2.1 Macroeconomic review and outlook

With the next 2–3 years since 2002 having been designated as a concentrated adjustment period, the Japanese government was making efforts to see bad loans written off and to undertake extensive structural reforms in the fiscal and regulatory structures.

However, the outlook for the Japanese economy even following these reforms is not necessarily rosy. The ageing of society and a shrinking population are expected to put some constraints on the Japanese economy in the near future. According to studies conducted by the National Institute of Population and Social Security Research, the population of Japan, which was 126.9 million in 2000 (National Census), is going to increase moderately until it hits its peak in 2006 of 127.7 million. After that, there will be a long-term population decline. At the same time, the ageing of society will progress at an accelerated pace, such that the population of senior citizens (aged 65 and older) will continue to increase rapidly from the current level of 22 million people to over 30 million by 2013 and 34.2 million by 2018. The ageing of society will inevitably increase the burden on working people in terms of taxes and social security. The savings rate is also expected to decline, thereby reducing the society's overall resources available for investment. In 2010, the pace of population decline will still be moderate, but if no extraordinary technological innovations or new means of increasing productivity are unearthed, the Japanese economy will have a difficult time getting back on a track of positive growth over the mid- to long term.

The aging of society will result in an increased reliance on fiscal measures, especially social security expenses, and will make fiscal reform even more difficult. Once the economy gets back on the track of autonomous recovery, fiscal reform must urgently be promoted by rationalizing and streamlining government administration.

On the other hand, the globalization of the world economy is also going to have an impact on Japan's future economy. In a post-Cold War environment where most countries have adopted a market economy and the information revolution has intricately affected the global market, Japan has been able to retain its high level of international competitiveness. This has made it essential to create a framework for implementing deregulation, promoting the IT revolution, and building infrastructures for transportation and telecommunications networks. In terms of its role within Asia, Japan needs to promote technology transfers to other countries and the liberalization of the Japanese market, while also playing a more active role in pursuing economic ties with its Asian neighbours.

Another imminent concern is the increased impact that global environmental problems are going to have on the economy. Cross-national efforts in areas such as global warming have already begun, but Japan is pouring its efforts into creating an international framework for dealing with environmental problems and providing government development assistance in the environmental sector. Domestically, the government is promoting a shift to a recycling-oriented social economy that addresses waste management and recycling issues.

As can be seen from the above, the outlook for the Japanese economy over the next 5 years since 2002 is not one of unbridled optimism, but instead one that recognizes the various problems that remain to be addressed. Increased wisdom and effort will be required if the Japanese economy is to continue to thrive now that it has shifted from a growing to a mature economy. Tables 7.6–7.8 show the major macroeconomic, demographic and financial indicators.

7.2.2 Construction market review and outlook

Table 7.9 shows the trends in construction investment and Figure 7.7 shows the estimate of construction investment. Recent trends and forecasts for construction investment based on the Construction Investment Outlook Using a Construction Economy Model issued by the Research Institute of Construction and Economy (RICE) show that in fiscal 2001 there were serious declines in all three sectors of construction investment. Government construction investment fell for the third consecutive year by a nominal 9.5 per cent, and private residential investment likewise fell by 8.8 per cent due to a worsening of the employment and income situations. Private non-residential construction investment also fell, dropping 12.2 per cent due to the deteriorating investment environment. Overall construction investment fell by a nominal 9.2 per cent from the previous fiscal year.

In fiscal 2002, overall construction investment fell by a nominal 6.8 per cent over the previous year, sinking below the ¥60 trillion mark for the first time in 16 years to settle at ¥56.3 trillion. In government construction investment, there was a 10.7 per cent decline in public works project expenditures in the initial budget, and without a doubt there was a noticeably large decline in independent local projects. Government construction investment overall fell by 9.4 per cent (nominal terms), about the same level of decline as in 2001. Private residential investments were supported by tax reductions on new home loans, but because of the time it takes

Table 7.6 Major macroeconomic, demographic and financial indicators

	1998	*1999*	*2000*	*2001*	*2002*
GDP and components (monetary unit: billion yen)					
GDP at real prices	5,172,204	5,269,504	5,356,903	5,265,241	5,264,866
GDP growth (%)	0.8	1.9	1.7	1.7	0.0
GDP at current market prices	5,132,448	5,143,487	5,130,060	4,972,186	4,915,429
Agriculture, forestry, fishing, mining and quarrying	89,932	82,382	76,320	–	–
% growth	2.0	8.4	7.4	–	–
Manufacturing sector	1,134,722	1,109,888	1,109,269	–	–
% growth	5.0	2.2	0.1	–	–
Services sector	3,752,703	3,752,353	3,777,580	–	–
% growth	0.5	0.0	0.7	–	–
Construction sector	397,399	384,955	376,358	–	–
% growth	3.8	3.1	2.2	–	–
Demographic indicators					
Population (1000s)	126,420	126,665	126,892	127,100	127,286
Population growth rate (%)	0.20	0.19	0.37	0.34	0.31
Total labour force (1000s)	67,930	67,790	67,660	67,520	67,080
Labour force growth rate (%)	0.09	0.21	0.40	0.40	0.86
Unemployment rate (%)	4.1%	4.7%	4.7%	4.8%	5.0%
Financial indicators					
Short-term interest rate (%)	0.66	0.17	0.23	0.13	0.13
Long-term interest rate (%)	0.972	1.836	1.663	1.311	1.399
Changes in consumer price index (%)	101.0	100.5	99.9	98.9	98.4
Base lending rate (Commercial banks)	1.5	1.375	1.5	1.375	1.375
Base lending rate (Finance companies)	2.2	2.2	2.1	1.85	2.1
Change against US$	130.90	113.91	107.77	121.53	133.52

Sources: Construction Economy Forecast (RICE, May 2002); Annual Report on National Accounts (Cabinet Office, March 2002); Institute of Population Problems Internet home page; Economic Data from Overseas (compiled by the Cabinet Office, April 2002); the Ministry of Public Management, Home Affairs, Posts and Telecommunications Internet home page; Bank of Japan Monthly Bulletin (compiled by the Research and Statistics Department, May 2002).

Notes

1 GDP and gross production figures are for the fiscal year. GDP figures for 2002 are estimates. Real figures are based on FY 1995 prices.
2 All gross productivity figures are in nominal terms.
3 Populations figures are estimates.
4 Figures for labour force population and unemployment rates are seasonally adjusted. Figures for 2002 are current as of May.
5 The consumer price index assumes a baseline figure of 100 for 2000.
6 Interest rate figures are from year-end statistics (except for 2002 figures which reflect interest rates as of March 31).
7 Short-term interest rates reflect the average interest rates published in the domestic commercial papers.
8 Long-term interest rates reflect interest rates on long-term (10-year) government bonds.
9 The consumer price index for 2002 is as of 30 April.
10 The base lending rates (Commercial Banks) reflect the short-term prime rates.
11 The base lending rates (Finance Companies) reflect the long-term prime rates.
12 Exchange rate figures are annual averages (except for 2002 figures, which reflect the average for February).

to change the consumer mentality, they still fell for the third consecutive year, by 6.6 per cent (nominal terms). There is little good to add in the areas of general building and civil engineering, as private non-residential construction investment continued to decline, falling 2.0 per cent.

Table 7.7 Main macroeconomic indicators (per cent change at constant price)

	1997[1]	*1995 (%)*	*1996 (%)*	*1997 (%)*	*1998*[2] *(%)*	*1999*[2] *(%)*
GDP	504,565.6	2.8	3.2	0.7	0.6	0.2
Private final consumption	305,745.1	3.1	2.8	1.2	0.6	1.7
Government final consumption	49,204.9	2.7	1.0	0.0	1.1	1.2
Public fixed capital formation	39,568.5	7.9	2.7	7.2	4.1	13.3
Private fixed capital formation	101,278.3	3.8	10.2	4.5	6.5	0.6
• Private residential buildings	22,416.0	6.7	13.7	21.1	3.4	5.4
• Private non-residential buildings, other construction and land improvement, etc.	78,862.3	7.4	9.1	0.7	7.3	2.2
Exports of goods and services	56,744.2	4.6	5.1	9.2	2.0	3.0
Imports of goods and services	49,235.3	15.2	8.7	2.9	3.4	1.2

Notes
1 Nominal price, billion yen.
2 1998,1999 are forecast.
Years are fiscal year (from April to March).

Table 7.8 Main demographic indicators

	1995	*1996*	*1997*
Population	125,570	125,864	126,166
Households	43,447	–	–
Rate of unemployed person (%)	3.2	3.4	3.4
Consumer price index (% change)	0.1	0.1	1.8
Construction price index (% change)	0.1	0.8	1.2
Short-term prime lending rate (%)	1.625	1.625	1.625
Long-term prime lending rate (%)	2.6	2.5	2.3
Annual average exchange rate of Yen per $US	102.91	115.98	129.92

Forecasts indicated that in fiscal 2003, overall construction investment would fall by a nominal 3.9 per cent from the previous fiscal year, dropping to ¥54.2 trillion. Government construction investment would fall by a nominal 6.9 per cent from the previous year since, with a stable budget and no additional measures in the revised budget, there is nothing to carry forward from the revised budget. Private residential investment was expected to bottom out in some sectors, such as owned houses and condominiums, but this would not amount to a full-blown recovery, as investment would continue to decline, falling 1.4 per cent. The declining trend in private non-residential construction investment would also continue, as forecasts indicate that it will fall by 1.3 per cent.

These construction investment forecasts did not make predictions for what would happen beyond fiscal 2003, but it was expected to be quite some time before economic recovery begins in earnest and before that movement is then reflected in the private construction market. With a decline in public works projects, construction investment would also inevitably decline over the next 5 years since 2002. Even over the mid- to

Table 7.9 Trends in construction investment

Construction investment (unit: local currency at current price) (100 million yen)

	1998	*1999*	*2000*	*2001*	*2002**1*	*2003**1*
Residential construction (A)	209,924	217,955	212,500	194,600	180,819	177,914
public	12,303	10,717	10,900	10,900	9,244	8,707
private	197,621	207,238	201,500	183,700	171,575	169,207
Non-residential construction (B)	139,191	128,580	123,700	105,600	101,175	100,139
public	36,472	34,725	30,300	23,700	20,775	20,820
private	102,719	93,855	93,400	82,000	80,400	79,319
Civil engineering construction (C)	365,154	338,504	328,800	303,900	281,328	263,488
public	291,155	273,937	263,200	240,900	219,578	202,838
private	73,999	64,567	65,600	63,000	61,750	60,650
Total construction investment (A + B + C)	714,269	685,039	665,000	604,100	563,322	541,541
Total public	339,930	319,379	304,400	275,500	249,597	232,365
Total private	374,339	365,660	360,600	328,600	313,725	309,176

*1 2002, 2003 are forecast.

(unit: change at local currency at constant price)

	1998 (%)	*1999* (%)	*2000* (%)	*2001* (%)	*2002**2 (%)	*2003**2 (%)
Residential construction (A)	−10.0	4.7	−2.5	−7.4	−6.3	−1.7
public	−8.3	−12.0	1.6	0.9	−15.1	−5.5
private	−10.1	5.8	−2.7	−7.9	−5.7	−1.5
Non-residential construction (B)	−11.6	−6.6	−4.0	−13.8	−3.7	−1.0
public	−8.4	−3.7	−12.9	−21.1	−12.3	1.0
private	−12.7	−7.6	−0.7	−11.4	−1.3	−1.5
Civil engineering construction (C)	5.3	−6.3	−3.2	−6.7	−7.3	−5.9
public	7.6	−5.0	−4.3	−7.6	−8.9	−6.9
private	−2.8	−11.7	1.3	−3.0	−1.5	−2.1
Total construction investment (A + B + C)	−3.1	−3.1	−3.1	−8.3	−6.4	−3.7
Total public	5.0	−5.1	−5.0	−8.6	−9.4	−6.2
Total private	−9.5	−1.3	−1.5	−7.9	−3.8	−1.6

*2 2002, 2003 are forecast.

long term, negative factors such as fiscal stringency and the ageing and shrinking of the population would make it impossible to avoid further contraction.

Some figures up to 1997, will forecasts for 1998, 1999 are available from the Japan Research Institute indicating the relative sizes of different sectors of the construction market. These are shown in Table 7.10.

Private construction

The number of private housing starts in FY 1998 was expected to remain basically unchanged, while investment in private non-residential construction would likely be weak. Private housing starts would almost certainly decline in the first half of

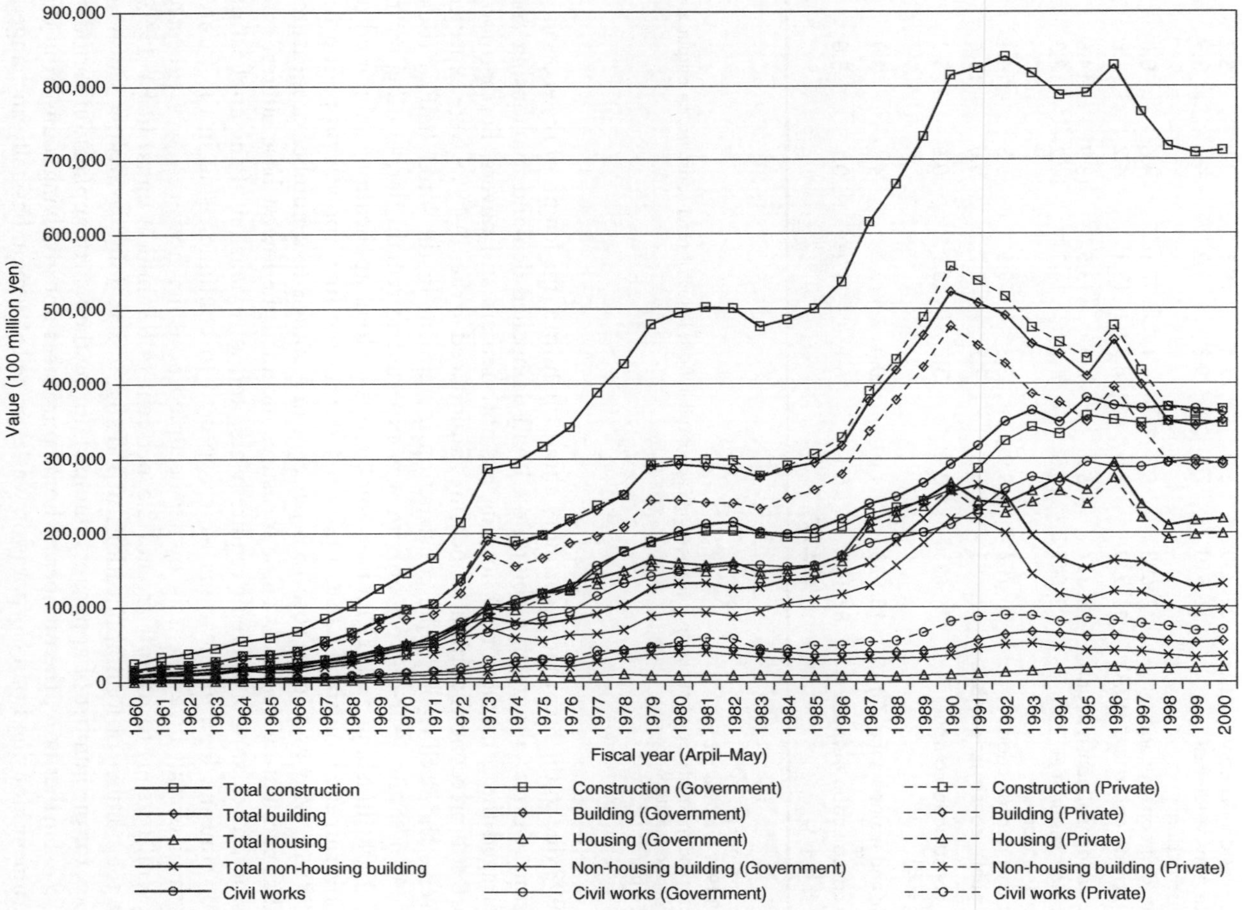

Figure 7.7 Estimate of construction investment.

Table 7.10 Construction market by type (per cent change at constant price)

	1997[1] (billion ¥)	1995 (%)	1996 (%)	1997 (%)	1998[2] (%)	1999[2] (%)
Private residential construction	22, 330	6.7	13.4	19.5	3.8	5.2
Private non-residential construction	11,910	7.1	9.3	3.2	4.6	3.9
Private civil construction	7, 690	5.3	4.1	7.1	4.6	0.0
Total private construction (A)	41, 930	4.6	8.9	13.1	4.2	1.6
Public residential construction	2, 337	6.3	6.4	13.5	13.9	14.6
Public non-residential construction	3, 649	10.5	1.3	14.2	11.2	18.3
Public civil construction	26, 707	9.9	3.3	9.5	5.5	14.4
Total public construction (B)	32, 693	6.8	2.5	7.7	4.0	13.4
Total construction output (A + B)	74, 623	0.2	3.8	10.9	0.6	5.3
Repair and maintenance (C)	15, 938	1.0	20.0	12.2	4.1	3.9
Total construction output (A + B + C)[3]	87, 552 (673.9)[4]	0.1	6.1	11.2	0.1	3.8

Notes
1 Nominal price.
2 1998,1999 are forecast.
3 Repair and maintenance of public civil engineering is counted in both items of total construction output and repair and maintenance.
4 US$4 billion.

FY 1998 due to the fact that uncertainty about the future (the length of the economic recession, worries about job stability, etc.) had dampened demand for housing and discouraged the construction of housing for both renters and buyers. Furthermore, the ¥16 trillion economic stimulus measures announced earlier in 1998 were expected to reverse the decline in public construction investment in the latter half of fiscal 2002. As a result, year-on-year growth was expected to remain flat; housing starts in FY 1998, finishing at 1.33 million units. Housing loan applications submitted to the Government Housing Loan Corporation during the first application period for FY 1998 (in May and June) rose 13.3 per cent in response to a number of measures aimed at stimulating demand. These measures included the lowest base interest rate in history, a relaxation of credit requirements, and an extension of the application period. Housing loan applications were expected to continue rising in the future. Growth in overall investment in owner-occupied housing, however, was expected for several reasons to top off around 2.5 per cent (4.62 million units) in FY 1998. While the number of loan applications was rising, it was doing so from a low base number, for the number of applicants slumped during the fourth application period in FY 1997. Furthermore, the number of home purchases not involving loans from the Government Housing Loan Corporation was expected to remain basically unchanged. As for rental housing, it appeared that the problem of oversupply was about to start clearing up. Housing starts in this category were expected to start bottoming out perhaps around mid-2002 before rising by 2.8 per cent to 530,000 units, particularly on the strength of brisk starts for up-market rental housing in prime locations. With regard to homes put on sale by builders, housing starts were forecast to drop by 9.9 per cent to under 316,000 units due to the following facts: (1) interest rates were not generally expected to rise; (2) a large amount of construction in the previous year and a rise in inventory have resulted in an oversupply which is affecting the market; and (3) builders are approaching any new land purchases with great caution.

Housing starts in FY 1999 were forecast to post a modest 5.6 per cent rise to approximately 1.405 million units because of the expected waning of the recession (refer Table 7.11). Another factor was the strong demand generated by an increase in the total number of households. Owner-occupied housing starts were expected to rise by 12.4 per cent to about 519,000 units. Borrowing by these owners from the Government Housing Loan Corporation, in particular, was expected to continue recovering because (1) many owners would probably tear down and rebuild; and (2) prices on residential land appeared to have bottomed out. Housing starts for rental homes were forecased to rise by 0.6 per cent to 533,000 units, supported largely by increased investment in prime locations. Until the time of writing, considerable progress towards inventory adjustment would have been made by FY 1999, the oversupply problem in the rental market would probably have tapered off, and land owners would continue to be very interested in rental property investments due to changes in inheritance taxes. Housing starts of homes for sale by builders were expected to rise, especially in prime locations, by 4.0 per cent to around 328,000 units thanks to: (1) continued low interest rates and other favourable investment factors; and (2) the fact that demand had bottomed out.

Private capital investment was on an unmistakable downward trend, falling 5.6 per cent in January–March 1998 over 2001. This marked the first decline since July–September 1994. Capital investment for FY 1998 in both the manufacturing and non-manufacturing sectors was expected to decline. Orders for machinery, which functioned as a leading indicator, had remained stagnant since October–December 1997, clearly indicating that capital investments were affected by stock adjustment. Due to slumping private consumption, it appeared that there would be a long wait before the government's ¥16 trillion comprehensive economic measures take effect, spurring private demand and bringing about spontaneous economic recovery. In view of these circumstances, private capital investment was expected to decline.

Investment in private non-residential construction (buildings and civil engineering projects) was also forecast to post real declines in FY 1998 and FY 1999 due to sluggish activity in the private sector.

Table 7.11 Housing starts

FY	*1995*	*1996*	*1997*	*1998*	*1999*
Owner-occupied	550.5	636.3	447.8	462.2	519.5
(growth rate)	5.2%	15.6%	29.6%	2.5%	12.4%
Rental	563.7	616.2	519.4	530.3	533.5
(growth rate)	1.8%	9.3%	15.7%	2.8%	0.6%
For sale by builder	344.7	352.0	349.7	315.9	328.5
(growth rate)	8.7%	2.1%	0.7%	9.9%	4.0%
Total	1,484.7	1630.4	1340.3	1330.8	1405.6
(growth rate)	−4.9%	9.8%	17.8%	0.8%	5.6%

Investment in private non-residential buildings would also drop by in FY 1998 and FY 1999, thus falling for 3 straight years after the short-lived recovery of 1996. Floor space for private non-residential building starts was expected to post a sharp 7.3 per cent drop in FY 1998 before rising very meagrely in FY 1999. Further breaking down these floor space figures by usage category, we find that office floor space building starts were estimated to be set to drop 13.9 per cent in FY 1998 and 1.1 per cent in 1999 due to the economic recession. Commercial floor space building starts, fuelled by the construction of large commercial facilities on the outskirts of metropolitan areas, should post the only gain to be seen in FY 1998, followed by a very small gain in FY 1999. As for factories, weak capital investment in the manufacturing sector should result in a large drop in FY 1998. The doldrums were expected to continue in FY 1999, with floor space for factory building starts expected to post a meagre rise of 0.7 per cent. Investment in private civil engineering was forecast to fall in FY 1998, thus marking the third consecutive decline. The growth rate for FY 1999 was expected to be zero.

Public construction

There were worries about the economic impact of a decline in public construction investment in FY 1999. Public construction investment in FY 1998 was forecast to post positive growth for the first time in 3 years. Although expenditures in this area were cut drastically in the main budget, large supplementary spending was provided for in the comprehensive economic measures. Due to the stimulus package and a push to carry out contracts ahead of schedule, public construction investment was expected to post a year-on-year gain in the second quarter of FY 1998. For the entire year, a real growth of about 4.0 per cent was forecast.

Public construction investment was expected to plummet in FY 1999. The Fiscal Structural Reform Law requires that ordinary public works expenditures in the FY 1999 main budget be less than those in FY 1998. If main-budget public works spending is reduced by 1 per cent at the national, prefectural, and local levels, public construction investment in FY 1999 can be expected to fall 13.7 per cent, even if part of the FY 1998 comprehensive economic measures are carried over to FY 1999 because the stimulus measures would have largely run their course before FY 1999 is out. This sharp drop in demand is likely to have a distinctly negative impact upon the Japanese economy.

Repair and maintenance

The repair and maintenance market is expected to achieve stable growth. The repair and maintenance market in FY 1996 was abnormally large due to the Great Hanshin–Awaji earthquake of January 1995. That explains why the market shrank in FY 1997 from the previous period, but over the medium to long term, this market will continue to grow. Building stock continues to grow every year, so there is a good reason to infer that the repair and maintenance market will continue to post stable growth.

Repair and maintenance is not limited to ensuring that homes and buildings remain in their original condition. There is also strong demand for modifications to increase the functionality of existing structures. Non-residential structures are frequently modified for the sake of office automation, or for upgraded air-conditioning systems.

They are also remodelled to increase guest-handling capacity and to achieve an enhanced image. Private homes are sometimes remodelled to accommodate two generations in a single household, to suit the needs of smaller family units, or simply to make better use of space. Table 7.12 shows the major projects programmed and underway at the time of writing.

Table 7.12 Major projects programmed and underway at the time of writing

Name	*Total cost*	*Construction period*	*Explanation of project*
Joban New Line	Est. ¥1.05 trillion	1993–2005	New railway which was expected to connect Tokyo and the Tsukuba Science City (58.3 km)
Makuhari New City Centre Construction	Est. ¥3.0 trillion (including the cost of building private-sector buildings)	1972–2000	Project area covers a total of 522.2 ha
No. 2 Tomei–meishin expressway	Est. ¥5.6 trillion (for approved 303 km)	1994–unsettled	490 km, 6-lane expressway
Hokuriku–Shinkansen (Takazaki–Komatsu) (Komatsu–Nanetsu)	Est. ¥3.4 trillion (est. ¥1.41 trillion) (est. ¥260 billion)	1989–	Total length of nearly 690 km (for the Tokyo to Takazaki portion between Tokyo and Osaka, the Joetsu Shinkansen is used. The Takazaki to Osaka portion is nearly 590 km)
Technoport Osaka	Est. ¥2.2 trillion	1983–2010	Construction of an international information city on the Osaka coastline. Covers a total of 775 ha
Shikoku–Odan expressway	Est. ¥522.0 billion	1985–	Total length of 443 km
Kyushu–Odan expressway	Est. ¥991.0 billion	1972–	Nagasaki–Oita: 254 km
Kansai International Airport (phase 2)	Est. ¥1.6 trillion	1998–	Runway 4,000 m reclaimed land 530 ha
Chubu New International Airport	Est. ¥700 billion	Scheduled to open in 2005	Two 4,000 m runways
Linear Central Shinkansen	Undecided	At the start of the twenty-first century	Tokyo–Kofu–Nagoya–Nara–Osaka
International exposition	Undecided	2005	Preparations for holding this exposition based on the 2005 International Exposition Treaty are proceeding
Relocation of capital functions	Est. ¥14 trillion	Undecided	Under revision by the Committee on Relocation of the National Diet, etc.

Construction cost

Tables 7.13–7.16 provide basic data on costs per cubic metre of building types, material prices, labour wages and sectoral wage costs.

7.2.3 Property market review and outlook

Trends in demand, supply, and price of property

LAND PRICES

According to officially announced government figures, land prices changed on a national average at about the same rate in 1997 as they did during the previous year. The price of residential land fell 1.4 per cent (versus −1.6 per cent in FY 1996), while that of commercial land declined by 6.1 per cent (versus −7.8 per cent in FY 1996).

Table 7.13 Construction costs (unit: 1,000 yen/m^2)

	Residential	*Offices*	*Stores*	*Industrial sites and factories*	*Schools*
1997	169	195	117	115	220
1998	167	193	115	116	218
1999	165	205	112	108	216
2000	162	187	100	104	202
2001	159	197	116	103	205

Source: Monthly of Construction Statistics, Ministry of Land, Infrastructure and Transport.

Note
Construction unit cost = Value of construction by investors reported per unit of area.

Table 7.14 Average construction material price (unit: yen)

	Cement in bulk (per ton)	*Steel bars* (per ton)	*Aggregate* (per m^3)	*Sand* (per m^3)	*Ready-mixed concrete* (per m^3)	*Precast concrete piles* (per pile)
1997	9,090	34,600	4,350	4,800	12,200	32,100
1998	8,850	29,700	4,190	4,710	12,000	32,100
1999	8,730	25,100	3,900	4,560	11,700	31,140
2000	8,880	26,300	3,700	4,350	11,700	29,000
2001	8,650	26,800	3,640	4,150	11,600	28,850

Source: Prices of Construction Materials and Wages, Construction Research Institute.

Note
All prices are for materials procured in Tokyo.

Types and specifications

Cement	Regular portland cement
Steel bars	D19 deformed bars (SD295A)
Aggregate	20–50 mm crushed stone for concrete
Sand	Coarse, washed
Ready-mixed concrete	Regular concrete (nominal strength: 18 N/mm^2; slump: 18 cm; maximum coarse aggregate size: 25 mm)
Precast concrete piles	PHC piles (type A, ϕ350 x 10 m)

Table 7.15 Construction industry salaries and wages (unit: yen)

	1997	*1998*	*1999*	*2000*	*2001*
Class I chartered architect (monthly salary)	452,800	443,900	471,000	451,000	430,600
Skilled worker (daily wage)	15,180	15,060	14,640	14,550	14,620
Unskilled worker (daily wage)					
Common labourer	13,190	13,010	12,910	12,840	12,860
Labourer, heavy	14,510	15,940	14,760	13,810	13,440
Labourer, light (male)	10,990	10,810	10,770	10,850	10,650
Labourer, light (female)	7,710	7,780	7,550	7,720	7,670

Sources: Figures for class I chartered architects are based on the average monthly salaries listed in the Basic Wage Structure Report, Ministry of Health, Labour and Welfare; and figures for skilled worker, common labourer, labourer (heavy) and labourer (light) are based on the Classified Outdoor Labour Wage Survey, Ministry of Health, Labour and Welfare.

Note

The "skilled worker" category includes carpenters, scaffold workers, masons, plasterers, electricians, plumbers, painters, truck drivers, sheet metal workers, welders, machine operators, steel workers, steel frame workers, drilling workers, blasters, tile setters, bricklayers, cement and concrete finishers, form builders, joiners, roofers, caisson workers, and boring machine operators.

Table 7.16 Average sectoral wage costs (per month) (unit: yen)

	1997	*1998*	*1999*	*2000*	*2001*
Construction	386,555	374,424	377,894	380,209	370,278
Manufacturing	375,612	371,437	366,793	371,881	368,200
Services	379,436	378,320	371,942	369,572	366,366
All industries	371,670	366,481	353,679	355,572	350,009

Source: Compiled from the total cash wages paid to full-time employees by companies with five or more employees, based on the full-time employees reported in the Monthly Labour Survey, Ministry of Health, Labour and Welfare.

Averaging together the three major metropolitan areas of Tokyo, Osaka, and Nagoya, for 1997 the decline in the price of commercial land slowed to single-digit percentages after 6 consecutive previous years of double-digit contraction. The price of residential land also fell by 2.2 per cent (versus an average annual decline of 2.8 per cent during the previous 6 years), thus showing a continued downward but stable trend.

In the Tokyo area, the price of residential land fell 3.0 per cent (3.4 per cent annually over the previous 6 years), while that of commercial land dropped 8.2 per cent (13.2 per cent annually over the previous 6 years). The rate of decline in prices for both residential and commercial land slowed down most notably in the 23 wards of Tokyo itself. At the time of writing, the standard price of commercial land had fell below where it was in 1983 when the bubble economy began, but the prices of some prime commercial lots rose slightly for the first time in 7 years.

In both the Osaka and Nagoya area, the price of residential land in 1997 fell 1.5 per cent (2.2 per cent annually over the previous 6 years), while that of commercial land dropped 6.8 per cent (9.9 per cent annually over the previous 6 years); the rate of decline had slowed down.

In other areas, at the time of writing, the price of residential land had fallen between 0.3 and 0.6 per cent annually over the past 4 years, while that of commercial land dropped between 5.1 and 5.8 per cent annually over the same period. Land prices had thus declined less here than in the three major metropolitan areas, but the downward trend continues.

THE REAL ESTATE MARKET

Trends in land prices in major cities According to the Posted Land Prices compiled by the MLIT, 2002 average land prices nationwide fell for the 11th consecutive year since the collapse of the bubble economy. The average price for residential land fell by 5.2 per cent (4.2 per cent in 2001), while that of commercial land fell by 8.3 per cent (7.5 per cent in 2001). Land prices in major cities as well as smaller localities fell by greater amounts than in 2001, and there was no end in sight for where they would finally hit bottom (Table 7.17). The trends in residentials and commercial land prices are shown in Figures 7.8 and 7.9 respectively.

Trends in rents in major cities The office vacancy rate in Tokyo's 23 wards was 4.5 per cent as of March 2002, an increase of 0.9 per cent over the same year-ago period. Due to the uncertain future of the economy overall and people who are waiting to rent in anticipation of future supply increases, projected increases in office demand have not materialized. Average advertised rents fell by 1.5 per cent from

Table 7.17 Officially announced land prices since 1997 (unit: per cent change from previous year)

Region	*Land use*	*1997*	*1998*	*1999*	*2000*	*2001*	*2002*
National	Average for all land uses	−2.9	−2.4	−4.6	−4.9	−4.9	−5.9
	Commercial land	−7.8	−6.1	−8.1	−8.0	−7.5	−8.3
	Residential land	−1.6	−1.4	−3.8	−4.1	−4.2	−5.2
Greater Tokyo	Average for all land uses	−5.1	−3.9	−7.1	−7.4	−6.4	−6.4
	Commercial land	−13.2	−8.2	−10.1	−9.6	−8.0	−7.4
	Residential land	−3.4	−3.0	−6.4	−6.8	−5.8	−5.9
Greater Osaka	Average for all land uses	−3.4	−2.3	−5.9	−6.9	−7.4	−9.1
	Commercial land	−9.9	−6.8	−9.6	−11.3	−11.0	−11.3
	Residential land	−2.2	−1.5	−5.2	−6.1	−6.7	−8.6
Greater Nagoya	Average for all land uses	−3.0	−1.9	−4.9	−3.0	−2.8	−5.3
	Commercial land	−8.5	−6.2	−11.2	−7.3	−5.6	−8.1
	Residential land	−1.7	−0.8	−3.3	−1.8	−1.9	−4.4
3 main metropolitan areas	Average for all land uses	−4.3	−3.2	−6.4	−6.6	−6.1	−6.9
	Commercial land	−11.5	−7.5	−10.2	−9.6	−8.3	−8.5
	Residential land	−2.8	−2.2	−5.7	−5.9	−5.6	−6.5
Non-metropolitan Japan	Average for all land uses	−1.6	−1.7	−3.0	−3.4	−3.8	−5.0
	Commercial land	−5.4	−5.1	−6.8	−7.0	−7.0	−8.1
	Residental land	−0.4	−0.6	−1.9	−2.3	−2.8	−4.0

Source: Posted Land Prices, Ministry of Land, Infrastructure and Transport.

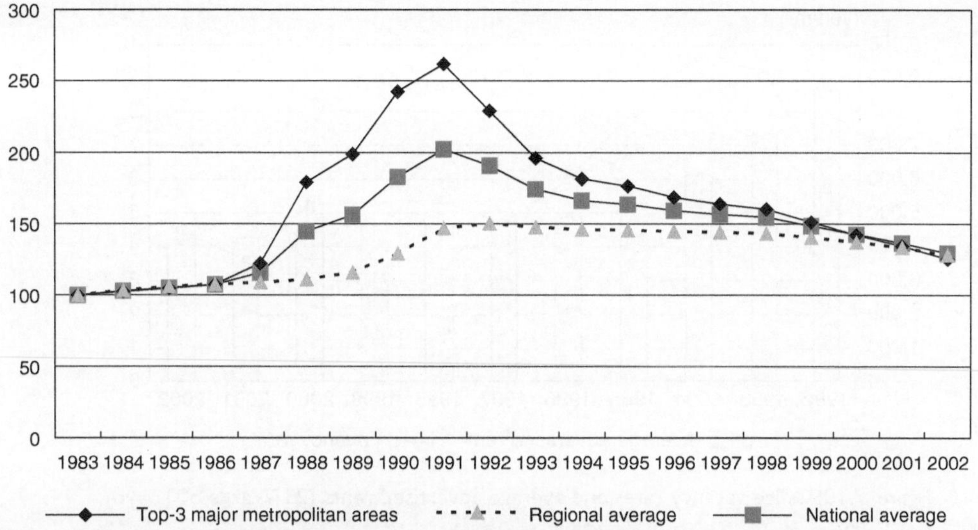

Figure 7.8 Trends in residential land prices (indexed).

Source: Posted Land Prices, Ministry of Land, Infrastructure and Transport.

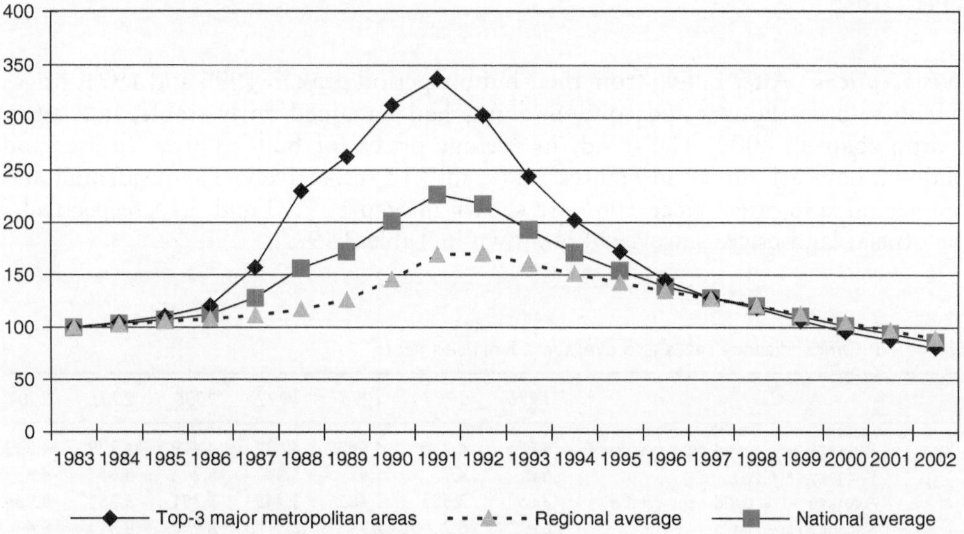

Figure 7.9 Trends in commercial land prices (indexed).

Source: Posted Land Prices, Ministry of Land, Infrastructure and Transport.

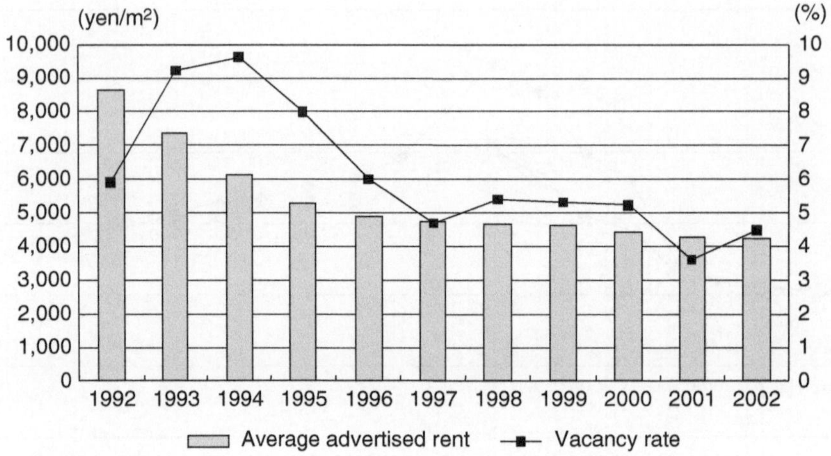

Figure 7.10 Office vacancy rates and average advertised rents (23 Wards of Tokyo).
Source: Office Market Report (Ikoma/CB Richard Ellis).

the previous period to ¥4,233 per square metre and rates were falling at a moderate pace.

Movements in office demand had been slow, with vacancy rates having deteriorated by 0.9 points versus the year-ago period in Osaka, and by 0.3 points in Nagoya. Average rents were continuing to fall as in the past (refer Figure 7.10 and Table 7.18).

Housing prices After falling from their bubble-period peak in 1990 and 1991, prices of built-to-order houses and condominiums had remained fairly stable, but began to drop again in 2000. The trends in average prices for built-to-order houses and condominiums are shown in Figures 7.11 and 7.12 respectively. The residential and commercial land prices since 1983 are shown in Figures 7.13 and 7.14 respectively. The official land prices since 1994 is shown in Table 7.19.

Table 7.18 Office vacancy rates and average advertised rents

		1996	*1997*	*1998*	*1999*	*2000*	*2001*	*2002*
Tokyo	Average advertised rent (yen/m^2)	4,885	4,730	4,648	4,627	4,418	4,279	4,233
	Vacancy rate (%)	6.0	4.7	5.4	5.3	5.2	3.6	4.5
Osaka	Average advertised rent (yen/m^2)	3,691	3,555	3,485	3,445	3,391	3,255	3,064
	Vacancy rate (%)	6.4	5.7	7.3	7.5	8.9	8.5	9.4
Nagoya	Average advertised rent (yen/m^2)	3,191	3,197	3,188	3,139	3,112	3,082	3,018
	Vacancy rate (%)	4.5	3.7	4.8	5.1	5.8	6.1	6.4

Source: Office Market Report (Ikoma/CB Richard Ellis).

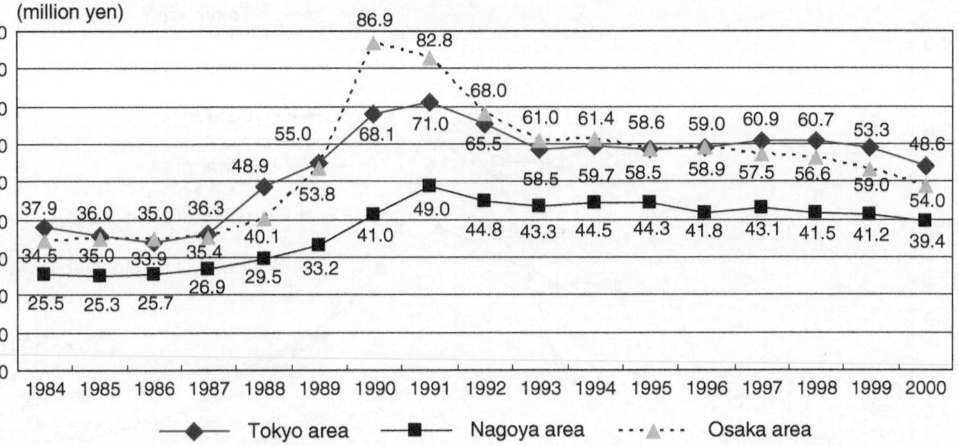

Figure 7.11 Trends in average prices (built-to-order houses).

Source: Report on the residential land and structure supply performance study, Urban development association.

Notes

The regional breakdown is as follows:

Tokyo region – Tokyo, Kanagawa, Saitama, Chiba, Ibaraki, and Tochigi prefectures.

Nagoya region – Aichi, Mie, and Gifu prefectures.

Osaka region – Osaka, Hyogo, Kyoto, Nara, Shiga, and Wakayama prefectures.

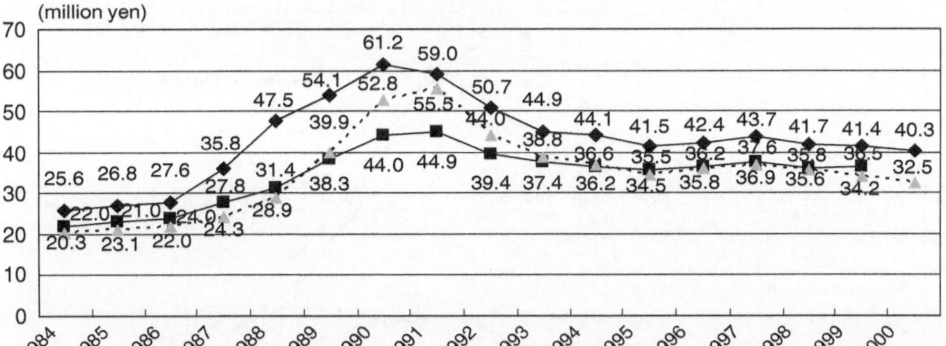

Figure 7.12 Trends in average prices (condominiums).

Source: Nationwide Condominium Market Trends, Japan Real Estate Institute.

Notes

1 The regional breakdown is as follows: Tokyo region – Tokyo, Kanagawa, Saitama, and Chiba prefectures.
Kinki region – Osaka, Hyogo, Kyoto, Nara, Shiga, and Wakayama prefectures.

2 Does not include resort condominiums.

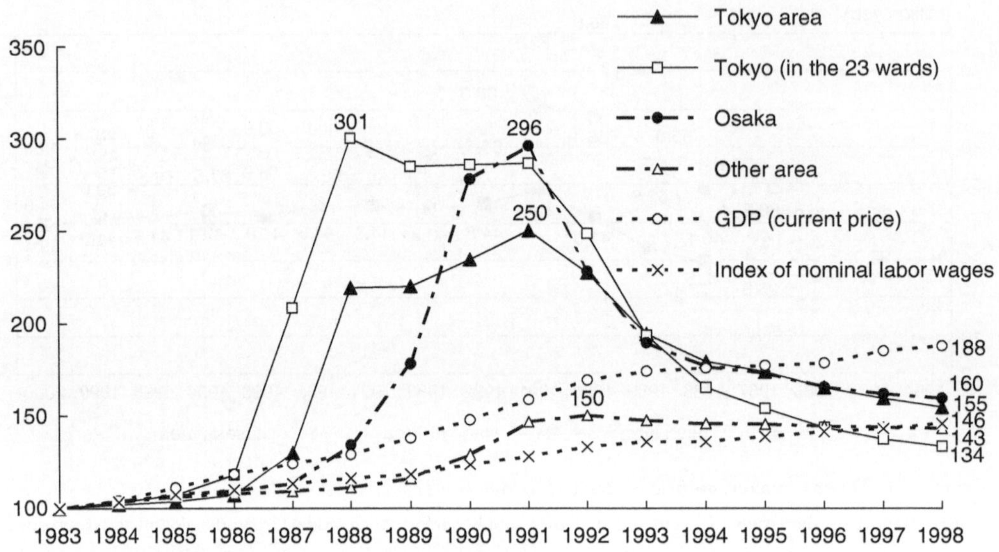

Figure 7.13 Residential land prices since 1983 (indexed).
Source: Official land prices and the Japan Statistical Yearbook.

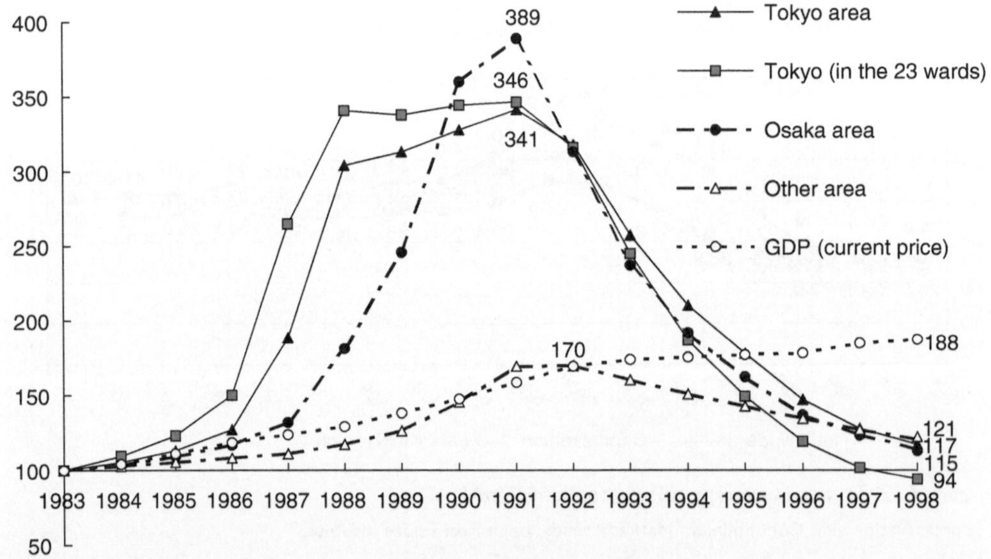

Figure 7.14 Commercial land prices since 1983 (indexed).
Source: Official land prices and the Japan Statistical Yearbook.

Table 7.19 Official land prices since 1994 (per cent change)

		1995	*1996*	*1997*	*1998*
Total	Total	3.0	4.0	2.9	2.4
	Commercial land	10.0	9.8	7.8	6.1
	Residential land	1.6	2.6	1.6	1.4
Tokyo area	Total	5.0	7.0	5.1	3.9
	Commercial land	15.4	17.2	13.2	8.2
	Residential land	2.9	5.0	3.4	3.0
Osaka area	Total	4.0	6.0	3.4	2.3
	Commercial land	15.3	15.8	9.9	6.8
	Residential land	1.9	4.3	2.2	1.5
Nagoya area	Total	5.6	5.2	3.0	1.9
	Commercial land	12.7	12.6	8.5	6.2
	Residential land	4.0	3.6	1.7	0.8
Three major areas	Total	4.8	6.4	4.3	3.2
	Commercial land	14.8	16.0	11.5	7.5
	Residential land	2.8	4.6	2.8	2.2
Other areas	Total	1.2	1.8	1.6	1.7
	Commercial land	5.5	5.8	5.4	5.1
	Residential land	0.3	0.6	0.4	0.6

Source: Official land prices and the Japan Statistical Yearbook.

LAND TRANSACTIONS

In FY 1997, 1.85 million land transactions occurred (i.e. registered transfer of a property deed in connection with a sale) in Japan (refer Figure 7.15). This figure was 6.6 per cent lower than that for the previous year, and was the first such drop in 4 years. (The year before this had seen an increase of 5.9 per cent.) While transactions in the first half of the fiscal year only dropped 1.7 per cent from the year-ago period,

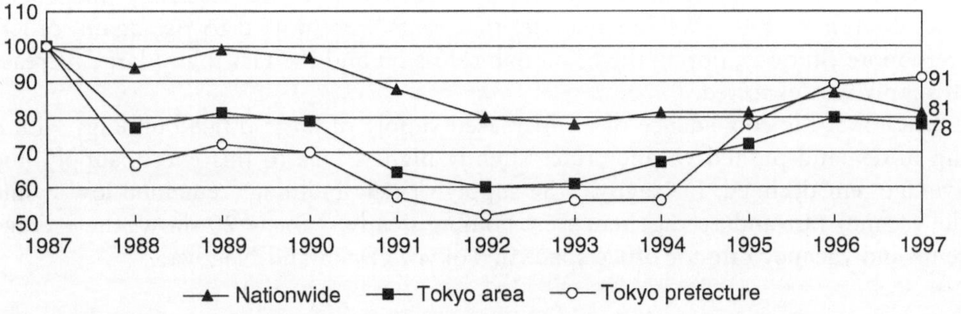

Figure 7.15 Number of land transactions since 1987 (indexed).
Source: Ministry of Justice Monthly Statistical Review.

Note
"Tokyo area" includes Tokyo Prefecture, Kanagawa Prefecture, Saitama Prefecture, and Chiba Prefecture.

those in the latter half plummeted by 9.8 per cent. In Tokyo prefecture, the number of transactions rose 1.6 per cent from the previous year, but year-on-year growth of 7.4 per cent in the first half of the fiscal year was followed by a decline of 4.4 per cent in the latter half. Strong demand for housing in the first half of the year was fuelled by dropping land prices and low interest rates, but this demand wilted in the latter half in the face of a consumption tax hike and the worsening of the economic recession beginning in the autumn.

PROPERTY MARKET TRENDS

Housing New housing starts in Japan in FY 1997 dropped 17.7 per cent to 1.34 million units due to (1) the fact that a sudden upward spike in demand ahead of the April 1997 consumption tax hike was followed by a big slump once the higher tax took effect and (2) worries about the economy.

New housing starts in the Tokyo prefecture fell 8.9 per cent. In the category of owner-occupied housing, in particular, the figure fell a whopping 27.7 per cent. The only bright spot, once again, was homes for sale by the builder, where housing starts rose 9.3 per cent. Average floor space, however, fell 2.2 per cent to 79.7 square metres the first such drop since 1990.

Sales of new apartments in greater Tokyo area dropped 14.8 per cent in 1997, and have continued spiralling rapidly downward in 1998. Only about 70 per cent of all apartments were being sold within a month after being put up for sale, and there are nearly 10,000 unsold units on the market. In addition, sales ratios for previously occupied apartments and detached houses were also on the decline, and prices were tumbling.

Office buildings In the 23 wards within Tokyo, the 6 month vacancy rate for office space stands at 4.7 per cent, virtually the same as for September–December 2001 (refer Figure 7.16). In five centrally located wards, vacancy rates have climbed upwards for the first time in 3 years, with even large buildings in prime locations showing signs of higher vacancy. In addition, the average rent has remained basically unchanged since March of 2001. While some feel that rents were poised to rise again, others were more cautious, noting the economic recession and the fact that a large increase in supply was expected.

In Osaka, the coincidence of an increased supply of large office buildings with a dip in demand pushed vacancy rates slightly higher. Due to this excess supply, the average rent declined. In Nagoya, the supply of new buildings remained low, while the vacancy rate and average rent were holding steady. Table 7.20 shows the average rents and vacancy rate for office space in Tokyo, Osaka and Nagoya.

Recent policy on city, land and housing

Table 7.21 shows the recent policy on city, land and housing. A comprehensive plan to create liquid markets for bad loans and associated real estate collateral (facilitating

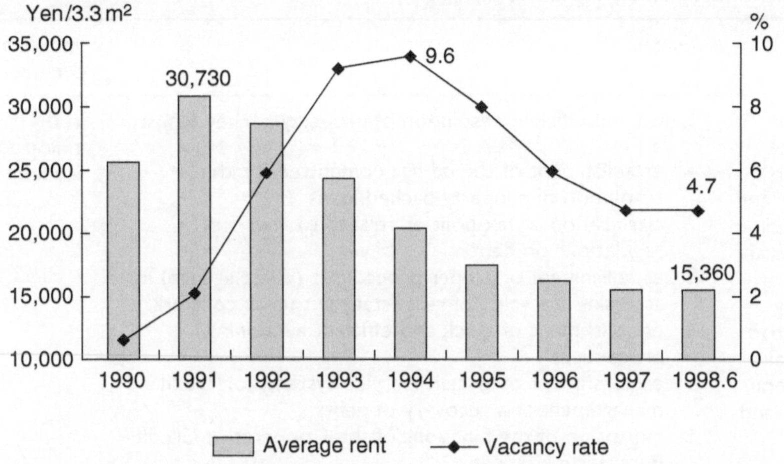

Figure 7.16 Average vacancy rate and rent for office space in Tokyo (23 wards).

Source: Office Market Report (Ikoma Data Service System), Vacancy rates are for December (except 1990, when it was for May).

Table 7.20 Average rents and vacancy rate for office space

	1994	*1995*	*1996*	*1997.6*	*1998.6*
Tokyo					
Average rents[1]	6,132	5,248	4,876	4,683	4,646
Vacancy rate (%)	9.6	8.0	6.0	5.0	4.7
Osaka					
Average rents	4,029	3,796	3,684	3,536	3,488
Vacancy rate (%)	6.8	7.0	6.4	5.6	6.3
Nagoya					
Average rents	3,394	3,255	3,185	3,143	3,104
Vacancy rate (%)	7.2	6.6	4.5	4.1	3.9

Source: Ikoma data service system.

Note

1 yen per square metre.

circulation of lands and securities in the market and promoting effective use of land) has been prepared (Table 7.22).

According to a 1991 study, there was a total of approximately 700 hectares of underused and unused land in eight of Tokyo's most centrally located wards. Of this land, some 80 per cent consisted of small lots measuring 300 square metres or less. The occupancy of the floor area (approximate actual floor area ratio/designated average floor area ratio) for the 23 wards as a whole was still short o 50 per cent.

Table 7.21 Recent policy on city, land and housing

Name of policy	*Content*	*Implementation date (period)*
Comprehensive plan to create liquid markets for bad loans and associated real estate collateral (creation of active trading markets for bad loans and real estate collateral, and efficient land use)	Quick and efficient resolution of property-backed loans: 1 establishment of the *ad hoc* committee for the resolution of property-backed loans 2 clarification of tax policies related to laws and regulations on debt 3 establishment of proper procedures (due diligence) for assessing the value of real estate put up as collateral 4 establishment of quick and efficient auctioning procedures 5 establishment of system (service system) for credit management and recovery of debts 6 expansion of the functions of the Cooperative Credit Purchasing Company 7 expansion of the functions of the auctioning subsidiaries of financial institutions 8 creation of secondary markets for asset-backed securities (ABS). Promotion of land consolidation and urban redevelopment: 1 system for urgent promotion of redevelopment • early implementation of high-priority urban redevelopment projects 2 active use of the Housing and Urban Development Corporation • establishment of an agency for promotion of effective land use • establishment of ¥30 million fund for purchase of underused and unused land. Consolidation of land and development of infrastructure to promote effective land use 3 establishment of system to promote effective use of the Organization for Promoting Urban Development • establishment by the Organization for Promoting Urban Development of a Centre for Promotion of Urban Redevelopment and Effective Land Use. Support of efforts to provide information and commentary on the real estate market, and development of the same • expansion of land purchasing fund to ¥50 million to promote private-sector development projects. Creation of public-sector demand for land: 1 Purchase of public-use lands to improve disaster preparedness, upgrade social services infrastructure, and revitalize downtown areas.	In the indicated sequence

Revitalization of downtown neighbourhoods	Passage of the Law on Downtown Areas (June 1998). Formulation of basic plan to develop urban neighbourhoods and revitalize commercial districts in accordance with policies adopted by local governments. Support by the national government for projects included in the basic plan.	Beginning in 1998

Table 7.22 Distribution and size of construction companies

Company size	*FY 1998* *No. of companies*	*FY 1999* *No. of companies*	*Share of total in 1999* (in %)
Fewer than 5	48,238	47,719	26.9
5–49	126,446	121,455	68.5
50–99	5,208	4,812	2.7
100–299	2,394	2,314	1.3
300–999	631	638	0.4
1,000+	257	242	0.1
Total	183,174	177,180	

Source: Annual Report on the Construction Works Survey, 1998 and 1999, Research and Information Division, Economic Affairs Bureau, Ministry of Land, Infrastructure and Transport.

Note
These figures are based on a sampling of Japan's construction firms. Sample data was extrapolated to obtain an approximation of the total numbers.

According to a recent study by the Organization for Promoting Urban Development, there are 167 pieces of property (approximately 24 hectares) in a state of total disrepair in a four-ward area. Some of this land was purchased by real estate developers and later abandoned. In other cases it was put up as collateral for a loan subsequently gone sour, or is otherwise encumbered.

Financial institutions, real estate companies, and construction firms are making little progress towards clearing up non-performing assets and loans, or towards creating active trading markets for property that has been put up as collateral on loans gone bad. The resulting bottleneck has partially squelched the business activities of business firms and other players in the Japanese economy.

In the comprehensive economic measures of April 1998 and the comprehensive plan for revitalization of the financial system (the latter of which was subsequently adopted by the government and ruling party in July), many measures are included as part of a total plan to inject liquidity into property and loans, and to encourage effective use of land. Legislation was being formulated at the time of writing to support the execution of these measures.

7.3 The construction industry

7.3.1 *State of the construction industry as of 2002*

Number of construction firms

TOTAL NUMBER OF CONSTRUCTION FIRMS

There were 568,548 firms licensed to engage in construction as of 31 March 1998, up 0.7 per cent from the previous year. The number of licensed construction firms began falling in March 1985 and bottomed out at 508,874 in 1990. Since then it has climbed steadily upwards (Figure 7.17).

Although the number of licensed construction firms has grown, Japan's economic recession and the contraction of the construction market have also led to an increase in the number of bankruptcies in the industry (Figure 7.18).

About 95 per cent of all licensed general contracting firms are small firms with less than 50 employees.

The contraction in the construction market since the 1990s has been described, but figures show that there was a serious supply surplus in the market. Construction investment (nominal value) per licensed construction company had dropped by 40 per cent from ¥160 million in 1991 to ¥99 million in 2002 and the construction investment per construction worker had dropped by 30 per cent from ¥13.6 million in 1991 to ¥9.6 million in 2002. Also, while future contractions in the construction market will inevitably focus on the government sector, adjustments to the numbers of construction companies and construction workers to suit the size of the market, through bankruptcies and worker lay-offs, for example, are not likely to be avoided (Figure 7.19).

Mirroring the situation regarding the number of construction firms, the number of bankruptcies had also been on the rise since bottoming out at 1,385 cases in 1990. The figure reached 4,785 in 1997. A string of bankruptcies among publicly listed

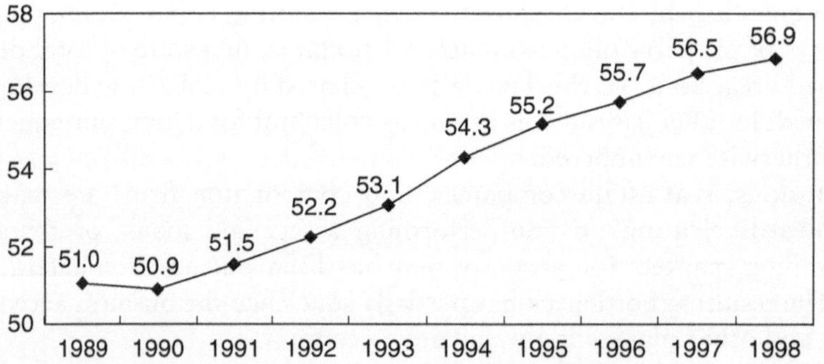

Figure 7.17 Number of construction firms since 1989.

Unit: 10,000 firms

Source: Ministry of Construction, Economic Affairs Bureau, Construction Contractors Division.

Note

Each year's figures are for 31 March.

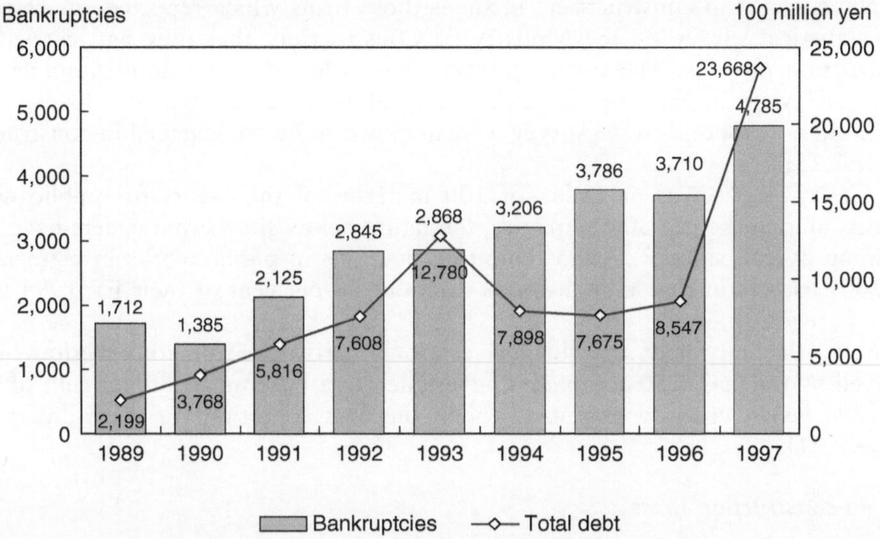

Figure 7.18 Construction sector bankruptcies and debts since 1989.

Source: Teikoku Data Bank Ltd.

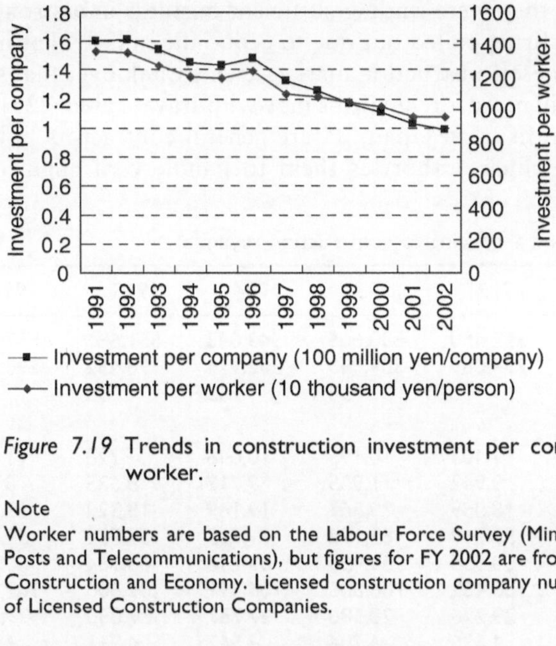

Figure 7.19 Trends in construction investment per construction company and construction worker.

Note

Worker numbers are based on the Labour Force Survey (Ministry of Public Management, Home Affairs, Posts and Telecommunications), but figures for FY 2002 are from estimates issued by Research Institute of Construction and Economy. Licensed construction company numbers are from the Survey of the Number of Licensed Construction Companies.

construction firms in 1997 had a serious impact upon total construction sector debt, pushing the figure up to ¥2.3668 trillion.

Both bankruptcies and the number of licensed construction firms were rising at the same time. However, the percentage of licensed construction firms actually engaged in construction projects has gone down every year since 1993. The category of "firms actually engaged in construction" includes those firms which reported, in response to a statistical survey by the Ministry of Construction, that they had carried out construction projects. The survey subjects were selected by random sampling, and the results were applied to the entire body of licensed construction firms. Firms which did not respond to the survey were assumed to be not engaged in construction (Table 7.23).

If FY 1994 is set as an index of 100 in terms of the orders for public works projects among the top 50 companies, the figure below shows that orders have been declining overall since FY 1995. Due to reductions in public works investments in FY 2001, orders in that year dropped to about 60 per cent of their FY 1994 levels (see Figure 7.20).

Due to the impact of the sluggish economy, private-sector construction orders received by the major 50 companies have fallen to a little under 80 per cent of their FY 1994 levels, creating extremely tough business conditions in the industry (see Figure 7.21).

Foreign construction firms

Foreign construction firms licensed to operate in Japan numbered 71 as of 31 March 1998, nine fewer than had been operating here one year earlier (11 firms withdrew from the market, while two new ones entered). Of the 11 firms that withdrew from the Japanese market, six were from the United States and three were from South Korea. The US firms withdrew because they were finding sufficient business in a strong US market, while the South Korean firms pulled out due to economic upheaval at home (Table 7.24). However, the Japanese construction market itself shrank over that same period. If this slump lasts long, still more foreign firms may withdraw (Table 7.25).

Of the 71 foreign construction firms in Japan, 24 are general contractors holding a specified contractor's license which authorizes them to handle civil engineering

Table 7.23 Construction firms (classed according to construction revenue)

Fiscal year	*1992*	*1993*	*1994*	*1995*	*1996*
Licensed construction firms	522,450	530,665	543,033	551,661	557,175
Firms engaged in construction	294,069	309,949	302,974	298,452	296,241
% of firms engaged in construction	56.3	58.4	55.8	54.1	53.2
Construction revenue					
Under ¥1 million	11,403	10,195	10,046	7,728	11,355
¥1–10 million	9,942	11,955	9,712	8,635	8,535
¥10–20 million	18,039	19,567	19,169	18,322	19,053
¥20–50 million	57,219	62,192	61,705	57,878	57,563
¥50–100 million	60,555	63,459	63,238	63,164	60,447
¥100–500 million	100,482	106,806	104,391	107,600	102,788
¥500 million–2 billion	29,226	28,586	27,757	27,830	29,006
¥2–5 billion	4,636	4,704	4,563	4,911	4,923
¥5–50 billion	2,357	2,269	2,182	2,170	2,335
¥50 billion and up	209	216	210	213	234

Source: Ministry of Construction, Economic Affairs Bureau, Research and Information Division. "Annual Report on the Construction Works Survey".

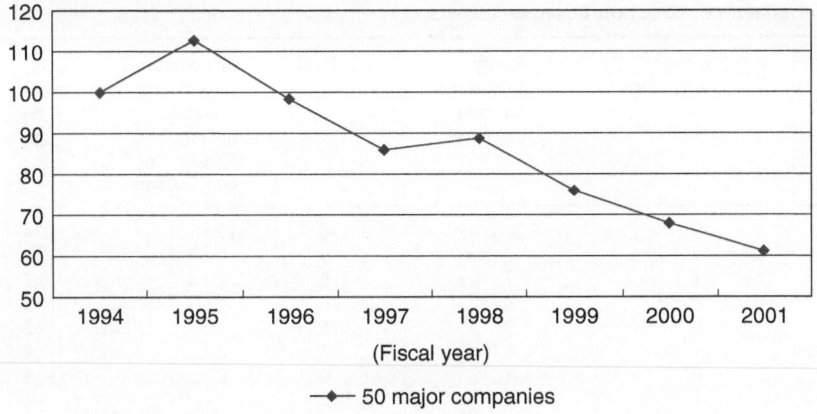

Figure 7.20 Trends in public-sector construction orders received by the major construction companies.

Source: Fifty major companies of orders received, Ministry of Land, Infrastructure and Transport.

Note
1994 = 100.

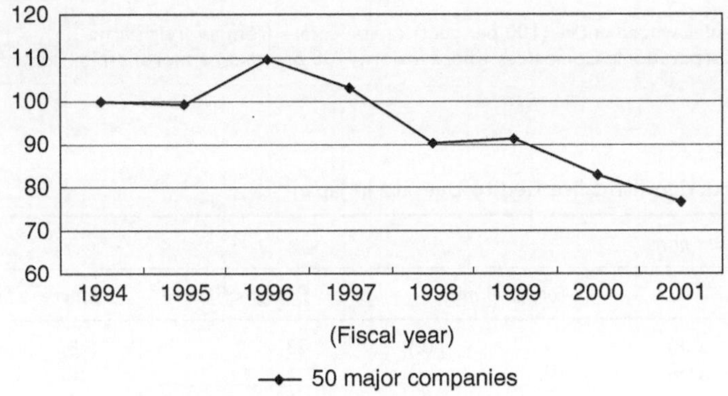

Figure 7.21 Trends in private-sector construction orders received by the major construction companies.

Source: Fifty major companies of orders received, Ministry of Land, Infrastructure and Transport.

Note
1994 = 100.

projects and to erect buildings (Table 7.26). The rest are licensed as specialists in such fields as electrical contracting, plumbing, etc. A construction firm must have specified contractor's license in order to accept directly from the end client all or part of a contract that it shall perform by issuing a subcontract of ¥20 million or greater.

Legal systems and procedures for establishing a business office; for a foreign corporation establishing a local office in Japan, any licenses or permits need not be obtained, nor must it register with the Legal Affairs Bureau. When a foreign corporation establishes a branch office (or sales office) or corporation in Japan, it is

Table 7.24 Number of foreign corporations and foreign-capitalized Japanese corporations that have obtained construction business licenses

Country	*Foreign companies*	*Foreign-capitalized Japanese companies*	*Total*	*By permit*	
				Minister of Land, Infrastructure and Transport	*Prefectural governor*
USA	10	26	36	12	24
South Korea	7	2	9	0	9
Germany	1	7	8	1	7
Holland	0	6	6	1	5
Switzerland	0	5	5	3	2
United Kingdom	1	6	7	3	4
Sweden	0	4	4	0	4
France	1	1	2	1	1
Finland	0	1	1	0	1
Singapore	0	1	1	0	1
Canada	0	1	1	0	1
Luxemburg	0	1	1	0	1
Total	20	61	81	21	60

Source: Ministry of Land, Infrastructure and Transport.

Notes
As of 31 March 2002.
Foreign corporations: Companies whose entire (100 per cent) capital comes from a foreign firm.
Foreign-capitalized Japanese corporations: Companies whose majority (50 per cent or more) capital come from a foreign firm.

Table 7.25 Foreign construction firms licensed to operate in Japan

	No. of licensed firms					
	Total		*Foreign firms*		*Foreign-affiliated*	*Japanese firms*
United States	32	(38)	9	(10)	23	(28)
South Korea	11	(14)	10	(14)	1	(0)
Sweden	6	(7)	0	(0)	6	(7)
Switzerland	6	(6)	0	(0)	6	(6)
Germany	5	(4)	0	(0)	5	(4)
France	3	(3)	1	(1)	2	(2)
Netherlands	3	(3)	1	(1)	2	(2)
England	2	(1)	0	(0)	2	(1)
Liechtenstein	1	(1)	0	(0)	1	(1)
Denmark	1	(1)	0	(0)	1	(1)
China	1	(1)	0	(0)	1	(1)
Malaysia	0	(1)	0	(1)	0	(0)
Total	71	(80)	21	(27)	50	(53)

Source: Ministry of Construction, Construction Contractors Dvision, Office for Construction Market Access.

Notes
Foreign-affiliated Japanese corporations: Japanese corporations in which foreign firms hold greater than 50 per cent ownership.
As of 31 March 1998.
Figures in parentheses are for 31 March 1997.

Table 7.26 Foreign firms holding a specified contractor's license (for civil engineering projects and the erection of buildings)

South Korea 10 firms	
	HYUNDAI Engineering & Construction Co., Ltd
	SamHwan Corporation
	Ssangyong Engineering & Construction Co., Ltd
	Kukdong Engineering & Construction Co., Ltd
	DAEWOO Corporation, Engineering & Construction
	LG Construction Co., Ltd
	Lotte Construction
	Dong Ah Construction Industrial
	Samsung Corporation
	Tong Yang Cement Corp. Engineering & Construction Group
US 8 firms	
	Schal Bovis Inc.
	Overseas Bechtel Inc.
	Parsons Brinckerhoff International
	GE Yokogawa International Ltd
	Flour Daniel Japan
	IBM Japan Ltd
	ServiceMaster Japan
	Fuji Xerox Co., Ltd
Netherlands 2 firms	
	Ballast Nedam International B.V.
	Hunter Douglas Japan
France 2 firms	
	GTM International S.A.
	DIC Degremont
Switzerland 1 firm	
	ABB
China 1 firm	
	Huacheng Construction
Total 24 firms	

Source: Ministry of Construction, Construction contractors division, Office for construction market access.

Note
As of 31 March 1998.

required to: (a) register the branch office (corporation) in accordance with commercial laws; (b) submit reports or documents in accordance with the foreign exchange and international trade laws; and (c) submit various documents to the tax authorities.

A company operating a construction business in Japan must obtain a construction business license in accordance with the Construction Business Law. A permit is not required if the company is only a subcontractor for a minor construction project as defined by government ordinance. But if the company engages in any other kind of construction work, it is required to obtain a construction business license regardless of whether it is a domestic or foreign firm. One condition for obtaining the license is that the company employ a person with management experience in Japan, but special approval may be obtained from the Minister of Land, Infrastructure and Transport for persons with the equivalent degree of experience in a foreign country if the Minister recognizes that person as having an equal or higher level of capabilities as that needed to fulfil this requirement. Likewise, the requirement of employing specialized engineers with certain qualifications or with actual work experience in Japan may be able to be met by a person with qualifications or actual work experience overseas authorized by the Minister of Land, Infrastructure and Transport.

Consultants

The number and value of contracts in fiscal 2001 for the major construction consulting (civil engineering), surveying, and land quality assessment firms are shown in Tables 7.27 and 7.28. Double-digit rates of decrease from the previous fiscal year were recorded for both the number and value of contracts, indicating that the market has cooled even further. However, while the land area for which designs were actually launched, as reported by architectural design firms, grew significantly in fiscal 1999, it fell again in fiscal 2000, and dropped sharply in fiscal 2001 by 17 per cent from the previous year (refer Tables 7.29 and 7.30).

TRENDS IN THE NUMBER OF REGISTERED CONSULTING FIRMS

The number of registered land quality assessment, surveying, and construction consulting (civil engineering) firms is increasing over time (see Table 7.31).

Table 7.27 Trends in the number and value of contracts of land quality assessment firms

	FY 1996	*FY 1997*	*FY 1998*	*FY 1999*	*FY 2000*	*FY 2001*
No. of contracts	33,482	30,327	30,575	30,246	29,177	25,203
Growth vs previous year (%)	–4.1	–9.4	0.8	–1.1	–3.5	–13.6
Contract amount (million yen)	161,046	139,206	144,809	139,191	124,525	105,955
Growth vs previous year (%)	–9.4	–13.6	4.0	–3.9	–10.5	–14.9

Source: Survey of Trends in Construction-related Industries, Ministry of Land, Infrastructure and Transport.

Note
These results are based on a survey of 50 selected land quality assessment firms.

Table 7.28 Trends in the number and value of contracts of surveying firms

	FY 1996	*FY 1997*	*FY 1998*	*FY 1999*	*FY 2000*	*FY 2001*
No. of contracts	40,135	36,513	33,462	30,002	33,993	34,339
Growth vs previous year (%)	1.7	–9.0	–8.4	–10.3	13.3	1.0
Contract amount (million yen)	143,384	127,446	126,704	112,776	109,539	93,587
Growth vs previous year (%)	–5.5	–11.1	–0.6	–11.0	–2.9	–14.6

Source: Survey of Trends in Construction-related Industries, Ministry of Land, Infrastructure and Transport.

Note
These results are based on a survey of 50 selected surveying firms.

Table 7.29 Trends in the number and value of contracts of construction consulting firms

	FY 1994	*FY 1995*	*FY 1996*	*FY 1997*	*FY 1998*	*FY 1999*	*FY 2000*	*FY 2001*
No. of contracts	43,432	47,646	47,526	45,410	46,875	45,432	48,210	52,467
Growth vs previous year (%)	–	9.7	–0.3	–4.5	3.2	–3.1	6.1	8.8
Contract amount (million yen)	468,716	523,596	516,205	484,146	506,773	465,843	466,852	445,133
Growth vs previous year (%)	–	11.7	–1.4	–6.2	4.7	–8.1	0.2	–4.7

Source: Survey of Trends in Construction-Related Industries, Ministry of Land, Infrastructure and Transport.

Note
These results are based on a survey of 50 selected construction consulting firms.

Table 7.30 Trends in the area of designs actually implemented by architectural design firms (FY1996–2001) (unit: 1000 m^2)

	FY 1996	*FY 1997*	*FY 1998*	*FY 1999*	*FY 2000*	*FY 2001*
Area of designs actually implemented	29,683	29,995	25,045	27,420	26,884	22,302
Growth vs previous year (%)	–3.1	1.1	–16.5	9.5	–2.0	–17.0

Source: Survey of Trends in Construction-related Industries, Ministry of Land, Infrastructure and Transport.

Note
These results are based on a survey of 90 selected architectural design firms.

Table 7.31 Trends in the number of registered consulting firms (FY 1994–2000)

	1994	*1995*	*1996*	*1997*	*1998*	*1999*	*2000*
Boring surveying	994	1,040	1,103	1,147	1,209	1,238	1,297
Growth vs previous year (%)	5.1	4.6	6.1	4.0	5.1	2.4	4.8
Land surveying	12,554	12,913	13,310	13,689	14,003	14,325	14,427
Growth vs previous year (%)	2.7	2.9	3.1	2.8	2.3	2.3	0.7
Construction consulting	2,619	2,720	2,893	3,076	3,277	3,426	3,686
Growth vs previous year (%)	5.6	3.9	6.4	6.3	6.5	4.5	7.6

Source: Ministry of Land, Infrastructure and Transport.

CERTIFICATION SYSTEMS FOR CONSTRUCTION PROFESSIONALS

Licensed surveyors To maintain the accuracy of basic surveys and public surveys, Japan's Land Surveying Law establishes a certification system for technicians (licensed surveyors or licensed assistant surveyors) engaged in the planning and implementation of surveys, and aims to maintain and improve surveying techniques. The number of licensed surveyors and licensed assistant surveyors comes from records of those who passed the required exam. Annual trends are shown in Table 7.32.

Licensed Architects The act on Architects sets out the qualifications for engineers engaged in the design and construction supervision of buildings. Trends in the number of licensed architects are shown in Table 7.33.

Material and machinery suppliers

Demand for major construction materials peaked in fiscal 1996 before starting to decline, but demand recovered in fiscal 1999 for all products except regular steel. However, demand in fiscal 2001 fell from the previous year for all major products.

Also, in fiscal 2000 the liquidation values of major construction machinery rose slightly over the previous year, while the number of pieces of major construction machinery fell slightly (refer Tables 7.34 and 7.35).

Construction sector employment

The number of workers in the construction sector began rising from 1987 after construction demand revived, and this number has continued to rise even with the industry in a slump following the collapse of the bubble economy. Total construction employment reached 6.85 million jobs in 1997 (Figure 7.22). Construction sector employment has been accounting for a steadily growing percentage of total employment in the economy as a whole, rising from 9.4 per cent in 1990 to 10.4 per cent in 1997. During that period, the number of workers in the construction sector increased by 656,000.

Several factors have been mentioned as possible explanations for the growing number of total workers while construction investment has been falling. One explanation is that growth in the number of workers did not keep pace with the rapid growth of construction investment in the late 1980s, thus creating a time lag. A second explanation is that pressure to cut costs has resulted in lower per-unit construction costs, so that demand for labour has not fallen in step with the rate of decrease in construction investment. Third, some argue that a contraction in the average size of

Table 7.32 Trends in registered licensed surveyors and assistant surveyors by fiscal year

	1994	*1995*	*1996*	*1997*	*1998*	*1999*	*2000*
Licensed surveyors	1,967	2,042	2,434	5,020	5,000	7,408	6,142
Licensed asst. surveyors	11,400	13,100	14,100	13,700	12,800	14,500	14,500

Source: Ministry of Land, Infrastructure and Transport.

Table 7.33 Trends in the number of licensed architects (based on registration)

	1951	*1970*	*1985*	*1992*	*1993*	*1994*	*1995*	*1996*	*1997*	*1998*	*1999*	*2000*
Level 1 licensed architects	15,819	152,017	197,507	243,906	250,688	257,466	264,398	271,231	278,184	285,255	292,620	299,247
Level 2 licensed architects	16,199	407,203	469,136	531,840	541,256	553,999	566,791	581,261	595,836	610,686	625,719	637,850

Source: Ministry of Land, Infrastructure and Transport.

Table 7.34 Domestic demand for key construction materials since 1996

		Unit	*FY1996*	*FY1997*	*FY1998*	*FY1999*	*FY2000*	*FY2001*
Cement	(Sales volume)	1,000 t	82,417	76,573	70,719	71,515	71,435	62,222
Ready-mixed concrete	(Shipment volume)	1,000 m^3	180,255	167,621	153,310	151,167	149,501	128,182
Lumber	(Shipping volume, lumber product)	1,000 m^3	24,395	21,103	18,924	18,369	17,282	14,011
Common steel	(Orders from construction firms)	1,000 t	30,659	28,642	25,715	26,863	28,024	23,829
Asphalt	(Sales to construction firms)	1,000 t	4,266	4,117	3,777	3,823	3,804	3,075

Source: Ministry of Land, Infrastructure and Transport.

Table 7.35 Key construction equipment: output value and unit production since 1994

	FY1994	*FY1995*	*FY1996*	*FY1997*	*FY1998*	*FY1999*	*FY2000*
Output (by value) of key construction equipment	1,442,494	1,424,173	1,542,086	1,385,569	1,021,517	1,000,688	1,043,409
Output (by of unit) of key construction equipment	462,775	433,061	464,550	403,603	310,165	324,623	322,781

Source: Japan Construction Equipment Manufactures Association.

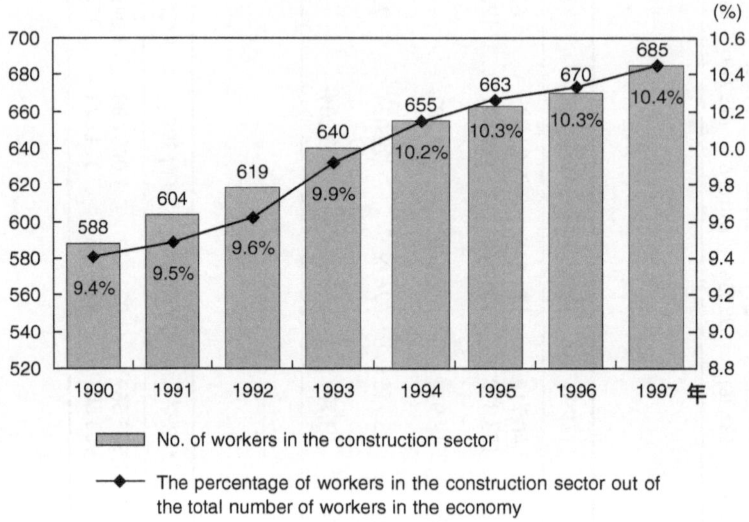

Figure 7.22 Number of workers in the construction sector since 1990.

Source: Management and Coordination Agency, Annual Report on the Labour Force Survey.

construction projects has lowered labour productivity, so that labour demand per unit of construction investment has gone up.

In the malaise that has followed the collapse of Japan's bubble economy, the construction industry has taken up the slack as the manufacturing sector and other parts of the economy sought to pare down the work force. However, construction demand was in a slump at the time of writing. The skilled labour shortage rate had declined in 1998, dropping to a record low of 1.1 per cent in May. The number of workers in the construction sector has been falling since November 1997 (refer Figure 7.22).

PROFESSIONAL AND TECHNICAL PERSONNEL AND WORKERS

The number of workers in the construction industry has been falling since 1997 (refer Table 7.36). However, the ratio of specialized technicians and labourers to workers in the construction industry has also been falling since 1997. The ratio of workers in the construction industry to those in all industries has remained stable at 10.1–10.4 per cent.

Enhancing the skills of the workforce

The Ministry of Health, Labour and Welfare has implemented a skill assessment system for the purpose of improving workers' skills and their interest in skill acquisition. This system comprises a four-level skill assessment for 38 types of construction-related jobs such as rebar work, and workers are assigned to a skill level based on their ability to pass both written and hands-on exams.

Table 7.36 Trends in the number of workers and technical workers/labourers in the construction industry (unit: 1,000 people)

	1995	*1996*	*1997*	*1998*	*1999*	*2000*
No. of workers in the construction	6,630	6,700	6,850	6,620	6,570	6,530
Growth vs previous year (%)	1.2	1.0	2.2	−3.5	−0.8	−0.6
1 No. of specialized technicians	430	430	410	430	420	420
2 No. of workers	4,380	4,420	4,550	4,340	4,320	4,320
The ratio of 1 and 2 to the no. of workers in the construction industry	66.1	66.0	66.4	65.6	65.8	72.6
No. of workers in all industries	64,570	64,860	65,570	65,140	64,620	64,460
Ratio of workers in the construction industry to workers in all industries (%)	10.3	10.3	10.4	10.2	10.2	10.1

Source: Annual Report on the Labour Force Survey, Ministry of Public Management, Home Affairs, Posts and Telecommunications, Statistics Bureau.

The Ministry of Land, Infrastructure and Transport is also working to develop training programmes for "multi-skilled workers" who can work in several different sector of construction, as well as for "core workers" who can devise efficient construction methods for contracting engineers (supervisors) and who can coordinate work with the managers of other sectors. In April 1997 the MLIT opened the Cross-Technique Training School to further these goals. To promote the training of core engineers, the MLIT is supporting the establishment of a system for evaluating them, and plans to investigate evaluation systems for companies that utilize core engineers, such as the public evaluation system, the "specialized construction companies strength index".

Also, the Ministry of Land, Infrastructure and Transport is conducting a preliminary training programme for new hires in the construction industry for small- and medium-sized companies and specialized construction companies who hire few new people and thus cannot efficiently implement their own independent training programmes. It is also planning to join construction companies to enhance construction skill training by providing industrial high schools with tools and materials, teachers, and volunteer lecturers.

Improving supervisory skills

Improving the skills of supervisors is largely being pursued through private instruction. Each company holds its own regular technology exchange meetings, provides quality control and safety and sanitation education, and has safety and sanitation coordinators conduct construction site safety patrols. In recent years, an increasing number of companies have been striving to elevate their technological capabilities by actively supporting the certification training various levels of construction managers and professional engineers, and by sharing construction techniques that use information technology.

Quality management

Significant changes are taking place in the public works environment, as tendering and contracting procedures are being reformed, globalization is progressing in the construction market, the need to reduce construction costs is rising, and companies are responding to global environmental problems. Given these conditions, many efforts are being taken to raise and maintain product quality at ever higher levels. The MLIT is investigating ways to apply management systems, such as the ISO 9000 series of international standards for quality management systems and the ISO 14001 series of international standards for environmental management systems (EMS), to public works projects. Aiming to further improve the level of quality assurance in public works projects starting in fiscal 2000, the government took a step further in its pilot projects and decided to try testing the effectiveness of a system of implementing ISO 9000 series compliant construction, in which ISO 9000s Certification was established as a competitive bidding prequalification criterion, to a certain range of construction projects. Since the rate of ISO 9000s Certification among companies prequalified by the MLIT in fiscal 2000 increased overall from fiscal 1999, the MLIT decided to expand the projects to which the trial implementation of ISO 9000 series compliant construction would be applied to 155 construction projects and 21 service projects in 2001. The government also decided to use the results of its study to promote further investigation into how the responsibilities for improving the efficiency of supervisory activities and other tasks should be divided between the construction company and the buyer. In the area of environmental management systems, the government's perspective is that efforts must be taken on the part of the company to fulfil their social obligations, and that the construction work buyers, who are comprised of organizations that commission public works, must actively contribute to those efforts. From this perspective, the MLIT began implementing ISO 14001 model projects since fiscal 1997 in 12 of the construction projects that fell under its direct administration. In addition to developing more model projects in fiscal 2001, the MLIT planned to conduct empirical studies to ascertain the effects of and problems involved in introducing EMS in public works projects.

To ensure the quality of construction, it is necessary to avoid various risks, such as delays due to the failure of the construction company or other problems. Several policies have been enacted in the Japanese construction industry to avoid these risks.

The prepayment guarantee system is a system in which the partial payment of construction fees paid in advance to the contractor by the organization commissioning a public works project (usually about 40 per cent of the contract amount) is guaranteed by a guarantee company. Specifically, if the contractor fails to complete the work in accordance with the terms of the contract, or the performance of guaranteed obligations is requested by the construction completion guarantor, the guarantee company will reimburse the losses up to the amount prepaid by the buyer. Because the funds used for public works projects are especially valuable public funds that come from taxes and other public sources, and thus need to be guaranteed accordingly, the Law Concerning Advance Payment Guarantee Undertakings for Public Works was enacted. Guarantee companies (eastern Japan, western Japan, and Hokkaido) are authorized to act as advanced payment guarantee companies. A guarantee company, in addition to accumulating guarantee capital that can be used to pay settlements, is required to strictly manage the use of the prepaid funds to ensure that they are used appropriately for the construction work to which they have been allocated. The advantages of this system are that the buyer can expect the construction work to be completed as planned because the advance payment necessary to implement the work has been paid, and the contractor does not have to worry about start-up capital for the project because they have access to the needed funds up front.

The performance guarantee system is a way of ensuring that construction contracts are carried out properly. Japan's Accounts Law and Local Government Law require that contractors pay a contract deposit, but negotiable securities such as government bonds, the guarantee of a guarantee company, or performance bonds may be offered in place of the contract deposit. Guarantees under the performance guarantee system can be categorized as "cash guarantees" in which losses due to contractor's non-performance of the contract are reimbursed to the buyer in cash, or "labour guarantees" in which a guarantee is made to complete the construction work. The contract deposit constitutes a cash guarantee, so if the contractor fails to complete the work in accordance with the terms of the contract, the guarantee company will pay the buyer the amount needed to cover the parts of the contract that were not fulfilled, up to the amount of the deposit. The contract deposit is set up as a special inclusion in the advance payment guarantee contract, and the work to which it applies is limited to the work for which the advance payment is to be made.

The MLIT is also in the process of studying the introduction of new guarantee systems, such as bid bonds, which require companies participating in the bidding process to obtain the guarantee of a financial institution in advance.

Improving quality assurance

The Ministry of Land, Infrastructure and Transport (MLIT) has been working since 1996 on about 50 pilot projects to improve quality assurance through ISO 9000 series compliance and to promote a better understanding of the ISO 9000 series, and to engage a wide range of interested parties in a discussion concerning the benefits and problems associated with the application of ISO 9000 standards to public works projects. As a result, the MLIT has determined that there is a high potential for the application of ISO 9000 series standards to effectively serve as a means of improving quality assurance standards on public works and other construction projects. In fiscal 2000, in the light of its findings, the MLIT was attempting 35 ISO 9000 series compliance projects that required project bidders to obtain ISO 9000 series certification, primarily projects with an especially high degree of difficulty. Based on the results of this attempt, in fiscal 2001 the MLIT was expected to expand the range of construction projects that require ISO 9000 compliance, increase the efficiency of supervisory activities, and simplify the documents that have to be submitted.

Japanese construction companies first began working in 1995 to achieve ISO 9000 series certification. Since then, the number of certified companies has risen

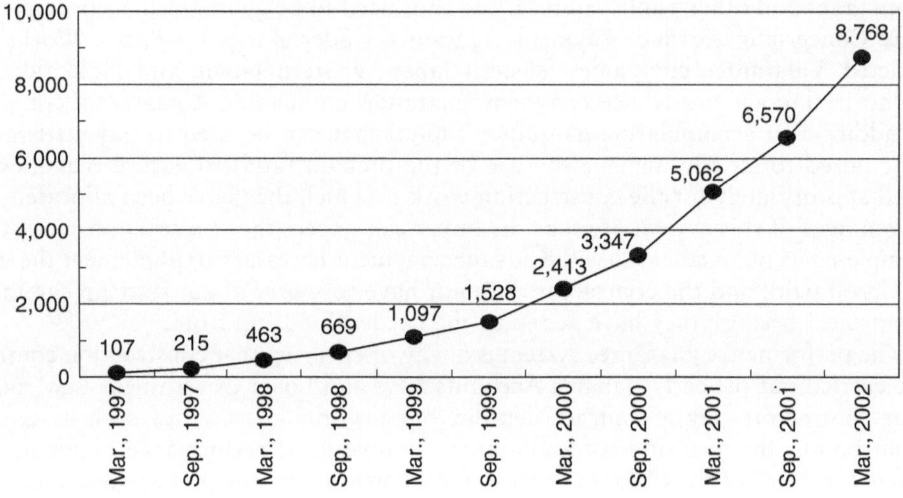

Figure 7.23 Registrations since March 1997.

rapidly. As of 31 March 2002, 8,768 construction companies had registered with the Japan Accreditation Board for Conformity Assessment (JAB) for certification in the construction sector. Most large and medium-sized general contractors had already obtained certification, and at the time of writing more local general contractors and specialty contractors were also obtaining certification (Figure 7.23).

Improving construction safety

In the area of safety, the MLIT has thus far made efforts to enact safety policies for construction work by providing the "Technical Guidelines for Construction Machinery and Construction Work Safety", but significant changes are taking place in the construction site environment as construction conditions are becoming more complex and workers are growing older. More than 20 per cent of the fatal incidents in the construction industry continue to be caused by construction machinery.

Thus, the Construction Work Safety Measures Investigation Commission was established in fiscal 2000 to analyse the causes of accidents and to advance the study of safety measures. In the future that commission will promote construction project safety policies and work to ensure that they are reflected in the "Technical Guidelines for Construction Machinery and Construction Work Safety". In fiscal 2001, an unmanned construction system (UCS) was adopted to implement rapid restoration activities on Miyakejima Island, where disaster restoration activities had been delayed due to continued volcanic activity since an eruption in July 2000 that made it dangerous for people to go in and do construction work. A UCS guidebook was created, and construction systems that utilize this device were prepared by each Regional Development Bureau.

Construction companies engaged in actual construction work periodically hold safety conferences and training sessions that are attended by construction contractors and specialized construction companies, and make efforts to ensure that all workers are thoroughly trained and have a full knowledge of safety issues. Individual work sites also take measures to maintain a constant awareness of safety among workers, including posting banners with safety slogans around the site and implementing call-out confirmation procedures.

Aiming to ensure the safety of construction work though public–private sector cooperation, the MLIT established the Comprehensive Safety Policy Project in May 1992. The basic aim of this project was to move away from promoting stronger regulations to emphasizing the importance of having each and every person involved in the construction industry, from those who order and manage work to the on-site skilled workers, be responsible for paying attention to safety issues. Its main policies were to coordinate the placement of responsible engineers through the improvement of construction system record keeping, to bolster education for construction workers using educational videos and other materials, and to offer seminars for engineers on relevant safety topics.

To further promote the causal analysis of worker accidents and to develop accident prevention measures, the MLIT announced plans in May 2000 to continue the performance of effective safety control activities at construction sites, to promote proper safety education for supervisors and on-site workers, and to construct an information exchange network on safety activities. It was also working on analyzing the causes of disasters attributable to human error and drafting measures to prevent such incidents (Table 7.37).

A growing number of companies are now performing safety management in accordance with the Occupational Health and Safety Management System (OHSMS).

Project inspection and acceptance system

The inspection of completed public works project is governed by the Public Works Standard Contract Agreement. When a project is completed, the contractor notifies

Table 7.37 Accident frequency rates (construction industry)

	General contractor	*Specialty contractor, plant and equipment contractor*
1997	1.11	0.83
1998	1.32	0.84
1999	1.44	0.74
2000	1.10	0.76
2001	1.61	0.81

Source: Survey of Occupational Accidents and Injuries, Ministry of Health, Labour and Welfare.

Note
The "frequency rate" is the frequency of accidents, expressed as the number of deaths or injuries from work-related accidents per one million labour hours.

the buyer. The buyer has 14 days from the day of notification to inspect the completed project in the presence of the contractor and to notify the contractor of the inspection results. If the project passes inspection, the buyer must take delivery of the finished structure. If the project does not pass inspection, the contractor must correct the deficiencies and have the project reinspected.

Inspection methods are stipulated in the design documents. National government contracts require that an inspection report be created following the inspection, but this is not always legally required of local government organizations. According to the Accounts Law, there are two kinds of inspections: inspections during construction and inspections of completion that are needed to initiate the payment process. The former are inspections of construction materials while the latter constitute completion inspections and inspection to confirm that certain portions of the construction have been completed so that payment for those portions can be made. According to the Public Works Standard Contract Agreement, the former are conducted by an inspector while the latter are conducted by the buyer.

Once the project has passed inspection and completion has been confirmed, the contractor can offer the delivery of the finished structure to the buyer. Once that notification is received, the buyer must immediately take delivery of the structure. If the contractor does not make an offer of delivery even though the buyer has confirmed the project completion, the buyer may ask the contractor to deliver the structure when the final payment is made.

Technical certification

To raise the level of building construction skills, the Construction Industry Law requires that workers in the building construction industry be certified annually. The types of technical certifications and the number of certified personnel are shown in Table 7.38.

Mutual recognition with foreign countries on technical certification

In recent years until 2002, the internationalization of Japan's construction market, including construction consulting services, proceeded at a rapid pace, and marked by reforms to the bidding and contracting system for public works projects and the issuance of a new WTO Agreement on Government Procurement. Even within the APEC region, which includes Japan, studies are being conducted on a framework for mutual recognition of certification systems and on the specific procedures and institutions that would be required for that purpose.

Establishing a mutually recognized system of technical certification that is internationally acceptable and that could assess technicians on some sort of common basis is important from the perspective of maintaining the quality of construction consultants working on public projects, as well as in terms of facilitating the overseas activities of Japanese construction consultants.

Thus, the Ministry of Land, Infrastructure and Transport established a Technician Certification Investigation Committee to study ways of implementing and policies for using a system of mutual recognition of technician qualifications.

Table 7.38 Number of workers passing the technical certification

	Construction machinery projects (FY 1960)		*Civil engineering project management (FY 1969)*		*Pipe construction project management (FY 1972)*		*Landscape project management (FY 1975)*		*Architectural project management (FY 1983)*		*Electric works project management (FY 1988)*	
	Level 1	*Level 2*	*Level 1*	*Level 2*	*Level 1*	*Level 2*	*Level 1*	*Level 2*	*Level 1*	*Level 2*	*Level 1*	*Level 2*
1995	768	6,246	13,437	36,902	3,781	10,982	4,758	5,810	7,306	22,720	7,312	6,215
1996	794	6,514	21,064	39,888	6,046	15,568	6,189	7,141	9,831	24,890	10,049	4,714
1997	991	6,700	21,641	46,072	6,952	13,408	4,566	7,882	10,321	35,850	6,559	5,439
1998	922	6,566	18,970	39,798	6,642	16,512	4,670	6,802	7,078	23,881	6,533	5,398
1999	871	6,839	23,896	49,495	5,511	28,370	3,835	5,993	10,809	20,414	5,798	6,059
2000	867	5,671	25,312	45,640	9,073	22,302	5,310	6,925	7,939	31,890	8,558	6,755
Total number of licensed workers after system implementation	16,190	108,789	464,608	1,118,760	106,984	292,593	78,119	137,580	170,420	306,816	119,655	162,836

Source: Ministry of Land, Infrastructure and Transport.

Note
Years shown in parentheses are the years in which each system was established.

Value-added per employee

Table 7.39 shows the value-added per employee.

Physical measurements of construction productivity

Table 7.40 shows physical measurements of construction productivity.

Structure of construction sector employment

Skilled labourers and other on-site employees comprise approximately two-thirds of all construction sector workers (Figure 7.24). This proportion has been falling slowly since 1980. By contrast, off-site technicians and office personnel have been increasing in both absolute numbers and in terms of percentage. There has been no significant change in the actual numbers of skilled workers since 1995, but since 2001 the overall numbers had been falling, with the exception of management staff (Table 7.41).

Table 7.39 Value-added per employee

	1997	*1998*	*1999*	*2000*
Construction (million yen)	7.52	7.48	7.00	7.11
Year-on-year change (%)	−1.31	−0.53	−6.42	1.57
Manufacturing (million yen)	8.01	7.65	7.50	7.81
Year-on-year change (%)	−0.87	−4.49	−1.96	4.13
Services (million yen)	5.97	5.70	5.87	5.72
Year-on-year change (%)	2.75	−4.52	2.98	−2.56
All industries (million yen)	7.34	7.12	6.94	7.02
Year-on-year change (%)	0.00	−3.00	−2.53	1.15

Source: Monthly Economic and Fiscal Statistics, Ministry of Finance.

Table 7.40 Physical measurements of construction productivity (unit: m^2/manday)

	1995	*1996*	*1997*
Residential	0.361	0.351	0.363
Commercial	0.316	0.372	0.401
Industrial	0.613	0.525	0.277

Source: Cost Analysis Information for Building Works, Management Research Society for Construction Industry.
However, Cost Analysis Information for Building Works is no longer being published, so statistics are only available through 1997.

Note
The "Residential" category comprises multiple-unit apartment buildings and dormitories that companies provide for their unmarried employees. The "Commercial" category comprises office buildings and retail stores.

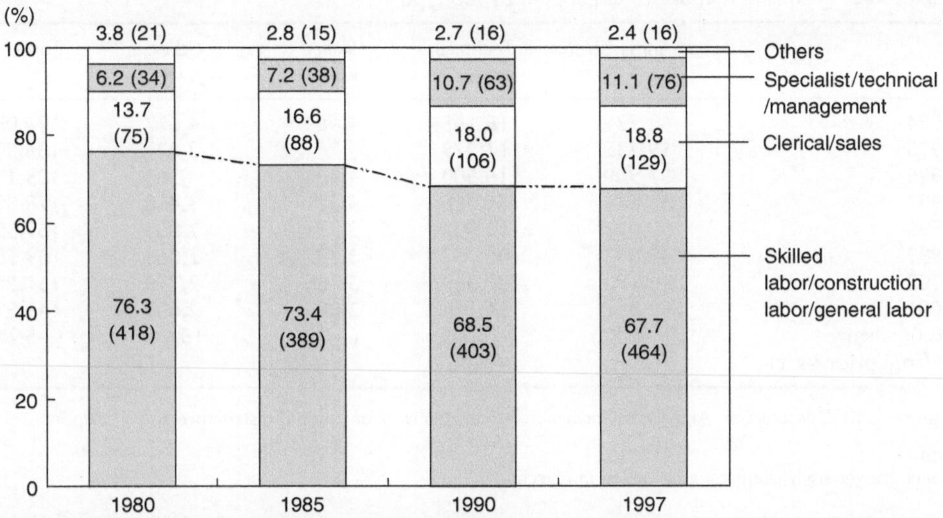

Figure 7.24 Structure of construction sector employment.
Source: Management and Coordination Agency, Annual Report on the Labour Force Survey.

Note
Numbers in parentheses are actual (not projected) figures.

Table 7.41 Breakdown of construction industry employees (unit: 1,000 people, avg. per year)

Type of work	*1990*	*1995*	*1999*	*2000*	*2001*
Specialists and technicians	290	430	420	420	390
Management staff	340	350	340	330	340
Clerical staff	840	930	940	930	910
Sales staff	220	290	320	340	330
Workers	3,950	4,380	4,320	4,320	4,150
General laborers	80	90	80	70	80
Others	160	150	150	140	120
Total	5,880	6,630	6,570	6,530	6,320

Source: Annual Report on the Labour Force Survey, Ministry of Public Management, Home Affairs, Posts and Telecommunications.

According to a survey conducted by the MLIT among the leading 36 general contractors, the number of staff in clerical, technical, and skilled work positions had fallen, and the total number of workers in all job types was down by 5.2 per cent until 2001 (the seventh consecutive year since 1994 that the number has fallen) as shown in Table 7.42.

In terms of age distribution, employees over the age of 50 in 1980 accounted for 23.4 per cent of the construction sector labour force, and were outnumbered by co-workers in their 30s and 40s. By 1997, however, this figure had risen to 35.5 per cent. The over-50 group had become the largest single age bracket. Thus the construction sector labour force was clearly ageing.

Table 7.42 Trends in number of employees by job type

	Administrative work	*Technical work*	*Skilled work*	*Other*	*Total*
1994	62,731	118,145	4,484	4,677	190,037
1995	59,735	117,979	4,751	3,871	186,336
1996	57,909	116,500	4,403	4,360	183,172
1997	55,831	114,768	4,371	3,408	178,378
1998	50,093	112,907	3,751	2,803	169,554
1999	49,053	104,732	3,732	2,068	159,585
2000	46,182	100,369	3,135	2,198	151,884
2001	44,341	94,165	2,884	2,603	143,993
(2001 change from prior year)	(−4.0%)	(−6.2%)	(−8.0%)	18.4%	(−5.2%)

Source: 2001 Construction Activity Performance Survey, Ministry of Land, Infrastructure and Transport.

Note

Years are generally the fiscal year ended in March.

The 30–39 age bracket, by contrast, declined from 28.6 per cent in 1980 to 16.8 per cent in 1997. Workers in this age bracket were at the peak of their careers, and their decreasing representation in the labour force gave cause for worry about the possibility of declining productivity.

The percentage of workers in the 20–29 age bracket declined during the bubble economy, but was on the rise in the 1990s. As for the reason, it was thought that the economic recession made employment hard to find and driven an increasing percentage of young work seekers to the construction industry (Figure 7.25).

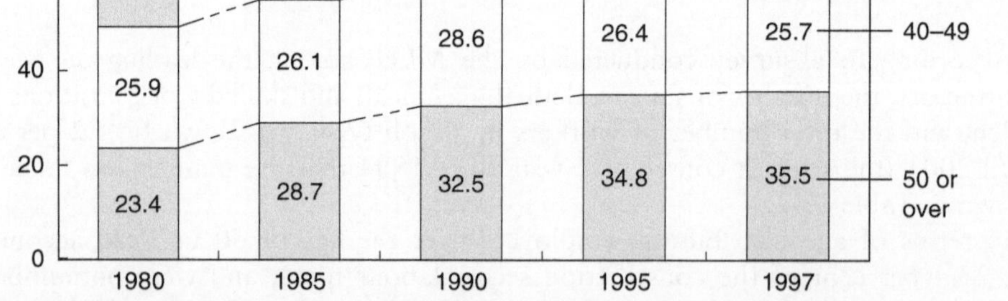

Figure 7.25 Age distribution of the construction sector labour force.

Source: Management and Coordination Agency, Annual Report on the Labour Force Survey.

Table 7.43 Construction sector wages since 1991

	1991	*1992*	*1993*	*1994*	*1995*	*1996*	*1997*
Average daily cash wages per person for outdoor construction personnel (unit: yen)	12,100	12,750	13,150	13,220	13,600	13,820	14,100
% rise	7.4	5.4	3.1	0.5	2.9	1.6	2.0
Monthly wages per permanent employee in the construction sector versus average wages in the economy overall	97.2	98.2	98.1	99.3	97.9	98.7	n.a.

Source: Ministry of Labour, Minister's Secretariat, Office of Policy. "Annual Report on the Wage Survey of Outdoor Workers by Occupation," "Annual Report on the Monthly Labour Survey".

Note
Average wages for the economy as a whole for each year are indexed to 100.

Wages have continued to rise steadily in spite of a slumping construction market and an increasing number of workers (Table 7.43). Since 1994, however, they have risen slowly, posting a 2.0 per cent rise in 1997 over the previous year. Wage differentials between the construction sector and the average for total industry have not made significant changes.

For data on average sectoral wages and employee wages and salaries see Tables 7.16 and 7.17 above.

Demand and supply for construction materials

CONSTRUCTION MATERIAL INVENTORY AND PRODUCTION INDICES IN RECENT YEARS

The market for construction materials has been affected by inventory adjustment since May 1997. Apart from lumber, for which inventory levels have been falling, inventories in all other categories have remained high, and production has not posted year-on-year gains in any category (Figure 7.26).

Material and machinery suppliers

Demand for major construction materials peaked in fiscal 1996 before starting to decline, but demand recovered in fiscal 1999 for all products except regular steel. However, demand in fiscal 2001 fell from the previous year for all major products.

Also, in fiscal 2000 the liquidation values of major construction machinery rose slightly over the previous year, while the number of pieces of major construction machinery fell slightly (Tables 7.44 and 7.45).

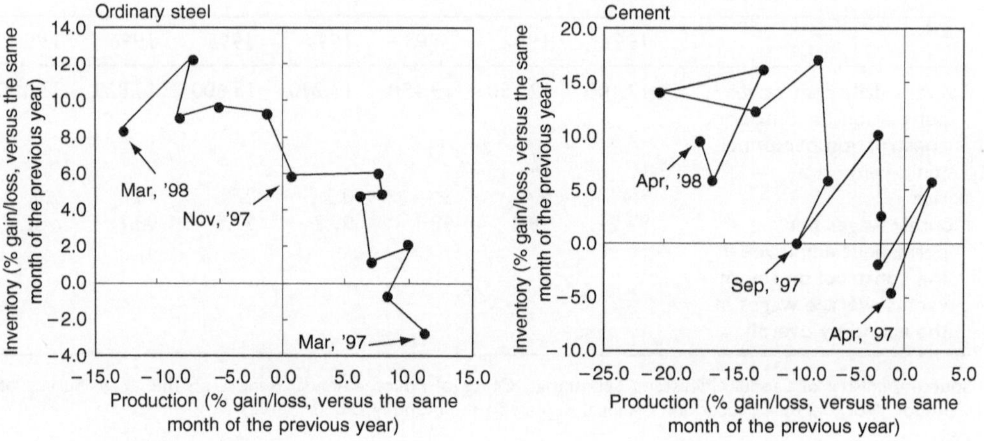

Figure 7.26 Construction material inventory and production indices in recent months.

Source: Ministry of Construction, Economic Affairs Bureau, Office of Labour and Materials, Monthly of Construction Labour and Materials.

Table 7.44 Domestic demand for key construction materials since 1996

		Unit	*FY1996*	*FY1997*	*FY1998*	*FY1999*	*FY2000*	*FY2001*
Cement	(Sales volume)	1,000 t	82,417	76,573	70,719	71,515	71,435	62,222
Ready-mixed concrete	(Shipment volume)	1,000 m^3	180,255	167,621	153,310	151,167	149,501	128,182
Lumber	(Shipping volume, lumber product)	1,000 m^3	24,395	21,103	18,924	18,369	17,282	14,011
Common steel	(Orders from construction firms)	1,000 t	30,659	28,642	25,715	26,863	28,024	23,829
Asphalt	(Sales to construction firms)	1,000 t	4,266	4,117	3,777	3,823	3,804	3,075

Source: Ministry of Land, Infrastructure and Transport.

Table 7.45 Key construction equipment: output value and unit production since 1994

	FY1994	*FY1995*	*FY1996*	*FY1997*	*FY1998*	*FY1999*	*FY2000*
Output (by value) of key construction equipment	1,442,494	1,424,173	1,542,086	1,385,569	1,021,517	1,000,688	1,043,409
Output (by of unit) of key construction equipment	462,775	433,061	464,550	403,603	310,165	324,623	322,781

Source: Japan Construction Equipment Manufactures Association.

WHOLESALE PRICE INDICES FOR PRINCIPAL CONSTRUCTION MATERIALS

Wholesale price indices for principal construction materials have basically been on a downward trend since 1993. Although price indices for almost all materials rose in 1997, that was due to the higher consumption tax.

Because of excessive inventories, wholesale price indices for almost all construction materials have been declining recently from month to month. Inventory adjustment is expected to continue, and it will probably be a long time before wholesale prices for construction materials begin to rise.

AVERAGE CONSTRUCTION MATERIAL PRICE

Tables 7.46 and 7.47 show the average construction material price and wholesale price indexes for principal construction materials.

Table 7.46 Average construction material price (unit: yen)

	Cement in bulk (per ton)	*Steel bars* (per ton)	*Aggregate* (per m^3)	*Sand* (per m^3)	*Ready-mixed concrete* (per m^3)	*Precast concrete piles* (per pile)
1997	9,090	34,600	4,350	4,800	12,200	32,100
1998	8,850	29,700	4,190	4,710	12,000	32,100
1999	8,730	25,100	3,900	4,560	11,700	31,140
2000	8,880	26,300	3,700	4,350	11,700	29,000
2001	8,650	26,800	3,640	4,150	11,600	28,850

Source: Prices of Construction Materials and Wages, Construction Research Institute.

Note

All prices are for materials procured in Tokyo.

Types and specifications

Cement	Regular portland cement
Steel bars	D19 deformed bars (SD295A)
Aggregate	20–50 mm crushed stone for concrete
Sand	Coarse, washed
Ready-mixed concrete	Regular concrete (nominal strength: 18N/mm^2; slump: 18cm; maximum coarse aggregate size: 25mm)
Precast concrete piles	PHC piles (type A, ϕ350 x 10)m

Table 7.47 Wholesale price indices for principal construction materials since 1993

Average for year	*1993*	*1994*	*1995*	*1996*	*1997*
Construction materials	103.3	100.6	100.0	100.2	102.1
Sawing lumber	104.0	104.5	100.0	100.9	103.1
Cement	108.7	102.6	100.0	100.0	103.4
Ordinary steel, other steels	107.4	101.6	100.0	98.7	100.3
Metal building materials	102.5	100.2	100.0	99.9	101.2
Stone, gravel, sand, etc.	90.8	97.1	100.0	95.1	94.8

Source: Ministry of Construction, Economic Affairs Bureau, Research and Information Division, Ministry of Construction Statistics Report.

7.3.2 *Major construction firms: status as of 2002 and future strategies*

Business climate in the construction industry

MARKET COMPETITION INTENSIFIES

Japan's construction market has been in a slump since the collapse of the bubble economy. In recent years, especially, economic recession has shrunk private construction investment in Japan. At the same time, the worsening financial situation of local governments and the implementation of fiscal structural reforms have also caused public construction investment to slide toward contraction. As a result, competition in the construction market is heating up, and profit margins are declining for all the major construction firms. The profitability of private-sector construction projects, in particular, has plummeted, and the number of money-losing projects has increased (refer Figures 7.27 and 7.28).

SURVIVAL STRATEGIES: LOCAL CONSTRUCTION COMPANIES

If the numbers of construction companies and construction workers are inevitably going to be reduced, then companies that focused on public works projects during the favourable business environment of the 1990s will be especially hard hit in the future, and will have to adopt business strategies aimed at survival.

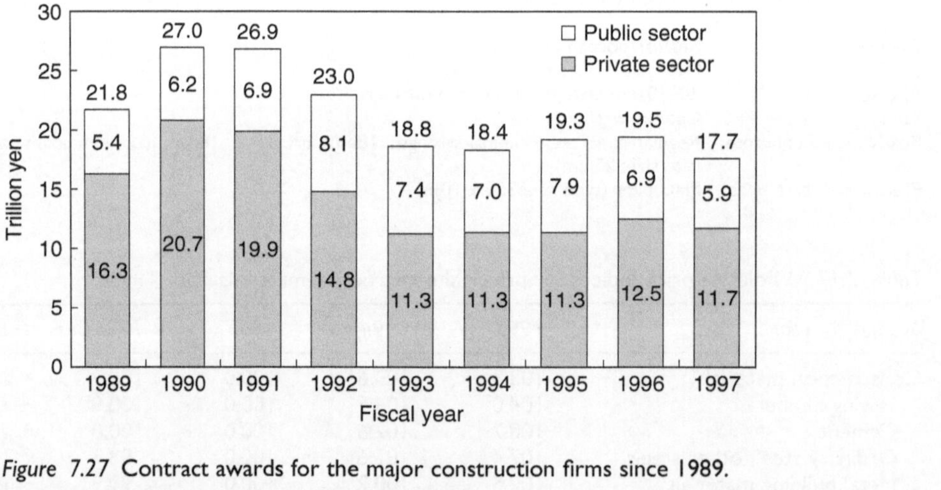

Figure 7.27 Contract awards for the major construction firms since 1989.

Source: Japan Federation of Construction Contractors.

Note

These figures indicate the combined totals of domestic contracts awarded to the 64 member firms of the Japan Federation of Construction Contractors.

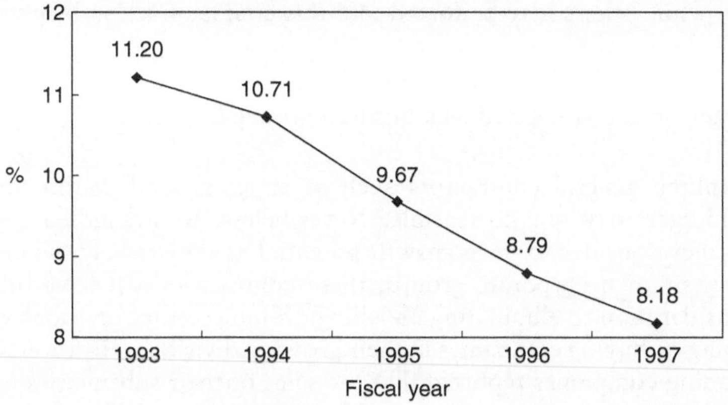

Figure 7.28 Gross profitability of projects completed by major construction firms since 1993.

Source: Japan Federation of Construction Contractors.

Note

These figures indicate weighted averages for the 64 member firms of the Japan Federation of Construction Contractors.

1 If businesses want to survive, they will have to move into construction and construction-related fields outside of the public works sector. If they hope to be able to expand, they will have to carve out a path in industry fields with growth potential.
2 A merger may be a potential growth strategy for survival. Specifically, a company could negotiate a merger with a company in a construction field or business area that would provide mutual benefits to both parties. One option would be to negotiate a merger with a company that focuses on private construction projects.
3 A company could also form alliances to facilitate the joint purchase of materials, the joint use of heavy machinery and assets/technologies, or the exchange of engineers.

Top national construction companies: business focus, range and organizational strategies

National general contractors need to seek out possibilities for partnerships and alliances that do not generate sentiments of resistance. However, as mentioned above, the industry was starting to see movements towards business consolidation. In April 2001, the businesses of Nitto Daito Kogyo and Mitsui Fudosan, though ranked lower than the top-50 companies, were consolidated (forming the Mirai Group with sales of ¥100 billion) and discussions of business consolidation have begun between top-50 companies Mitsui Construction and Sumitomo Construction.

Amidst the anticipated contraction of the construction market, the major construction companies are naturally reviewing the focus and range of their businesses and considering such options as alliances with other companies. In the long term, however, they may be left with little choice but to develop their businesses through mergers, sales, and acquisitions.

Under these conditions, some companies may select an aggressive merger or sale as part of its strategy, while others may be forced into this course of action by external factors.

1 When the merger or sale is selected as a business strategy:

   a For top-ranking general contractors such as super general contractors, a full-scale merger may not be feasible. Nevertheless, to strengthen certain fields that they consider to have growth potential as they review the organization of their own corporate groups, these contractors will have to give careful consideration to eliminating or selling off unnecessary divisions while strengthening or buying companies in their growing divisions. However, even the five leading companies reported that the sales of their subsidiaries (averaging about 60 companies as of the end of fiscal 2000) account for only about one-fourth of their total sales. This shows that the construction industry has a structure centred on the parent company. Corporate groups also have the option of downsizing as the construction market contracts, but if they want to try to maintain or expand their corporate group as a whole over the mid- to long term, they may also choose, regardless of what the rest of the market is doing, to take aggressive action in the sectors where they expect to see future growth and to develop into a corporate group that has mixed interests in the construction industry as well as in non-construction industries.
   b Middle and lower ranking companies (companies in the middle and lower levels of the top 50, lower-ranked listed companies that have faced or are soon likely to face difficulties in sustaining their businesses at the national level due to declining sales in recent years, and that have had problems sustaining their top rankings in the public works sector) may choose to downsize their businesses and transform themselves into local companies, or, if they want to continue to develop on a national scale, may choose to negotiate mergers or purchase other companies. The business consolidation between Nitto Daito Kogyo and Mitsui Fudosan mentioned above was actually chosen as a means for each of the companies, whose sales individually were less than ¥100 billion, to develop their business on a national scale.

2 When merger or sale is necessitated by external factors
   The major construction companies covered by this institute's "Financial Analysis" have been reduced by three due to legal measures, leaving 48 companies. Some companies have been replaced, but seven companies reported a consolidated capital ratio (to total capital) of less than 1 per cent (one of which reported a negative ratio) at their fiscal 2001 interim closing. Even when the newly added companies are included, however, there are still six companies reporting less than 1 per cent (with one reporting a negative ratio). As sales continue to decline and construction gross profit rates fall, some companies will find themselves in a difficult position come fiscal 2005 when changes will be introduced in the way companies report the value of tangible fixed assets. Main banks are expected to take the initiative in promoting efforts to revise this system.

Sales for 8 of the 13 largest construction firms have fallen from the previous year 2001.

Sales and number of employees by sector for 15 major general contractors (FY 2001)

The sales and number of employees for major construction firms for different types of construction projects are shown in Table 7.48.

AFTERMATH OF THE BUBBLE

In addition to the declining profitability of construction projects, some of the major construction firms are further burdened by payments for interest-bearing debts acquired during the bubble period. Japan's 53 top construction firms held a combined total of ¥8.8 trillion worth of interest-bearing debt in FY 1997. While this figure is down from an earlier peak, it remains 3.5 times greater than where it stood in 1983, and the debt-to-sales ratio is more than twice as high as that for 1983. Most of the construction firms that have been buffeted worst have adopted strategies which focus on reducing these interest-bearing debts, but the intensely competitive construction market of recent years has cut into profits and prevented them from making much progress in this regard (Figure 7.29).

FEWER OVERSEAS CONSTRUCTION PROJECTS

With the growth of the Asian economy, Japanese construction firms have been winning a steadily increasing number of overseas construction contracts since 1992. The Asian economic crisis at the time of writing however, was having an enormous impact on the Japanese construction industry. The percentage of total contracts accounted for by overseas contracts was a very low 6.4 per cent even in 1996, when this figure peaked. The Japanese construction industry had been looking forward to steady growth in overseas contracts, especially in Asia. This unexpected decline makes the then current Asian slowdown especially painful and was forcing construction firms to change their strategies (Figure 7.30).

Trends in overseas construction orders received by the big 50 general contractors (1986–2001) are shown in Figure 7.31.

OVERSEAS CONSTRUCTION CONTRACTS IN FISCAL 2001 (RATIO OF ORDER VALUES BY REGION)

Figure 7.32 shows two sets of statistics on overseas business.

RESTRUCTURING EFFORTS ATTAIN LIMITED SUCCESS

To prevent the slump in the construction market from cutting into profitability, construction firms adopted various restructuring schemes. In order to put themselves back on a firmer business footing, some firms wrote off large extraordinary losses to reduce the amount of interest-bearing debts and unproductive assets on their books.

Table 7.48 Sales and number of employees for major construction firms for different types of construction projects (FY 1997)

Firms	Sales (million yen)				No. of employees
	Total	Building construction	Civil engineering	Real estate, etc.	
Kajima Corporation	1,550,317	1,072,703	387,321	90,292	10,761
Obayashi Corporation	1,286,544	934,059	321,145	31,338	10,685
Shimizu Corporation	1,285,444	1,017,089	244,406	23,948	9,876
Taisei Corporation	1,241,392	944,578	272,899	23,914	9,989
Takenaka Corporation	1,031,793	994,492	27,843	9,458	8,134
Kumagai Gumi	650,352	321,943	302,689	25,718	4,825
Toda Construction	550,628	386,737	159,291	4,598	4,709
Nishimatsu Construction	513,527	271,866	227,245	14,416	4,633
Fujita Corporation	441,688	287,902	142,982	10,804	3,581
Penta-Ocean Construction	396,742	155,648	238,582	2,511	3,655
Maeda Corporation	389,752	235,801	153,950	0	3,810
Mitsui Construction	380,810	250,392	124,048	6,369	2,771
Hazama Corporation	377,470	209,337	162,332	5,801	3,089
Tokyu Construction	347,138	223,158	116,746	7,233	2,805
Tobisima Construction	321,589	160,836	158,521	2,231	2,463

Firms	Sales								No. of employees	
	Total	% change from previous term		% change from previous term	Civil engineering	% change from previous term	Real estate, etc.	% change from previous term		% change from previous term
Kajima Construction	15,128	−5.6	10,638	−0.6	3,634	−12.1	856	−27.9	13,588	−2.1
Shimizu Corporation	14,738	0.2	11,383	1.7	2,952	−4.2	403	−7.0	11,499	−2.7
Obayashi Corporation	14,652	−3.0	10,060	−5.6	4,109	2.5	483	10.2	11,721	−1.1
Taisei Corporation	13,848	−11.6	10,183	−8.7	3,256	−7.4	409	−58.7	12,204	−3.1
Takenaka Corporation	12,624	0.2	12,103	−0.1	291	−14.2	230	58.8	9,452	−1.9
Kumagai Gumi	10,133	8.9	6,745	4.6	3,224	27.2	164	−48.9	7,991	−3.1
Nishimatsu Construction	7,117	−2.2	3,791	14.4	3,100	−20.8	226	332.8	5,654	−3.4
Fujita Corporation	6,976	−1.7	4,371	−2.5	1,961	−5.2	644	17.9	5,103	−6.2
Toda Construction	6,650	−3.5	4,828	−5.1	1,744	1.5	78	−12.0	5,417	−4.3
Penta-Ocean Construction	5,672	3.0	2,446	15.1	3,192	−4.7	34	10.0	4,671	−8.1
Sato Kogyo	5,144	−8.8	2,980	−8.7	2,064	−3.1	101	−58.6	4,635	−8.6
Hazama Corporation	5,111	−9.7	3,077	−4.5	1,970	−13.3	65	−61.0	4,809	−4.1
Maeda Corporation	5,053	2.4	2,943	6.0	2,110	−2.2	0	0.0	4,323	0.4

Source: Each firm's financial statement.

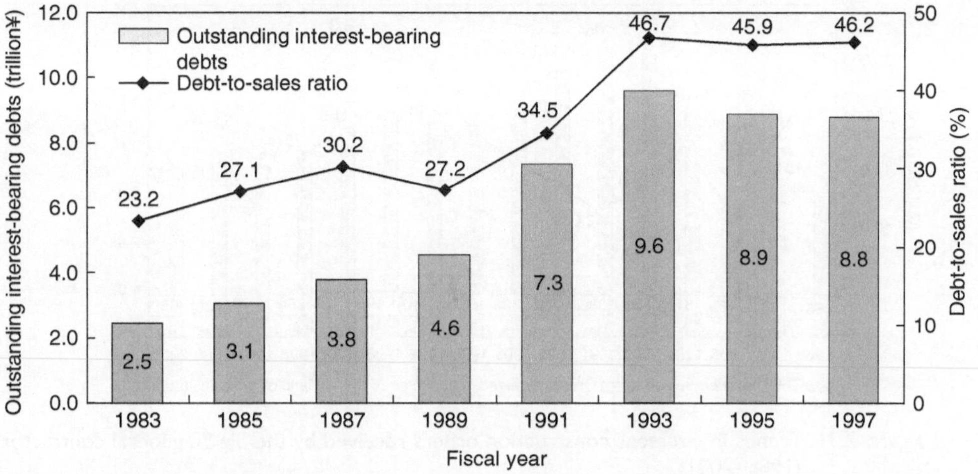

Figure 7.29 Interest-bearing debts of the major construction firms since 1983.

Source: Financial statements of the top 53 construction firms, and surveys conducted by the Research Institute of Construction and Economy.

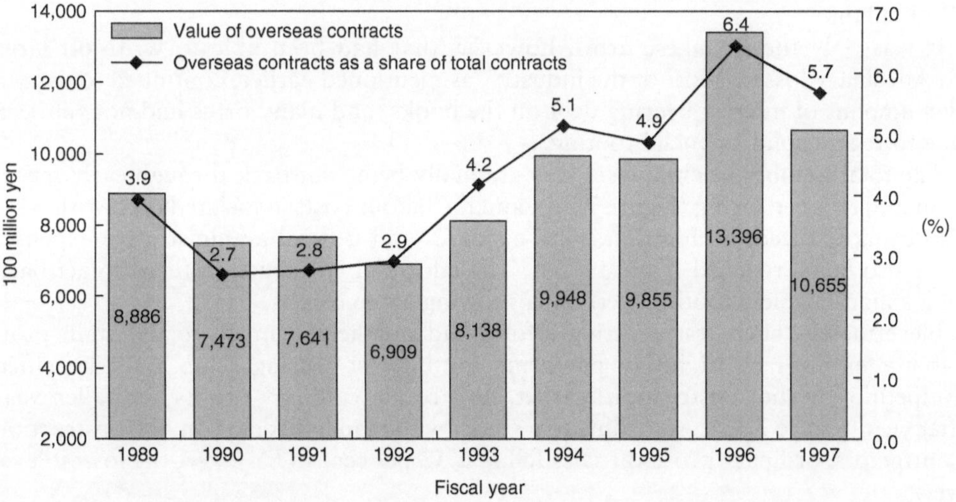

Figure 7.30 Overseas contracts for major construction firms since 1989.

Source: Japan Federation of Construction Contractors.

Note

These figures indicate the combined totals of overseas contracts awarded to the 64 member firms of the Japan Federation of Construction Contractors.

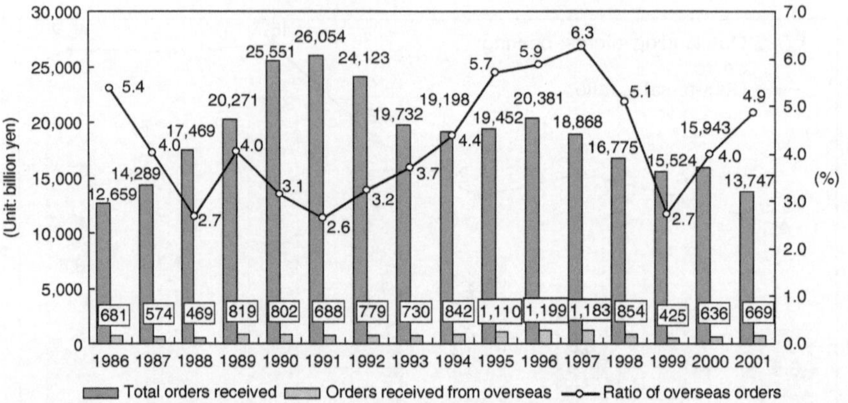

Figure 7.31 Trends in overseas construction orders received by the big 50 general contractors (1986–2001).

Source: Survey A of Orders Received, Construction Research and Statistics Division, Information and Research Department, Policy Bureau, Ministry of Land, Infrastructure and Transport. Survey conducted among the big 50 general contractors.

Notes
1 The number of overseas construction orders does not include those received by local corporations.
2 Figures shown here differ from figures reported by the Overseas Construction Association of Japan due to differences in the companies surveyed.

According to the Research Institute of Construction and Economy, the combined total of such write-offs for the top 53 construction firms in FY 1997 came to approximately ¥750 billion.

It was only the strongest firms, however, that had been able to write off large extraordinary losses. Most of the industry, as mentioned earlier, continued to carry a high amount of interest-bearing debt on the books, and many firms had not yet been able to regain solid financial footing.

The total number of employees was gradually being cut back through early retirement and reduced hiring (Figure 7.33), and the labour costs associated therewith were also being reduced. In addition, sales expenses and general administrative expenses were also being reduced (Figure 7.34). The efforts of construction firms to get back on a sound business footing were thus showing some results.

Nevertheless, such restructuring efforts had not helped much to maintain profitability in the face of heavy payments for interest-bearing-debts and intensified competition in the construction market. As a result, ordinary profits had fallen year after year (Figure 7.35). According to a study by the Japan Federation of Construction Contractors, ordinary profit on sales fell to 1.52 per cent in FY 1997, the lowest level ever.

Recent strategy of construction firms

Construction firms are faced with an extremely harsh business climate. In addition to restructuring, construction firms are working to formulate new business strategies. These firms have also sought actively to diversify their business during the bubble

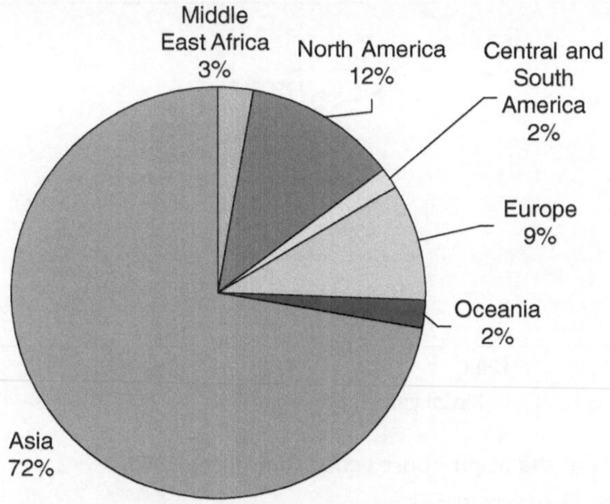

Source: Overseas Construction Association of Japan. Totals for the 48 member companies.

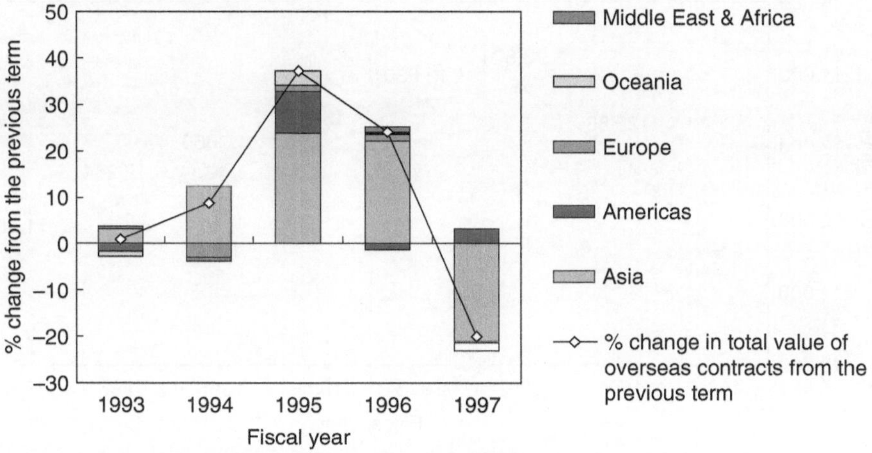

Figure 7.32 Two sets of statistics on overseas business.

Source: The Overseas Construction Association of Japan, a survey for growth in overseas construction contracts.

Note

These figures indicate the combined totals of overseas contracts awarded to the 56 member firms of the Overseas Construction Association of Japan.

economy, but they branched out into "expansion-oriented construction activities", such as real estate, hotels, and resorts. Now, by contrast, they are concentrating their efforts on "core construction activities" aimed at spurring greater demand for peripheral construction activities. Below we shall discuss some of the principal areas they are now concentrating on.

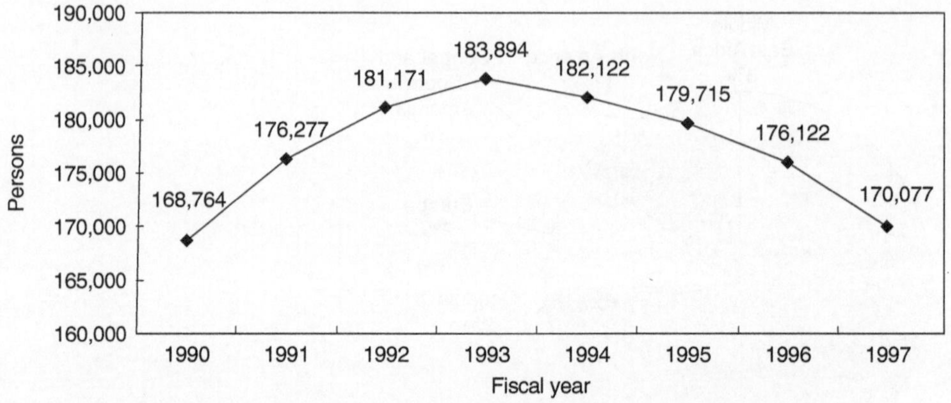

Figure 7.33 Number of employees at the major construction firms since 1990.

Source: Financial statements of the top 53 construction firms.

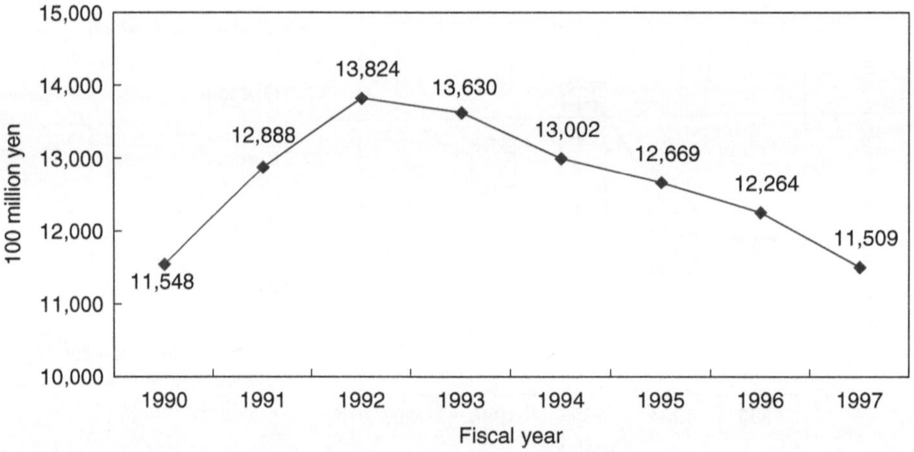

Figure 7.34 Sales and general administrative expenses at the major construction firms since 1990.

Source: Financial statements of 53 major construction firms.

PRIVATE-SECTOR PARTICIPATION IN PUBLIC WORKS PROJECTS

Some Japanese construction firms are involved in build-operate-transfer (BOT) projects in Hong Kong, Australia, and other overseas markets. At the time of writing, however, the Japanese government was also moving to address its financial problems by adopting a programme modelled on Britain's Private Finance Initiative (PFI). The Diet was discussing a bill which included measures to provide support for PFI contractors, and it was expected to be passed into law in autumn of 2002. Prefectural and local governments, as well, are considering the possibility of adopting

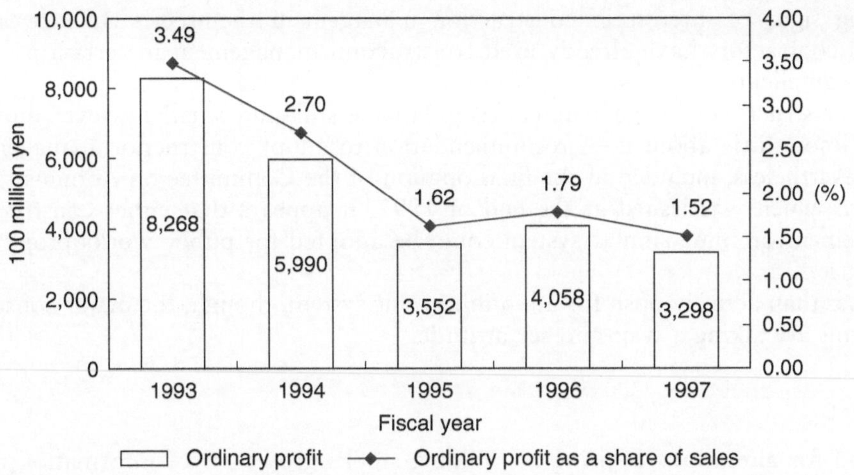

Figure 7.35 Ordinary profits of the major construction firms since 1993.

Source: Japan Federation of Construction Contractors.

Note

These figures indicate the combined totals for the 62 member firms of the Japan Federation of Construction Contractors.

a PFI system for the construction of toll roads, government buildings, and other such infrastructure. PFI projects and similar undertakings have the potential to become an important market in the future. With an eye on this important potential, major construction firms have established project teams and committees to study and collect information on PFI procedures, and to consider the possibility of taking part in such projects. (For further information regarding PFI, see Section 3.3.2, recent R&D projects.)

INTRODUCTION OF CONSTRUCTION MANAGEMENT (CM)

The national government adopted the Action Plan for Reduction of Public Work Construction Costs in April 1997. Construction firms have also been moving to cut costs in order to deal with the slumping private-sector construction market. Of the various measures taken towards this end, construction management has generated some of the keenest interest.

Major construction projects in Japan have always been awarded in total to a single general contractor. While it is convenient for the end client to award a contract in this manner, it is difficult for the client to know anything about the costs associated with the project. Construction management is drawing attention because it introduces greater cost transparency into the system, thus making it possible to achieve cost savings. Construction management made great inroads into the industry. Firms specializing in construction management have been established, and consulting firms and design offices have begun to be involved in construction management.

This trend poses a threat to the major general contractors, and the latter are showing signs of interest in construction management themselves. In fact, some general contractors have already used construction management in certain projects for foreign clients.

The construction management market in Japan is still quite small, however, and end clients know little about it. A recommendation to adopt construction management was, nevertheless, included in the final opinion of the Committee on Administrative Reform, which was issued at the end of 1997. It appears that either construction management or some similar system could be adopted for public works projects in the future.

Rather than actively push for adoption of this system, though, the major construction firms are taking a wait-and-see attitude.

RENEWAL

Demand for alterations is rising as building stocks grow, use of information technology increases, and the population ages. As a result, the renewal market is expected to take on increased significance in the near future. The Ministry of Construction has carried out a study to develop an estimate of future demand for renewal services, breaking the renewal market down into three segments: maintenance, repair, and remodelling. The Ministry concluded that the renewal market would grow at an average annual rate of 2.2 per cent, to become 1.5 times larger in the year 2010 than it was in 1995 (Figure 7.36). It thus appears that the renewal market will become too big for construction firms to ignore. Major construction firms have recently begun working actively to develop new technologies and organizational modes to meet the needs of the renewal market.

Even if a large renewal market takes shape, however, it is not certain how much of that market would be captured by the major construction firms. Major construction

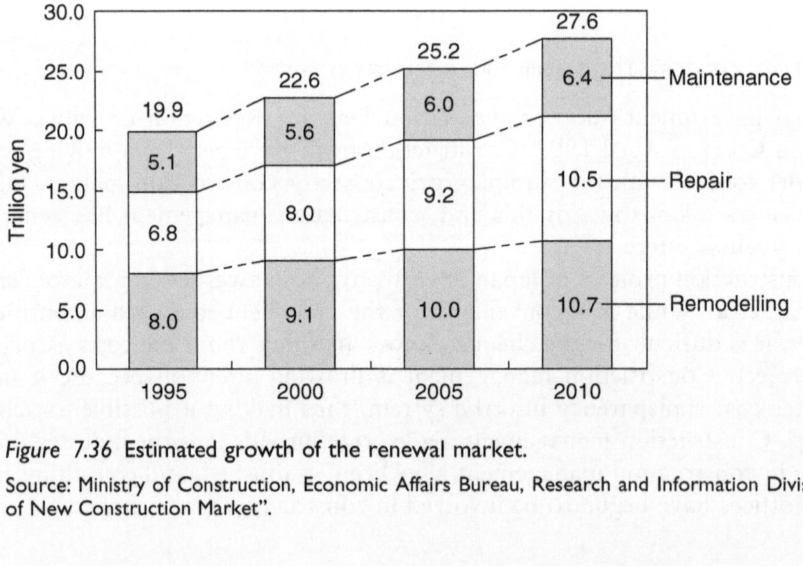

Figure 7.36 Estimated growth of the renewal market.

Source: Ministry of Construction, Economic Affairs Bureau, Research and Information Division. "Forecast of New Construction Market".

firms have been able to take advantage of their special strengths in large renewal projects, such as the ones at Kasumigaseki Building and the Trade Centre Building. It is not clear, however, what percentage of renewal projects in the future will offer the large scale that suits major construction firms.

ENHANCEMENT OF ENGINEERING CAPABILITIES

The number one business of Japanese construction firms has always been, and continues to be, total subcontracting, but what "total subcontracting" entails has been gradually changing. The most common practice is for the general contractor to build the superstructure of a building while subcontractors install equipment and systems. Especially in the case of factories, specialist engineering firms have a particularly important role to play. More recently, however, general contractors have begun branching out into the engineering sector, offering one-stop service that covers everything from planning to design, construction, and user support.

Examples include: (1) the adoption of the Hazard Analysis and Critical Control Point (HACCP) approach at food production facilities; (2) the adoption of automated warehouses at physical distribution facilities; and (3) the adoption of project management to provide such facilities as turnkey power plants overseas.

In Europe and North America, there are many engineering, procurement, and construction (EPC) contractors capable of handling everything from construction to engineering. Japanese general contractors have begun taking steps to acquire the capability to offer EPC-like services, but the path to establishment of an EPC firm in Europe or North America differs greatly from the Japanese pattern.

In the West, almost all EPC firms have evolved as a result of the vigorous M&A activities so prevalent in the West, that is, construction firms have acquired engineering firms, and vice versa. In Japan, firmly established engineering firms already exist, and the competition between them is fierce. General contractors, for their part, prefer to add engineering capacity by strengthening their own in-house technological expertise. As such, it is difficult to imagine that Japan's general contractors will undergo a quick transformation into EPC firms.

ENVIRONMENTAL TECHNOLOGY AS A NEW BUSINESS

With today's emphasis on global environmental preservation, construction firms are creating new lines of business for themselves. Many construction activities have a negative environmental impact. Hazardous wastes are generated, and significant carbon dioxide emissions are connected with the production of construction materials. For this reason, environmental issues have always been of great relevance to the construction industry. The industry has long been called upon to take active steps to address environmental issues, and now these issues are actually presenting construction firms with business opportunities (Table 7.50).

To capitalize on these business opportunities, it is advantageous to obtain ISO 14000 series certification, and increasing numbers of construction firms are working to do just that. Six major construction firms had obtained this certification as of July 1998.

DOWNSIZING OF MANAGEMENT

The measures described above are long-term strategies, however, aimed at markets that are expected to emerge in the future. As such, they will not improve the bottom line in the short term.

To quickly bolster the bottom line during the current economic recession at the time of writing, construction firms must downsize and improve management. They must also take steps to enhance productivity.

7.3.3 Research and development in the construction sector

Current status of R&D

R&D EXPENDITURES IN RECENT YEARS

The R&D expenditures of the major construction firms are on a downward trend in absolute terms, but among the top five firms, R&D expenditures continue to account for 1 per cent of total sales. The combined R&D expenditures of the construction industry as a whole in FY 1996 came to ¥224.5 billion, 32.2 per cent of which was spent by the top five firms. This percentage had been dropping in recent years, however (Figures 7.37 and 7.38). Various reasons have been advanced to explain this trend. First, it may be that the sophistication and complexity of today's technologies have forced the top five firms to carry out joint R&D projects in cooperation with various specialist construction firms. A second possible explanation is the decreasing number of large-scale construction projects. R&D expenditures are often included as part of the expenses associated with individual projects, but economic recession has meant fewer large construction projects of the type that generate a need to develop highly sophisticated technologies. The result has been less spending on R&D.

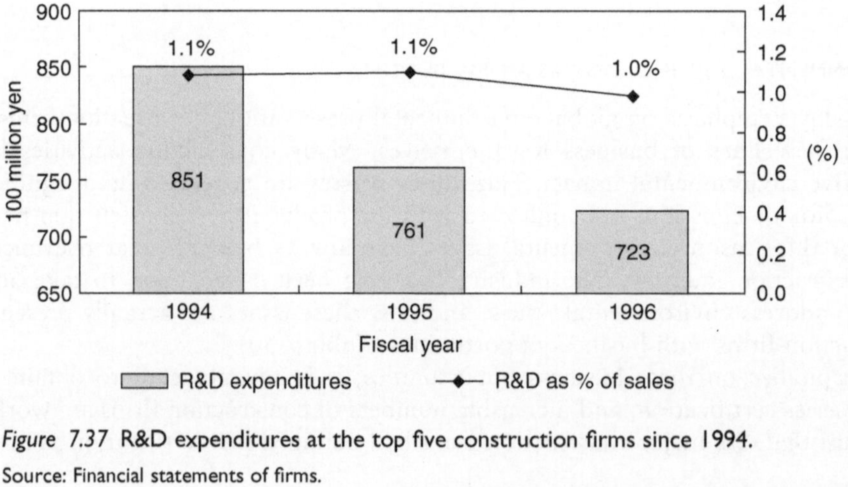

Figure 7.37 R&D expenditures at the top five construction firms since 1994.

Source: Financial statements of firms.

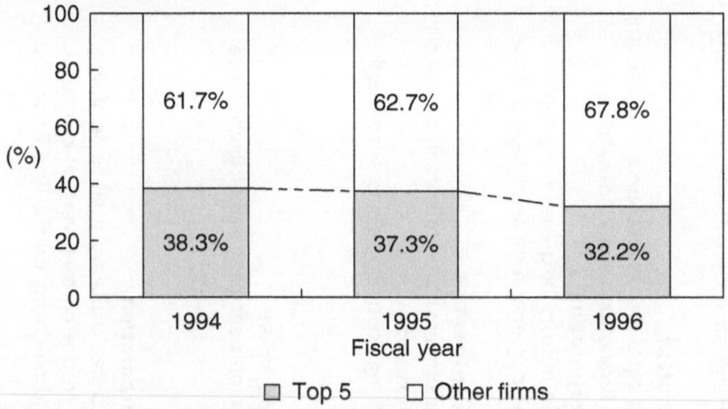

Figure 7.38 R&D expenditures: top five versus the rest.

Source: Financial statement of firms, and the Report on the Survey of Research and Development (Management and Coordination Agency, Statistics Bureau).

Recent R&D projects

In addition to the usual projects aimed at increasing productivity, the main construction firms have recently been devoting considerable R&D resources to environmental technology, information technology, and earthquake-resistant construction (Tables 7.49 and 7.50).

Use of prefabricated components

Table 7.51 shows the level of prefabrication from 1995 to 2000.

7.3.4 Bidding systems

According to Japan's Accounts Law, open and competitive bidding is generally the bidding system to be used for public works projects, but in recent years different commissioning bodies, such as local governments, have begun to use different bidding methods. In fact, there have been many cases in which designated competitive bidding has been used. Still, recognizing the trend towards economic liberalization as symbolized by the WTO, more construction service buyers have begun employing open and competitive bidding in recent years, and are also accepting new bidding methods such as the value engineering (VE) system in which technical proposals for construction methods and other elements of the project are accepted during the bidding process, and the integrated evaluation bidding system in which the function and quality of the proposed structure are comprehensively evaluated during the bidding process.

In February 2001, the Act for Promoting Proper Tendering and Contracting for Public Works was enacted to secure the national citizens' trust towards public works and to promote the sound development of the construction industry that performs public works projects commissioned by the national government, quasi-governmental agencies, and local governments. The act is based on (1) ensuring transparency in

Table 7.49 Principal R&D projects recently carried out by the major construction firms

Environmental technology

The Obayashi BIMA System, an ecological recycling system — ***Obayashi Corporation***

The Obayashi BIMA System entails the churning of organic waste material in a special fermentation vat. The gas (methane) produced during this process can be used to generate power, while the solid waste thus produced can be used as compost.

The Step Revegetation Method — ***Sato Kogyo***

Large planter boxes built in a terraced structure are used to capture fallen leaves, snow, and other biomass so the nutrients in it can be slowly returned to the soil. These planters are good for preserving the ecological balance of steep terrains, maintaining harmony with the surrounding environment.

Building technique for preserving the natural ground water flow — ***Taisei Corporation***

By installing water-permeable mats to underground walls to accumulate water, which is then routed through pipes along the course of its natural flow, large architectural structures can be built without their underground portions interrupting the natural flow of underground water.

CLIS-HR (Crystal Liquid Ice Thermal/Storage and Heat Recovery System) — ***Takenaka Corporation***

This air conditioning system uses the spontaneous flow of the coolant to refrigerate air without the use of a motor. Inexpensive nighttime electricity is used to run an ice heat-storage system. The coolant used is R-22, which contains no CFCs. The conversion of latent heat and the use of spontaneous coolant flow provide a very economical air conditioning system.

Thermal core type solar heating system — ***Nishimatsu Construction***

Soil is used as a long-term heat storage material, thus enabling surplus heat generated by solar systems in the summer to be utilized in the winter.

ECOGRAB — ***Maeda Corporation***

Electrolytic precipitation and gravel filtration can be used (instead of chemicals) to clean the water by removing floating particles, suspended organic matter, nutrient salts, metals, etc. from ponds, moats, and other bodies of standing water.

Information technology

Shimiz Cyber Building — ***Shimizu Corporation***

An electromagnetic radiation shield is used to prevent electronic devices from malfunctioning, enhance electronic security, and facilitate more effective use of radio systems.

Earthquake-resistant construction

Vibration isolation for wooden structures on the list of cultural properties — ***Obayashi Corporation***

This is a vibration system designed for wooden structures on the list of cultural properties. Steel frames and steel rods are used to reinforce wooden structures. New concrete beams are added beneath the flooring on the first floor, and the weight resting on each vertical column is transmitted via these concrete beams to a rubber vibration damper. A high-quality precast form is used for the flooring on the first floor, and precast concrete beams are used beneath this flooring, thus providing a crack-free, long-lasting structure.

Active Vibration Control System *Kajima Construction*

This is a conventional vibration isolation system "with a brain". Sensors detect earthquake vibrations and transmit signals to an actuator, which moves the floor to offset the effect of the earthquake.

T-LAMDA (Taisei Linear motor type Active Mass Damper) *Taisei Corporation*

A counterweight hanging from a special support mechanism is moved by a linear motor to counteract vibrations caused by wind and small- to medium-sized earthquakes. This is Japan's first active vibration isolation system. It can counteract the effect of 50 metres per second winds and earthquakes measuring up to 4 on the Japanese earthquake intensity scale.

CFT concrete-filled steel tube structural system *Takenaka Corporation*

This is a completely new structural system which makes use of steel pipes filled with high-strength concrete. By combining the special strengths of steel pipe and concrete, this structure provides rigidity, strength, and flexibility for outstanding resistance to earthquakes, fire, and other natural disasters.

Others

V-Block Method – New Technology for Immersed Tunnel Construction *Penta-Ocean Construction*

This construction method is used to build underground tunnels. A wedge-shaped block called a V-block is inserted into the end joint of an underground tunnel, and the forces (water pressure, weight of the tunnel) acting upon the joint are utilized to create a tight seal between the two ends of the V-block and the previously sunk caisson, thereby completely preventing water seepage. Once the water seal is established, the sea water inside the joint is pumped out, and workers enter the air-tight section of tunnel to complete construction of a strong joint.

FACES Future Automated construction Efficient System (all-weather automated construction method) *Penta-Ocean Construction*

With this construction method, a building under construction is completely covered by a temporary roof, and on-site conveyance equipment enables work to proceed as if the workers were inside a factory. Once the first floor is completed, the temporary roof is raised automatically to make way for construction of the second floor.

Segment of high flowing concrete (SHFC) *Sato Kogyo*

Manufacturing of the segments accounts for a high proportion of the total costs generated by shield tunneling. This construction method uses high flowing concrete to manufacture the segments. With conventional super-hard mixing concrete, a vibrator has to be used to pack down the concrete, and the surface has to be finished by hand. In addition, the framework has to be moved after the completion of each segment. The SHFC method simplifies the manufacturing of segments, requires less equipment, and achieves significant cost savings.

Table 7.49 (Continued)

SMART: Shimizu Manufacturing system by Advanced Robotics Technology (all-weather automated building construction method) — ***Shimizu Corporation***

This method is an improved version of the pioneering Smart System. Usable for the construction of buildings of up to 30 stories in height, it makes use of a welding robot (development is now complete), an automated process for straightening a crooked steel frame, and other automated devices.

Branch shield construction method (underground branch construction method) — ***Nishimatsu Construction***

This construction method is used to construct a T-joint to connect a branch tunnel to the main tunnel. The shield for the branch tunnel is branched off from within the shield for the main tunnel. The branch tunnel can be built without the use of pillars.

Artificial upwelling fishery (creating an upwelling to form a new fish habitat) — ***Hazama Corporation***

Carbon blocks made from the large amounts of fly ash generated by coal-fired power plants are placed on the sea floor and piled high into an artificial mountain, which creates an artificial upwelling. This current carries up nutrient salts, thus cultivating phytoplankton and turning the area around the artificial mountain into a gigantic fishery.

Geo-cycle (mechanized underground bicycle parking lot) — ***Hazama Corporation***

The same technology that goes into automated warehouses can be used to build a mechanized bicycle parking lot 20 to 30 metres underground. These lots occupy an area of approximately 330 square metres, and can accommodate 1,000 bicycles. Through the use of ID cards, these lots can operate 24 hours a day, and the property above can be used for many different purposes. It only takes about 20 seconds for the user to park or retrieve a bicycle.

REED building method — ***Maeda Corporation***

This high-speed method for building bridge footings makes use of knobby H-beams and SEED forms, a type of high-strength caisson. The knobby H-beams which are used in place of the usual molded steel bars offer superior bonding strength. By assembling the SEED form within these H-beams and using a crane to assemble steel-reinforced precast concrete caissons, the assembly of forms and precast caissons is finished with extreme rapidity. Furthermore, there is no need to strip the form or cure the concrete, which cuts the time required to complete the latter part of the project.

Table 7.50 New construction technologies

Brief description of new technologies	*Developer*
Flat slab base isolation SYSTEM	
Comprised primarily of multilayered earthquake-resistant walls and flat slabs (floor panels), this new earthquake resistant structural system for high-rise condominiums and apartment buildings offers an optimal combination of earthquake resistant components that make it possible to lengthen the building's natural period to 4–6 seconds and thus improve it's seismic resistance. Without beams or pillars protruding into the room, this structural system achieves simple, open spaces that allow boundless layout options. Also, the structure allows the use of large windows, both hanging and sliding, to create bright and open rooms with good views.	Takenaka Corporation
Tunnel wall tapping sound-resonating diagnostic system	
A system for automatically diagnosing the soundness of a tunnel lining using the SonicMeister, a wall tapping sound-resonating diagnostic device. This device allows more accurate operations at 3–5 times the speed of conventional methods by manpower. Diagnostic results can be immediately confirmed on the spot, and the recorded data can be saved for use in diagnosing deterioration over time.	Taisei Corporation
Cyber building	
This is a building that can comprehensively resolve problems involving electromagnetic waves, such as malfunctioning IT equipment, information leaks, and interference affecting wireless systems using electromagnetic wave shield technology. Shimizu Corporation predicts whether problems will occur using electromagnetic environmental measurement and simulation technologies, and performs appropriate electromagnetic environmental engineering on the entire process from planning to design, construction, and maintenance.	Shimizu Corporation
Hybrid slip form method	
This is an extremely fast method of constructing tall bridge piers that uses steel pipes for the reaction abutments and raises a slip form (PC form). Steel pipes are used as the main reinforcement, and a complex structure using a line from the PC strand is used in the tie. As a result, Obayashi Corporation has improved construction safety, reduced the energy needed for the process of rebar fixing, and increased the speed of construction.	Obayashi Corporation
WATAR (wave trapping artificial reef construction method)	
A method of constructing breakwaters submerged in the ocean, WATAR is a method of controlling waves and combating shoreline erosion. In contrast to other artificial reefs, WATAR is more effective at reducing wave height and offers lower construction costs because of its compact structure. Because part of the structure can be prefabricated, construction times can be shortened. Also, because it is a submerged breakwater, it does not interfere with the coastal scenery.	Nishimatsu Construction
Parent–Child shield construction method	
By allowing cross-sectional changes, this shield construction method is a technology for economically and efficiently building tunnels that can be adapted to the tunnel's purpose. One shield can excavate different tunnel radiuses. There is a child-to-parent method, in which, after excavating with the child shield, the parent shield is attached to continue excavation, or a parent-to-child method, wherein the child shield is attached to the parent	Sato Kogyo

Table 7.50 (Continued)

Brief Description of New Technologies	*Developer*
shield ahead of time, and then once excavation with the parent shield is finished, the child shield is taken off before continuing with excavation work.	
Rapid construction method for high seismic-resistance bridge piers (REED construction)	
This is a structural shaping and construction method for reinforced-concrete complex-structure bridge piers that uses studded H-shaped steel (that can withstand tension) and a prefabricated high-durability form (that can withstand compression). As a result, this method enables fast-paced construction through the repetition of simple tasks, improved earthquake resistance through the use of highly rigid H-shaped steel, and enhanced beauty and durability through the use of pre-cast form.	Maeda Corporation
Pneumatic capsule material transport system	
This system is used to transport and unload earth excavated during underground construction by placing it in a capsule and using a low-pressure air current in a vertical pipeline to transport it above ground without ropes. The system can be used in vertical holes with depths of 30–1000 metres and 5–15 metres in diameter, and features a transport cycle time that is about three times faster than conventional wire rope systems. Processes from piling to unloading are completely automated. Once construction work is completed, the system can be used for vertical distribution.	Kajima Corporation
4-Centered shield machine construction method	
This method of constructing underground stations using a 4-centered slurry shield machine can be used in locations with a lot of underground structures or where excavation is difficult. The 4-centered slurry shield machine has two horizontal rotating cutters and two centered vertical rotating cutters on a single surface.	Hazama Corporation
Digital camera earth transport volume measurement system	
This is a system of automatically measuring soil volumes accumulated in soil transport barges with a stereo image measurement device that uses two digital cameras. While offering a high level of measurement precision, this system is much less expensive that conventional mechanical methods. Also, because the parts that comprise the system are sold on the market, the system is inexpensive and easy to maintain. Absolutely no additional equipment is needed on the soil transport barge.	Penta-Ocean Construction

the tendering and contracting procedures, (2) promoting fair competition between bidders, (3) completely eliminating improper behaviour, and (4) ensuring that public works projects are performed properly – and makes it compulsory for commissioning organizations to (a) announce their projected orders each fiscal year, (b) announce information related to tendering and contracting procedures, (c) establish measures to eliminate improper behaviour, and (d) establish proper construction systems. The act was applied to tendering and contracting procedures starting in fiscal 2001.

Table 7.51 Level of prefabrication

	Portion of total sales by domestic cement firms accounted for by finished cement products (%)
1995 FY	14.5
1996 FY	14.5
1997 FY	14.6
1998 FY	14.1
1999 FY	13.9
2000	13.6

Source: Cement Handbook (2001 edition), Japan Cement Association.

Bidding system for public works projects

SYSTEM FOR ASSESSING A CONSTRUCTION FIRM'S CAPABILITIES

The capabilities of Japanese construction firms are assessed on the basis of several different factors. Under this assessment system, it is the government official from which a firm has received its construction license (e.g., Minister of Construction, or a prefectural governor) who is responsible for assessment of the firm, which is awarded points on the basis of the amount of business handled, its financial situation, its technical expertise, etc. This assessment process is mandated by law. When a company is being evaluated to determine whether it is qualified to bid for public works projects, this system is used to assess its credit, technical capabilities, and on-site construction expertise. The basic elements of the system are summarized in Table 7.52.

Table 7.52 Trends in number of construction business license holders (1975–2002)

	Total	*Growth rate (in %)*	*Permit from Minister*	*Growth rate (in%)*	*Permit from governor*	*Growth rate (in %)*
1975	350,817		6,331		344,486	
1980	488,520	39.3	7,465	17.9	481,055	39.6
1985	518,964	6.2	8,337	11.7	510,627	6.1
1990	508,874	−1.9	8,944	7.3	499,930	−2.1
1991	515,440	1.3	9,022	0.9	506,418	1.3
1992	522,450	1.4	9,124	1.1	513,326	1.4
1993	530,665	1.6	9,332	2.3	521,333	1.6
1994	543,033	2.3	9,619	3.1	533,414	2.3
1995	551,661	1.6	9,871	2.6	541,790	1.6
1996	557,175	1.0	10,062	1.9	547,113	1.0
1997	564,849	1.4	10,485	4.2	554,364	1.3
1998	568,548	0.7	10,742	2.5	557,824	0.6
1999	586,045	3.1	10,815	0.7	575,230	3.1
2000	600,980	2.5	10,899	0.8	590,081	2.6
2001	585,959	−2.5	10,877	−0.2	575,082	−2.5
2002	571,388	−2.5	10,909	0.3	560,479	−2.5

Source: Ministry of Land, Infrastructure and Transport.

Note
Numbers of companies are as of the end of March in each year.

CONSTRUCTION BUSINESS LICENSING SYSTEM

There are two types of construction licenses, national licenses granted by the MLIT, and prefectural licenses granted by prefectural governors. There are also two types of construction firms, special and ordinary construction firms.

A single construction firm cannot simultaneously hold both a national and prefectural construction business license, nor can a single firm be categorized simultaneously as both a special and ordinary construction firm.

When a company intends to establish offices and do business in two or more prefectures, it must obtain a national construction license. If a company intends to establish an office and do business in only one prefecture, it must obtain a prefectural license from the prefecture that has jurisdiction over that location.

All construction firms – both general contractors and subcontractors – must obtain an ordinary construction license unless they only intend to perform extremely small jobs. A special construction license, however, must be obtained to engage in general contracting where contracts worth ¥45 million or more are received directly from the end client, or to engage in subcontracting involving contracts worth ¥30 million or more.

The special construction license is part of a system designed to protect the interests of subcontractors and to ensure that projects are carried out properly. The required qualifications for a special construction license are more stringent than those for an ordinary license.

CORPORATE CAPABILITY EVALUATION SYSTEM

The principal system used in Japan for evaluating a construction firm's capabilities is the Business Evaluation. For construction firms that receive direct contracts for public works projects, business evaluations are carried out by the Minister of Land, Infrastructure and Transport or a prefectural governor. These evaluations entail an examination of a specifically formulated list of objective criteria for evaluating the business of a construction firm. The evaluation results are used to assign the company a point score, which is publicly disclosed (see Table 7.53).

Organizations that place orders for public works projects assign each prospective contractor a ranking, taking into consideration their score on the Business Evaluation (objective evaluation) as well as the results of the organization's own evaluation of the contractor's construction performance (subjective evaluation) (Table 7.54).

The criteria for the subjective evaluation include construction performance, history of instructions to halt work, history of legal violations of the Construction Business Act, awards for outstanding construction projects, etc.

THE TYPICAL BIDDING PROCEDURE

Until recent years till 2002, public works contracts in Japan were usually awarded through a selective bidding process. The Cabinet, however, adopted the Action Plan on Reforms of Bidding and Contracting Procedures of Public Works, which provided for the introduction in January 1994 of an open bidding system. There were a number of developments which led to this Action Plan. One was a recommendation put forward in December 1993 by the Central Council on Construction Business entitled Reforms of Bidding and Contracting Systems for Public Works. Another was a GATT

Table 7.53 Total points on the business evaluation $(P) = 0.35X_1 + 0.10X_2 + 0.20Y + 0.20Z + 0.15W$

		Evaluation criteria
Scale of operation	X_1	Average annual construction revenue, classed by type of project
	X_2	Shareholder's equity, number of employees
Business condition	Y	Ratio of revenues to operating profit; Ratio of capital to ordinary profit; Ratio of cash flow to revenue; Ratio of require working capital to monthly turnover; Ratio of account receipt and work in process to annual sales; Ratio of accounts receivable to monthly turnover; Equity ratio; Ratio of interest-bearing debt to monthly turnover; Net interest payments ratio; Ratio of shareholder's equity to fixed assets; Ratio of fixed assets long-term capital; Ratio of value-added output to fixed assets
Level of technical expertise	Z	Number of technical employee, classed by type of business
Other evaluation criteria	W	Non-wage benefits; Working days per year; On-site safety record; Number of construction industry accountants on staff

Table 7.54 Outline of assessment system-items upon which a construction firm is assessed

		Item assessed
Business size	X_1	Average construction revenue in each category of construction
	X_2	Equity capital, number of employees
Financial situation	Y	Ordinary profit on construction revenue; ordinary profit on total assets; break-even ratio; current ratio; quick ratio; months of operating capital on hand; construction revenue per employee; value-added per employee; total assets per employee; ratio of fixed assets to net worth; equity ratio; ratio of fixed liabilities to net worth
Technical capabilities	Z	Number of technical personnel in each category of construction
Other items assessed	W	Fringe benefits; years in business; construction safety record; number of construction bookkeepers

Formula for calculating composite score $(P) = 0.35X_1 + 0.10X_2 + 0.20Y + 0.20Z + 0.15W$

government procurement agreement. As a result of the Action Plan's adoption, the national government and other government-related organizations began using the open bidding system to award contracts for projects with a pro forma value of 4.5 million SDR or greater (then equivalent to \730 million). There were 393 such projects in FY 1994. These open bidding procedures were equally open to all construction firms both domestic and foreign. For projects worth less than 4.5 SDR, a modified version of the selective bidding system was implemented. (Modifications included the adoption of public invitations and project interest registration.)

The WTO government procurement agreement took effect in January 1996. The provisions of this agreement apply to construction services for all projects worth over 4.5 million SDR (then the equivalent of \650 million) that are awarded by the national government, and to all projects worth over 15 million SDR (then the equivalent of \2,160 million) that are awarded by government-related organizations, prefectural governments, and ordinance-designated cities. The implementation of the different bidding procedures by different types of governmental organizations is shown in Table 7.55.

Tables 7.56 and 7.57 show an overview of the bidding systems and the contract award criteria respectively.

PUBLIC WORKS CONTRACTS

Public works contracts are characterized by the following special features:

a Types of contracts – In principle, public works projects are divided into a design contract, a construction contract, and an equipment contract.
b Scope of contracts – The end client usually awards the entire construction project to a single general contractor.
c Contract price – The overall cost of the project is usually included in a single lump-sum contract.

ADOPTION OF THE DESIGN-BUILD CONTRACTING SYSTEM

The design-build system is a method in which client institutions accept design and other technical proposals at the time of the bidding, and contract both the design and construction work together.

On some projects that require advanced or specialized technological capabilities and for which there is considerable development of construction technologies, it may be better to combine special design and construction technologies developed by separate companies than to use the design and construction technologies developed separately. (In FY 2001, this system was used for 14 projects under the direct supervision of the Ministry of Land, Infrastructure and Transport.)

As stated above, a separate contract for design is usually awarded for public works projects, but discussions are now under way regarding the possibility of adopting a design-build system, and this system is already being used on an trial basis. The basic facts concerning a design-build contract are included in Table 7.58.

Table 7.55 Implementation of different bidding systems (1996)

Client	*Open bidding (in %)*	*Selective bidding (in %)*		
		Public invitation	*Project interest registration*	*Other*
National government	100.0	91.7	66.7	–
Prefectural governments	100.0	70.2	4.3	25.5
Ordinance-designated cities	100.0	75.0	–	25.0
Cities, towns, villages	15.1	7.4	4.0	–
Public corporations	100.0	94.1	17.6	–

Source: Research Institute of Construction and Economy (1997). Survey and Report on Fundamental Issues Related to Procurement, Contracts and the Guarantee System.

Note

The figures in this table indicate what percentage of a given type of government organization has implemented the type of bidding system in question.

Table 7.56 Overview of the bidding systems

Open bidding

Size of project

- Awarded by the national government: over \720 million (4.5 million SDR)
- Awarded by government-related organizations, prefectural governments, or a ordinance-designated city: over \2,430 million (15 million SDR)

Qualifications for bidding

- To obtain bidding qualifications, a firm must be awarded a specified number of points in a business assessment
- The firm must have an established track record handling the type of project to be awarded
- The firm must be able to assign certified technical professionals to the project in question

Procedures

1. Invitation to bid is published, and design drawings are made available
2. Firms wishing to bid provide documentation of their bidding qualifications
3. Client confirms bidding qualifications, notifies prospective bidders
4. Bidding commences
5. Contract is awarded

Selective bidding

Selective bidding by public invitation

Size of project

- In principle, used for projects worth \200–650 million

Selection procedure

1. Client ranks the prospective bidders
2. Client requests technical documents from the prospective bidders
3. Approximately ten firms are selected on the basis of the technical documents submitted

Bidding procedure

1. Selected bidders are notified
2. Design drawing are delivered to the bidders
3. Bidding commences
4. Contract is awarded

Selective bidding by project interest registration

Size of project

- In principle, used for projects worth \100–200 million

Selection procedure

1. At the beginning of the year, companies register to indicate the type of projects they wish to bid for
2. As each individual project arises, 10–20 registered firms are requested to submit technical documents
3. Approximately ten firms are selected on the basis of the technical documents submitted

Bidding procedure

Same as for public invitation selective bidding

Discretionary contract awards

Type of project

- Projects which require special technology, machinery, or equipment, or which can only be awarded to a particular construction firm
- Emergency projects for which there is not time to go through a competitive bidding process. Determination of contract price
- The contract price must be lower than the projected price. If it is higher, the bidding procedure must usually be repeated

Table 7.57 Contract award criteria

- A contract is usually awarded to the firm that tenders the lowest-priced bid
- There is a system for investigating bids that appear to be priced so low as to make successful performance of the contract highly unlikely. The purpose of the investigation is to determine the ability of such a bidder to perform the contract
- Many prefectural and local governments use a minimum bid system in which the lowest allowable bid price is specified beforehand. A contract may not be awarded to any bidder whose bid price falls below this minimum
- The possibility of adopting a comprehensive evaluation system is also being discussed. Under this system, contracts would be awarded on the basis of a comprehensive evaluation of technical documents and price estimates submitted to the client

Table 7.58 Design-build contracts now under consideration

Design-build contract system

Type of project

Design-build contracts are used for projects which require special, highly sophisticated technology and an especially large R&D effort, and which by their nature call for a single contractor to handle both design and construction.

Contract award procedure

The companies to participate in the bidding are selected upon the basis of design documents submitted by a pool of candidates. The contract is either awarded to the lowest bid, or to strongest bid as determined upon the basis of a comprehensive evaluation of both design documents and price.

Competitive bid plus partial design

Type of project

This type of system is used for projects which can be most effectively carried out when a single construction firm uses its design expertise and equipment to develop design details for a specific part of the overall project.

Contract award procedure

A single contract covers both construction and the design details prepared on the basis of a previously completed basic design.

INTRODUCTION OF VALUE ENGINEERING (VE)

There are two types of value engineering (VE) systems: a bidding VE system, where technical proposals on construction methods are accepted in the bidding stage, and a post-contract VE system, in which technical proposals on construction methods are accepted in the construction stage.

The VE system can be utilized in conjunction with either open or designated competitive bidding, or a negotiated contract system (such as by the MLIT, local housing supply corporations, and local public works organizations). There are plans to expand the post-contract VE system by making it applicable to all construction projects using open and competitive bidding and public-invitation designated competitive bidding.

The VE systems described in Table 7.59 are now under discussion, and are being implemented on a trial basis.

Table 7.59 VE systems now under consideration

VE during bidding	
Contract award based on price competition	
	Construction firms which are able to propose viable cost reduction technologies are chosen from a pool of candidates. Selected firms submit bids based on their technical proposals, and the contract is awarded to the lowest bidder.
Contract award based on price and technology	
	Bidders are first selected by examining the past track record of a pool of candidates in the type of construction in question. Bidders propose both technologies and cost control methods for reducing costs. The contract is awarded on the basis of a comprehensive evaluation of both proposals.
VE after bidding	
	The firm that has been awarded the contract develops a technical proposal for reducing costs. If the proposal is accepted, the design documents and contract are revised accordingly. When the contract is revised, part of the cost savings reverts to the construction firm.

Bidding systems for private-sector projects

With public works projects, separate contracts are generally awarded for construction and equipment installation. With private-sector projects, however, there are two systems – construction-only contracting, and design-build contracting. In FY 1996, design-build contracts accounted for 41.8 per cent of the total value of private-sector projects handled by the top five construction firms. For the top 54 firms (including the top five just mentioned), this figure was 26 per cent (according to a survey conducted by *Nikkei Construction*).

Supervision of construction work is usually carried out by either the end client or the design firm. Certified quantity surveyors are not used for the checking of construction quantities, unit costs, or the amount of work done.

Private-sector contracts are usually awarded on a lump-sum basis. Payment is most often made in instalments, with an agreed upon sum paid, for example: (1) when construction begins; (2) at the halfway point; and (3) after completion.

The use of CM contracting has been on the rise in recent years for private-sector projects, including projects awarded by foreign clients, but the total number of such projects remains low.

7.3.5 Market access in the construction sector

Market access initiatives

CONSTRUCTION SERVICES PROCUREMENT: EQUAL OPPORTUNITY FOR DOMESTIC AND FOREIGN CONSTRUCTION FIRMS

The World Trade Organization (WTO), which was founded in January 1995, established the General Agreement on Trade in Services (GATS) in order to deal with government measures that have an impact upon trade in the service sector, which includes the construction industry. Japan became a party to the WTO Agreement on Government Procurement in 1996. In accordance with the GATS agreement, Japan has pledged not to set numerical limits upon foreign participation in the construction industry and five other industries, and not to adopt other measures that would have a discriminatory effect against foreign firms in these industries.

In addition, the terms of the WTO government procurement agreement reached in January 1996 require equal opportunity for domestic and foreign firms to bid for projects worth over 4.5 million SDR (currently equivalent to ¥720 million) that are awarded by the national government, and for projects worth over 15 million SDR (currently equivalent to ¥2,430 million) that are awarded by government-related organizations, prefectural governments, and ordinance designated cities.

MAJOR PROJECTS ARRANGEMENT (MPA)

In 1989, Japan and the United States agreed upon special measures to allow foreign construction firms to register for business in Japan and bid on 17 major construction projects. The purpose of this agreement was to give foreign firms an opportunity to become familiar with the Japanese construction market. The special measures are as follows:

- Notices of selective bidding contracts were published in daily construction industry newspapers. In the case of construction contracts, the period for specifying qualified bidders was lengthened to 30 days, and the period from specification to deadline for submission of bid was lengthened to 40 days.
- An independent, third-party grievance body (the Construction Procurement Review Board) was established to handle grievances arising in connection with procurement activities carried out under the terms of the special measures.

Of the 42 projects carried out thus far in accordance with MPA provisions, 22 have been completed and seven were in progress at the time of writing. The then most recent addition to the MPA line-up is the Chubu New International Airport.

FOREIGN FIRMS AND THE MANAGEMENT ASSESSMENT SYSTEM

In order to properly assess the management of foreign construction firms, the assessments are carried out as follows.

- Assessment of construction revenues, equity capital, number of employees, and financial situation is based on the firms' operations outside Japan.
- The number of technical personnel and years in business are based upon the firms' operations outside Japan, subject to approval by the Minister of Construction.
- When the Minister of Construction has determined that the subsidiaries of a foreign construction firm are all engaged together with the parent company in integrated construction-related operations, the firm shall be assessed on the basis of the operations of its entire group.

CONCESSIONARY POLICIES FOR FOREIGN CONSTRUCTION FIRMS

There are no special concessionary policies or regulative policies for foreign construction firms.

Brief statistics on foreign participation in government procurement The national government and various government-related organizations awarded three contracts worth a combined total of ¥3.4 billion to foreign construction firms in FY 1996. In FY 1997, these figures rose to six contracts worth a combined total of ¥5.8 billion. This figure, however, only represents 0.37 per cent of contracts awarded through open bidding.

7.3.6 Construction industry policy

Measures to improve the business climate in the construction industry (January 1998)

In response to a rapidly deteriorating business climate in the construction industry, the Ministry of Construction in January 1998 adopted the Measures to Improve the Business Climate in the Construction Industry. The total package comprises 27 measures aimed at the needs of eight different segments of the construction industry, and is intended to achieve three main goals: (1) to help construction firms overcome the current credit crunch by ensuring an adequate supply of funds; (2) to support the efforts of construction firms to improve their financial systems and strengthen management; and (3) to provide a better environment for the awarding of construction contracts.

RECOMMENDATIONS BY THE CENTRAL COUNCIL ON CONSTRUCTION BUSINESS (FEBRUARY 1998)

Competition in the construction industry is expected to grow increasingly fierce due to slumping construction investment, changing trends in public-sector investment, and globalization of the construction market. To build a competitive and transparent marketplace in which firms with the best technology and management will rise to the top, the Central Council on Construction Business in June 1996 established the Fundamental Issues Committee. The work of this committee led to the issuance in February 1998 of a set of recommendations entitled "How the Construction Industry Should Respond to Restructuring in the Construction Market". These recommendations call for the following concrete measures to be taken:

1 Further improve bidding and contract systems by tapping the technical expertise of the private sector, achieve greater transparency in bidding and contract procedures, and further improvements at the prefectural and local levels.
2 Promotion of structural reforms in the construction industry includes:

- revision of industrial classifications used in the construction licensing system
- review of the management assessment system and other arrangements for the evaluation of construction firms
- elimination of weak construction firms
- promotion of tie-ups among construction firms to improve managerial and technical capabilities
- establishment of proper business relations between general contractors and subcontractors.

3 Action plan for measures to reduce the cost of public works (April 1997) – In order to prepare for the graying of Japan's population, make the best possible use of limited financial resources, and efficiently carry out public works projects so as to make solid progress in the building of public infrastructure, the national government has formulated the Action Plan for Measures to Reduce the Cost of Public Works.

This Action Plan sets specific numerical targets for cost reductions in public works projects. Furthermore, the various ministries and agencies in the national government are carrying out action plans of their own to achieve numerical targets based on those in this Action Plan.

4 Fiscal Structural Reform Law (June 1997) – Under the terms of the Fiscal Structural Reform Law, which is intended to reduce the government's ¥500 trillion cumulative fiscal deficit, the period covered by the Basic Plan for Public Investment has been extended by 3 years. (It originally called for ¥630 trillion of government spending between FY 1995 and FY 2004.) In addition, the five-year plans of each government ministry and agency have been extended by 2 years (and all ten-year plans have been extended by 4 years). The law also requires that public works expenditures in FY 1998 be cut by 7 per cent from FY 1997.

Since the passage of that law, however, the extended economic recession has prompted a change in the direction of economic policy, with top priority now being placed on spurring economic recovery in Japan. To achieve this goal, the measures called for by the Fiscal Structural Reform Law have been put on hold to allow the floating of deficit-covering bonds. Tax reductions have been enacted, and public investment has been boosted.

7.4 Infrastructure development

7.4.1 Current status, long-term plans

Current infrastructure development efforts

Infrastructure development has not been carried out for a very long time in Japan. It started with the redevelopment of land that had been abandoned during World War II, and then jumped into high gear to meet new demand as the economy grew and the standard of living rose. Although infrastructure development has proceeded steadily since the war, the level of development still lags in certain areas and for certain types of infrastructure (Table 7.60).

Long-term plans for infrastructure development

Long-term plans for infrastructure development in Japan comprise comprehensive plans, and sectoral development plans. Comprehensive infrastructure development plans are long-term plans that cover every type of infrastructure. Examples of such plans include the National Comprehensive Development Plan, various economic plans, and the Basic Plan for Public Investment.

Table 7.60 Principal infrastructure: targets and current levels of development

Type	*How it is measured*	*Current level of development*	*Target for early twenty-first century*
Roads	Length of high-grade arterials	7,265 km (as of 31 March 1998)	approx. 14,000 km
Sewage lines	Percentage of households and businesses with service	55% (as of 31 March 1997)	approx. 90%
Urban parks	Park area per person (for the population targeted by the plan)	Nationwide: 7.3 m^2 23 wards of Tokyo: 2.9 m^2 (as of 31 March 1997)	approx. 20 m^2
River improvement (large rivers)	Flood prevention rate	65%	Ability to prevent large floods that come every 30 or 40 years
(medium- and small-size rivers)	Flood prevention rate	44%	–

The National Comprehensive Development Plan is a long-term infrastructure development plan formulated from an overarching, national perspective. In March 1998, a new National Comprehensive Development Plan, entitled A Comprehensive Nationwide Land Development Plan for the 21st century – Promoting Regional Independence and Beautification Efforts for National Land, was adopted. The plan is for the years of 2010–2015 and aims to decentralize the formulation of national land planning in order to maximize the discretionary authority of local communities, and to encourage these communities to form links amongst themselves as they participate in national land planning. The plan includes four strategic approaches: (1) creation of residential areas with more natural scenery; (2) urban renewal; (3) projects carried out by local communities, which work together amongst themselves; and (4) the formation of broad international people-to-people contacts.

The plan also calls for preservation and management of land, development plans for different fields (such as traffic, telecommunications, etc.), and basic plans for different local communities.

Economic plans include the Social and Economic Plan for Structural Reforms Toward a Vital and Secure Life, which was formulated in 1995. This plan includes specific targets for different types of infrastructure development to bring about the achievement of the following three policy objectives: (1) creation of a pleasant environment for everyday activities; (2) creation of an environment in which everyday activities can be carried out with safety and peace of mind; and (3) establishment of the necessary conditions for further development of the Japanese economy.

The goal of the basic Plan for Public Investment which was formulated in 1994 is to "ensure that all necessary infrastructure has basically been completed by the early part of the twenty-first century". The plan provides for the expenditure of ¥630 trillion during the ten-year period from 1995 through 2004. Priority has been given to the development of infrastructure that will directly improve the quality of living. The plan states that "the government shall raise to approximately 60 per cent the proportion of total investments directed toward projects related to the everyday living environment, social services, and cultural facilities".

The deficit reduction measures adopted by the national government in June 1998 decree that the ¥630 trillion slated to be spent under the provisions of deficit reduction measures shall be spent over a period of 13 years, rather than the originally scheduled 10 years. In effect, spending on infrastructure development has been slashed.

Sectoral development plans also exist covering most construction of roads, sewage lines, and other types of infrastructure, which are carried out under long-term plans covering periods of 5–10 years. There are currently long-term development plans for 16 different projects, each of which has been approved by Cabinet resolution. Each plan sets forth the size and specific goals of each project (Table 7.61).

The period covered by all these plans has been extended by 2 years in accordance with the provisions of deficit reduction measures adopted in June 1998, which amounts to a reduction in spending on infrastructure development.

Citizens now seek increased participation in infrastructure development, and new efforts are being made to ensure that these long-term plans fully meet the needs of local communities.

In connection with the formulation of a new five-year road improvement program, which was completed in May 1998, the following measures were taken from early on in the process so that as many citizens as possible were actively encouraged to contribute their ideas.

- A public involvement system was used to solicit the views of 130,000 citizens. This system, which relied on questionnaires, facilitated public involvement during the formulation of the plan and during the final decision process.
- 222 round tables in 64 communities attracted a total attendance of 3,000.

7.4.2 Organization and funding of infrastructure development

Organizations in charge of infrastructure development

In addition to governments at the national, prefectural, and local levels, infrastructure development in Japan is also carried out by public corporations, local public utilities, local public corporations, public–private partnerships, and private companies (Table 7.62).

The development of electric power and telecommunications infrastructure has been transferred to the private sector, as has much of the responsibility for the development of railways. As for harbour infrastructure, a large portion of the development and management process has been privatized in many cases at the level of individual wharves.

Where feasible, facilities are developed by public corporations which charge user fees. To a certain degree, this introduces the market mechanism into the process.

Responsibility for funding of infrastructure development

Infrastructure development projects fall into a number of different categories. Public works projects are not usually funded just by the organization in charge of the project.

Table 7.61 Long-term public works plans

Sector	*Plan*	*Period*	*Budget* (trillion yen)	*Purpose*
Roads	Five-Year Road Improvement Program	1998 ~ 2002	78	• Provide support for economic structural reform (high-grade arterials, etc.) • Provide support for community and urban development (bypasses, burying of power lines, etc.) • Provide a better living environment (wide sidewalks, noise barriers, etc.) • Achievement of an environment in which one can feel secure (improve maintenance, provide evacuation routes for disasters, etc.)
Ports	The Ninth Seven-Year Investment Plan for the Improvement of Ports and Harbours	1996 ~ 2002	7.49	• Formation of physical distribution network that will support an internationally competitive economy (international sea terminals, etc.) • Creation of a safe environment (more earthquake-resistant seawalls, etc.) • Provide support for community development (harbour redevelopment, etc.)
Airports	The Seventh Seven-Year Investment Plan for the Improvement of Airport	1996 ~ 2002	3.6	• Build primary airports for metropolitan areas • Improve the environment near airports • Build air traffic safety facilities
Traffic safety	The Sixth Seven-Year Traffic Safety Facilities Improvement Program	1996 ~ 2002	2.69	• Traffic safety for pedestrians and bicyclists • Accident prevention near schools • Prevention of vehicular accidents • Traffic control centres
Housing	The Seventh Housing Construction Five Year Plan	1996 ~ 2000	730 million dwelling units	• Construction of high-quality housing stock that meets the needs of residents • Safe, pleasant urban environment • Development of an environment which supports an active old age • Development of housing and neighbourhoods which support a vital community
Sewage	The Eighth Seven-Year Sewage Improvement Program	1996 ~ 2002	23.7	• Development of sewer service in small- and medium-sized towns where it is most lacking • Development of advanced waste water treatment • Efficient use of sewerage infrastructure

Table 7.61 (Continued)

Sector	*Plan*	*Period*	*Budget (trillion yen)*	*Purpose*
Waste facilities	The Eighth Plan for Developing Waste Disposal Facilities	1996 ~ 2002	5.05	• Increase recycling • Development of circulative disposal process • Treat household wastewater
Parks	The Sixth Seven-Year Program for Developing City Parks	1996 ~ 2002	7.2	• Provide space for refuge in times of natural disaster • Provide an opportunity to come in contact with nature • Provide a place for exercise • Provide a green buffer space
Improvements to mountain areas	The Ninth Seven-Year Plan for Improvement of Mountain Areas	1997 ~ 2003	3.77	• To build a safe country that is resistant to natural disaster • To improve the water-holding capacity of headwater areas • To build an environment of plenty
Flood control	The Ninth Seven-Year Flood Program	1997 ~ 2003	24.0	• To build a safe environment for society • To resolve the problem of frequent water shortages • To provide clean water and beautiful waterside scenery • To support the development of unique, vibrant communities (by providing places for community activities, etc.)
Improvement of steep slopes	The Fourth Five-Year Program for the Prevention of the Collapse of steep slopes	1998 ~ 2002	1.19	• To prevent landslides • To promote the planting of greenery on steep slopes
Coastal improvement	The Sixth Seven-Year Program for Coastal Improvement	1996 ~ 2002	1.77	• To build a safe, high-quality coastline • To build a coastline where nature thrives • To build a beautiful coastline that people can easily visit
Land improvement	The Fourth Long-term Land Improvement Plan	1993 ~ 2006	41.0	• To ensure the productivity and profitability of farm land • To build vibrant rural communities • To ensure efficient use of land
Fishing ports	The Ninth Long-term Fishing Port Improvement Plan	1994 ~ 2001	3.0	• To ensure adequate fishing industry output • To build safe, pleasant fishing harbours • To build vibrant fishing communities
Coastal fisheries	The Fourth Coastal Fisheries Development Plan	1994 ~ 2001	0.6	• To build fish habitat • To build breeding facilities • To build coastal fishing areas
Forest improvement	The Second Forest Improvement Plan	1997 ~ 2003	5.38	• To improve the capacity of forests to conserve water and soil • To promote coexistence between man and forest • To promote the reuse of forest resources • To build vibrant mountain communities

Table 7.62 Principle organizations in charge of infrastructure development

Type of infrastructure	*Organization*
Roads	National government; prefectural and local governments; public (koudan) corporations (Japan Highway Public Corporation, etc.); local public utilities; public (kousha) corporations; public-private partnerships (Trans Tokyo-Bay Highway Corporation)
Railways	A special corporation (Japan Railways group); a public (koudan) corporation (Japan Railway Construction Public Corporation); local public utilities; public-private partnerships; private companies
Airports	National government; prefectural and local governments; a public (koudan) corporation (New Tokyo International Airport Authority); a special corporation (Kansai International Airport Company); private companies
Harbours	National government; prefectural and local governments; public (kousha) corporations (Foreign Trade Port Corporation, etc.); public–private partnerships; private companies
Sewage service	Prefectural and local governments
Electric power	A special corporation (Electric Power Development Company); private companies
Telecommunications	Special corporations (NTT, KDD); private companies

Rather, the burden of funding is shared among several related entities, including the national government, prefectural and local governments, public corporations, and those who benefit from the infrastructure.

How the burden of funding is divided among these entities is decided on a case-by-case basis. In principle, each different entity takes responsibility for directly funding part of a project (Table 7.63).

When a project is carried out by a public corporation, that corporation usually funds the project by issuing public debt. It then covers this debt by charging a user fee.

One question which always arises concerns the proportion of funding to be provided by different public entities (the national government, on the one hand, versus prefectural and local governments, on the other). It is difficult to give a general rule of thumb regarding the rate of subsidization granted by the national government to prefectural and local governments. It varies from project to project depending on such factors as: (1) the importance to the national government of the project in question; (2) the size of the project; (3) the scope of the project's impact; and (4) the degree of priority accorded to the project.

Sources of funding for infrastructure development

Next, there is the question of where the national government, prefectural and local governments, public corporations, and other organizations obtain the funds which they disburse for infrastructure projects.

National government obtains most of its funds for public works projects by issuing bonds and using specially earmarked revenues (Table 7.64).

Table 7.63 Division of funding responsibility for infrastructure development

Organization in charge of project	*Organization(s) responsible for funding*	*Principal projects*
National government	National government only	National government hospitals; national education institutions; class I airports
	National government; prefectural and local governments	Roads (under direct government control); harbours (under direct government control); class 2 airports; mountain improvement projects (under direct government control); flood control projects (under direct government control); national parks; nature parks
	National government; prefectural and local governments; beneficiaries of the infrastructure	Agricultural infrastructure (government-run projects)
Prefectural and local governments	National government; prefectural and local governments	Roads; harbours; class 3 airports; mountain improvement projects; flood control projects; traffic safety facilities; waste treatment facilities; urban parks; nature parks; social service facilities; medical facilities; educational facilities
	National government; prefectural and local governments; beneficiaries of the infrastructure	Sewage service; public housing; agricultural infrastructure; forest roads; fishing ports
	Prefectural and local governments only	Projects listed under item which do not receive subsidies from the national government
	Prefectural and local governments; beneficiaries of the infrastructure	Projects listed under item which do not receive subsidies from the national government
Public corporations, etc.	National government; public corporations	Toll roads; water service; apartments built by the Japanese Housing and Urban Development Corporation; railways
	National government; public corporations; prefectural and local governments	Toll roads; container wharves

Note
The term "public corporation", as used here, includes public (koudan) corporations, public (kousha) corporations, local public utilities, and local public corporations (chihou kousha).

Table 7.64 General account revenues and expenditures (main budget, FY 1997)

Revenues	(unit: 100 million yen)
Taxes and stamp revenues	21,296
Taxes accounted for by gasoline tax, LPG tax, etc.	20,600
Other revenues	2,294
Bond revenue	92,370
Carryover from previous term	179
Funds appropriated from 2001 budget	984
Total	117,123
Expenditures	
Public works expenditures	91,419
Expenditures for facilities and equipment	11,235
Investments and loans	12,123
Other	2,337
Total	111,123

GOVERNMENT BONDS

Japan's national government is not officially authorized to issue bonds to raise revenue, but it is allowed to issue the so-called construction bonds to finance public works projects. The reason for this exception to the restriction against the use of government bonds is that the funds thus raised are used to build infrastructure which effectively becomes the assets of all citizens, who thus benefit from such bonds.

Two types of tax revenues are earmarked for the funding of public works projects. Gasoline taxes and other related taxes are used to fund road projects, and aircraft fuel taxes are used to fund airport construction. Eighty per cent of the combined total of all gasoline tax revenues, 1/2 of LPG tax revenues and a high proportion of motor vehicle tonnage tax revenues provide fund for road projects, 11/13 of aircraft fuel tax revenues are used to help fund airport project.

Prefectural and local governments obtain most of their funding for infrastructure development from general revenues, local government bonds, earmarked revenues, and national treasury disbursements.

General revenues are generated by: local taxes; and local allocation taxes from the national government (a type of revenue sharing scheme). Prefectural and local governments, like the national government, are only authorized to issue bonds for the financing of capital expenditures.

There are two sources of earmarked revenues. One is local transfer taxes (another type of revenue sharing scheme, in which all or part of certain national taxes are transferred to prefectural and local governments to be used for a specific purpose. The local road transfer tax is an example of this type of tax). The other is earmarked taxes (taxes levied by prefectural and local governments for a specific purpose, such as urban planning taxes).

An examination of revenue structure in at the time of writing revealed that general revenues are accounting for a falling percentage of infrastructure development funding, while reliance upon local government bonds is on the rise. The national

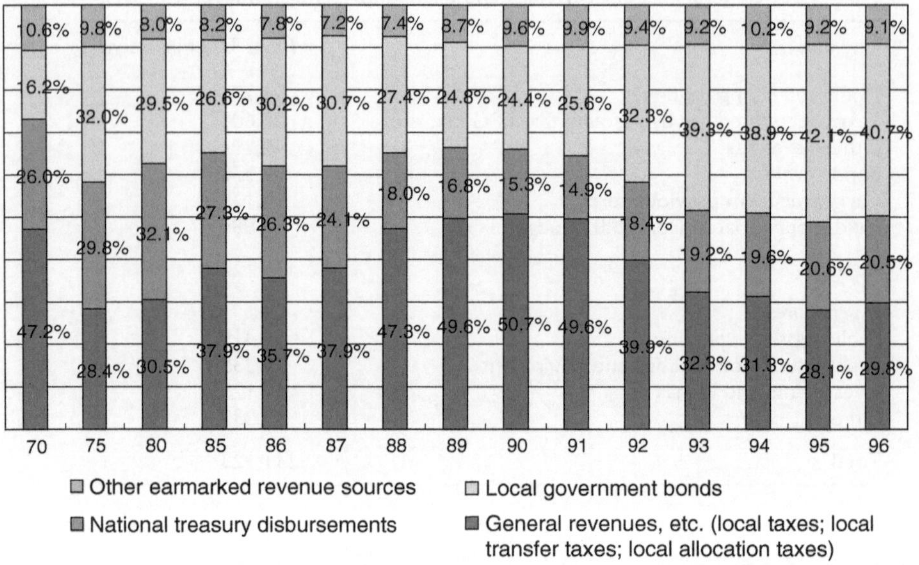

Figure 7.39 Structure of funding since 1970 for ordinary construction projects carried out by prefectural and local governments.

government relies more heavily than do prefectural and local governments on specially earmarked revenues (Figure 7.39).

As for public corporations, they obtain funding through debt in principle, most of which takes the form of loans following from the fiscal loans and investment. Some projects are also funded by private placement bonds, or loans from private banks. User fees generate the revenue to repay this debt.

7.4.3 *Tapping private-sector resources for infrastructure development*

Private-sector involvement to date

Amidst political, social, and economic change the burden of funding and developing the infrastructure has shifted back and forth at the time of writing between the public and private sectors, but the general trend is towards privatization and greater reliance upon the resources of the private sector.

Particularly in the years after Second World War, as Japan went through reconstruction and into a phase of extremely rapid economic growth, a critical lag in infrastructure development coincided with a shortage of public-sector funds. In order to promote infrastructure development, the government sought actively during those years to tap into private-sector funds, technology, and know-how (Table 7.65).

Infrastructure development in Japan at that time was characterized by two especially noteworthy features: the existence of the Fiscal Loans and Investment and the many different types of entities in charge of infrastructure projects.

Table 7.65 Entities in charge of infrastructure projects

	Economic and financial background	*Trend in types of entities*
1950s	In the latter half of the 1950s Japan entered a period of rapid economic growth. As a result, roads and other forms of infrastructure became overburdened.	To support the development of the country's industrial and social infrastructure, various special public and semi-public corporations were established to help carry out infrastructure development. Public corporations (koudan) were a completely new type of entity. Set up as self-supporting accounting units, they were needed in order to actively funnel private-sector funds into infrastructure projects, thereby overcoming restrictions upon the use of special account funds and local government revenues for infrastructure development.
1960s	The shortage of public-sector funds became acute in the latter half of the 1960s, as evidenced by the issuing of government bonds.	The government stepped up its efforts to encourage private-sector involvement in infrastructure development. However, there were many barriers (permit systems, etc.) preventing private enterprise from engaging in infrastructure development on its own. Many "third-sector" (i.e. semi-public) corporations were established during this period to enable the public and private sectors to work together, particularly in the area of community development.
1970s and 80s	The two oil crises of the 1970s ushered the Japanese economy into a period of more moderate economic growth. Due to a large balance of outstanding government bonds, the government began to suffer from structural deficits.	Japan's bubble economy began in the latter half of the 1980s. In 1986, the Law on Temporary Measures to Encourage Private-Sector Involvement in the Development of Specified Facilities was passed, thus paving the way to increased private-sector participation (via semi-public entities) in the development of infrastructure needed for R&D activities, telecommunications, international exchange activities, etc. Some of the key entities involved in infrastructure development underwent very significant change during this period. Nippon Telegraph and Telephone Public Corporation and Japan National Railways (both of which had been converted after the war from state-run enterprises into public corporations) were both privatized. Privatization of the Nippon Telegraph and Telephone Public Corporation (now known more commonly as NTT) was carried out in April 1985 in order to: (1) spur the development of more advanced telecommunications technology; (2) respond to the needs of an increasingly complex and diversified market; and (3) inject greater efficiency and vitality into the telecommunications industry by allowing new competitors to enter the market. Japan National Railways was broken up into six privatized regional passenger companies and one privatized freight company in April 1987 in order to: (1) resolve the company's chronic debts; and (2) downsize the organization as a means of securing greater discretionary freedom for the company's management.

Table 7.65 (Continued)

	Economic and financial background	Trend in types of entities
		Another major trend during this period was the establishment of such corporations as the Trans Tokyo-Bay Highway Corporation (1986) and the Kansai International Airport Company (1984). These corporations were given a specific mission to accomplish – to actively tap into private-sector management know-how and funds for the construction of the Trans Tokyo-Bay Highway and the Kansai International Airport.

Below are two of the main large-scale projects which have been carried out with significant private-sector involvement.

TRANS TOKYO-BAY HIGHWAY

The Law on Special Measures to Construct the Trans Tokyo-Bay Highway was passed in 1986 so that the private sector could contribute funds as well as managerial and technical expertise to the Trans Tokyo-Bay Highway project, thereby commencing the project as quickly as possible and lightening the burden shouldered by the national government. The project was to be carried out jointly by the Japan Highway Public Corporation and a corporation which was newly founded (Table 7.66).

KANSAI INTERNATIONAL AIRPORT

The Law on the Kansai International Airport Co., Ltd. was passed to facilitate efficient construction and management of the Kansai International Airport. The Kansai International Airport Company was established in 1984, with investments from the private sector, prefectural and local governments, and the national government. The company commenced the airport project that same year.

In Phase 2 of the project, a decision was made to establish two separate entities for construction of the facility and preparation of the island on which the airport stands. The Kansai International Airport Land Preparation Co., Ltd was established in 1996, and is currently working on the airport island.

Future private-sector involvement in infrastructure development

The national government has serious financial problems due to the accumulation of many long-term outstanding bonds. Infrastructure development must be carried out more efficiently. At the same time, the fiscal system must be restructured to reduce public works expenditures.

The national government's economic stimulus package includes measures aimed at promoting infrastructure development. For example, the BOT system and Britain's PFI programme have been adopted in order to tap the funds as well as the managerial and technical expertise of the private sector.

Table 7.66 Trans Tokyo-Bay highway project

Outline	The highway is a 15-kilometer four-lane toll bridge and tunnel spanning Tokyo-Bay. It runs from Kawasaki City (Kanagawa Prefecture) to Kisarazu (Chiba Prefecture), and opened to traffic in December 1997
Total cost	Approx. ¥1,440 billion (¥1,239.2 billion invested by the Trans Tokyo-Bay Highway Corporation; ¥200.8 billion invested by the Japan Highway Public Corporation)
Organization in charge	Trans Tokyo-Bay Highway Corporation (a semi-public corporation formed by investment from the private sector and the Japan Highway Public Corporation)
How the project has proceeded	• After receiving authorization to build a toll road, the Japan Highway Public Corporation signed a construction agreement with the Trans Tokyo-Bay Highway Corporation which specified the division of responsibilities for this project. Responsibilities of the Japan Highway Public Corporation: Basic survey; design; land acquisition Responsibilities of the Trans Tokyo-Bay Highway Corporation: Detailed design; construction. • After construction was completed, the Trans Tokyo-Bay Highway Corporation handed over control of the highway to the Japan Highway Public Corporation, for which it is scheduled to receive payment from the Japan Highway Public Corporation in instalments over an extended period of time. • Ultimate responsibility for management of the highway rests with the Japan Highway Public Corporation, but actual management is carried out by the Trans Tokyo-Bay Highway Corporation under the terms of a separately concluded contract. The Japan Highway Public Corporation pays an annual management fee to the Trans Tokyo-Bay Highway Corporation.
Government support measures	• Appropriations from the Japan Highway Public Corporation and various prefectural and local governments • Government guarantees for bonds issued to fund the project • Low-interest loans

Making use of the funds as well as the managerial and technical expertise of the private sector enables more efficient development of infrastructure. At the same time, it also relieves the burden that must be shouldered by government at the national, prefectural, and local levels.

In addition, the adoption of these new systems has drawn considerable attention as a means of promoting economic restructuring. For example, they will afford private companies greater opportunity to create new lines of business.

The Diet is now deliberating a bill for promotion of the use of private-sector funds for the construction of public facilities (Table 7.67).

The Ministry of Construction has established a Committee to establish a New Infrastructure Development Model Aimed at Encouraging Private-Sector Investment. This committee (whose membership comprises persons from academia, the construction industry, financial institutions, etc.) completed the Guidelines for a Japan-style PFI in May 1998 (Table 7.68). These guidelines have laid clear ground rules concerning the promotion of a Japan-style PFI. In addition to setting forth a new perspective on promoting private-sector involvement, and new ideas concerning the

Table 7.67 Bill promoting the use of private-sector funds for the development of public facilities

Purpose
To promote the use of the funds as well as the managerial and technical expertise of the private sector for efficient and effective development (construction, maintenance, operation, and planning) of public facilities.

Types of facilities
Roads, harbours, airports, water and sewage services, government buildings, public housing, waste treatment facilities, medical facilities, alternative energy facilities, recycling facilities, heat generation facilities, telecommunications facilities, research facilities, etc.

Formulation of basic policy and detailed policy
- Basic policy is determined by the prime minister (e.g., basic guidelines for the selection of projects and contractors, etc.).
- Detailed policy is determined by the minister with relevant jurisdiction, or by the prefectural governor or local mayor where the facility is to be built (e.g., solicitation of contractors for a specific project).

Selection of projects and contractors
- Selection of private contractors through public invitation, etc.
- Objective assessment of project effectiveness and efficiency, and public announcement of assessment findings.

System for promotion
- The Committee to Promote Projects that Make Use of Private-Sector Funds has been established by the Office of the Prime Minister.
- This committee is responsible for formulating basic policy, taking follow-up measures, soliciting public opinion, and communicating the views of the prime minister.

Support measures
- Extension of the period of government responsibility for funding (up from 5–30 years).
- Provision of government allocations.
- Guarantees (by governments at the national, prefectural, and local levels) for bonds.
- Interest-free loans from the Japan Development Bank.
- Use of government land free of charge, or with reduced fees.

basic nature of the task at hand, the guidelines also. Name various types of projects and infrastructures to which the new infrastructure development policies could apply, make important points about the financing of infrastructure development initiative and clarify the manner in which public assistance is to be provided.

The Ministry of International Trade and Industry has established a group to study various issues related to infrastructure development. This group released a set of recommendations in June 1998 concerning the adoption of a PFI programme suited to Japanese conditions, and it also called for action to develop alternative energy resources, an example being the construction and operation of power generation facilities fuelled by waste material.

Prefectural and local governments also have high hopes for what can be achieved through a PFI programme. Many are actively studying ways to take advantage of such a programme and the construction and financial industries are also looking into how they can benefit from a PFI programme in their respective fields.

Table 7.68 Guidelines for a Japan-style PFI

Purpose

- To enhance infrastructure development by making effective use of government spending.
- To reassess the proper roles of the public and private sectors.
- To generate new projects for the private sector.

Basic approach

- The Japan-style PFI is a means of promoting infrastructure development by using private-sector involvement and market mechanisms to achieve the efficient development of infrastructure and to spend government money most effectively.
- This system is targeted towards infrastructure projects that in the past would normally have been carried out by the government (national, prefectural, and local) or some other public entity.
- The Japan-style PFI is to be carried out under the managerial leadership of the private sector, with appropriate public-sector participation, so as to benefit from the creativity of the private sector.
- The following shall be decided on a case-by-case basis after consultation with private-sector contractors and managers: (1) the proper roles and responsibilities of the public and private sectors; (2) the risks to be assumed by the public and private sectors; and (3) the degree of public support to be provided.

Possible PFI projects

Type 1 (projects for which a user fee is charged) Projects for which a private-sector contractor recovers its investment by charging a user fee or by deriving income from operations related to the project.

Type 2 (joint public-private projects)

Projects in which a public facility can be developed more efficiently in conjunction with the development of a private facility (i.e. the burden shouldered by the government is lightened).

Type 3 (purchase of public services from the private sector)

Projects in which the infrastructure in question can be developed and managed efficiently by the private sector. The government can then buy the services provided by the private-sector organization.

Thus, it is not only the government, but also the construction industry, financial sector, and others in the private sector that are engaged in discussions concerning private-sector involvement in infrastructure development. In the future, the national government, prefectural and local governments, the business community, and other parties must work closely together to overcome problems and create conditions conducive to the formulation of concrete projects.

7.5 International exchange and cooperation in the construction sector

7.5.1 What Japan can do to promote international exchange and cooperation

Official Development Assistance (ODA)

Due to depreciation of the yen, Japan's ODA assistance declined in FY 1996 to US$9.608 billion, but this figure still made Japan the world's leading aid donor for the sixth year in a row. This aid consisted of grant assistance, technical cooperation, loans, and funding of international organizations. While funds thus provided were used for a wide variety of purposes, the highest proportion of this money went for infrastructure development in developing countries.

Infrastructure development is the single biggest target of bilateral ODA assistance, 40.7 per cent of which was used to support the development of roads, harbours, power plants, airports, and other types of economic infrastructure (Table 7.69). ODA assistance in 1995 was distributed to 161 countries and union territories. The largest share, as in previous years, went to Asian countries (Table 7.70).

In addition, large amounts have been provided to developing nations by the Japanese government and private-sector organizations in the form of export credits and direct investments (Table 7.71).

Technology transfer

Technology transfer in the construction sector is carried out by the Japan International Cooperation Agency (JICA). Under its "project-type technical cooperation" program, JICA: (1) dispatches experts abroad; (2) runs training programs; and (3) donates equipment and materials (Table 7.72).

In addition, the Ministry of Construction is also carrying out the Program for Development of Construction Technology Overseas, which is focused on developing construction technologies that are suited to the natural, social, and economic conditions of developing nations (Table 7.73).

The Ministry of Construction regularly receives requests from developing nations for construction technology assistance. In response, the Ministry formulates plans for technologies suited to conditions in the country concerned, and works to ensure that the technology transfer is carried out in a smooth and effective manner (Overseas Construction Technology Cooperation Program) (Table 7.74).

Dispatching experts abroad

Countries throughout Asia have high praise for the technology and know-how at the disposal of government agencies and public corporations in Japan. In response to requests from developing countries and international organizations, the Ministry of

Table 7.69 Types of ODA assistance

Type of aid	*Amount in 1996* (million dollars)	*% change from previous year*	*Primary aid provider*
Grant aid	2,396	−19.4	Ministry of Foreign Affairs, Ministry of Finance, JICA
Technical cooperation	3,181	−8.1	JICA
Government loans	2,780	−32.6	OECF
Multilateral aid	1,252	−70.0	Ministry of Finance, Ministry of Foreign Affairs
Total ODA assistance	9,608	−34.8	

Source: 1998 Diplomatic Bluebook (Ministry of Foreign Affairs).

Table 7.70 ODA assistance to Asian countries (1996) (million dollars)

Type of aid	*Indonesia*	*Vietnam*	*Singapore*	*Thailand*	*S. Korea*	*Philippines*	*Malaysia*	*India*	*Sri Lanka*	*China*
Grant aid	64.41	46.37	None	1.86	None	91.14	0.69	35.18	52.39	24.99
Technical cooperation	163.31	46.67	8.54	135.41	95.00	94.34	69.91	21.83	34.16	303.73
Government loans	737.81	27.81	None	526.73	−222.94	228.96	−553.11	522.26	87.39	533.01
Total ODA assistance	965.53	120.86	8.54	664.00	−127.94	414.45	−482.51	579.26	173.94	861.73

Source: 1998 Diplomatic Bluebook (Ministry of Foreign Affairs).

Table 7.71 Flow of funds to developing nations other than ODA

Type	*1996* (million dollars)	*% change from previous year*
Government funds		
Export credit	−185	−119.7
Financing of direct investments	1,930	−49.1
Financing of international organizations	−67	−106.6
Total government funds	1,678	−70.8
Private-sector funds		
Export credit	−1,427	−146.3
Financing of direct investments	8, 338	−12.2
Bilateral securities investments	22,061	111.4
Financing of international organizations	−581	***
Total private-sector funds	28,392	23.6
Donations from private non-profit organizations	232	7.7

Source: 1998 Diplomatic Bluebook (Ministry of Foreign Affairs).

Table 7.72 JICA project-type technical cooperation (recent projects)

Country	*Project description*	*Project period*
Philippines	Construction productivity enhancement plan	1993/4–1998/3
Turkey	Earthquake preparedness centre	1993/4–1998/3
China	Automated command system for the national flood prevention centre	1993/6–1998/5
Indonesia	Development of technologies for construction of multi-unit dwellings	1993/11–1998/10
Kenya	College for surveying and cartography	1994/10–1999/9
Thailand	Sewerage research centre	1995/8–2000/7
China	Centre for new housing technologies and personnel training	1995/9–2000/8
Sri Lanka	Centre for training in equipment operation	1996/10–2001/9

Source: International Exchange in the Construction Sector (Ministry of Construction).

Table 7.73 Program for development of construction technology overseas

Country	*Project description*	*Project period*
Indonesia	Development of housing construction technology	1979–1981
Thailand	Development of road construction technology for laterite soils	1973–1983
Indonesia	Development of construction technologies for fire-resistant buildings	1982–1984
Indonesia	Development of flood warning technologies	1984–1986
Indonesia	Development of construction materials technology	1985–1987
Peru	Development of satellite cartography	1987–1990
Thailand	Development of model housing technologies for developing nations	1989–1990
South East Asian countries	Development of materials and equipment for disaster response	1991–1993

Source: International Exchange in the Construction Sector (Ministry of Construction).

Construction works through JICA and other channels to dispatch experts in rivers, roads, housing, urban planning, and other specialties. These experts work together with counterparts in other countries to plan and carry out development projects (Table 7.75).

Seventy seven per cent of all experts dispatched in FY 1996 went to Asian countries (Figure 7.40).

Training

In response to requests from developing countries, JICA and other entities work through such organizations as the Interchange Association and the Infrastructure Development Institute to provide training programmes in Japan for personnel engaged in a broad range of different specialties (refer Table 7.76). In FY 1996, 54 per cent of all trainees came from Asian countries (Figure 7.41).

Table 7.74 Overseas construction technology cooperation program (recent projects)

Project name	*Project description*	*Project period*
Construction technology cooperation projects		
Preliminary Study for Development of a Master Plan for Environmental Improvement in Metropolitan Areas	In order to work toward environmental improvement in large cities in developing nations, this project was launched to develop basic concepts regarding a model environmental improvement programme in Jakarta. This project also included recommendations concerning funding of the programme.	1993–1995
Formulation of a Policy Concerning Assistance for the Formulation of a Master Plan for Development of Sewer Systems in Developing Nations	The purpose of this project was to support efforts to formulate a master plan for the development of sewer systems in developing nations, which are working to improve the quality of residential environments by cleaning up the water supply.	1993–1995
Discussion of Methods for Comprehensive Environmental Surveys	The purpose of this programme was to study the comprehensive surveys that are required before concrete action can be taken in the field of environmental protection. On the basis of this study, an effort was made to develop an integrated policy to address water and sewer systems, river improvement, waste treatment, etc.	1995–1997
Preliminary Study for the Formulation of a Policy on the Use of Private-Sector Funds for International Infrastructure Development Projects	The purpose of this project was to study the use of private-sector funds for infrastructure development, a funding method that is expected to become increasingly important in the future. On the basis of this study, the project sought to formulate policy on this issue.	1995–1996
Study on Methods of Environmental Improvement in the Metropolitan Areas of Developing Nations	The purpose of this study was to study the problems associated with efforts to improve the environment in the metropolitan areas of developing nations. On the basis of this study, concrete ways were discussed in which Japan can cooperate with developing nations in their efforts to improve residential environments, and a manual was prepared.	1996–1998
Preliminary Study to Formulate a Policy on the Improvement of Sewer System Management in Developing Nations	In order to help developing nations keep their sewer systems running at peak performance, this project was organized to formulate a policy on maintenance and management techniques suited to the natural, social, and economic conditions of developing nations.	1996–1998
Projects to promote the transfer of construction technology		
Project for Promotion of International Exchange	The purpose of this project is to provide training in Japan for persons from overseas (i.e. to carry out efficient training programmes in the most efficient possible manner).	1991

Table 7.74 (Continued)

Project name	*Project description*	*Project period*
Project to Promote Transfer of Land Readjustment Technology	The purpose of this project was to: (1) prepare a manual of land readjustment technologies suited to conditions in developing nations; and (2) discuss technology transfer measures.	1991–1993
Project for OJT Technology Transfer Aimed at Job Creation	The purpose of this project was to prepare for the Project for OJT Technology Transfer Aimed at Job Creation, which will include efforts to create more jobs and provide on-the-job training in the construction sector in developing nations.	1993–1997
Project to Promote the Transfer of Construction Industry Know-How	The purpose of this project is to: (1) transfer the managerial, organizational, and business know-how of the Japanese construction industry; and (2) carry out seminars on methods of establishing construction industry systems. Part of this project also involves the preparation of seminar materials, which are targeted to the specific needs of individual countries.	1995–
Project for Transfer of Planning and Management Technologies for Public Facilities	The purpose of this project was to: (1) formulate policy regarding methods for transferring public facility planning and management technologies connected with bidding procedures that incorporate proposals; and (2) carry out case studies to determine the effectiveness of previous technology transfer efforts.	1996–1997

Source: International Exchange in the Construction Sector (Ministry of Construction).

Table 7.75 Number of experts dispatched in recent years

Fiscal year	*1993*	*1994*	*1995*	*1996*
Long-term assignments (one year or longer)	136 persons, 22 countries	136 persons, 22 countries	136 persons, 22 countries	136 persons, 22 countries
Short-term assignments	136 persons, 22 countries	136 persons, 22 countries	136 persons, 22 countries	136 persons, 22 countries
Total	136 persons	136 persons	136 persons	136 persons

Source: International Exchange in the Construction Sector (Ministry of Construction).

Foreign direct investment

Japan's overseas direct investment (all figures converted to dollars) rose for 3 consecutive years from FY 1992 through FY 1995 before dipping by 5.3 per cent to US$48.019 billion in FY 1996. This decline, however, was primarily due to the depreciation of the Japanese yen versus the dollar (the yen-based figure rose in FY 1996 by 9.1 per cent to ¥5.4094 trillion).

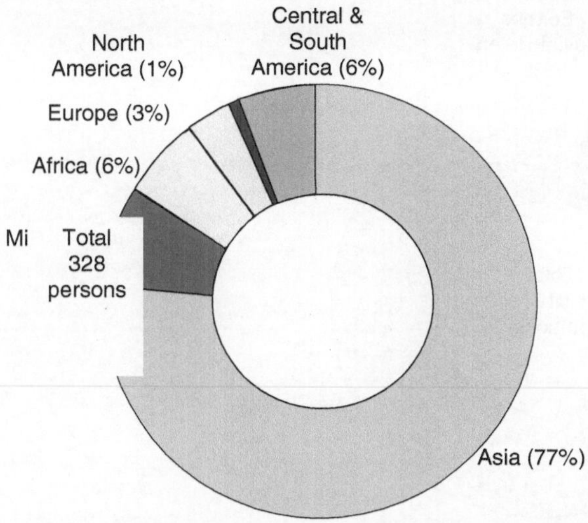

Figure 7.40 Destinations of experts dispatched in FY 1996.
Source: International Exchange in the Construction Sector (Ministry of Construction).

Table 7.76 Number of training courses/trainees in recent years

Fiscal year	*1993*	*1994*	*1995*	*1996*
Group training courses	20 courses; 233 persons	26 courses; 277 persons	25 courses; 265 persons	27 courses; 315 persons
Individual training courses	131 courses; 357 persons	106 courses; 404 persons	116 courses; 387 persons	99 courses; 304 persons
Total	704	681	652	619

Source: International Exchange in the Construction Sector (Ministry of Construction).

Note
Group training courses: Not restricted to persons from any particular country.
Individual training courses: Designed for specific individuals.

The number of investments fell 12.6 per cent to 2,501 (Table 7.77). Investment in the manufacturing sector rose 8.8 per cent in FY 1996 to US$20.258 billion, fuelled primarily by a 94.7 per cent jump in transportation equipment facilities (US$3.873 billion), and a 23.2 per cent rise in electrical equipment facilities (US$6.513 billion). Investments in textile and machinery facilities declined.

Investment in the Asian manufacturing sector rose 41.6 per cent in FY 1994, then jumped to 55.5 per cent growth in FY 1995 before going down to a 17.7 per cent rise in FY 1996. These investments continue high in dollar-denominated terms (US$6.628 billion), and have contributed greatly to construction demand and employment in Asia (Table 7.78).

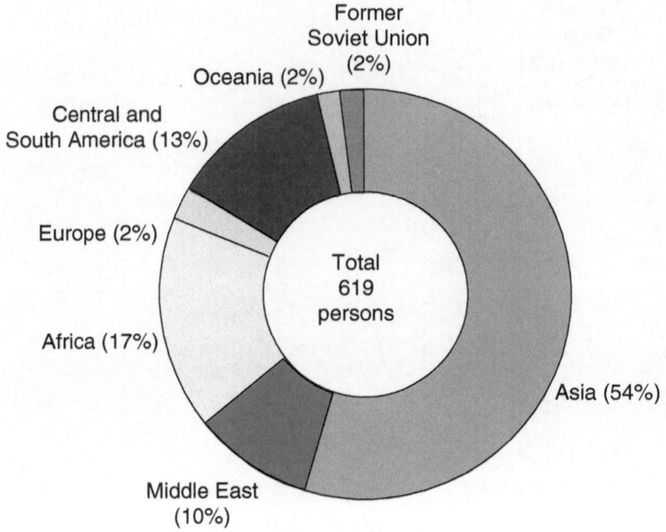

Figure 7.41 Where trainees in FY 1996 came from.

Source: International Exchange in the Construction Sector (Ministry of Construction).

Table 7.77 Japanese foreign direct investments during FY 1994 to FY 1996 (units: $m, per cent)

	FY 1994	*FY 1995*	*FY 1996*
Overall			
(amount)	41,051	50,694	48,019
(% change)	14.0%	23.5%	–5.3%
N. America			
(amount)	17,823	22,761	23,021
(% change)	16.6%	27.7%	1.1%
Europe			
(amount)	6,230	8,470	7,372
(% change)	–21.5%	35.9%	–13.0%
Asia			
(amount)	9,699	12,264	11,614
(% change)	46.1%	26.4%	–5.3%
Central and South America			
(amount)	5,231	3,877	4,446
(% change)	55.2%	–25.9%	14.7%

Source: JETRO White Paper (Japan External Trade Organization).

Participation in BOT projects

Many developing nations, especially in Asia, are actively promoting BOT (build-operate-transfer) projects as a means of tapping into private-sector resources for infrastructure development. Many Japanese construction firms are taking advantage of their advanced technology, long experience, and strong funding to take part in these BOT projects (Table 7.79).

Table 7.78 Japanese foreign direct investment in recent years (manufacturing sector) (units: $m, per cent)

	FY 1994	*FY 1995*	*FY 1996*
Overall			
(amount)	13,784	18,623	20,258
(% change)	23.8%	35.1%	8.8%
N. America			
(amount)	4,763	7,326	8,792
(% change)	14.9%	53.8%	20.0%
Europe			
(amount)	1,855	1,995	2,871
(% change)	−9.1%	7.5%	43.9%
Asia			
(amount)	5,181	8,058	6,628
(% change)	41.6%	55.5%	−17.7%
Central and South America			
(amount)	1,159	320	1,489
(% change)	218.4%	−72.4%	364.9%

Source: JETRO White Paper (Japan External Trade Organization).

Table 7.79 Participation of Japanese firms in infrastructure development abroad

Country/territory	China, Hong Kong, Malaysia, Philippines, Australia, Indonesia, Turkey, etc.
Type of project	Power plants, tunnels, water systems, expressways, etc.
Form of participation	Investments, subcontracting, supplying parts

Source: Ministry of Construction.

In November 1997, the Ministry of Construction established a BOT Support Team, whose membership included knowledgeable parties from outside the Ministry. In addition to discussing various BOT-related issues (e.g., the proper responsibilities of the public and private sectors; land acquisition; income guarantees; risk reduction; etc.), the BOT Support Team is also charged with responsibility for keeping abreast of project opportunities and devising technical support measures.

Market access and foreign direct investment in Japan

In accordance with the provisions of the WTO agreement, Japan has pledged not to set numerical limits upon foreign participation in the construction industry and five other industries, and not to adopt other measures which would have a discriminatory effect against foreign firms in these industries. (See "Brief statistics on foreign participation in government procurement" p.296)

In addition, the Government Procurement Agreement, which took effect in January 1996, stipulates the types of government organizations and the size of projects that are subject to the terms of the WTO agreement (Table 7.80).

Foreign construction firms are expected to develop a greater presence in Japan's domestic construction market over the medium to long term, but the number of such

Table 7.80 WTO government procurement agreement: requirements in the construction services sector

	National government (31 organizations)	*Prefectural and local governments (47 prefectures and 12 cities)*	*Government-related organizations (84 organizations)*
Construction projects	4.5 million SDR (¥720 million)	15 million SDR (¥2.43 billion)	15 million SDR (¥2.43 billion)
Consulting	4.5 million SDR (¥720 million)	15 million SDR (¥2.43 billion)	4.5 million SDR (¥720 million)

Source: Japan Federation of Construction Contractors.

firms licensed to operated in Japan changed little from 1992 through 1997. In 1998 the number has fallen due to a shrinking domestic construction market (Figure 7.42).

Total foreign direct investment (all sectors) jumped sharply upward by 78.3 per cent to US$6.841 trillion. (In terms of Japanese yen, these investments rose 108.5 per cent to ¥770.7 billion.) Investments in both the manufacturing and non-manufacturing sectors rose, but in the construction sector they fell (Table 7.81).

Foreign direct investment in Japan from every one of the world's regions rose in FY 1996. Investments from Asia, in particular, posted a spectacular 374.4 per cent rise (US$1.217 billion) (Table 7.82).

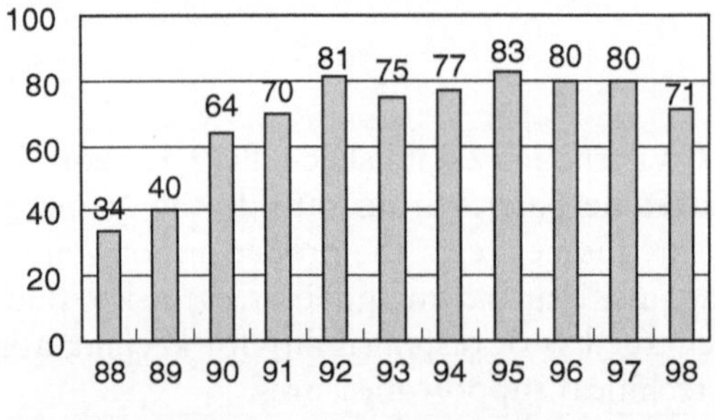

Figure 7.42 Number of foreign construction firms in Japan since 1988.
Source: Japan Federation of Construction Contractors.

Table 7.81 Foreign direct investment in Japan (based on officially reported figures) (unit: $ m)

	FY 1995	*FY 1996*	*% change from previous year*
Total	3,837	6,841	78.3
Manufacturing	1,464	2,762	88.7
Non-manufacturing	2,373	4,079	71.9
(Construction)	1	0	−22.4

Source: JETRO White Paper.

Table 7.82 Foreign direct investment in Japan (based on figures reported to the management and coordination agency) (unit: $ m)

	FY 1995	*FY 1996*	*% change from previous year*
Total	3,837	6,841	78.3
N. America	1,857	2,171	16.9
Europe	1,318	1,954	48.3
Asia	257	1,217	374.4
Central and South America	147	582	294.8

Source: JETRO White Paper.

Procurement of construction materials

As for construction materials (other than lumber), the Japanese construction industry does not depend upon foreign resources very much (Table 7.83). This is generally because of dissatisfaction with foreign materials over issues related to quality, delivery time, and selection. However, imports by major construction firms have been on the rise until 2002.

The top import item, stone, accounts for about 40 per cent of the total. In terms of origin, about half of all construction material imports come from other Asian countries (Table 7.84). According to a survey conducted by the Japan Federation of Construction Contractors, these materials were imported for a number of reasons, including: price; client request; unavailability on the Japanese market; and quality.

Table 7.83 Construction material imports by the principal construction firms in recent years

Fiscal year	*1993*	*1994*	*1995*	*1996*	*1997*
Total value	¥27.0 billion	¥34.2 billion	¥45.7 billion	¥44.1 billion	¥48.5 billion

Source: Japan Federation of Construction Contractors (survey of 35 member firms).

Table 7.84 Construction material imports

Item	*Value* (million yen)	*% of total imports*	*% from Asia*
Stone	20,388	42.1	47.5
Wooden fittings	1,967	4.1	60.4
Glass	1,857	3.8	8.7
Furniture	1,679	3.5	17.3
Tile	1,584	3.3	40.7
Electrical equipment	1,491	3.1	11.8
Lumber	1,457	3.0	20.6
Total	48,470	100.0	44.0

Source: Japan Federation of Construction Contractors (survey of 35 member firms).

Supply of construction equipment

Almost all construction equipment used in Japan is produced in Japan (Table 7.85). Very little is imported. Japan exports large amounts of construction equipment, especially to other Asian countries. The Japanese construction industry, working through various international cooperation schemes, is capable of providing a stable supply of high-quality, low-cost construction equipment to countries throughout the world (Table 7.86).

7.5.2 *International exchange and cooperation*

Market access

The total value of overseas construction projects awarded to Japan's 56 top construction firms declined in FY 1997 for the first time in 4 years, to US$10.403 billion (including the portion of such projects performed by overseas affiliates). Although the post-bubble economic recession and the resulting slowdown in the domestic construction market have spurred Japanese construction firms to make concerted efforts to win overseas contracts, the economic doldrums stemming from last year's Asian currency crisis have sharply reduced construction orders from Asia and caused an overall decline in overseas construction contracts (Table 7.87).

This cause-and-effect chain shows that strong ties exist between construction markets in Japan and the rest of Asia, and that these markets have a strong influence upon each other. Of the above total, 36.2 per cent of all contracts went to overseas Japanese affiliates in FY 1995, 36.3 per cent in FY 1996, and 38.3 per cent in FY 1997. These rising percentages show that Japanese construction firms are establishing an increasingly strong presence overseas.

Japanese construction firms active in Asia and Oceania give generally high marks to countries in these regions for ease of market access, but calls have been made for

Table 7.85 Selected construction equipment: production and domestic demand (1995)

Type	*Units sold*	*Units produced*
Earth-moving equipment	124,050	124,982
Levelling and rolling equipment	4,185	5,213
Concrete mixers, asphalt equipment	6,455	2,130

Source: Annual Machinery Statistics (Ministry of International Trade and Industry), Survey of Construction Equipment Trends (Ministry of Construction).

Table 7.86 Construction equipment exports (1996) (unit: million dollars)

Region	*Total*	*US*	*EU*	*Asian NIEs*	*ASEAN7*	*China*
Value	4,011	928	194	599	1,518	178
Growth	−2.4%	−13.0%	−20.8%	−0.9%	7.0%	−12.0%

Source: JETRO White Paper.

Table 7.87 Overseas construction contracts awarded to Major Japanese construction firms (including overseas affiliates)

	Total	*Asia*	*N. America*	*Europe*	*Other*
1995	8,188	6,619	589	36	943
	100.0%	80.8%	7.2%	0.5%	8.7%
1996	10,148	8,439	407	195	1,106
	100.0%	83.2%	4.0%	1.9%	10.9%
1997	7,876	6,655	216	11	993
	100.0%	84.5%	2.7%	0.1%	12.6%

Source: Overseas Construction Association of Japan (survey of 56 member firms).

Notes
Upper row: 100 million yen.
Lower row: share of total overseas contracts.

improvement for specific items. Problems mentioned by Japanese firms are summarized in Table 7.88.

Foreign firms (in all sectors) operating in Japan also report market access problems. The problems mentioned are summarized in Table 7.89.

Table 7.88 Access problems in overseas construction markets

Problem area	*Description of problems*
Legal systems	• Corporate taxes are levied regardless of whether or not a corporation has made a profit. • Value-added taxes lack transparency, and taxes are not always refunded when they should be. • There are limits on foreign ownership of overseas affiliates. Foreign firms are not allowed to have wholly owned subsidiaries. • Laws are frequently amended, and enforcement is not uniform.
Government oversight	• Permit applications, tax breaks, and payments for construction services provided are often delayed due to disagreements among different government ministries and agencies. • It takes a long time to apply for development permits, construction permits, etc.
Economic infrastructure	• Power outages are commonplace. • Due to inadequate water, sewer, and phone services, companies are forced to build their own.
Technical capability	• Contractors do not always have the necessary expertise. • There are quality problems with locally produced construction materials.
Labour markets	• Insufficient supply of skilled labour. • Job hopping is commonplace. • It is difficult to hire qualified employees.
Social customs	• After-sales services are not sufficient. • A large volume of documentation (contracts, related documents) is required. Preparation and checking of this documentation is expensive. • There is corruption in government and the business community.

Source: Survey conducted by the Research Institute of Construction and Economy.

Table 7.89 Market access problems mentioned by foreign firms in Japan

Rank (top 6)	*Description*
1	High business costs
2	Exacting demands of users
3	Difficulty of hiring
4	Closed nature of Japanese market
5	Restrictions, regulations, inspections, commercial practices
6	Difficult country for foreigners to live in

Source: Opinion Poll of Foreign Firms Concerning Direct Investment in Japan (JETRO).

Note
Multiple responses were allowed. The six items above were the ones most frequently mentioned as problems.

Supply of construction materials

As mentioned in "procurement of construction materials" (section 5.1.8), more and more of the construction materials used in Japan are being imported. It is necessary that we take steps to ensure a steady supply of inexpensive, high-quality materials from other Asian countries, which is where over 40 per cent of these imports currently come from.

Imports of stone, lumber, processed materials and temporary support materials (scaffolding, etc.) for interior and exterior work, and other materials are expected to increase for a number of different reasons, including the fact that some are: (1) unavailable in Japan; (2) of unique design or workmanship; or (3) lower-priced because of inexpensive labour.

At the same time, there are other factors which inhibit the import of construction materials, including: (1) loss of the cost merits caused by the depreciation of the yen; (2) worries about quality; (3) inadequate after-sales service; (4) late delivery; and (5) storage yard fees. Asian countries would be well advised to take advantage of their competitive strengths to overcome these problems, thus ensuring their ability to continue providing Japan with a steady supply of construction materials.

7.6 Concluding remark

The outlook for the Japanese economy over the next 5 years from 2002 is not one of unbridled optimism, but instead one that recognizes the various problems that remain to be addressed. Increased wisdom and effort will be required if the Japanese economy is to continue to thrive now that it has shifted from a growing to a mature economy.

Notes

1 "Mid-term policies concerning government development assistance", 10 August, 1999.
2 Overseas Construction Association of Japan. Totals for the 48 member companies.
3 Overseas Construction Association of Japan. Totals for the 48 member companies.

Chapter 8

Malaysia

Construction Industry Development Board, Malaysia and Edwin H.W. Chan

8.1 Introduction

8.1.1 Background

Malaysia was created in 1963 through the merging of Malaya and the former British Singapore, both of which formed West Malaysia, while Sabah and Sarawak in north Borneo composed East Malaysia. Singapore separated from the union in 1965. The first 3 years of independence of Malaysia were marred by hostilities with Indonesia (Abacci Atlas: Malaysia Atlas Entry).

Malaysia had a population of 22,229,040, with the population growth rate at 1.96 per cent (July 2001 est.). The 2001 statistics showed the proportion of males slightly higher than that of females (1.01 male/female). Malaysia is a diverse country with a great degree of cultural variation among the nation's three major ethnic groups. The indigenous Malay made up 58 per cent of the total population while Malaysia had a considerable number of ethnic Chinese (27 per cent) and Indians (7 per cent). Diversity also exists in religion, with Islam, Buddhism, Daoism, Hinduism, Christianity and Sikhism being practised over the whole country, and Shamanism practised in East Malaysia. The official language of Malaysia is Bahasa Melayu, but English, Chinese dialects (Cantonese, Mandarin, Hokkien, Hakka, Hainan and Foochow), Tamil, Telugu, Malayalam, Punjabi and Thai are also spoken in the country. Several indigenous languages are spoken in East Malaysia and the most popular ones are Iban and Kadazan. Almost 83.5 per cent of the population in Malaysia is literate (Abacci Atlas: Malaysia Atlas Entry).

Malaysia, including the peninsula and northern one-third of the island of Borneo, is located in Southeast Asia, near Indonesia and the South China Sea, south of Vietnam at geographic coordinates of 230 N Latitude and 11,230 E Longitude. The country has a total area of 329,750 square kilometres consisting of 328,550 square kilometres land area and 1,200 square kilometres water area. Malaysia controls a coastline of 4,675 kilometres (Peninsular Malaysia 2,068 km, East Malaysia 2,607 km), with 2,669 kilometres of land boundaries. Bordering countries include Brunei, 381 kilometres, Indonesia, 1,782 kilometres, and Thailand, 506 kilometres. The terrain of Malaysia is distinguished by coastal plains, which rise to hills and mountains, with elevation extremes ranging from the lowest point at Indian Ocean (0 m) to the highest point Gunung Kinabalu (4,100 m). In terms of climate, Malaysia has a tropical marine climate with two distinct monsoon seasons – annual southeast

(April–October) and northeast (October–February). The principal natural resources of Malaysia are tin, petroleum, timber, copper, iron ore, natural gas and bauxite. Other natural resources include arable land (3 per cent of total lands), permanent crops (12 per cent), forests and woodland (68 per cent) and others (17 per cent). Although, there are no permanent pastures in Malaysia, according to the 1993 estimates, a total of 2,941 square kilometre area of the country was irrigated land (Abacci Atlas: Malaysia Atlas Entry).

8.1.2 Political structure

Malaysia is a constitutional monarchy. Malaya (what is now Peninsular Malaysia) was first formed on 31 August 1957. Federation of Malaysia (Malaya, Sabah, Sarawak and Singapore) was formed on 9 July 1963, and Singapore later became independent from the federation on 9 August 1965. The federation is nominally headed by the paramount ruler and a bicameral Parliament, which consists of a non-elected upper house and an elected lower house. Although Peninsular Malaysian states have hereditary rulers in all, the governors of Melaka, Penang, Sabah and Sarawak are appointed by the Malaysian government. The federal constitution determines the powers of state governments, but under terms of the federation, Sabah and Sarawak retain certain constitutional prerogatives (e.g. the right to maintain their own immigration controls). In the House of Representatives, Sabah and Sarawak each holds 20 seats and 28 seats respectively. The foreign affairs, defence, internal security and other powers of both Sabah and Sarawak are delegated to the federal government (Seekins, 1991; Encyclopaedia: Malaysia).

8.1.3 Development process in Malaysia: through the years

Malaysia, being one of the second-tier NICs (Newly Industrialized Countries), is a fascinating case with a dramatic history that challenges many conventional models of national development. It serves as an interesting example of how a frontier area, even within its own region, became one of the key economic prizes of the British colonial empire. A few decades early in the twentieth century, British Malaya was transformed from a colonial backwater into the largest dollar-earner in the empire. In the early twentieth century, after decades of trying one plantation crop after another without much success, Malaya hit the jackpot with rubber trees. Rubber from colonial Malaya met the almost insatiable demand for tyres created by the development of the automobile industry in industrial countries, especially in the United States. The rubber industry, combined with expansion of the tin-mining industry, made Malaya one of the most prosperous economies in the colonial world, though dependence on any primary export product does create a boom and bust cycle. By 1929, British Malaya had already achieved the highest per-capita GDP of any country or territory in Asia. The colonial system was stabilized with the British policy of granting pensions to the Malay ruling class in return for their withdrawal from political and economic governance, except for symbolic roles that legitimated the colonial system. The system supplements the rubber boom. In spite of many emergent problems of the colonial period with booms and busts, a measure of social progress was made possible by the enormous profits of rubber and tin.

Though most of the benefits of colonial prosperity were sent home to Britain, the spillover was enough to benefit the Malayan economy. A reasonably good physical infrastructure and some elements of a modern educational system and civil service were created for Malaya. During the colonial period, there was massive immigration from China and India that provided cheap labour required by the expanding export economy. The economic activities of the immigrants developed into ethnic tensions of the "plural society". For most of the colonial era, the latent problems were largely circumvented through institutional and geographical segregation (Drabble, 2001).

This was beyond the expectation of the colonial government, as it thought that the Chinese and Indians arriving in Malaysia for economic reasons would eventually return to their homelands. With the ethnic groups that are deeply divided along economic, social, cultural and religious lines, the major challenge is to create a common nation that belongs to the major ethnic groups. The political stability of Malaysia has always been depending on the balance among the three ethnic groups (Malay, Chinese and Indian), particularly between the Chinese and the Malays. Politically the Malays have control, but the Chinese are by far the wealthiest group. The process of income re-distribution had been extremely slow. The pre-existing formula for peaceful co-habitation of the three ethnic groups collapsed. It led to racial conflicts on 13 May 1969 and riots started in Kuala Lumpur when Malays demanded the expulsion of non-Malays. Consequently, the Malaysian government reorganized the economic and political foundations of the country, which aimed at eradicating ethnic tensions. It is recognized that for harmony to be re-established, economic and political equity among the three ethnic groups is necessary. During 1971–1990, the Malaysian government implemented the New Economic Policy (NEP), which had two goals: the eradication of poverty among all Malays and a reduction in the economic disparities among the ethnic groups (Jesudason, 1997; Drabble, 2001).

It was the historical events in the years before and after the Second World War that had led to the formation of the Federation of Malaysia in 1963. The changes in the politics and power sharing in the ethnic groups were prompted by the Second World War, which shattered the colonial world forever. Southeast Asia was conquered by the Japanese in the opening weeks of the Pacific War and the British were driven out for a period. The British colonial regime returned after the end of the Second World War, to face a new international environment and a resurgent nationalism which strived to put an end to colonialism. With the defeat of a communist guerrilla insurgency, successful local elections were held in the mid-1950s, political stability at the time provided the opportunity of a peaceful transition to independence for Malaya in 1957. In the 1950s and the 1960s, by a successful programme of replanting rubber estates and smallholdings with more productive varieties of rubber trees, the traditional export economy was re-habilitated. In spite of the introduction of new synthetic rubber and declining world prices of natural rubber, the rubber industry has been able to flourish. As Malaysia became the world's largest producer of palm oil, the export economy was revived again in the 1960s and 1970s. In the 1970s and 1980s, the development of a substantial oil and natural gas industry became the most important source of Malaysian economic growth and the developments in manufacturing, particularly in electronics and textiles, had been providing new jobs in the urban sector (Drabble, 2001).

In the 1960s, the Malaysian government provided price supports to subsidize rice production. With the intention of making Malaysia self-sufficient in rice production, the government made massive investments in irrigation facilities. However, Malay rice farmers have remained among the poorest segment of the population. The goal of self-sufficiency has not been achieved and, increasingly, rural Malays left agriculture to move to cities to find better economic prospects. Such a trend persisted in the 1970s and 1980s and had also inspired migration from Indonesia to the Malaysian rural economy. The New Economic Policy (1970–1990) of Malaysia was intended to reduce the ethnic economic gap and to reduce poverty. Unfortunately, the policy has created a Malay rentier class based on political and government patronage. The idea of a prosperous peasantry was an ideal that generated considerable support in colonial and postcolonial circles. It heralded the rural Malay community as the source of legitimacy for the state (Drabble, 2001).

The role of the government in the post-independence economy has been one of the most controversial issues in Malaysian economic debates. There are ups and downs with the Malaysian economy. Effects of economic mistakes seem to have been absorbed by the infusion of oil revenues, which has kept the economy growing. The government has invested heavily in physical infrastructure and also in education and health. Over the last four decades, modern Malaysia has weathered many economic and political storms. Malaysian economic growth perked along at 5–7 per cent per year from the early 1960s until the late 1990s. Malaysia's economic growth rates have not been equal to the East Asian NICs' (South Korea, Taiwan and Hong Kong), but sustained growth rate at such levels has narrowed the economic gap with the industrial world (Drabble, 2001).

8.2 The national economy

With its economy grown, social standards improved and the intellectual levels on the rise, Malaysia has become an emerging market. Three decades ago, it was classified as a third-world country, but has now shrugged off its dependency on tin mines and rubber plantations. Instead, Malaysia had become one of the world's largest semiconductor producers and a major maker of cellular phones and stereos. Under the prudent guidance of the Malaysian government, it had become an economic powerhouse by the middle of the twentieth century. Its gross domestic product (GDP) was growing at an average rate of 9.0 per cent until the Asian financial crisis hit the country in 1997. However, with near double-digit phenomenal economic growth rate, Malaysia met with two trouble periods in the run-up to the twenty-first century. The first economic crisis came in 1985, bringing about the collapse of the prices of all the commodities including rubber, tin, palm oil and petroleum, all of which the Malaysian economy was based on. Under such conditions, the conception of the two-pronged strategy to invite foreign investors to open factories in Malaysia, easing restrictions on foreign ownership was introduced. At the same time, most of the government projects and enterprises were placed in private hands. The choices given by the government were the emerging market mix of utilities, resource companies, builders, land developers and banks. They flocked in multitudes and have changed everything in Malaysia. With total returns to investors of 24 per cent in 1987, 42 per cent in 1988 and 94 per cent in 1989, Capital's Emerging Markets Growth

Fund did exceptionally well. The results provided the platform for the irrational rush of 1993, such that the investment amount of an entire year could amount in a day. In 1993, US$15 billion in net foreign investment flowed into emerging Asian stock markets. This figure is up from the previous record high of US$2 billion the year before. At the same time, Malaysia's stock market went up 98 per cent that year. The economic performance was probably never heard of before by any group of economies of similar size. Impressive macroeconomic performance was backed by moderate inflation, an absence of significant fiscal imbalances and high savings. The economic performance was so successful in the Asia-Pacific region that it had attracted large inflows of foreign portfolio funds into the region. This success became a root cause for the currency crisis for the region. In the second half of 1997, the region's economic slowdown had brought about a scurry of foreign funds fleeing. Unlike neighbouring countries, Malaysia is fortunate enough to be performing under stable political conditions (Encyclopaedia: Malaysia).

8.2.1 Macroeconomic review and outlook

Since 1988, Malaysia has experienced an average economic growth rate of 9 per cent per annum. An overall GDP growth of an average 8.7 per cent per annum was recorded during the Sixth Malaysia Plan (6MP) 1990–1995. In 1995, per capita gross national product (GNP) increased from RM4426 in 1990 to RM5815, and was targeted to increase to RM7213 in 2000. Price stability and virtual full employment, since 1993, have made the high growth rate of the economy possible. The benefits of strong overall economic growth were widely distributed which was evidenced with the sharp decline in poverty and the high attainment levels reflected in the social indicators. Large foreign direct investment (FDI) inflows as well as investments in physical infrastructure, technology, education and skill training contribute to the expansion of real output. The strong overall growth had induced substantial structural transformation to the Malaysian economy during the 6MP. Manufacturing sector increased sharply from 26.9 per cent of the GDP in 1990 to 33.1 per cent in 1995. However, for the agriculture sector, the share of GDP declined from 18.7 to 13.6 per cent during the same period (Table 8.1). The shares of GDP in 2000 were expected to record further decline with agriculture to 10.5 per cent, but an increase to 37.5 per cent with manufacturing. Among the other sectors, during the 6MP, for the construction sector, the share of GDP increased from 3.6 to 4.4 per cent, while the share of the services sector increased from 40.6 to 41.9 per cent (Asian Development Bank, 1997, 1999, 2000). The composition of employment experienced similar changes as the structural changes in output. This was reflected in an increase from 19.9 per cent of total employment in 1990 to 25.9 per cent in 1995 in the share of employment in manufacturing. However, for agriculture, the share of employment declined from 26 per cent in 1990 to 18 per cent in 1995. In anticipating a further decline in the importance of agriculture in output and employment in 2000, the shares of manufacturing and agriculture in employment were estimated at 28.9 and 13.1 per cent respectively. During the 6MP period, the share of private investment in GNP increased from 20.7 per cent in 1990 to 32 per cent in 1995, whilst during the period the share of public investment grew at 14.7 per cent. During same period, the share of public investment in GNP increased from 12 to 13.4 per cent (Asian Development Bank, 1997, 1999, 2000).

Table 8.1 Gross domestic product by industrial origin at constant 1978 prices (per cent share in GDP) sector/industry 1985, 1990, 1995, 2000

	1985	*1990*	*1995*	*2000*
Agriculture, forestry and fishing	20.9	18.7	13.6	10.5
Industry	36.7	42.2	47.3	50.7
Mining and quarrying	10.5	9.8	7.4	5.7
Manufacturing	19.7	26.9	33.1	37.5
Construction	4.8	3.6	4.4	4.8
Electricity, gas and water	1.7	1.9	2.3	2.7
Services	41.8	40.6	41.9	43.1
Wholesale and retail trade, hotels and restaurants	12.1	11.1	12.1	12.7
Transport, storage, and communication	6.4	6.9	7.3	8.3
Finance, insurance, real estate, and business services	8.9	9.8	10.7	11.9
Government services	12.2	10.6	9.7	8.1
Other services	2.3	2.1	2.0	2.1
Plus: Import duties	3.9	3.7	4.2	3.3
Less: Imputed bank service charges	3.2	5.1	7.0	7.5
GDP at market prices	100.0	100.0	100.0	100.0

Source: Government projections under the 7MP 1996–2000. Bank Negara Malaysia, Annual Report 1995, Ministry of Finance, Malaysia, Economic Report, various issues.

The strong growth performance sustained for nearly a decade. Such performance has also resulted in fundamental structural change in the economy and led to the corresponding change in the composition of output and employment. Manufacturing increased its contribution to the total GDP sharply from 20 per cent in 1985 to 33.1 per cent by the end of the 6MP (1991–1995). During the same period, contribution of agriculture to the GDP declined to 13.6 per cent from 21. As illustrated in Table 8.1, the composition of total employment also transformed in line with the composition change of output. In terms of its ability to attract inflows of private FDI, Malaysia ranked among the highest in the world. This helped the growth of export-oriented manufactures, which was the underlining factor for the growth of the manufacturing sector. Government involvement in the economy was substantially cut down through the privatization programme, which has helped the government to transform a significant fiscal deficit into a surplus by 1993. Malaysia has made great strides towards realizing its vision of a modern industrial society. The country has been a top economic performer in Asia with the excellent overall economic and social progress of the past decade (Asian Development Bank, 1997, 1999, 2000).

For nearly a decade, Malaysia had been experiencing a very high rate of economic growth. However, its overall GDP growth rate declined from 9.5 in 1995 to 8.2 per cent in 1996. In comparison with the high rates achieved in 1995, growth in all sectors was lower in 1996, which truly reflected the growth potential of the economy. The main impetus for overall economic expansion was still attributed to the manufacturing sector, which was growing at a double-digit rate in 1996. Nevertheless, the sector also experienced a drop from 14.5 in 1995 to 12.2 per cent in

1996, as the country's manufacturing sectors were heavily relying on the export-oriented electronics industry, which has come under pressure from increasing global competition, slower demand growth and a build-up of inventories. Due to reduced public-sector investment, growth of the construction sector declined to 13 per cent in 1996 from 17.3 in 1995, and contributed 4.5 per cent of GDP in 1996.

Supported by the growth of palm-oil output, growth of the agriculture sector increased from 1.1 per cent in 1995 to 2.4 in 1996. With such low growth, contribution of the agricultural sector output was reduced to 12.7 per cent of GDP in 1996. For the other sectors, growth of the mining sector declined sharply to 3.9 in 1996 from 9 per cent in 1995; and a decline of growth from 9.4 in 1995 to 9.1 per cent in 1996 was also recorded in the services sector (Asian Development Bank, 1998). The Asian financial crisis had severely affected Malaysia with severe economic recession by August 1998. As output of all major sectors of the economy declined, the real GDP contracted by 6.2 per cent in 1998. The construction and manufacturing sub-sectors contracted sharply and due to supply constraints, agricultural sector also declined. In 1998, the regional economic crisis brought difficulties that are more significant to Malaysia's two million migrant workers and urban communities. Unemployment in urban areas was particularly noted and unemployed figures climbed by 109,000–342,000. The numbers for poverty increased from 6.8 in 1997 to 8 per cent in 1998, which again affected mainly the cities (Encyclopaedia: Malaysia).

The Malaysian economy made the turnaround in the second quarter of 1999. The return of confidence, and a more stable economic environment through selective capital controls and the pegging of the ringgit provided the driving factors for the growth. Initially, the economy was externally led, but later became broader-based as domestic demand picked up as it was driven by expansionary fiscal and accommodative monetary policies. All sectors in the economy registered positive growth except for mining. Strong real GDP growth of 8.5 per cent was recorded in 2000. The construction sector also made a turnaround in 2000 after consecutive falls in two preceding years (1998 : –23.0 and 1999 : –5.6 and recorded 1.1 per cent growth. The implementation of several large infrastructure projects and new investments in the residential sector, particularly low- and medium-cost houses, were the main factors helping the turnaround of the construction sector (Malaysian Property Market 2000 in Overview).

The high growth rate was attained in an environment of strong domestic demand, and low inflation and unemployment. The per capita gross national income (GNI) increased from US$3,238 in 1999 to 3,516 in year 2000. The dramatic recovery of the economy can be attributed largely to the National Economic Recovery Plan. The economic recovery plan contained over 580 recommendations to stabilize the currency, restore market confidence, maintain financial market stability, as well as address medium-term issues such as improving economic fundamentals and addressing the equity and socio-economic agenda. The stability accorded by the pegged ringgit facilitated the business decision-making process that enabled Malaysia to capitalize on the recovery in external demand. This resulted in unprecedented levels of current account surpluses of 13.9 per cent of GNP for 1998, 17.1 per cent of GNP for 1999 and 10.0 per cent of GNP for 2000. The surpluses contributed to the sharp rise in international reserves (1998, US$26.2 billion; 1999, US$30.8 billion; and 2000, US$29.9 billion), increased the monetary base and provided the liquidity to support economic growth. The collateral values of loans were strengthened with the improving stock market prices, thereby facilitating the resolution of non-performing loans (NPLs) of the banking system.

Expenditure on emoluments, supplies and services of the federal government rose strongly in 2000 due to the upward salary adjustment and increase in housing allowances for selected categories of civil servants. The higher spending on supplies and services was for professional services, supplies of materials, maintenance and renovation works. Nevertheless, growth in total public consumption moderated due mainly to lower spending for defence.

Private sector expenditure rebounded strongly in 2000 to record double-digit growth of 15.7 per cent (3.9 per cent in 1999). Private consumption was sustained at a high rate, reflecting improved consumer confidence in an environment of higher disposable income, low interest rates as well as favourable employment prospects. Major consumption indicators such as sales of passenger cars, imports of consumption goods, loans disbursed for consumption credit and activities in the wholesale, retail, restaurants and hotels, all improved. Private investment activities increased by 26.7 per cent after 2 consecutive years of contraction (1998: −55.2 and 1999: −21.8 per cent). In line with the strengthening of investment activities, loans disbursed by the banking system also increased in 2000, mainly due to fund manufacturing activities and property development. In tandem with the higher capacity utilization, strong demand in the export-oriented industries, strengthening investor confidence and the introduction of policy measures aimed at attracting foreign investors through the pre-packaged incentive scheme for strategic projects, contributed to higher investment in the manufacturing sector. Manufacturing sector was estimated to have increased by 36.8 per cent in 2000. Table 8.2 shows the main macroeconomic indicators.

8.2.2 Economy in the first half of 2001

Real GDP registered a growth of 3.2 per cent in the first quarter of 2001. Growth was broad based, with all sectors of the economy recording positive growth rates. In an environment of less-favourable external developments in the first quarter of 2001, growth was increasingly supported by domestic demand. Growth in the first quarter of 2001 was supported by expansion in domestic-oriented industries, the semiconductor sub-sector and strong growth in both the agriculture and the services sectors. This diversified economic base has cushioned the marked slowdown in the electronics equipment sub-sector. Domestic demand was also supported by the higher government expenditure. The pre-emptive pro-growth policies led to the strong increase in gross fixed capital formation which helped contain the impact of the slowdown in the US economy. The fiscal stimulus was also supported by major administrative changes undertaken to ensure that the counter-cyclical fiscal policy would be effective in stimulating growth. Steps were taken to ensure the full utilization of allocation to achieve maximum impact on the economy.

Looking forward, the fiscal stimulus package, including the additional spending of RM 3 billion (US$0.8 billion) announced on 27 March 2001 was targeted to facilitate economic growth through higher domestic demand, shorter gestation period and low import content. The impact was expected to be felt in the economy within

Table 8.2 Main macroeconomic indicators

	1996	*1997*	*1998*	*1999*	*2000*	*2001 (f)*
GDP and components						
GDP at real price (RM billion)	183.9	196.7	182.2	192.8	209.3	219.8
GDP at current prices	253.7	281.8	284.5	300.3	339.4	353.2
GDP growth (% change)	8.6	7.3	−7.4	5.8	8.5	5.0
Manufacturing sector (RM billion)	53.3	58.8	50.9	57.8	69.9	75.8
% growth	18.2	10.1	−13.4	13.5	21.0	8.5
Service sector (RM billion)	93	102.2	101.4	104.6	109.9	115.7
% growth	8.9	9.9	−0.7	3.3	4.9	5.2
Construction sector (RM billion)	8.6	9.5	7.3	6.9	67.5	75.7
% growth	16.2	10.6	−23.0	−5.6	3.1	5.5
Demographic indicators						
Population – (million persons)	21.2	21.6	22.2	22.7	23.3	23.8
Population growth rate (% change)	2.4	1.7	1.7	1.8	2.3	1.7
Total labour force (million persons)	8.6	9.0	8.9	9.2	9.5	9.8
Labour force growth rate (% change)	2.4	4.6	−2.5	1.7	2.1	n.a
unemployment rate (% of labour force)	2.5	2.4	3.2	3.4	3.0	2.9
Financial indicators						
Changes in consumer price index (2000 = 100)	3.5	n.a	5.3	2.8	1.6	1.5–2.0
Short-term interest rate (%) (3-month interbank rate)	–	8.7	6.46	3.18	3.25	n.a.
Long-term interest rate (5) (12-month fixed deposits bank rate)	–	9.33	5.74	3.95	4.08	n.a.
Annual average exchange rate with US$	1.41	2.81	3.92	3.80	3.80	3.80
Change against US$	0.5	35.0	2.3	00	00	00

Sources: Bank Negara Malaysia Annual Report, 2000.
The Malaysian Economy in Figure 2001, by EPU, Prime Minister's Dept.
Economic Report 2000/2001. Statistics Handbook Malaysia 2001 by Dept. of Statistics, Malaysia.

Note
base year 1987 (f) forecast.

a shorter time span. The expenditure, which was largely focused on education and training, will not only stimulate economic growth in the near term, but will also strengthen the long-term objective of manpower development. It was likely that the full impact of these measures was expected to be more visible in the second-half of the year 2001.

During the first quarter, monetary policy remained accommodative amid a benign inflation situation. The prevailing ample liquidity together with historically low interest rates has continued to provide a positive environment for private-sector investment activities. The monetary aggregates continued to expand, albeit at a slower rate, reflecting continued expansion in business activities. Lending activities were sustained with loans outstanding increasing by 5.8 per cent at the end of March 2001. Including Private Debt Securities (PDS) held by the banking institutions, the total financing provided by the banking system increased by 7.1 per cent. Total financing including PDS held by non-bank private sector rose by 8.1 per cent as at the end of March 2001.

8.2.3 *Role of construction/property market in national economy*

The construction industry is an integral part of Malaysian economy. The growth rate of the construction sector over the last decade has been considerably high with a peak level (19.5 per cent) in 1990 (Table 8.3). The construction industry consistently attained a double-digit growth rate throughout the 1990s until the Asian financial crisis hit the country's economy in 1997. The construction sector recorded a growth of 1.1 per cent after contracting by 10 per cent in 1998 and by 5.6 per cent in 1999. The small growth in 2000 was attributed primarily to government spending under the fiscal stimulus programme, privatized infrastructure projects and residential development, especially affordable housing. The construction activities in the non-residential sub-sector, however, were subdued. Activities in this sub-sector were supported mainly by ongoing projects that had commenced prior to 1999. The unsold units of residential property stood at 45,500 units and the occupancy rate for commercial space at about 75 per cent. As of mid-2000 it was estimated there were still 2,773,186 square metres of vacant completed office spaces. On top of this incoming supply of 2,44,601 square metres of office spaces and 1,492,555 square metres of retail spaces are expected to enter the market in the next few years.

Act 520 that governs the activities of Construction Industry Development Board (CIDB) in managing the development of the construction industry makes it incumbent on contractors to inform the board of any construction activities exceeding RM500,000 two weeks after being awarded the project. Unfortunately, there appeared to be a tendency to delay reporting projects resulting in a time lag between the time the data on a project is acquired and the actual time of award. As a result timely acquisition of information has, as yet, not been effected. Although there are punitive provisions for such neglect CIDB, as of now, has not resorted to such measures preferring to harness the industry's cooperation in undertaking its responsibility.

Table 8.3 Growth trend of GDP and construction industry

Year	*% growth of construction industry*	*% growth of GDP*
1990	19.5	9.8
1991	14.6	8.7
1992	11.2	7.8
1993	11.2	8.3
1994	14.1	9.2
1995	17.3	9.5
1996	14.2	8.6
1997	9	7.8
1998	−10	3
1999	−5.6	
2000	1.1	

General trend of supply

Tables 8.4 and 8.5 show the quarterly trend of construction projects awarded for the period 1997–2000 (first half) based on the information gathered as of March 2001. In 1997, 5,982 projects worth about RM58.6 billion were awarded. In 1998 the value declined to RM29.5 billion representing 3,243 projects. The reduction was the direct result of the fallout of the financial crisis of September 1997. In 1999 the figure climbed back slightly to RM50.4 billion involving 4,844 projects. The first half of 2000 recorded 1,898 construction projects awarded with a value of about RM16.0 billion based on data as of March 2001 down by RM11.5 billion from RM27.5 billion of the previous half. This indicates that there was a decline in the value of construction projects being awarded for the first half of 2000. Year upon year the decline was about RM7.0 billion from RM22.9 billion to RM15.9 billion.

Although the data collected on projects included information on their clients, analyzing the data to determine the extent of the public-sector's involvement in the construction sector was not conclusive. This was because public-sector projects included only those undertaken and implemented by government departments but

Table 8.4 Total volume and number of projects in 2000 (as of 4 September 2001)

	Volume in RM	*No. of projects*
q1-2000	8,965,444,658	892
q2-2000	9,396,026,025	1,217
q3-2000	11,600,416,881	1,025
q4-2000	9,427,444,946	1,028
Total	39,389,332,510	4,162

Table 8.5 Quarterly project values and numbers (projects recorded as of 10 April 2001)

	Value of projects (billions)	*Quarterly totals*	*Numbers*	*%*
q1-1997	12.20		1,335	
q2-1997	16.98		1,637	39.19
q3-1997	18.58		1,744	9.43
q4-1997	10.87	58.62	1,266	−41.51
q1-1998	6.27		620	−42.32
q2-1998	6.02		818	−3.97
q3-1998	5.95		958	−1.17
q4-1998	11.30	29.54	847	89.95
q1-1999	7.87		781	−30.34
q2-1999	14.99		1,201	90.49
q3-1999	9.71		1,381	−35.22
q4-1999	17.78	50.36	1,481	83.01
q1-2000	8.02		822	−54.86
q2-2000	7.97	16.00	1,076	−0.64

precluded projects undertaken by corporatized and privatized public-sector entities as well as non-financial public enterprises, which were considered part of the private sector. Further, in rejuvenating the economy, when necessary, the government had, in most cases, resorted to the use of the privatization mode as the vehicle for implementation since it was deemed to be more effective and in tune with the concept of making the private sector the engine of growth for the nation. As such, stratifying projects into public or private domain did not provide for a true picture of the degree of involvement of the public sector in the construction industry. (A major setback was the absence of information on clients' profile in a large part of the data.)

Beginning 1999 and extending into 2000 the quarterly value and volume of construction projects awarded were more volatile showing a roller-coaster trend across each quarter. Year-on-year the first half of 2000 showed a sharp decline from RM22.9 billion in the 1999 to RM16.0 billion in 2000. Quarterly trend also indicated a decline in the first quarter of 2000 from RM17.8 billion in fourth quarter of 1999 to RM8.0 billion in the first quarter 2000 and RM8.0 billion in the second quarter.

Quarterly project profiles

Table 8.5 gives the value profile of projects for 1997 until the middle of 2000. It can be seen that the majority of projects are those that fall between RM500,000 and RM20 million with a maximum of 95 per cent of projects in the fourth quarter 1999 and a minimum of 91 per cent in the first quarter of 1999. This was followed by projects within the RM20–RM40 million category and subsequently by projects in the RM40–RM60 million category. However, in the fourth quarter of 1999 projects with value exceeding RM100 million exceeded projects within the RM40–RM60 million category. In fact, projects exceeding RM100 million are more frequent than projects between RM60 and RM80 million in all cases except the second quarter of 2000.

The roller-coaster trend of the quarterly volume of construction awards can be attributed to the periodical entry of exceptionally large projects into the market. These are mostly projects in the oil and gas sector (e.g. the EPCC contracts for construction of petroleum by-products processing plants) and the implementation of infrastructure projects whether by the public sector or as privatized projects. From Table 8.4 it can be seen that the number of exceptionally large projects in terms of percentage to overall number of projects has relatively lower correlation (0.89) with projects that are below RM20 million in value unlike the other categories of projects with value below RM100 million which had correlation factor greater than 0.9. This may arise because projects exceeding RM100 million are grouped together and are thus not subjected to the RM20 million range as other project categories. Since projects grouped in the category exceeding RM100 million have lower correlation with the projects in the less-than RM20 million category as compared to the other categories, projects exceeding RM100 million are treated as distortions or outliers to the real trend of new project entry as well as the growth trend in this review. On the other hand, the projects below RM100 million can be viewed as reacting to the demand of the market and are thus demand-pull projects, whereas projects exceeding RM100 million can be considered as supply-push projects, usually government initiated, with the purpose of rejuvenating the economy during sluggish times.

As mentioned earlier, the payment of levies tended to lag the date of awards of projects sometime as long as nine months. For the purpose of evaluating the trend of construction starts it is not practical to wait for all the information to be collected before a reliable estimate can be done. Correlations can be obtained between data collected three months after the end of each quarter as against the data collected as of April 2001 for projects with value less than RM100 million. Based on values of projects the correlation factor was 0.97 and based on the percentages of change in quarterly figures the correlation factor was 0.95. This showed that using data collected three months after the end of a quarter is sufficiently accurate to predict the trend of construction starts.

Projects awarded to foreign contractors

A total of 120 projects valued about RM8.3 billion were awarded to foreign contractors in 1997. This was reduced to 58 projects worth about RM6.6 billion in 1998, which increased to 70 projects worth about RM8.5 billion in 1999. In 2000 as of March 2001, 59 projects worth about RM2.9 billion have been awarded to foreign contractors.

Japan has consistently been securing the highest number of projects (1997: 59, 1998: 34, 1999: 47, 2000: 52). The other countries have only managed to intermittently secure jobs in Malaysia. It is most likely that Japan has managed to secure many projects in tandem with the Japanese investment in Malaysia. It is worth noting that countries that have secured projects in Malaysia also happen to be the major foreign investors here. Japan, for example, was the second largest single investor in Malaysia in 2000 (see Economic Report 2000/2001).

Looking at the list of projects awarded to foreign contractors, it can be seen that the successful bidders for projects are basically established multinational corporations. The majority of contractors from other countries are leveraged on their special expertise in securing projects in this country. It is thus not surprising to note the involvement of foreign contractors in projects that usually require some special expertise in its implementation.

8.2.4 Properties sub-sector

The Asian financial crisis in 1997 has left a major impact on the property market in Malaysia, which has been thriving in the early part of 1990s. The currency and economic crisis affected the region since 1997. With decline in the GDP, office take-up slowed down and this was met with oversupply of office space, for example, the Petronas Tower accounted for 2.4 million square feet. The situation got worse in the first three quarters of 1998. The high cost of funds and tight liquidity situation contributed to the continuing weakness. Bank policy on the RM250,000 cap for funding of development on new residential and shophouse units was not helping the property market. Over the first half of 1998, prime office rents declined by 10 per cent and 6 per cent. It dampened the Klang Valley and Kuala Lumpur property market. The oversupply situation happened in almost all sectors. Lower-end residential sector involving units priced below RM150,000 were still in demand, which provide the only market activities. However, vacancy rates soon soared when the government

Table 8.6 Number and value of property transaction by sector

Property type	*1998*		*1999*		*Annual % change*		
	Number	*RM million*	*Number*	*RM million*	*1998*	*1999*	
Residential	122,881	13,949	157,082	18,539	−30.3	27.8	32.9
Commercial	12,085	4,916	15,902	6,204	−44.1	31.6	26.2
Industrial	4,862	2,838	8,330	3,599	−48.1	71.3	26.8
Total property transaction in 1999	186,077	27,911	225,901	34,423	−32.3	21.4	23.3

Source: Property Market Report, 2000

shifted administrative functions to the intelligent garden city of Putrajaya (Malaysian Property Market 2000 in Overview).

As shown on Table 8.6, the property transaction activity contracted significantly. Whilst overall prices fell in 1998, in 1999 prices were mixed. In 1998, it reduced by 32.3 per cent in number and 47.6 per cent. However, in 1999, the situation improved for all the residential, commercial and industrial sub-sectors, with 21.4 per cent and 23 per cent in numbers and value respectively.

Favourable lending conditions with reduced interest rate generated demands in the residential sector. The base lending rate (BLR) came down from a high of 12 per cent in June 1998 to 8.05 per cent by early November 1998. In addition, several schemes launched by government were positive, including re-classifying loans to the property sector as priority/non-priority loans instead of being considered non-productive. By 2000, the performance of the property market in Kuala Lumpur was rather mixed in the three major sectors. In the commercial sector, the office sub-sector saw occupancy rates reduced, but this was in contrast with the retail and hotel sub-sectors. In the residential sector, buyers avoided condominiums and apartments but continued to show interest in landed property in selected locations. However, those flats priced within the affordable range of less than RM200,000 were still attractive. There were no new significant supply in the industrial sector and it was rather quiet (Property Market Report, 2000).

Commercial properties

In 1995, the purpose-built office market performed better than anticipated. High take-up rate with the building put to market in 1995, such as the MNI Twins and Plaza Raja Chulan. Average vacancy rate was less than 10 per cent and the general rentals for Class One office remained at RM48–RM59 per square metre per month inclusive of service charges. Up to the third quarter of 1995, 220,000 square metre of purpose-built office space added to the market. The supply was expected to be in surplus in 1996 with large office buildings such as the Petronas Towers completed. In that year, new office building completed contributed 2.1 million square feet space making a total office stock of 33 million square feet. In Kuala Lumpur, the average vacancy rates increased from 3.4 per cent in 1996 to 11.4 per cent in 1997 and to 15 per cent in June. The situation was better with office within Kuala Lumpur Central

Table 8.7 Supply of office, retail and residential space in Klang valley

Year	*Office space*		*Retail space*		*Apartments*
	Square metre	*Occupancy rate*	*Square metre*	*Occupancy rate*	*Units*
1996	269,742	95.5	139,964	92.8	14,568
1997	869,394	94.9	362,574	90.5	5,473
1998	873,346	79.8	395,328	59.5	14,151
1999	795,750	76.2	58,486	76.6	9,547

Source: Property Market Report, 2000.

Area (KLCA). In the Klang Valley, the vacancy rate for purpose-built office was 16.5 per cent in June 1998 (Table 8.7). By the end of 1999, the take-up rate of office improved, when the average vacancy rates at Klang Valley and Kuala Lumpur stood at 23.6 per cent and 22.0 per cent respectively. There was a marked difference in the occupancy rate of purpose-built office in KLCA and Kuala Lumpur Metropolitan (KLM) areas. Over the period from 1999 to 2000, while the KLCA saw an increase in occupancy rate from 80.7 per cent to 84.8 percent, the KLM areas experienced a drastic drop from 70.3 per cent to 56.7 per cent (Property Market Report, 2000).

By 2000, the nationwide total stock of newly completed office space increased by 13.0 per cent with 1.19 million square metres. Klang valley contributed to 55.7 per cent of the new supply while Johor and Perak contributed with 9.6 per cent and 8.8 per cent respectively. The overall occupancy rate dropped to 76.9 per cent from previous year's 82.1 per cent. Due to the relocation of government departments to Putrajaya, Kuala Lumpur accounted for almost half of the unoccupied space. Due to the existing oversupply especially in Johore Bahru and Batu Pahat, the rentals in Johor reduced as much as 30.0 per cent. Generally, the rentals in the states of Kuala Lumpur, Selangor, Penang, Negeri Sembilan and Melaka were stable. In the first quarter of the year 2001, rentals of office market had suffered further reduction, by 5–10 per cent from 1999 rates. For the retail property, the growth of 3.8 per cent in total transactions and value decelerated in 2000. The most active transactions for retail property were from Perak, which made up 18.5 per cent of the total. This was followed by Selangor and Johor with market share of 18.2 and 15.4 per cent respectively. Positive growth in total volume was recorded in major states of Selangor, Pulau Pinang and Perak, while a drop was registered in Kuala Lumpur, Johor, Negeri Sembilan, Perlis and Kedah. Retail properties between RM250,000 and RM500,000 attracted the buyer's interests. For this price range of retail property, Selangor made up 33.6 per cent of state-wide total transactions. Prices for the conventional commercial property were mixed. There was mixed performance in Selangor and Pulau Pinang. Shophouses in Johore Bahru had a price increase of 10 per cent especially in city area. Generally, prices in Kuala Lumpur were stable but with substantial increase in some selected areas such as Sri Hartamas with 30 per cent increase. For shopping complexes, the overall occupancy rate dropped slightly from 75.7 to 73.2 per cent in 2000. For states in the country, the rate ranged from 58.2 per cent for Pulau Pinang to 91.4 per cent for Perlis. With the almost full occupancy rates for KLCC Suria and Mid-Valley Megamall, the rate was pushed up by 5.1–78.8 per cent. In Negeri

Sembilan, the situation of oversupply was significant. In Selangor, the rentals for ground-floor retail units of some established shopping complexes were either stable or improved up to 56.0 per cent due to the high demand. In Johor, rentals decreased with those in city fringe areas dropping up to 28.0 per cent, but rentals of retail building in the city centre managed to maintain the existing rate (Property Market Report 2000; Malaysia Country Guide).

Residential properties

In Malaysia, residential unit refers to a building for dwelling purposes. It does not include hostels, tenement houses and parsonages. The categories of residential units include detached houses, semi-detached houses, terrace houses, apartments and condominium units, cluster housing, townhouses, strata-detached and strata semi-detached housing. Condominium is a type of housing comprising a block or blocks of apartments with common recreational and other facilities for communal living. It is an apartment built by the private sector held with a stratatitle. A Management Corporation (MC) is elected amongst the owners, which runs the common facilities. To upkeep the facilities and provide a sinking fund for major repairs, monthly management fees are collected from the owners. Condominium caters to the needs of the middle and upper-middle income groups and is normally built in urban centres (Malaysian Property Market 2000 in Overview).

In 1995, the residential sub-sector, in particular for conventional housing, saw good demand and price. In 1996, about 10,603 transactions were recorded, while housing stock was at 172,731. There were a total of 9,281 condominiums and apartments entering the market in Kuala Lumpur in 1996 alone and another 11,694 were expected to be completed in 1997. With total number of transactions increasing by 3.3 per cent over the previous year, the growth continued into 1997 and the total value transacted rose by 15.2 per cent. In 1998, due to uncertainty of employment, demand for residential units was weak with few transactions taking place. The high-end residential sub-sector of the market was worse hit with their prices in some location falling by as much as 40–50 per cent in the year. In 1999, the growth in the manufacturing sector helped the economy move and improved the employment opportunities. With such background and with the success of the First Home Ownership Campaign held at the end of 1998, the residential sector became brighter in 1999, especially for landed properties, which have improved since then. However, such revival of confidence in the primary market did not benefit the secondary market. Activity in the secondary market was rather selective, such that Tun Dr Ismail, Damansara Heights and Bangsar continued to be in demand. Demand remained weak for apartment and condominiums in 2000. Generally, the residential property prices, including rentals, were stable or increased. The residential sector was moderately active in 2000 with transactions at 8.8 and 18.3 per cent growth in total volume and value compared to previous year's 27.8 and 32.9 per cent growth respectively. After experiencing negative growth in transaction volume for the states of Johor, Kedah and Kelantan against only Perak in 1999, there were 170,932 transactions recorded in 2000 and they represented RM21.94 billion in total value. The most popular residential properties are those priced below RM100,000, which accounted for 58.0 per cent of total transactions with a lower share compared to 62.0 per cent the previous year. In Kuala Lumpur unlike apartments and condominiums, which remain stable, the terrace houses in selected areas enjoyed 18 per cent increase. The same applied to Selangor and Johor. However, apartment transactions in Penang recorded price improvements in almost all areas. Due to the reduction of units under construction and in the pipeline, the supply was lower for the year 2000. The weak demand for apartments and condominiums was expected to continue in 2001. Overall, the exceptional low interest rate, employment stability, accelerating consumer spending and positive shifts in household asset allocation were factors that could improve the volume and push residential prices up in 2001 (Property Market Report 2000).

Industrial properties

Industrial properties include factories and warehouses. Factory refers to space used for industrial purposes, which include uses for the manufacturing, altering, repairing, ornamenting, finishing, cleaning, washing, packing, canning, breaking-up or demolition of any article or its parts and the processing and treatment of minerals. Factory spaces are further categorized into flatted factory space and non-flatted factory space. Warehouse refers to covered space used as storage area for raw materials, semi-finished or finished goods (Malaysian Property Market 2000 in Overview). Due to the scarcity of land, the Federal Territory has a very small industrial base. Industrial developments are very limited and the factories are terraced and semi-detached factories, which are mainly located in those old, established industrial areas such as Sungai Besi, Segambut, Jinjang, Old Klang Road, Cheras, Setapak, Jalan Genting Klang and Salak South. Industrial developments, were concentrated on the fringe of Kuala Lumpur and in Selangor in the past few years. The areas that performed better were new industrial developments, including areas along the NorthSouth Expressway, Shah Alam, Klang and new townships such as Puchong, Rawang, Bukit Beruntung, Subang Jaya, Bangi, etc. In 2000, the existing stock of industrial property was 75,322 units, and a total of 2,336 units of industrial property were completed. The largest share of 63.0 per cent (1,471 units) went to Selangor while Perak and Pahang each shared 435 and 203 units respectively. The total number of transactions for the year was 7,556. In 2000, although the total transactions of industrial sector fell by 9.3 per cent a significant rise of 100.9 per cent in total value was registered. Melaka recorded the largest percentage of improvement of 54 per cent. For capital and rental, some upward movement was recorded in Kuala Lumpur with prices in some selected areas appreciating by 10.0–25.0 per cent. Price movement varied in Selangor and prices in Johor generally declined by 10.0 per cent (Property Market Report, 2000).

8.2.5 Changes of economic policies due to the financial turmoil

Global economic and financial developments impinged significantly on the prospects for the region. The slower than expected growth in the industrial economies, the increase in the volatility in the major currencies and the sharp movements in the global stock markets have generated further weakness in external demand and resulted in an unstable financial market condition. Imbalances in the USD, euro and yen currency areas were significant and posed continued risk to global economic expansion. Further, the continued lack of constructive initiatives to strengthen the international financial architecture particularly in respect of the increased movement of speculative short-term capital remains a potential source of instability for international markets and the recurrence of financial and economic crises. The experience gained from the East Asian financial crisis in 1997/1998 was useful in providing the direction and thrust for Malaysia's future development. Some of the lessons learnt were,

- efforts to strengthen economic fundamentals must be continuous. Malaysia needs to further intensify the resilience of domestic economy and enhance its competitiveness, not only to mitigate adverse external economic development but also to reap the opportunities available;
- monetary and fiscal policies must be directed towards promoting sustainable growth, while keeping inflation at a low level;
- strong economic fundamentals must be supported by sound financial and corporate sectors;
- develop the capital market to diversify risks and lessen the reliance on bank financing; and
- strong corporate governance to increase resilience and sustain investor confidence in the event of crisis.

The major currencies experienced significant movement during the first quarter. In particular, the rapid depreciation of the Japanese yen raised concerns that there would be competitive depreciation of regional currencies. This led to market concerns that there would be a downward adjustment in the ringgit exchange rate to remain competitive. Such a depreciation for competitive purposes will not benefit Malaysia as any competitive advantage derived from the depreciation would only be temporary. Over-reaction to adjusting the exchange rate in response to these short-term movements would result in misalignments of exchange rates. The exchange rate should adjust to structural changes that result in shifts in the underlying fundamentals. For Malaysia, domestic macroeconomic fundamentals continued to remain consistent with pegged exchange rate system (RM 3.80 = US$1.00). The exchange rate continued to be supported by a strong current account surplus, projected at 6–7 per cent of GNP in 2001. Inflation was expected to remain low at about 2 per cent. External debt remained low with reserves exceeding short-term external debt by 5.7 times. The assessment was that the ringgit remained close to its fair value and was consistent with the fundamentals of the economy.

The US economy had entered a period of cyclical slowdown after a decade of rapid expansion. Given that the US accounts for slightly above one-fifth of world output and absorbs about 19 per cent of would exports, its slowdown was expected to result in a more moderate world growth. While there had been aggressive monetary easing by the US, it was likely that the full impact or recovery will only emerge towards the end of the year 2001 or early 2002. Before clear signs of a recovery in the US emerge the external environment can be expected to be vulnerable to volatility as markets react to specific development. This uncertain environment had raised the potential risks of over-reaction by the market and of further instability. Against this background, Malaysia had already experienced some moderation in the expansion of Malaysia's export and output in the first two months of 2001. Given this scenario, the strategy was to promote further the domestic sources of growth. While growth will moderate, the momentum of growth built up over 1999–2000, the preemptive measures introduced in the 2001 budget in October 2000 and the package announced on March 2001 were expected to mitigate the effect of the moderation.

8.2.6 Policies and strategies adopted to overcome the economic slowdown in the construction sector

Malaysia's performance will be influenced by developments in the external environment as it is highly integrated with the global economy. It will be exposed to the risks of economic and financial shocks, which transcend national borders and regions and are difficult to predict. The achievement of the growth and structural transformation targets will hinge on Malaysia's ability to strengthen its resilience and competitiveness. Efforts will be undertaken to strengthen indigenous capabilities and capacities to enhance national competitiveness in the world market. The sectoral thrusts will be actively pursued thus enhancing the services sector to be a key driver of economic growth and expanding the coverage of infrastructure and utilities as well as improving the efficiency and quality of services. The sectoral strategies will focus on achieving sustainable growth with resilience.

Public sector

The construction sector was expected to grow at 6.6 per cent per annum. The major contributor to this growth will be the ongoing infrastructure projects such as the Express Rail Link for KLIA, the LRT System 2 and the development of Putrajaya as well as other projects to strengthen the social infrastructure. Infrastructure development will focus on ensuring availability of supply, enhancing efficiency as well as improving the quality of services. The expansion of infrastructure facilities will take into consideration long-term demand, economic growth and competitiveness. The development strategies will also emphasize further expansion, upgrading and integration of roads and highways, railways, ports, airports as well as telecommunications will be undertaken in a more coordinated and integrated approach.

Development expenditure of the Federal Government increased by 110 per cent to US$1.4 billion during the first quarter of 2001, a portion of which was expended on construction of infrastructure projects including roads, schools and public utilities. As activity in the non-residential sub-sector remained subdued amidst the large oversupply of retail and office space as well as industrial units, the government announced three measures to reduce the property overhang and to stimulate economic growth. They were:

a to discourage further addition to the property stocks, local authorities would be directed not to approve any new construction office and commercial space in view of the existing large property overhang;

b to help reduce excess stocks in the property market, in particular residential units, incentives provided during the Housing Campaign in 1999 was reinstated for year 2001 with immediate effect. These incentives include the exemption of stamp duty, waiver of processing fee and higher margin of financing; and

c to attract more foreign property investors to come in and purchase local properties Foreign Investment Committee guidelines on foreign equity as well as property and asset acquisition by foreigners were liberalized, effective from 25 April 2001. This is to provide a more liberal environment for foreign investors and the inflows of FDI.

Housing development will continue to give emphasis to the provision of adequate, affordable and quality housing to all, particularly the low-income group. In this respect, focus will be on the provision of better quality houses in suitable locations and conducive living environment. Housing development will continue to be closely coordinated with industrial and overall development planning to ensure adequate and timely provision of houses including access to social services and amenities as well as an integrated and adequate public transport system. These efforts were to be undertaken within the framework of the National Spatial Plan.

Efforts will be undertaken to promote R&D activities related to production of cheaper building materials and better construction techniques with a view to improving the quality and safety of houses and ensuring their affordability. The housing sector will also address a more broad-based agenda such as fulfilling the needs of the elderly and disabled, energy-efficient design as well as implementation of the community-based, garden city concept. Emphasis will continue to be given to the maintenance of high-rise buildings especially low-cost and low-medium-cost flats.

The establishment of new integrated complexes and townships that are managed on an energy-efficient manner will also be encouraged. Towards this end, the Malaysian standards on efficient lighting, air-conditioning and building-envelope systems will be formulated and energy-efficiency provisions will be incorporated in the Uniform Building By-Laws.

The Government also launched a fully computerized second-generation monitoring system (PMS2) to closely monitor the implementation of all ongoing projects. In addition, other measures to ensure timely implementation of projects included the establishment of a flying squad and a review of government's procurement procedures (such as faster processing of tenders, fast-track implementation using turnkey, design-and-build concept and relaxation of guarantee procedures). A late-payments unit was also setup to handle late-payment grievances. In the first quarter 2001, the government also established the Project Implementation Coordination Unit under the Malaysian Industrial Development Authorities (MIDA) to expedite approval of building plans for manufacturing projects.

Private sector

The private sector will continue to play an increasingly important role in meeting the overall housing needs, while the public sector will concentrate on providing houses for the low-income group and public sector employees. The residential and commercial spaces are expected to experience slow growth mainly due to the supply overhang. The Real Estate and Housing Development Association Malaysia (REHDA) will organize a year-long campaign to promote home and property ownership by organizing a nation-wide property exposition. Promotional activities will include year-round TV, newspaper, radio, Internet and other media advertisement. REDHA will complement and coordinate governmental effort for maximum impact and mobilize its more than 800 members to contribute positively to this national effort.

8.2.7 Economic outlook in the next decade

The long-term outlook for the global economy was expected to be positive with real GDP growth projected to increase in all major regions. The world economy is expected to expand by 2–3 per cent per annum between 2000 and 2010 (Table 8.8). This improvement will be spurred by the continued growth of the US economy, robust expansion in Europe and recovery in Japan. The developing countries are expected to grow at an average rate of 3.7 per cent per annum following the stronger economic fundamentals in most countries, consolidation of recovery in Asia and Latin America and moderate growth in the Middle East and in Africa. The East Asia and Pacific region is estimated to expand at 5.4 per cent per annum between 2000 and 2010, boosted by increasing gross domestic investments at 32.6 per cent of GDP and exports at 8.5 per cent per annum. With stable growth in population and moderate inflation, the GDP and consumption per capita in this region are expected to increase by more than 5.0 per cent during the same period, providing an expanding market as well as opportunities for business and industry. During the next decade the volume of world merchandise trade is expected to grow at 6.8 per cent per annum in tandem with the increase in world output. At the regional level, the implementation of the ASEAN Free Trade Area (AFTA) will provide the impetus for trade expansion. The prospects for further trade liberalization remains good with the progress made in the implementation of the Common Effective Preferential Tariff (CEPT), the main instrument for realizing AFTA. AFTA, which was originally targeted for full implementation by 2008, was expected to be realized by 2003, with some flexibility given to its new members. Under the CEPT, ASEAN member countries will reduce intra-regional tariffs and eliminate quantitative restrictions and other non-tariff barriers.

Table 8.8 Economic growth forecast for the decade comparative to the region

	Year ended December	*% growth*
Malaysia	2001	3.3
	2000–2010	3.3
Total Asia-Pacific (average)	2001	3.6
	2000–2010	4.6
ASEAN (4)	2001	5.1
	2000–2010	5.3
OECD	2001	1.8
	2000–2010	2.7

Source: http://www.ey.com/global/vault.nsf/International/Asia_Pacific_Economic_Outlook_2000_–_2010/$file/AP_Economic.pdf.

The economic recovery in East Asia will depend on the commitment of the countries affected by the financial crisis to continue to undertake comprehensive restructuring of the economy, including their financial sector, as well as restore political stability. Malaysia's performance will be influenced by developments in the external environment as it is highly integrated with the global economy. It will be exposed to the risks of economic and financial shocks, which transcend national borders and regions and are difficult to predict. The achievement of the growth and structural transformation targets will hinge on Malaysia's ability to strengthen its resilience and competitiveness. The Eighth Five-Year Malaysia Development Plan (2001–2005) will incorporate the strategies, programmes and projects designed to achieve the National Vision Policy objectives of sustainable growth and strengthen economic resilience as well as create a united and equitable society.

The economy was expected to grow "reasonably strongly" in 2002, with the consensus forecast at 2.9 per cent, while a relatively high average growth of 6.5 per cent per annum expected over the next decade barring constraints of corporate restructuring and foreign investment returns. Indications were positive with exports and imports falling sharply and government reserves reaching more than $30 billion at the end of October.

8.3 The construction sector

8.3.1 Productivity in the construction industry

In 1998, productivity and employment in the construction sector declined by 12.7 and 13.5 per cent respectively (1997: 1.9 per cent, 8.6 per cent). With the decline in productivity and employment growth, output in construction declined by 24.5 per cent (1997: 10.5 per cent) (refer to Table 8.9). Due to economic slowdown,

Table 8.9 Growth of productivity, GDP/output and employment by sector

	Productivity		*Employment*		*GDP*	
	1997	*1998e*	*1997*	*1998e*	*1997*	*1998P*
Malaysia	5.6	1.79	2.2	−4.95	7.8	−6.70
Sector	1997	1998c	1997	1998c	1997	1998e
Agriculture	4.57	−0.66	−1.57	−3.37	3	−4
Mining and quarrying	0.98	1.35	2.02	−0.52	3	0.82
Manufacturing	5.78	−7.01	6.73	−3.41	12.46	−10.17
Construction	1.86	−12.71	8.62	−13.45	9.5	−24.45
Electricity	251	2.03	7.79	4.05	10.3	6.15
Transport	8.42	0.97	2.22	0.18	8.5	1.16
Commerce/Trade	6.36	−0.24	2.14	−1.8	8.5	−2.04
Finance	5.82	1.81	5.18	2.49	11	4.35
Other services	−0.1	−0.54	−1.57	1.63	5.69	1.08
e-Estimates						

Source: National Productivity Corporation, Malaysia.

output in the construction sector is expected to decline by 1.7 per cent while employment is expected to decline by 1.5 per cent. The construction sector productivity is expected to decline by 0.2 per cent. Productivity in terms of value-added per employee for the construction sector grew at 10.6 per cent during the period 1988–1993 following rapid industrialization and modernization of the country. However, the low generation of added value content was mainly due to the high import content and the inefficient management of resources in construction projects. Competitiveness in terms of labour cost was growing at 1.2 per cent in added value per labour cost. Similarly, wages for the sector were increasing at the rate of 9.3 per cent in labour cost per employee annually. The sector had also registered a significant shift from labour-intensive to capital-intensive construction operations as indicated by the increase of 12.2 per cent in fixed assets per employee. The acquisition of the physical assets includes the heavy construction equipment such as the travelling cranes, large earth-moving equipment and modern construction machinery equipment.

To improve productivity of the construction industry, it is imperative to strategize the development of the construction industry. The strategies should include enhancing managerial skills and technical capabilities of contractors. CIDB is in the process of developing and formulating training programmes to this effect. Another strategy is the enhancement of skills of workers, which is discussed in Section 8.3.5.

Enhancement of technology is another important aspect of achieving the desired productivity. There has to be a shift from the traditionally labour-intensive to a highly mechanized industry. In attaining this, CIDB carries out promotional programmes on industrialized building, systematic embracing of prefabrication technology and modular coordination. Modular coordination is currently introduced into the housing industry as a start, which would be followed by other types of building. Modular coordination training programmes are also being undertaken. For effective enhancement of technology CIDB sets the technology, benchmarking, and monitors the technological advancement in the industry. Any new technology would be promoted to the industry. Recognizing the importance of efficiency and productivity enhancement through computerization, CIDB is earnestly pursuing its efforts in this direction. Parallel with the nation's IT agenda, CIDB is paving the path for the implementation of e-construction. This involves linking CIDB to the City Hall of Kuala Lumpur and the local authorities, which are on the GIS, thus giving an added value in both directions.

8.3.2 Construction cost

Unit construction cost

Records indicate that there has not been much increase in the construction cost of construction works. Except for the rise in cost from 1995 to 1996, there was no increase from 1996 to 1998 (refer to Table 8.10).

Average construction material price

Prices of several major building materials were adjusted downwards amid a contraction in demand during the year 1998. Analysing the price trend for the central region

Table 8.10 Price index for construction

Type of construction materials	*Base year*	*December*	*December*	*December*	*December*	*December*	*Percentage annual change*	
	1991	*1995*	*1996*	*1997*	*1998*	*1996*	*1997*	*1998*
Steel and metal sections	100	89.3	88.4	102	96	0.9	13.6	−6
Sand	100	130.6	143.9	146.9	145.1	13.3	3	−1.8
Timber	100	164.6	181	193	174.3	16.4	12	−18.7
Plumbing materials	100	100	100	101.3	100.7	0	1.3	−0.6
Bricks	100	108.8	111.6	112.6	110.4	2.8	1	−2.2
Floor and wall tiles	100	95.1	95.1	95.1	98.1	0	0	3
Plywood	100	169	204	187.9	184.3	35	−16.1	−3.6
Roofing materials	100	104.9	107.5	108.2	11.1	2.6	0.7	2.6
Ceiling materials	100	102.5	103.2	103.2	105.9	0.7	0	2.7
Aggregates	100	87.7	88.5	88.9	83.9	0.8	0.4	−5

Source: Department of Statistics Malaysia.

of Peninsular it was discovered the price of timber plunged by 18.7 per cent compared to an increase of 16.4 per cent and 12.0 per cent respectively for 1996 and 1997. In Malaysia prices for steel and metal section fell by 6 per cent while price for aggregates fell by 5 per cent. Prices of other major materials including sand, plumbing materials and bricks have fallen below 5 per cent. However, prices for floor and wall tiles, roofing and ceiling, materials have increased marginally by less than 4 per cent (refer to Table 8.11). Except for bitumen, there has not been much chance to the prices of major materials for civil works.

Table 8.11 Average construction material price

Year	*Cement in bulk* (RNI per tonne)	*Steel bars* (RM per tonne)	*200 mm aggregates* (RM per tonne)	*Concreting sand* (RM per tonne)	*Common bricks* (RM per thousand pieces)	*RMC Grade 30* (RM per in m³)
1995	198	1,119.20	20	10	200	n.a.
1996	198	1,119.20	20	12	220	n.a.
1997	198	1,119.20	21	17	220	123
1998	198	1,119.20	25	19	230	147
1999	198	1,119.20	27	21	240	160
RNIC: Ready-mix concrete						

Source: Public Works Department Malaysia.

Table 8.12 Construction industry salaries and wages

	1995	*1996*	*1997*	*1998*	*1999*
Professionals	n.a.	n.a.	n.a.	n.a.	n.a.
Technicians	n.a.	n.a.	n.a.	n.a.	n.a.
Skilled worker (daily wage)	58	59	60	62	65
Unskilled (daily)	33	33	35	35	40

Source: Public Works Department Malaysia.

Construction industry salaries and wages

Salaries and wages of construction workers have been increasing fairly. Records from the Public Works Department found that salaries and wages of skilled and unskilled workers have been increasing (refer to Table 8.12).

8.3.3 *Construction quality*

Prior to the economic recession, the pace of the construction activities was tremendous. This caused contractors to hurry for timely completion in anticipation of growth of other projects. This situation coupled with shortages of workers led to a deteriorating quality of workmanship. Realising the significance of quality construction, the CIDB has embarked on a few programmes to improve quality in the construction industry. Some of these programmes are as follows.

Standard development

The Construction Industry Development Board (CIDB), Malaysia has been appointed as a Standards writing Organisation to develop standards particularly Malaysian Standards for building and civil engineering works. At the time of writing there were 55 standards in the process of development, out of which eleven were already completed.

Quality assurance development

The Construction Industry Development Board, Malaysia has also embarked on a quality assurance programme covering the quality of construction materials such as ready-mixed concrete, precast concrete piles, timber piles and treated piles. In this respect CIDB has been authorized to certify the quality of all imported cement before acceptance into the country. To upgrade quality in the construction industry CIDB promotes the ISO 9000 certification scheme by conducting courses and embarking on the "Do-it-yourself Scheme" (DIY) for ISO 9000 certification to encourage industry players to develop their own quality assurance system for the certification towards ISO 9000 standards. CIDB would render its assistance to those who require it for the implementation of the "DIY Scheme". On quality assurance in construction practices,

CIDB prioritized the area of site investigation in cognizance of the importance of foundation stability. This program embraces the following:

- development of code of practice for site investigation;
- development of skill standards for drillers and master drillers;
- development of site investigation training modules;
- training and accreditation of site investigation personnel;
- development of a standard procurement document;
- accreditation of soil testing laboratories;
- nation wide seminar on reengineering of site investigation practice;
- regulation and quality assurance on site investigation practices.

Meanwhile focus is also given to construction safety, quality assurance in project management, guidelines on demolition works and guidelines on construction site facilities. To ensure quality assurance in construction testing, CDDB carries out registration of construction testing laboratories and is currently preparing guidelines for the accreditation of the laboratories.

Formulation of quality policy

To ensure proper direction in the implementation of the quality upgrading programmes CIDB together with the industry players has formulated a quality policy. This policy was launched by the prime minister during, the International Construction Week on 6 September 1999.

Qlassic

To promote and upgrade quality in the construction industry, CIDB has implemented quality assessment system or QLASSIC. QLASSIC is a method to measure construction quality and to benchmark the level of quality achieved in the construction project. QLASSIC enables quality of works of different construction projects to be objectively compared and provides a basis for the parties involved in the construction team to set quality attainment targets and measure their progress over time. The emphasis of QLASSIC is on the physical work by measuring dimensions against tolerances and inspecting, workmanship against a set of approved standards and specifications. One of the principles of QLASSIC is "first time inspection" which means that assessment is done prior to any rectification works to defects. The idea is to encourage contractors to do things right for the first time. QLASSIC differs from ISO 9000. ISO 9000 is an internationally recognized set of standards on quality management and assurance system that attempt to ensure a product's constantly being produced according to the required specifications. QLASSIC, on the contrary, is a quality assessment system to evaluate the end product based on approved standards and specifications.

QLASSIC was developed to achieve the following objectives:

- to evaluate the quality of workmanship in the construction project based on approved standards and specification;
- to have a standard quality assessment system for the construction industry;

- to be used for quality comparison between construction projects;
- to be used as a system to evaluate contractor's performance, on the quality of workmanship; and
- to be used for data compilation for statistical analysis in estimating, the quality of workmanship and productivity level of the construction industry.

QLASSIC sets out the standards for the various aspects of construction elements for general building works. The standards cover three main components: structural, architectural and external works. Assessments on the workmanship are done based on these standards and points are awarded if the workmanship complies with the standards. These points are then summed up to give a total quality score called the QLASSIC score for the building.

For simplicity of assessment, QLASSIC evaluations are done only for the superstructure components of a building and do not cover substructure works, such as piling and heavy foundation, and mechanical and electrical services which are heavy equipment-based and are usually called under separate, contracts or sub-contracts. Moreover, although workmanship is substantially important, the design elements are more critical in the substructure elements. The rapid construction of the substructure elements is also another factor to be considered in simplifying the assessment process.

Zero-Defect Programme

The "Zero-Defect Programme" is implemented with the support and cooperation of the Ministry of Housing and Local Government. The programme aims to provide quality houses to owners. Currently the programme is focused on low cost and medium-low housing arising from the many complaints regarding defects, poor workmanship and poor quality. The main activities under the programme are the zero-defect campaign and the implementation of a Pioneer zero-Defect Low Cost Housing Scheme. Both activities call for adherence to the Construction industry standard for low cost houses. This programme also embraces a standard procedure for issuance of a Certificate of Fitness, and approval of plans and planning, approvals for low cost houses amongst various local councils.

Various quality assurance programmes have been set up to support and facilitate Trade with a view that these programmes add value to the investors and exporters. The national standardization has been strongly orientated towards international participation and cooperation. It is made under the inception of the Law of Malaysia – the Standard Act. Assurance of compliance through the conformity assessment activities follows the international practices so as to maintain the credibility and competency of the schemes. This includes the establishment of accreditation and calibration activities. The quality assurance programmes include the publicity and awareness programme, the certification programme, conformity assessment programme, calibration programme and the competency programme. Standards and quality are one of the major ways in providing confidence in the marketplace besides giving added value that is acceptable to customers. Constant awareness through quality promotion programmes is important to create better and more competitive output. These are achieved in Malaysia through annual events like the Quality Control Conventions, Quality Conferences, November as the Quality Month, regional quality programmes, and competition and presentation of the Prime Minister Quality Awards.

The Construction Industry Development Board (CIDB) is a statutory body established under the Act of Parliament, viz. Act 520 Lembaga Pembangunan Industri Pembinaan Malaysia Act in July 1994. The objective of the board is to develop the construction industry as a major contributing sector to the national economy and capable of producing and delivering high quality construction works, with value for money and responsive to the nations needs. The Department of Standards Malaysia (DSM) appointed SIRIM Berhad as the agent to develop Malaysian Standards. Subsequently, SIRIM Berhad appointed CIDB as a Standards Writing Organisation (SWO) on 22 April 1997. The period of appointment was 5 years covering the scope of development of standards pertaining to the construction industry. CIDB became the sixth organization to be appointed as SWO. This is in line with one of the functions and roles of CIDB, which is to encourage the standardization and improvement of construction techniques and materials. CIDB's commitment in enhancing the development and the usage of standards in the construction industry has inspired DSM to appoint CIDB as a member of its standard advisory committees. Malaysian Standards are developed through consensus by committees which comprise a balanced representation of producers, users, consumers and others with relevant interests, as may be appropriate to the subject in hand. These standards were appropriate adaptations of international standards. Approval of a standard as a Malaysian Standard is governed by the Standards of Malaysia Act 1996 (Act 549). Malaysian Standards are reviewed periodically. The usage of a Malaysian Standard is voluntary and market driven except in so far as they are made mandatory by regulatory authorities by means of regulations, local by laws or any other similar ways. This directory shows as on 31 December 1997, about 650 Malaysian Standards relevant to the Construction Industry, which have been established on product specifications, codes of practice terminologies, guides and methods of test. The information provided in this catalogue of Malaysian Standards for the construction industry has been extracted from Malaysian Standards Catalogue 1998, produced by the Department of Standards Malaysia.

8.3.4 Construction safety record

Records from the Social Security Organization (SOCSO) showed that the number of accidents reported to the Society in 1996 for the construction sector is 11 per cent of the number of construction workers registered with them. In 1997, there has been a slight increase of 1 per cent in the number of accidents reported to the SOCSO (Table 8.13). Claims for permanent disability for 1995 are 10.6 per cent of accidents

Table 8.13 Number of accidents reported for the construction sector

Year	*No. of workers registered*	*No. of employers*	*No. of accidents*	*Claim for permanent disability*	*No. of fatalities*
1995	n.a.	n.a.	4,406	468	60
1996	2,245,734	25,861	5,302	476	116
1997	251,034	30,108	3,510	458	116

Source: Social Security Organization.

reported while in 1996 it reduced to 9.0 per cent. However in 1997 claims for permanent disability have increased to 13 per cent. Accidents involving facilities are small, which is recorded at 1.3 per cent, 2 per cent and 3.3 per cent respectively for 1995, 1996 and 1997.

The general perception of the construction industry in Malaysia is "Difficult, Dirty and Dangerous" (The 3-Ds Syndrome). It is an industry that contributes to high accident rates and very often, the employees are exposed to hazardous environments. Locals, especially the younger generations, are reluctant to take up a career in the construction sector for the hazards involved. As such, more than 70 per cent of the Malaysian construction industry manpower depends on foreign workers.

Construction Industry accidents are a relatively small part of the total accidents occurring in the all different sectors including the industry sector in Malaysia. However, the industry has been experiencing tremendous growth and hence there has been an increase in the construction activities which affects the safety and health of the public. Table 8.14 gives an overview of the accident frequency in the construction sector. The number of accidents in this sector that peaked in 1996 (5,302) has dropped by over 57 per cent by 2000. Despite the total construction volume peaking in 1997, there was a drop of over 20 per cent while the number of fatalities remained lower. The number of fatalities however has not been lowering consistently. In 1999, the number of fatalities was a record high at more than double the figure of 1995, when the number of accidents was nearly at the peak. The Government has shown its serious commitment to ensuring and improving safety practices in any workplace. Hence, much legislation, regulation, codes of practice, and guidelines were introduced and among these are:

- The Occupational Safety And Health Act 1994 (OSHA, 1994);
- The Safety Regulation on Building Operations and Works of Engineering Construction (1986), which was made under The Factories And Machinery Act 1967 (FM Act, 1967);
- Guidelines for Public Safety and Health at Construction Sites which was developed by Committees appointed by The Director General of Department of Occupational Safety and Health (DOSH), Malaysia;
- The Uniform Building By-laws, 1984.

Table 8.14 Yearly accidents frequency in construction industry compared to industries in general

Year	*Reported*		*PD reported*		*Fatalities reported*	
	Construction industry	*Total industry*	*Construction industry*	*Total industry*	*Construction industry*	*Total industry*
1995	4,406	114,134	468	10,356	60	952
1996	5,302	106,508	476	10,725	116	1,207
1997	3,510	86,589	459	11,184	81	1,307
1998	979	85,338	571	10,736	104	1,046
1999	4,747	92,074	610	10,613	146	909
2000	1,854	36,854	289	4,623	85	454

Source: http://www.ohseinstitute.org/new/ohse/Malaysia/statIndustrial_Accident.htm.

In spite of the above, the accident rate in the construction industry was still high. According to the statistic published in the SOCSO Annual Report, in 1997 a total of 3,510 accident cases in the construction industry were reported, out of which 459 were permanent disability cases and 81 deaths. The high accident rate is a clear indication that things are not right, and something has to be done about the safety at construction sites. It is imperative to pay attention to different aspects of construction in order to improve the safety standards at construction sites. These are:

i **Legislation and regulation** – The government should review the assisting legislation and regulation from time to time to ensure adequate coverage of safety and health in all the areas of construction industry. At the same time, the law must keep abreast of the needs of new construction methods and technologies.

ii **Roles of government agencies** – The government agencies, such as CIDB, DOSH, and Local authorities play important roles in the enforcement of laws. Regular inspection at construction sites should be carried out and stern action taken against any contractor not complying with the safety requirements.

iii **Cooperation between government agencies** – Under Malaysian law, all contractors carrying out construction works must register with CIDB, which is empowered to cancel, suspend or reinstate the registration. DOSH is the authority on safety and health. It issues a "Stop Work Order" to a construction site not complying with the safety and health regulations. A system can be worked out between these two government agencies to monitor the performance and safety standards of contractors. Contractors with high safety standards get better ratings, which would consequently be recorded and monitored by CIDB. The following actions will be taken against contractors with poor rating:

- Not considered for renewal upon expiry of registration with CIDB,
- Registration may be cancelled,
- Registration may be suspended,
- Class of registration may be downgraded.

Contractors with good records and ratings may be considered for a higher class of registration. Better records and ratings will mean better reputation and more contracts. A higher class of registration will entitle one to bigger contracts.

iv **By safety education** – Enforcing the laws alone will not overcome the safety problems. There is another aspect to be included, which is by safety education. Safety education should begin at the schools, institutions, colleges and universities. Today, there is very little emphasis on safety aspects in our education system. Besides acquiring knowledge in a particular subject, the education system should also instil safety culture from a young age.

v **Buy green card for safety** –

- Workers in the construction industry are encouraged to obtain a Green Card prior to entering a construction site and must carry the card wherever and whenever at the site. In order to get the card workers are required to undergo a one-day training in safety under the tutelage of an accredited safety officer.

- Workers who have successfully undergone the safety training will have to pay RM 50.00 to obtain the card. The card comes with a 24-hour insurance coverage in any part of the world. The card is valid for 2 years.
- Over and above the need to instil safety awareness among workers the card also is a means at registering construction industry workers.
- Although currently the Green Card is still optional in future it shall be made mandatory for all workers to have a card in order to practice.
- The card is equipped with two-dimensional bar code with safety encryption features.

8.3.5 Skill level of workforce and supervision level

One of the concerns in the construction industry is the increasing pressure on wages, especially before the economic turmoil, when the growth of the sector was tremendous, and growth of productivity was lagging. To increase productivity, the Construction Industry Development Board has embarked on programmes to uplift the skills of construction workers. These programmes are as follows:

i **Registration of workers** – The first level is to register skilled workers and site supervisors. The main aim is to keep a register of the number of available workers in the country and the skills that they possess, which would from a basis for development of policy regarding skilled workers and site supervisors. To encourage registration of workers, an insurance coverage is provided for all workers registered. The coverage is twenty hours and worldwide, covering death and personal injuries caused by construction-related activities or otherwise.

ii **Accreditation of skilled workers** – Then next level is to accredit workers for the skill that they have acquired. Accreditation for workers may be obtained by undergoing formal training or undergoing testing to demonstrate their skills. The formal training embodies theory and practical work while testing is for those workers who are already skilled and competent to demonstrate the skills possessed by them. The accreditation is focused on the existing work force in the country. However school leavers and fresh entries for training are also acceptable.

iii **Setting up of national construction academies** – To provide the infrastructure for the accreditation process, National Construction Academies are/would be set up to undertake training activities and testing beds for the accreditation of workers. Besides skills training, the academies are/would conduct training to upgrade professional and managerial level for industry players. An Academy each for Sabah and Sarawak will be set up to cater for the training needs of states and sub regions. Currently negotiation is being held with the state government on land acquisition for the development of the academy.

iv **Development of skills standard** – The basis for accreditation and skill training is the skill standard. Currently 24 skills standard and 6 for site supervisors have been prepared and 14 more are in the process of preparation. The attached appendix provides the list of standards already prepared.

8.3.6 *Research and development*

As the country moves into the industrialization era, it places greater demand on quality of construction works necessitating technological advancement towards greater mechanization. Depletion of natural resources and rising cost call for introduction of new materials and components. In subscribing to this demand, Malaysia has given emphasis to research and development in the area of science and technology. The allocation made by the government for science and technology development had increased to RM3.049 billion under the Seventh Malaysia Plan (1996–2000) from RM1.43b billion as allowed under the Sixth Malaysia Plan (refer to Table 8.15).

Construction research has been grouped under the funds allocated to the Identified Research on Priority Area (IRPA), where a committee was formed to select projects to be given priority for research funding. For the period from 1996 to year 2000, the allocation for IRPA research and development totalled RM685 million. To date, 2,839 projects have been identified by IRPA that would eventually channel RM540.5 million into the projects. In 1999 out of the 7 1.1 million granted for R&D support, 5.84 per cent of it, or RM4.2 million, was allocated to construction projects (refer to Table 8.15).

8.4 Construction outlook 2002–2010

8.4.1 *Business opportunities and major projects in pipeline*

Malaysia has been one of the attractive places for foreign investment in Asia. Statistics of the Malaysian Industrial Development Authority (MIDA) showed that the US was the largest foreign investor in the manufacturing sector in 2000 with approved projects valued at US$1,972 million. The second largest investor was Japan with US$758 million, followed by Netherlands, with US$572 million. In 2000, the three made up more than 63 per cent of total foreign manufacturing investments in Malaysia. These statistics only represent manufacturing investment; upstream oil and gas investments are not included. MIDA was established under the Federal Industrial Development Authority Act, 1965. One of the main tasks of the MIDA is to promote Foreign Direct Investment (FDI) in Malaysia and it has been successful. Under the challenge that overseas companies will increasingly look to China and Vietnam, the Malaysian government continues its efforts to attract more investment, both foreign and domestic. In 1992 alone, investment proposals totalling MYR27.4 billion (US$10.96 billion) were

Table 8.15 Development allocation for science and technology, 1991–2000 (RM m)

Programmes	*Sixth Malaysia Plan*		*Seventh Malaysia Plan*
	Allocation	*Expenditure*	
Direct R&D	629	567.1	1,000.00
Technology development for SMIS	–	–	100
Technology acquisition	–	–	100
Commercialization of technology	–	–	100
S&T infrastructure and development	807.7	629.2	1,749
Total	1,436.70	1,196.3	3,049.00

Source: Sixth and Seventh Malaysia Plan.

approved by the MIDA and 65 per cent of the sum came from abroad. Although Taiwan and Japan are the traditional major foreign investors, the US and Australia also make significant commitments. Historically, the largest European investor is Britain whose investment is valued at over £2 billion. Britain is the second largest investor overall, after Japan. In terms of human resource, if there is a lack of trained Malaysians to fill the positions, expatriates are welcome to work in the country. They are allowed to hold permanent key posts in order to safeguard the interests of foreign investors (source: http://pacificrim.bx.com/country/country_malaysia.html, and Country Profile of Malaysia; Malaysia Country Guide).

To facilitate foreign investment, it is important to have a reliable arbitral institute to resolve international trade dispute. The Regional Centre for Arbitration based at Kuala Lumpur provides reliable international arbitration in settling their commercial disputes. The Centre is a non-profit administering agency whose main function is to serve the needs of international organizations and business communities in the Asian and Australian region. Malaysia's liberal policy on equity participation allows up to 100 per cent foreign ownership in projects which export 80 per cent or more of their production. The policy has helped to increase foreign investment in the manufacturing sector. Up to 100 per cent foreign equity ownership is allowed for projects which export less than 80 per cent of their production, under the following conditions (Country Profile of Malaysia):

- foreign investment in fixed assets (excluding land) is at least RM50 million or when the project has at least 50 per cent value-added; and
- the company's products do not compete with products presently being manufactured for the domestic market.

After the Asian financial crisis of 1997/1998, the Malaysian government installed a set of tax measures to stop investors selling shares and taking their money out of the country. It imposes a 10 per cent tax, levied on investment profits. These measures have been scrapped lifting the last remaining controls over foreign investment in Malaysia's stock market. For export-oriented manufacturing investment, there used to be significant curbs to hinder manufacturing investment aimed at the domestic market. This major barrier has been removed when the government relaxed this policy in July 1998. All equity and export restrictions were dropped for new manufacturing projects approved through December 2000 up to December 2003. In the same year, foreign equity restrictions in the telecommunications, insurance, shipping and freight-forwarding sectors were also removed. With most sectors of the economy open to trade, doing business with Malaysia become more encouraging and US electronics, infrastructure and consumer products companies have been doing particularly well (Country Profile of Malaysia).

8.4.2 Areas where foreign investments were needed

The Malaysian government has invested heavily in world-class infrastructure. For the petrochemical, electronics and semiconductor-manufacturing industries Malaysia is an attractive foreign investment destination. The Multimedia Super Corridor (MSC) of the country is also an attractive place for R&D-based investment. Malaysia's MSC is designed to create an ideal environment for ICT-related production as well as to provide the backbone for an information superhighway. The network contains a high-speed link (10GB/s network) that connects the MSC to Japan, ASEAN, the US and Europe. The network is capable of supporting extensive public administration, education and business applications. The superhighway is intended to provide quality access to global information as quickly and easily as possible. At the same time, the Demonstrator Application Grand Scheme (DAGS) is expected to facilitate social and economic progress through the innovative use of ICT. Citizens are given access to the opportunities associated with the MSC and to be involved in multimedia development. The MSC is Malaysia's blueprint for developing a high-technology information-based research and manufacturing region. There is no restriction on currency exchange and expatriate employment for investments that have been approved for MSC status. Oil and gas, and manufacturing companies making semiconductors and other electronic products form the core of FDI in Malaysia. The principal foreign investors in the electronics and electrical sectors come from the US. According to Malaysian statistics of 2000, the US, with approved new manufacturing investment totalling RM7.5 billion (US$1.97 billion), was ranked first in FDI approved in Malaysia's manufacturing sector (Malaysia Country Commercial Guide 2002; Malaysia Country Guide).

Despite the economic slowdown since 1998, certain sectors of the Malaysian economy are continuing to attract FDI. The Ministry of International Trade and Investment (MITI) approved 296 manufacturing projects worth approximately US$3.12 billion during the first four months of 1998. An additional 212 applications were received – not approved until 2002 – worth about US$1 billion. The country's oil and gas sector remains the most attractive investment which was expected to absorb about $14 billion until 2003. Thirty-one per cent of the investment would come in the form of FDI. Many multinational firms, including Union Carbide and BASF, are planning to build large petrochemical plants in Malaysia. Other companies, such as Amareda Hess and Santa Fe, are also entering the exploration and production market. The oil and gas/petrochemical sector will also generate upstream and downstream activities. For example, a record RM15 billion (about US$3.75 billion) was earmarked for the related activities in 1998. With large-scale investment in the sector, the market for such items as industrial process controls is anticipated to grow in the near future. It will also stimulate importing business as much of the equipment will need to be imported from the US and elsewhere. The construction sector will benefit from the recovery in private-sector investment and fiscal expansion. The demand for housing, in particular low- and medium-cost units as well as infrastructure projects in the health and education sub-sectors and rural development will also contribute towards further growth in this sector. The construction sector will likely become an important sector for foreign investment in the near future. (PMO 2001 at http://www.smpke.jpm.my/NewSMPKE/economy/prospects.htm, and Malaysia Country Commercial Guide 2002; Malaysia Country Guide).

References

Abacci Atlas: Malaysia Atlas Entry (http://www.abacci.com/atlas/country.asp?countryID=257).

Asian Development Bank (1997, 1998, 1999, 2000) Country Operational Strategy Study Malaysia, 1997, 1999 and 2000 (http://www.adb.org/scripts/rwisapi.dll/@adb.env) and (http://www.adb.org/Documents/COSSs/mal.pdf).

Country Profile of Malaysia (http://www.bisnetworld.net/bisnet/countries/malaysia13.htm).

Drabble, John H. (2001) *An Economic History of Malaysia, c. 1800–1990: The Transition to Modern Economic Growth*, published by EH.NET (July2001) (http://www.eh.net/bookreviews/library/0371.shtml).

Encyclopaedia: Malaysia (http://www.nationmaster.com/encyclopedia/Malaysia).

Jesudason, James V. (1997) Chinese Business and Ethnic Equilibrium in Malaysia, *Development and Change* 28 (1), 119–141.

Malaysia Country Commercial Guide 2002 (http://strategis.ic.gc.ca/epic/internet/inimr-ri.nsf/en/gr-78786e.html).

Malaysia Country Guide (http://www.propertynetasia.com.my/exlink/investmentguide/malaysiacountryguide.asp).

Malaysian Property Market 2000 in Overview (http://www.propertynetasia.com.my/exlink/investmentguide/malaysiapropertyinvestment.asp).

Pacific Rim Review (http://pacificrim.bx.com/country/country_malaysia.html).

Prime Minister's Office, 2001 (Malaysia) (http://www.smpke.jpm.my/NewSMPKE/economy/prospects.htm).

Property Market Report 2000 (http://www.propertyzoom.com/pz/mktTrend/trend2000/start_2000.htm).

Seekins, Donald M. (1991) Malaysia: Chapter 4A. Government and Politics: Countries of the World, 1 January 1991.

Chapter 9

Singapore

Building and Construction Authority, Singapore and Eddie C.M. Hui

9.1 Introduction

9.1.1 Background

Singapore was founded as a British trading colony in 1819, which joined Malaysia in 1963, but withdrew 2 years later to become an independent city-nation. It subsequently became one of the world's most prosperous countries, with strong international trading links (its port arguably one of the world's busiest) and per capita GDP above that of leading nations of Western Europe.

Singapore is a small city-state with a population of 4,131,200 (2001 mid-year est.). Since 1987 the slogan "have three, or more (children) if you can afford it" has been the major theme of Singapore's new population policy. The policy, which may be described as "selectively pronatalist", represented a fundamental change in direction from the blanket "stop at two" policy which had been in effect for about two decades until the mid-1980s. The population growth rate during the time of writing was 2.8 per cent, owing perhaps to the government's drive to increase population. The proportion of males is slightly lower than that of females (0.96 male(s)/female).

Singapore is a predominantly Chinese country, but it has a sizeable Malay and Indian population. Ethnic groups comprise Chinese (76.7 per cent), Malay (13.9 per cent), Indian (7.9 per cent) and others (1.5 per cent).

Geography

Singapore is located in Southeastern Asia, between Malaysia and Indonesia, at geographic coordinates of 1 22 N Latitude and 103 48 E Longitude. Singapore has a total area of 647.5 square kilometres consisting of 637.5 square kilometres land area and 10 square kilometres water area. Singapore has a coastline of 193 kilometres with no land boundaries.

Singapore is a moderately uniform country in terms of climate, terrain and the government. The climate is generally tropical, hot, humid and rainy with two distinct monsoon seasons: north-east monsoon from December to March and south-west monsoon from June to September; inter-monsoon – frequent afternoon and early evening thunderstorms. Singapore's topography is mostly lowland, however, the gently undulating central plateau contains water catchment area and nature preserve. The elevation extremes range from the lowest at Singapore Strait (0 m) to Bukit Timah (166 m). Singapore's staple natural resources include fish, deep-water ports including arable land (2 per cent of total lands), permanent crops (6 per cent), forests and woodland (5 per cent), and others (87 per cent). There are no permanent pastures in the country.

9.1.2 *Political structure*

Singapore has a parliamentary type republican government, and a single party has ruled since its separation from Malaysia. There are no administrative zones due to its small size. And the central state is responsible for government and legal machinery and regulates growth.

9.1.3 *Development process: through the years*

Immediately after separating from Malaysia on 9 August 1965, Singapore was faced with uncertainties of survival. Hence, the government adopted a main political ideology, first, by creating a society that is flexible enough to accept new values, attitudes and perspectives, secondly, by maximum mobilisation of a tightly organised society and finally, by wider nation building. Although the separation had come with promises of cooperation in economic areas, Singapore's economy in its early years of freedom was hit due to the demise of a common market with Malaysia as Malaysia sought to protect its own economy by bypassing Singapore's ports and rising tariff walls. During the period between 1959 and 1970, the lion's share of the nation's per capita income fell to a small group of political, executive and commercial elites, and the real wages of workers only increased by 5 per cent. The breakdown of the common market forced the island to develop independently of its main land. The key was to go global. To do so, however, a conducive industrial climate had to be created, industrial peace restored and wages held down to attract business investment. History was made when the main party, the People's Action Party (PAP), swept up all 58 parliamentary seats and was returned as a single party government. With political stability at home, the Singapore government began revamping the regulatory and institutional framework, and subsequently redefining the country's role as an export-orientated economy. Foreign investors were attracted by the improved labour situation and by incentives such as tax relief for up to 5 years and unrestricted repatriation of profits and capital in certain government-favoured industries. US firms flocked to invest in Singapore, accounting for 46 per cent of new foreign capital invested in 1972. Companies from Western Europe, Japan, Hong Kong, Taiwan, Malaysia and Australia also invested capital, and by 1972 one quarter of Singapore's manufacturing firms were either foreign-owned or joint-venture companies.

Singapore's industrialisation came amidst the booming conditions as a result of the escalation of the Vietnam War and the exploration of oil in Indonesia. Singapore capitalised on this by offering a natural base for dozens of exploration, engineering, diving and other support companies for the petroleum industry in nearby Indonesia, as well as being the oil-storage centre for the region. By the mid-1970s, Singapore was the third largest oil-refining centre in the world. Singapore became a supply centre for American forces and provided shipping facilities. The country sailed through the oil crisis in the 1970s unscathed due to pragmatic strategies that sought to withdraw from excessive dependence on foreign capital and to generate contributions from

Table 9.1 GDP growth in selected years

Year	*Overall GDP* (S$m)	*% growth*
1971	15,622.5	12.5
1981	36,039.1	9.6
1991	71,151.3	7.1

employers and employees in lieu of property and infrastructure services. Thus, by the mid-1970s, the PAP's pragmatic strategy had succeeded in creating a suitable economic climate of necessary infrastructure services and financial facilities.

Singapore's pragmatic economic ideology and far-sighted approach was reflected in its GDP growth rate. In much of the last three decades until 2002, Singapore saw its GDP grow at a double-digit rate. The GDP of the country itself doubled within less than 10 years during the 1970s and tripled by the mid-1980s (refer Table 9.1).

9.2 The national economy

9.2.1 Macroeconomic review and outlook

The Singapore economy achieved a spectacular double-digit growth during most of the late 1980s and early 1990s. However, Singapore, like many other East Asian countries, was hit by the regional crisis in 1997. Despite its strong financial and economic fundamentals, the Singapore economy was adversely affected by the crisis. In the years preceding the financial crisis, the country's GDP growth rate was phenomenal at 12.8 per cent in 1993, 11.4 per cent in 1994, and comfortable at 8.1 per cent in 1995 and 7.5 per cent in 1996, which continued to look strong even in 1997 driven by the previous year's strong economic drive. With most of the major industries showing negative growth, Singapore's economic growth dipped sharply from 8.5 per cent in 1997 to a dismal −0.1 per cent in 1998.

Singapore is a highly developed and successful free-market economy, a remarkably open and corruption-free business environment, with stable prices, and the fifth highest per capita GDP in the world. Exports, particularly in electronics and chemicals, and services are the main drivers of the economy. Mainly because of robust exports, especially electronic goods, the economy grew 10.3 per cent in 2000 (Table 9.2). However, Singapore experienced a negative growth of 2.0 per cent in 2001 largely because of weaker global demand, especially from the US (Table 9.3). The government has continued to promote high levels of savings and investments through a mandatory savings scheme and spends heavily on education and technology. It also owns government-linked companies (GLCs) – particularly in manufacturing – that operate as commercial entities. In the face of increasing globalisation, Singapore has been positioning itself as the region's financial and high-tech hub. With a GDP of just S$159,888 m in 2000, the Singapore economy was small by world standards. However, 2000 per capita GNP stands at S$39,796 making Singapore one of the richest countries in the world.

Table 9.2 Main macroeconomic indicators

Year	*1995*	*1996*	*1997*	*1998*	*1999*	*2000*	*2001 (revised forecast)*
GDP at real prices (Base year = 1990) (S$ m)	102,859	110,730	120,191	120,081	128,405	141,572	138,683
GDP at current market prices	117,768	128,201	140,279	137,618	140,070	159,888	153,455
GDP growth (%)	8.0	7.7	8.5	−0.1	6.9	10.3	−2.0
Manufacturing sector (Base year = 1990) (S$ m)	26,406	27,182	28,399	28,239	32,089	36,993	32,720
% growth	10	2.9	4.5	−0.6	13.6	15.3	−11.5
Financial and business services sector (Base year = 1990) (S$ m)	24,536	26,591	30,176	29,363	30,653	32,480	33,307
% growth	5.3	8.4	13.5	−2.7	4.4	6.0	2.5
Construction sector (Base year = 1990) (S$ m)	7,459	9,193	10,688	10,992	10,027	9,854	9,646
% growth	9.6	23.2	16.3	2.8	−8.8	−1.7	−2.1
Primary sector (Agriculture, fishing and quarrying) (Base year = 1990) (S$ m)	203	210	210	196	194	183	174
% growth	3.1	3.5	0.1	−6.9	−1.1	−5.7	−4.9
Population – Singapore residents[1] ('000)	3,014.6	3,067.8	3,121.1	3,174.8	3,221.9	3,263.2	3,319.1
Population growth rate (%)	1.8	1.8	1.7	1.7	1.5	1.3	1.7
Total labour force ('000)	1,749.3	1,801.9	1,876.0	1,931.8	1,976.0	2,192.3	2,119.7
Labour force growth rate (%)	3.3	3	4.1	3.0	2.3	10.9	3.3
Unemployment rate (%) – Seasonally adjusted	2.0	2.2	1.8	2.4	3.5	3.3	2.6
Changes in consumer price index (Base period = Nov. 1997 to Oct. 1998, % change over previous year)	1.7	1.4	2.0	−0.3	0.0	1.3	1.0
Changes in GDP deflator (Base year = 1990, % change over previous year)	2.7	1.1	0.8	−1.8	−4.8	3.5	−1.9

Table 9.2 (Continued)

Year	*1995*	*1996*	*1997*	*1998*	*1999*	*2000*	*2001 (revised forecast)*
Short-term interest rate (%) (3-month SIBOR rate)	5.53	5.63	5.81	5.13	6.06	6.39	1.88
Long-term interest rate (%) (12-month fixed deposits bank rates)	4.01	3.99	4.41	2.51	2.46	2.42	1.63
Annual average exchange rate with US$ (Singapore dollar per US dollar)	1.4174	1.4101	1.4848	1.6736	1.6949	1.7239	1.7917

Source: Singapore Department of Statistics, Yearbook of Statistics.

Notes

1 Singapore residents refer to citizens and those who have been granted permanent residence in Singapore.

2 The industries are classified according to the Singapore Standard Industrial Classification 1996.

Table 9.3 GDP growth of each sector in 2001

	(%)
Owner-occupied dwelling	4.4
Manufacturing	−11.5
Construction	−2.1
Utilities	4.0
Other goods and services	−4.9
Wholesale and retail	−2.8
Hotel and restaurants	−2.6
Transport and communication	2.7
Financial services	2.2
Business services	2.8
Other services industries	6.2

Source: http://www.contactsingapore.org.sg/work/economy.htm.

9.2.2 *Economy in 2001*

The Singapore economy contracted by 2 per cent in 2001, after achieving 10 per cent growth in 2000. This sharp reversal in growth was primarily due to the slump in external demand. Weak household consumption and business investments also contributed to the contraction. All key economic sectors saw lower growth compared to a year earlier. The economy saw an improvement in the fourth quarter, with a growth of 5.6 per cent on an annualised quarter-on-quarter basis, after three consecutive quarters of decline.

All major sectors performed worse than the year 2000. In particular, the manufacturing sector fell by 12 per cent, a sharp reversal from the 15 per cent expansion in 2000. The construction, wholesale and retail trade, and hotels and restaurants sectors also declined in 2001. The manufacturing sector contracted by 12 per cent in 2001, after a 15 per cent expansion in 2000. The sharp contraction was a result of the adverse impact of lower demand for electronic products and the weak US economy. The construction sector contracted for the third consecutive year in 2001, by 2.1 per cent due to weak property market sentiments. As a result of the uncertain economic climate and weaker consumer sentiments both domestically and regionally, the wholesale and retail trade sector declined by 2.8 per cent, a sharp reversal from the 15 per cent growth in 2000. The hotel and restaurant sector also shrank by 2.6 per cent in 2001 as compared with the healthy growth of 8.9 per cent in 2000. The contraction was on the back of the dip in visitor arrivals and weak economic sentiments. Although the transport and communications sector continued to grow, growth was much lower at 2.7 per cent compared to the 8.5 per cent of the year before. The financial services sector moderated from a growth of 4.6 per cent in 2000 to 2.2 per cent in 2001. This was attributed to weaker performance in the Asian Dollar Market, insurance industry and investment advisory business. The business services sector rose by a moderate 2.8 per cent, down from a 7.3 per cent growth in 2000. The real estate services segment, which accounts for about half of the sector, reported slower growth on the back of weak economic conditions.

After soaring by 14 per cent in 2000, total demand dropped by 7.7 per cent, the first decline in 3 years. External demand was the major drag on growth, hit by weaker global economic conditions and the persistent electronics slump. Merchandise exports were dragged down by electronic products while services exports, especially transport-related ones, were adversely affected. Domestic demand also dipped with the weakening of private consumption and business investment. Inventory corrections were substantial. Total consumption expenditure grew by 1.8 per cent, a significant moderation from the 11 per cent in 2000.

Total fixed asset investment fell by 4.6 per cent in 2001, reversing the 6.3 per cent increase in the previous year. Impinged upon by the negative business sentiments, private investment in machinery and equipment, in particular, scientific and medical instruments, office machines and electric power machines, slumped by 18 per cent. Investments in residential and non-residential buildings fell significantly in tandem with the weak property market.

The labour market was affected by the deterioration in external environment in 2001. A rapid contraction in economic activity saw retrenchments hitting 25,600 in 2001, more than double the 11,624 seen in 2000. Employment in goods-producing industries experienced a sharp decline, but this was mitigated by employment growth in the services sectors. At the end of 2001, the overall unemployment rate rose to a 15-year high at 4.7 per cent. For the whole of 2001, the unemployment rate averaged 3.3 per cent, compared to 3.1 per cent in 2000.

As the outlook of the global economy was expected to pick up in 2001, the Ministry of Trade and Industry forecasted Singapore's 2002 growth to be between 1 and 3 per cent.

9.2.3 Role of construction/property market in national economy

Construction investment (or market) volume

Total contracts awarded in 2001 reached $13.1 billion, a substantial drop of 34.9 per cent over the preceding year. The fall was also accompanied by a 14.4 per cent contraction in the total number of construction projects awarded.

Construction demand for both public and private sectors declined due to cutbacks in HDB new flat construction, the softening of the private property market and weaker private industrial developments.

Total construction output in 2001 reached $15.2 billion, registering a drop of 4.5 per cent over 2000. In 2001, construction output was supported mainly by infrastructure projects such as the reclamation project at Jurong Island and Tuas View Extension, Mass Rapid Transit (MRT) extension to Changi Airport, North East Line, as well as Light Rapid Transit (LRT) projects at Sengkang and Punggol. Figure 9.1 shows the variation between contracts awarded and progress payments during the period 1984–2001.

Public sector

A total of $7 billion worth of public sector contracts were awarded in 2001. While industrial and institutional developments registered higher demand, the demand for residential, commercial and infrastructure developments contracted.

Residential construction demand continued its downward slide for the third straight year to $931 million in 2001. The decline of 42.2 per cent was primarily due to the reduction in the construction of new HDB flats in the last 2 years in response to the significant overhang of unsold flats and the weak property market. The Build-To-Order (BTO) scheme, under which flats will be built only if demand for flats in a new development site exceeds 70 per cent, was introduced by the HDB in 2001 to better balance the demand and supply of public housing.

Commercial construction demand fell by 53 per cent over the year to about $75 million. The drop was mainly due to the significantly lower demand for the redevelopment of government office buildings, as most of them, such as the Ministry of Education, Ministry of Foreign Affairs, Ministry of Home Affairs and HDB headquarters, had either been completed recently or their contracts were already awarded in previous years.

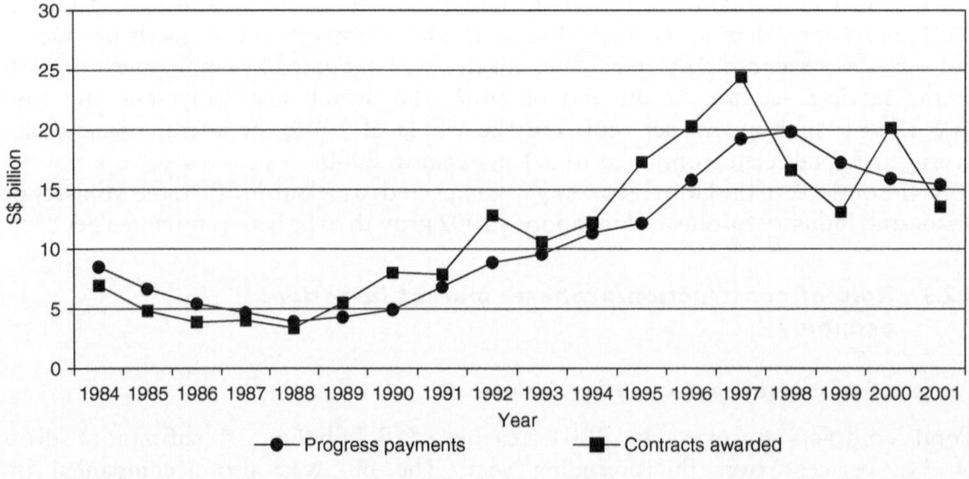

Figure 9.1 Contracts awarded and progress payments.

However, 2001 saw a pick-up of industrial construction demand by more than three times the level in 2000 to about $0.5 billion. This was fuelled by the award of JTC's second phase of Woodlands Spectrum, the development of Cargo Agents Building "B" for the Civil Aviation Authority of Singapore (CAAS) as well as the development of the Changi Water Reclamation Plant by the Public Utilities Board (PUB).

Institutional and other building construction demand also recorded a robust expansion of 35.3 per cent from $1.9 billion in 2000 to $2.6 billion in 2001, the highest in the past 20 years, until 2002. Major projects during 2002 included the construction of the national library building along Victoria Street, and the development of the mega-prison complex on 48 hectares of land to centralise all prison establishments.

Civil engineering construction demand dropped by 64.2 per cent from $8.3 billion in 2000 to $3 billion in 2001. Of this amount, 65.2 per cent was contributed by the construction of rail systems and road-related projects. In particular, in 2001, the Land Transport Authority (LTA) invested about $1 billion to develop the Marina MRT Line in its bid to provide better public transport services. Furthermore, LTA also invested heavily in the construction of the Kallang and Paya Lebar Expressway (from East Coast Parkway to Pan-Island Expressway) to ease heavy traffic flows.

Private sector

Total private-sector construction demand fell by 28 per cent to $5.8 billion in 2001. Construction demand for all development types shrank, ranging from 14.9 per cent for residential construction to 53 per cent for civil engineering construction. In line with lower demand, the total number of private-sector projects awarded dropped by 17.7 per cent over the year 2001.

Residential developments suffered a drop to $2.7 billion, on account of the conservative approaches amongst property developers amid economic slowdown. Condominium projects, which contributed 59 per cent to the overall private-sector residential construction demand, continued to be the mainstay of private residential property developments in 2001. Since the introduction of the Deferred Payment Scheme by developers which allows property buyers to defer up to half of the 20 per cent cash down payment, the private residential market had shown signs of picking up towards the end of 2001.

Commercial construction demand declined further to $0.4 billion in 2001, the lowest since 1989 as island-wide office occupancy levels in Singapore fell because of the weak market demand and an increase in premature lease terminations. Most of the commercial construction demand in 2001 stemmed from the construction of the Tekka Mall at Sungei Road, addition and alteration works of commercial developments such as the United Square Building at Thomson Road and the 16-storey office and cable car operation building at Telok Blangah Road.

About $2.1 billion worth of industrial projects were awarded in 2001. The strong industrial building construction demand of $3.1 billion in 2000 could not be sustained amid the deteriorating manufacturing sector caused by the continued slump in the global electronics industry. Nonetheless, private-sector construction demand was supported by the development of a Phase-2 Design and Build (D&B) project of the Senoko Power Station Stage-1, procurement and construction of a pharmaceutical plant at Tuas South, as well as erection of a wafer fabrication plant at Pasir Ris.

Institutional and other building construction demand registered contraction of 39.6 per cent, from $0.6 billion in 2000 to $0.4 billion in 2001. Projects awarded in 2001 included the development of a four-storey meditation hall and temple at Bright Hill Road and a church development at Woodlands Circle.

Civil engineering construction demand in 2001 recorded a drop of 53 per cent from the level in 2000 to $0.2 billion. Apart from the development of a transmission cable tunnel from Senoko Power Station to Gambas Avenue, most of the other infrastructure projects were small in scale and cost less than $10 million. Tables 9.4 and 9.5 show contracts awarded and progress payments by sector and type of work.

9.2.4 *Construction outlook in 2002*

In 2002, in view of the uncertain economic conditions foreseen, it was unlikely to see a rebound in construction demand. Total construction demand was expected to reach

Table 9.4 Contracts awarded by sector and type of work (S$ m)

	1994	*1995*	*1996*	*1997*	*1998*	*1999*	*2000*	*2001**
Both sectors	**12,227.01**	**17,287.37**	**20,290.51**	**24,399.60**	**16,636.69**	**13,096.07**	**20,162.79**	**13,567.87**
Building work	11,234.53	14,135.44	16,257.11	19,863.26	12,926.34	10,778.40	11,395.41	10,160.20
Residential	6,124.11	7,974.14	7,497.33	11,309.37	5,318.94	4,938.70	4,749.55	3,854.23
Commercial	1,637.35	1,460.17	2,320.35	2,222.05	2,482.56	882.24	826.56	492.93
Industrial	1,945.02	2,700.01	4,094.31	3,967.94	2,979.98	2,650.23	3,278.57	2,664.81
Institutional and others	1,528.05	2,001.12	2,345.12	2,363.90	2,144.86	2,307.23	2,540.73	3,148.23
Civil engineering work	992.48	3,151.93	4,033.40	4,536.34	3,710.35	2,317.67	8,767.38	3,407.67
Private sector	**6,648.46**	**8,251.28**	**9,653.51**	**12,669.79**	**5,973.01**	**6,651.50**	**8,054.42**	**6,158.99**
Building work	6,509.42	7,892.74	9,384.89	12,540.55	5,722.78	5,912.30	7,549.01	5,910.68
Residential	3,126.61	3,933.75	3,686.28	6,850.04	2,024.01	2,114.74	3,139.57	2,909.04
Commercial	1,468.07	1,356.54	2,001.99	1,433.63	1,011.06	762.83	667.78	416.77
Industrial	1,603.63	2,277.28	3,338.93	3,645.85	2,370.16	2,458.06	3,136.58	2,152.12
Institutional and others	311.11	325.17	357.69	611.03	317.55	576.67	605.08	432.75
Civil engineering work	139.04	358.54	268.62	129.24	250.23	739.20	505.41	248.31
Public sector	**5,578.55**	**9,036.09**	**10,637.00**	**11,729.81**	**10,663.68**	**6,444.57**	**12,108.37**	**7,408.88**
Building work	4,725.11	6,242.70	6,872.22	7,322.71	7,203.56	4,866.10	3,846.40	4,249.52
Residential	2,997.50	4,040.39	3,811.05	4,459.33	3,294.93	2,823.96	1,609.98	945.19
Commercial	169.28	103.63	318.36	788.42	1,471.50	119.41	158.78	76.16
Industrial	341.39	422.73	755.38	322.09	609.82	192.17	141.99	512.69
Institutional and others	1,216.94	1,675.95	1,987.43	1,752.87	1,827.31	1,730.56	1,935.65	2,715.48
Civil engineering work	853.44	2,793.39	3,764.78	4,407.10	3,460.12	1,578.47	8,261.97	3,159.36

Source: BCA as on 30 April 2002.

Note
* Forecast.

Table 9.5 Progress payments certified by sector and type of work (S$ m)

	1994	*1995*	*1996*	*1997*	*1998*	*1999*	*2000*	*2001**
Both sectors	**11,342.00**	**12,145.64**	**15,771.67**	**19,235.60**	**19,828.16**	**17,249.28**	**15,925.96**	**15,410.04**
Building work	9,469.24	10,343.86	13,596.73	16,294.00	16,510.35	13,848.81	12,438.11	11,070.65
Residential	4,530.41	4,932.18	6,989.82	8,717.91	8,760.33	6,935.29	5,099.96	4,050.28
Commercial	1,977.32	1,953.60	1,774.74	1,912.14	1,977.04	1,469.57	1,570.05	1,403.80
Industrial	1,317.82	2,079.43	3,296.79	3,392.28	3,181.25	2,647.81	3,093.21	2,676.04
Institutional and others	1,643.69	1,378.65	1,535.38	2,271.67	2,591.73	2,796.14	2,674.89	2,940.53
Civil engineering work	1,872.76	1,801.78	2,174.94	2,941.60	3,317.81	3,400.47	3,487.85	4,339.39
Private sector	**5,462.22**	**6,326.63**	**8,542.19**	**9,733.12**	**9,656.02**	**7,295.25**	**7,482.97**	**6,799.05**
Building work	5,313.22	6,170.33	8,208.98	9,350.31	9,150.19	6,854.69	7,000.84	6,226.85
Residential	1,851.82	2,446.26	3,590.25	4,452.67	4,209.51	3,251.07	2,680.53	2,385.60
Commercial	1,838.50	1,737.34	1,565.41	1,562.03	1,674.16	1,053.25	990.86	816.56
Industrial	1,118.37	1,754.90	2,678.07	2,931.34	2,791.42	2,174.27	2,890.24	2,472.70
Institutional and others	504.53	231.83	375.25	404.27	475.10	376.10	439.21	551.99
Civil engineering work	149.00	156.30	333.21	382.81	505.83	440.56	482.13	572.2
Public sector	**5,879.78**	**5,819.01**	**7,229.48**	**9,502.48**	**10,172.14**	**9,954.03**	**8,442.99**	**8,610.99**
Building work	4,156.02	4,173.53	5,387.75	6,943.69	7,360.16	6,994.12	5,437.27	4,843.80
Residential	2,678.59	2,485.92	3,399.57	4,265.24	4,550.82	3,684.22	2,419.43	1,664.68
Commercial	138.82	216.26	209.33	350.11	302.88	416.32	579.19	587.24
Industrial	199.45	324.53	618.72	460.94	389.83	473.54	202.97	203.34
Institutional and others	1,139.16	1,146.82	1,160.13	1,867.40	2,116.63	2,420.04	2,235.68	2,388.54
Civil engineering work	1,723.76	1,645.48	1,841.73	2,558.79	2,811.98	2,959.91	3,005.72	3,767.19

Source: BCA as on 6 March 2002.

Note
* Forecast.

$12.9 billion, marginally below the demand in 2001 (Figure 9.2). Excluding reclamation projects, the total value of contracts was also expected to fall to $11.7 billion, down by 10.2 per cent over the corresponding level in 2001, mainly due to weak private-sector demand. Public sector was expected to pump prime the economy through continued strong spending on institutional building and civil engineering projects.

Private-sector construction demand was expected to remain weak if the economy continues to be in the doldrums. Building construction demand was anticipated to fall by one-quarter in the total value on account of sluggish demand for property and industrial developments. Civil engineering construction demand, on the other hand, was likely to be higher with strong support from utility-related projects.

Public sector

The public sector was expected to award $8.4 billion worth of projects in 2002, registering an expansion of 19 per cent over 2001. Hence, the share of the public sector

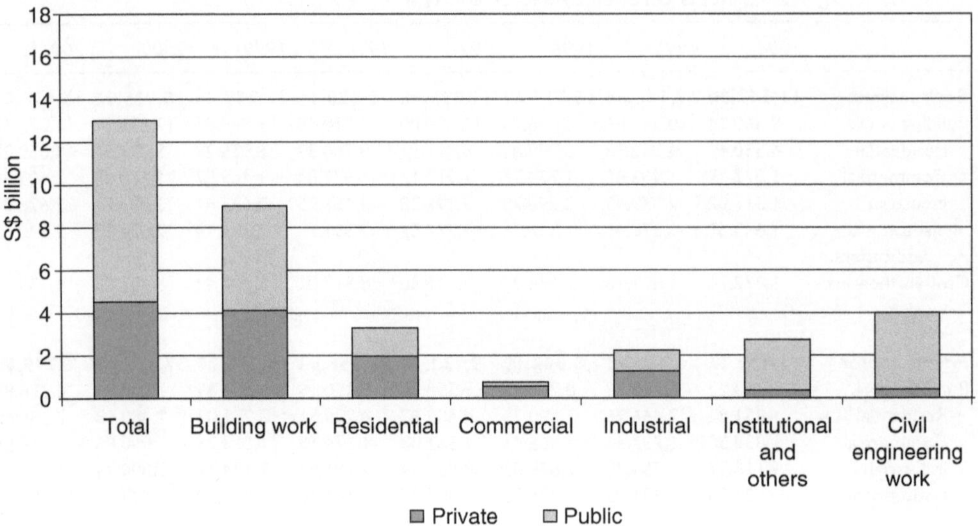

Figure 9.2 Construction demand 2002.

Source: Building and Construction Authority as on 30 April 2002.

construction demand was likely to expand from 54.8 per cent in 2001 to 64.9 per cent in 2002. Of these $8.4 billion, about 60 per cent was expected to be spent on building developments and the remaining was expected to be spent on infrastructure projects.

After experiencing a sharp downward slide in 2001 since the peak period in 1997, public residential developments are likely to reach only $1.2 billion[1] in 2002. With the inclusion of the Lift Upgrading Programme (LUP), spending on upgrading works was estimated to be higher in 2002. However, the news released in January 2002 revealed that HDB would stop building new flats under the Registration for Flats System (RFS) in order to clear its stock of some 17,500 unsold flats in the new towns. In view of this development, the demand for public residential construction was expected to be even lower if the economy continues to be weak.

Commercial construction demand was expected to enjoy about a two-fold expansion over the year 2002 to $0.2 billion, compared to a drop of 53 per cent in the preceding year. Similarly, the demand for industrial construction was likely to double to $1 billion with the award of various Changi Water Reclamation Plant projects. It was expected that further support would come from the development of a multi-user Biomedical Research and Development Complex at North Buona Vista, and HDB's Light Industrial Factory developments at various industrial estates.

Institutional and other building construction demand was expected to remain strong and reach $2.5 billion in 2002, a level comparable to that of 2001. As for 2001, Ministry of Education (MOE) was expected to invest in rebuilding and improving existing schools to provide teachers and students with more up-to-date learning facilities. The substructure works of the Singapore Management University (SMU) were also slated to commence in developing a city campus at Bras Basah Road. Other then upcoming major institutional projects in 2002 included the development of the Law Enforcement Academy building at Choa Chu Kang; construction of a new columbarium and a crematorium complex at Mandai Road; and the erection of the Basic Training School at Jalan Bahar.

Depending on the pace of the economic recovery, the institutional building construction demand was expected to reach a level of more than $3 billion if the construction of the Changi Airport's new Terminal 3 was awarded as scheduled. The construction of the new terminal was expected to be delayed in view of the 2002 economic downturn, which would affect the volume of passengers.

Civil engineering construction demand was likely to reach $3.5 billion in 2002. Highlights of infrastructure projects in 2002 included the development of the mass rapid transit (MRT) Outer Circle Stage 2; Paya Lebar Expressway; development of the Changi Effluent Outfall and the construction of Equalisation Basin at Bedok Water Reclamation Plant.

Private sector

Private sector construction demand was expected to reach only $4.5 billion in 2002, down by 22 per cent from $5.8 billion in 2001. With the exception of commercial and infrastructure construction demand, which would experience a growth of 45 and 68.2 per cent respectively, demand for all other development types shrinks.

Residential construction demand was expected to shrink from $2.7 billion to about $2 billion in 2002, on the back of the existed weak economic situation ensuing from the downturn in the global electronics cycle and the terrorist strikes in the United States. Condominium projects would have continued to dominate the private residential developments, with a share of 64 per cent. Major residential buildings to be awarded in 2002 include: The Blissville at Bukit Timah Road; Blue Horizon at West Coast Crescent; Hillview Regency at Hillview Avenue by Far East Organization Centre; a condominium development at Beatty/Sturdee Road by Allgreen Properties; and another condominium development at Draycott Drive by Wing Tai Property Management.

Commercial construction demand was anticipated to be at $0.6 billion. The main support was expected to come from the construction of a 26-storey office tower at Keppel Road by PSA Corporation and the development of an entertainment/shopping/office complex at Handy Road by Cathay Properties. Industrial building developments were forecasted to be close to $1.3 billion, down by 41.7 per cent from the $2.1 billion in 2001. The decrease was mainly caused by the lack of demand as the manufacturing sector had been badly hit by the global downturn.

Institutional construction demand was likely to remain comparable to the 2001 level at $0.4 billion. Apart from the redevelopment of the bus-depot at Bedok North by Waterbank Properties, 2002 was expected to see various developments of religious buildings.

Civil engineering construction demand was forecasted to expand by 68.2 per cent to $0.4 billion in 2002. This was mainly due to the development of utility tunnelling projects from Ayer Rajah to Paya Lebar and Seraya to Labrador by the PowerGrid.

As for 2002, construction output was forecasted to fall slightly to $15 billion due to the weak demand in 2001 and 2002. Given the weak performance of the other economic sectors, the industry's share of the national GDP for 2002 was expected to be somewhere between 6 and 7 per cent.

9.2.5 *Property prices in Singapore*

Property prices have fallen since the financial crisis of 1997, as a result of the slowdown of the economy. In 2000, the commercial property market improved, but it then slackened in 2001 due to weak market sentiments. Private residential and commercial property prices decreased by 10 and 8 per cent respectively. As for industrial properties, prices inched up only 1 per cent during the same year.

Residential property

According to DTZ Debenham Tie Leung, a property consultancy firm, the residential market softened further in 2001 as the economy slipped into recession. Yet, there were pockets of strengths. Demand for executive condominiums was strong in part due to the $30,000 grant for first-time home-buyers. Projects under construction, which offered deferred payment schemes, where the initial payment is only 10–20 per cent and the balance is paid upon project completion, were also enthusiastically received.

The well-being of the market was underpinned by local demand among households wishing to upgrade to private properties for owner-occupation. By contrast, demand from investors and foreigners was limited. The leasing market weakened considerably as expatriates were relocated out of Singapore and housing allowances were reduced. This was exacerbated to some extent by the corporate restructuring of some MNCs. Given the supply-demand imbalance, the residential market was expected to remain patchy. However, the suspension of Government Land Sales in 2002 would help mitigate the imbalance, leading to some market stability.

Commercial property

The office market had a relatively good start in 2001 with 2 million square feet scheduled for completion. However, the US technology crash happened in early 2001, followed by the rapidly deteriorating international political and economic environment. All this bad news filtered through rapidly to the office market. Demand for office space contracted by an estimated 914,000 square feet as leases were preterminated. Occupancy dropped to 90 per cent, compared to 95 per cent in 2001. Average rents also declined by 12 per cent, compared to 15 per cent in 2001. In a bid to lower business costs, the government relaxed planning regulations to allow home offices in some city areas. As the small-office-home-office (SOHO) concept was gradually introduced, commercial developments were exploring the viability of incorporating home offices. According to DTZ Debenham Tie Leung, the office market was expected to weaken further in 2002. New supply beyond 2002 was expected to be small, at only 1.9 million square feet per annum. And, it was expected that 2003 will see only 444,000 square feet due for completion. Such reductions in supply would conceivably help ameliorate the supply-demand imbalance in 2002.

Industrial property

Contrary to the slackened demand for conventional industrial space, demand for high-tech industrial space was strong. Overall, however, supply continued to outstrip demand, and thus the government suspended the sale of some 7 hectare of industrial land. Average industrial rents declined by between 5 and 11 per cent in 2001. The industrial market was expected to remain soft as the manufacturing sector takes its toll. Singapore's focus on the electronics industry suggests that demand for industrial properties was likely to recover only with an upturn in the global electronics market. The government was reviewing Singapore's reliance on the electronics industry, promoting other clusters such as biotechnology and life sciences. To facilitate this, JTC Corporation was developing One-north, a self-contained 200 hectare life sciences and IT hub comprising housing, research labs, business, financial and retail uses in Buona Vista near the National University of Singapore.

Table 9.6 shows the historic development of property market in Singapore.

Table 9.6 Historic development of property market in Singapore

	1996	*1997*	*1998*	*1999*	*2000*	*2001*	*Q1 2002*
Private residential							
Existing stock (thousand sq. m)	149.1	161.3	173.6	183.6	193.0	198	200
Occupancy rate (as at 4Q) (%)	93.8	91	91.0	92.0	93.1	91.8	92.4
Net change in existing stock	2,291	2,827	3,382	2,765	1,895	952	n.a.
Net change in occupied stock	1,578	2,539	4,263	1,712	3,726	1,711	n.a.
Price index[1] as at 4Q (4Q98 = 100)	173.1	151.6	100	134.2	132.8	117.2	115.9
Rental index as at 4Q (4Q98 = 100)	146.1	138.3	100	98.2	98.6	95.9	94.7
Office							
Existing stock as at 4Q (thousand sq. m)	5,107	5,481	5,733	5,879	6,137	6,250	6,366
Occupancy rate as at 4Q (%)	91.5	88.8	85.4	85.9	88.7	88.7	86.4
Net change in existing stock (thousand sq. m)	73	67	45	57	45	9	n.a.
Net change in occupied stock (thousand sq. m)	42	56	–5	83	71	64	n.a.
Price index[1] as at 4Q (4Q98 = 100)	160.3	143.2	100	98.7	125.9	98.8	93.6
Rental index[2] as at 4Q (4Q98 = 100)	128.2	120.7	100	90.7	102.1	94.1	91.9
Factory							
Existing stock (thousand sq. m) as at 4Q	19,482	21,113	22,575	23,188	24,021	24,638	24,872

Source: Urban Redevelopment Authority.

Notes

1 The price index base year has been re-based to Quarter 4 of 1998. HDB flats, shops and other public sector buildings are excluded in the compilation of the property price index.

2 Rental index of flatted factories/warehouse space.

9.2.6 Changes of economic policies

In October 2001, the then Prime Minister Goh Chok Tong announced the establishment of the Economic Review Committee (ERC) to fundamentally review our development strategy and formulate a blueprint to restructure the economy.

Singapore faces major new challenges. The world economy is undergoing dramatic changes driven by rapid technology advances and globalisation. In the aftermath of the September 11 events, the international situation has changed even more radically, affecting trade and investment. China's growing importance as a major competitor increases pressure to shift out of those industries in which Singapore is assumed to be no longer competitive, and to nurture and grow new and promising sectors. Uncertainties in the region call for new strategies to develop Singapore as a hub, for the region and the broader Asian economy. In short, Singapore wishes to upgrade, transform and revitalise its economy.

The ERC has the following terms of reference:

a to review government policies with economic implications, including taxation, the CPF system, land allocation, and the framework for wage competitiveness;
b to recommend ways to strengthen the spirit of entrepreneurship and innovation among Singaporeans, and foster the growth and internationalisation of Singapore-based companies;
c to recommend measures to enhance our human capital, to support all sectors of the economy;
d to recommend ways to upgrade the manufacturing sector and move into higher value-added, knowledge intensive activities including R&D and biomedical science, to build on the successful record of MNC investments in Singapore;
e to recommend strategies to promote service industries where we enjoy or can develop a competitive edge, including information and communications technology (ICT), education, health care, tourism, financial services, logistics and legal services;
f to recommend measures to upgrade the operations and raise the productivity of businesses that serve mainly the domestic market, especially the smaller ones; and
g to assess the impact of economic restructuring on Singaporeans, and recommend measures to prepare workers to adapt to the changes, particularly through education, skills upgrading and training, to ensure continued employability.

Seven sub-committees were set up, one for each term of reference, to review the specific issues in greater detail and seek views from the public. The sub-committees were chaired by ERC members and involve further representation from both private and public sectors, so as to achieve wider participation in the Committee's work.

9.3 The construction sector

9.3.1 Productivity

Labour productivity

Growth in labour productivity in the first half of 2001 resulted in a 0.2 per cent gain for the whole year (Table 9.7). This improvement was attributed primarily to the further tightening of Man-Year-Entitlement (MYE) formula, thus a decrease in

Table 9.7 Labour productivity growth (per cent change)

Year	*1995*	*1996*	*1997*	*1998*	*1999*	*2000*	*2001*
Construction	−3.8	−2.9	−3.2	−4.4	−2.2	−0.5	0.2
Manufacturing	6.5	3.0	5.5	1.4	17.6	11.6	−13.5
Financial	−1.4	0.2	9.3	−9.8	4.0	−3.5	−3.8
Business services	–	–	−0.4	−5.3	−2.5	−2.7	−6.7
Transport and communication	6.5	3.8	4.6	4.7	6.7	3.5	−1.3
Wholesale and retail trade	5.9	3.5	3.9	−3.3	8.8	10.9	−6.2
Hotels and restaurants	–	–	1.9	−9.0	3.9	6.5	−4.4
Other services industries	–	–	0.2	−2.5	2.4	1.9	0.7

Source: Yearbook of Statistics, Singapore.

employment figures overall. Among other contributing factors were more buildable designs and more skilled workers.

Physical measurement of construction productivity (production)

The physical measurement of construction productivity showed improvement in 2001 except for the HDB and private residential sectors which registered decreases. The overall construction industry average increased by 2.6 per cent in 2001, when compared to 2000 (Table 9.8).

9.3.2 *Construction cost*

Construction cost indices

The aggregate Basic Material Cost Index in 2001 was 59.7, a decrease of 5.3 per cent from 63.0 in 2000, a further weakening from the 3.4 per cent decline in 2000. This was due to the softening prices of basic materials. On the other hand, the aggregate Labour Cost Index fell moderately by 0.7 per cent in 2001, after the 4.4 per cent increase registered in 2000 (see Figure 9.3).

Despite the fall in aggregate Basic Material Cost Index and Labour Cost Index, total construction cost, as measured by the aggregate Tender Price Index for All Buildings, recorded a marginal increase of 0.2 per cent in 2001. This was due to the

Table 9.8 Average productivity (m²/man-day)

Year	*1994*	*1995*	*1996*	*1997*	*1998*	*1999*	*2000*	*2001*
HDB	0.66	0.66	0.66	0.66	0.69	0.66	0.74	0.71
Residential (landed)	0.22	0.18	0.21	–	0.21	0.21	0.22	0.15
Residential (non-landed)	–	–	–	–	0.22	0.23	0.24	0.21
Commercial	0.24	0.25	0.22	0.27	0.26	0.23	0.22	0.29
Industrial	0.36	0.37	0.4	0.40	0.45	0.45	0.43	0.45
Institutional	0.22	0.23	0.24	0.29	0.26	0.26	0.31	0.33
Industry average (m²/md)	0.31	0.32	0.34	0.36	0.37	0.37	0.38	0.39

Source: Building and Construction Authority.

Figure 9.3 Cost indices (base year 1990).
Source: Building and Construction Authority.

higher construction cost of public housing posted in the second half of the year as a result of higher risk allowances.

Unit construction cost

Unit construction cost in Singapore has changed very little when compared to the prices of the finished product (Table 9.9). However, unit construction costs for almost all kinds of buildings rose to its peak in 1997 and have come down since then to 1995–1996 levels. As the labour and material index remained relatively stable, the rise and fall seen in the unit construction cost over the years could be attributed mainly to the changes in operational and transaction costs in the industry.

The construction cost serves as a guide for preliminary cost appraisals and budgeting. According to Davis Langdon and Seah, an international cost consultant, construction costs in 2002 were expected to remain soft due to the keen tendering environment. As demands for construction materials and manpower were likely to drop, overall construction cost was expected to fall marginally in 2002.

Average construction material price

Apart from a brief rise, prices of most construction materials in 2001 remained low. In fact, prices of cement, concrete and steel bars fell to a ten-year low in 2001 (Figure 9.4). Economic uncertainties clouded the industry and kept prices down. Suppliers were generally pessimistic about the market outlook and resorted to price cutting to maintain their market share. It was even more evident from the tender prices submitted to HDB for supply of cement (such prices indicate the future trend), where prices dropped by nearly 20 per cent. The shrinking market and intense competition

Table 9.9 Unit construction cost (S$ per sq. m of gross floor area)

Year	*Residential (high rise)*		*Commercial office (high rise)*		*Industrial*		*Hotel*
	Standard	*Luxurious*	*Standard*	*Prestige*	*Light*	*Heavy*	*5 Star*
1995	1,200–1,400	1,550–1,750	1,550–1,900	1,950–2,650	850–1,350	1,000–1,600	2,850–3,450
1996	1,300–1,450	1,700–2,100	1,700–1,900	2,100–2,650	900–1,350	1,000–1,600	2,900–3,450
1997	1,402.5	1,980	1,905	2,433	1,155	–	3,159
1998	1,100–1,250	1,600–2,100	1,500–1,800	2,100–2,500	850–1,200	1,000–1,500	2,600–3,300
1999	1,100–1,250	1,650–2,100	1,500–1,800	2,000–2,500	800–1,000	1,000–1,400	2,600–3,300
2000	1,100–1,250	1,650–2,100	1,500–1,800	2,000–2,450	800–950	1,000–1,400	2,600–3,300
2001	1,183	1,792	1,604	2,159	878	1,174	3,073

Source: Davis Langdon and Seah Handbook 2002.

Note
Costs for 2001 are as at 4Q 2001 levels, excludes basement carparks.
Conversion rate for 2001 was S$1.7917–US$1.

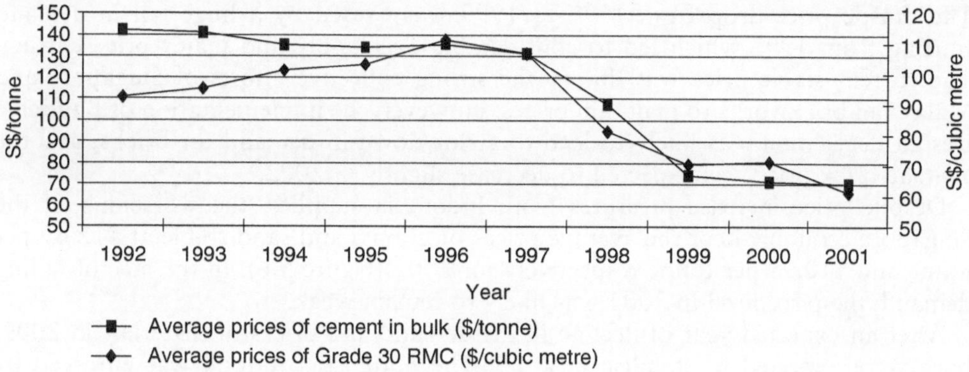

Figure 9.4 Average prices of cement and ready-mixed concrete Grade 30.
Source: Building and Construction Authority.

among suppliers in 2001 affected every player in the market, including NatSteel, a local steel supplier. The latter announced its plans to cut down its production capacity by 35 per cent to maintain competitiveness in July 2001.

Looking at the prospects for 2002, demand and price trends for basic construction materials were projected to fall further. In addition to the slowdown in construction activities which affect the demand and prices of construction materials, other major factors would also influence the price trend of the materials.

The reported average price of ordinary Portland cement (in bulk) was $70 per tonne in 2001. The highly competitive cement market and free trade environment in Singapore led prices to fall to the lowest in 10 years, until 2001. In fact, the price level in 2001 was not even half of that in 1991. Similarly, the average prices of bagged cement and clinker declined to $84 per tonne and $39 per tonne respectively in 2001. Comparing the whole year, the price slide was more substantial in the second half of last year. The dip in the HDB tender award price of supply for cement from $72.30 per tonne (awarded on September 2000) to $58 per tonne (awarded on August 2001) further confirmed the possibility of a future decline in cement prices.

The average price of ready-mixed concrete in 2001 plunged by 15 per cent to $61 per cubic metre for Grade 30 (the benchmark used for monitoring concrete price trends). The average prices of Grades 35 and 40 concrete also dived by an average of 14 per cent over the year 2001 to $64 per cubic metre and $67 per cubic metre respectively. In the current cut-throat atmosphere, prices of ready-mixed concrete in 2002 were expected to be low.

Due to weak demand and excess supply, the average price of rebars fell from $459 per tonne to $433 per tonne in 2001. Moreover, the successful HDB bulk contract price to supply rebars for HDB projects also dropped to an all-time low of $418.50 per tonne (contract commenced in September 2001). Singapore was not the only country experiencing oversupply of rebars as it was an international trend. To clear the stock, producers worldwide were likely to lower prices. This means the rebar price was expected to remain relatively low in 2002.

The average price of bricks increased slightly to $147 per thousand pieces compared to the rock bottom price of $135 per thousand pieces seen in 1999 (refer Figure 9.5). The drastic price drop from 1997 to 1999 was caused by a huge fall in demand starting from 1997 which led to considerable oversupply and major price cutting. The current stable price was due to the strong collective efforts of Singapore and Malaysian brickworks to maintain prices. However, the implementation of Buildable Design Legislation was likely to lead to a slowdown in demand for bricks, and the average brick price was expected to decrease slightly in 2002.

Despite price increase pressures from Indonesia supplies, the worsening of the Singapore economy kept the average prices of granite and sand stable at $12.69 per tonne and $10.95 per tonne respectively for 2001 (Figure 9.6). In the face of falling demand, the price level in 2002 was likely to remain weak.

After an expected year of decline in the demand and prices of materials in 2002, both were expected to stabilise in 2003. The optimistic outlook was enforced by

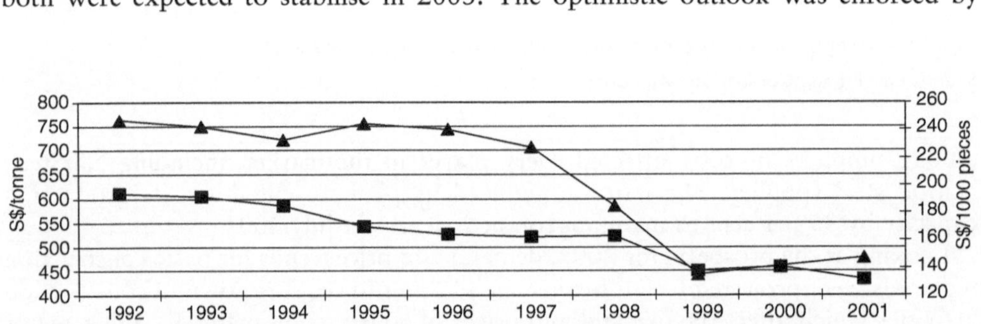

Figure 9.5 Average prices of steel bars and bricks.

Source: Building and Construction Authority.

Note

Prior to 1995, the price of steel bars represented NatSteel's listed price of 20 mm high tensile steel bars. With effect from 1Q1995, the price refers to the average selling price of imported and locally produced steel bars.

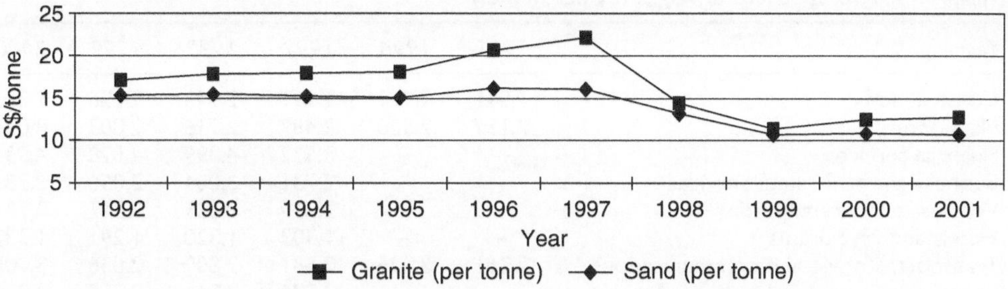

Figure 9.6 Average prices of granite and sand.

Source: Building and Construction Authority.

government's off-budget package which was expected to take effect in the first half of 2003. Moreover, producers were expected to then be able to adjust their production which would lead to less oversupply and possible recovery in prices.

Construction industry salaries and wages

With higher construction cost and anticipating uncertain prospects, the construction industry salaries for professionals and technicians showed a decrease of 10.1 and 7.0 per cent respectively in 2000 (Table 9.10). However, the figures had been steadily rising despite the economic crisis and peaked in 1999 to 3,592 and 2,721 for professionals and technicians respectively. In 2001, the daily wages of skilled workers suffered a drop of about 2.7 per cent from that of 2000. Similarly, the daily wages of unskilled workers fell by about 1.9 per cent from 2000.

Average sectoral wages

Basic wages in the construction sector had been steadily rising during the past decade. However, the rise was small compared to a more significant rise in the basic wages of the manufacturing sector, which saw a rise of over 50 per cent during the same period. Since 1995, most other sectors of the economy such as the financial, business,

Table 9.10 Construction industry salaries and wages

Year	*1995*	*1996*	*1997*	*1998*	*1999*	*2000*	*2001*
Professionals (S$ per month)	–	3,302	3,474	3,576	3,592	3,228	n.a.
Technicians (S$ per month)	–	2,498	2,551	2,660	2,721	2,531	n.a.
Skilled workers (S$ per day)	29.1–64.8	33.9–69.5	33.5–72.7	32.0–69.7	22.8–62.5	23.4–61.4	23.2–59.3
Unskilled worker (S$ per day)	16.9–32.6	16.9–31.6	17.7–33.6	17.0–30.6	16.9–33.9	17.1–35.6	17.0–34.7

Sources: Report on Wages in Singapore 1997–2000, Ministry of Manpower.
Building and Construction Authority.

Table 9.11 Average sectoral wages (S$ per month)

Year	*1995*	*1996*	*1997*	*1998*	*1999*	*2000*
Construction	1,941	1,971	1,991	2,143	2,226	2,333
Manufacturing	2,157	2,320	2,487	2,716	2,803	3,036
Financial services	–	–	3,522	4,399	4,528	4,931
Business and real estate services	–	–	2,731	3,004	3,056	3,281
Wholesale and retail trade	–	–	2,319	2,539	2,561	2,721
Hotels and restaurants	–	–	1,303	1,320	1,291	1,332
Transport, storage and communications	2,261	2,425	2,581	2,802	2,836	3,105
Community, social and personnel services			2,576	2,811	2,955	3,336

Source: Yearbook of Statistics, Singapore.

Note
The industries are classified according to the Singapore Standard Industrial Classification 1996.

transport and hotel industries saw a higher increase in basic wages compared to the construction sector.

In line with the strong economic expansion in 2000, total wages increased by 6.6 per cent in 2000. Basic wages increased by 4.9 per cent for all employees, substantially higher than the 2.1 per cent registered in 1999 (Table 9.11).

9.3.3 *Construction quality*

Quality certification schemes

In order to maintain and enhance construction quality, BCA launched the ISO 9000 Certification Scheme as early as 1991 and made it a mandatory requirement in July 1999. This scheme applies to contractors with a registration grade of G6 and above in BCA's Contractors Registry undertaking general building, civil engineering and piling works. It also includes consulting firms engaged in architectural, engineering and quantity surveying services who undertake public construction projects valued at more than S$30 million. The BCA ISO 9000 Certificate is awarded to companies in the architectural, engineering, contractor, property and building materials and related fields. It recognises companies that have demonstrated continuous and effective operation of a quality management system.

The BCA also administers the BCA ISO 14000 (Environmental Management System, EMS) Certification Scheme and the BCA OHSAS 18000 (Occupational Health and Safety Management Scheme) Certification Scheme. Contractors awarded the ISO 14000 EMS certification will have the management system that enables the firms to control the impact of their activities, products and services on the environment. Contractors awarded the BCA OHSAS certification will become more aware and self-regulatory in promoting safety at their construction sites, leading to better safety standards for their projects (see Table 9.12).

CONQUAS score

The Construction Quality Assessment System (CONQUAS) was officially developed by the then Construction Industry Development Board in 1989, in conjunction with

Table 9.12 Total number of certified firms

Types of certification	*No. of firms*
BCA ISO 9000 certification	327
Contractors	211
Consultants/public agencies	97
Education and training for construction industry	2
Developers/property services/project management	7
Others	10
BCA ISO 14000 certification	29
BCA OHSMS certification	27

Source: Building and Construction Authority as on 25 April 2002.

the major public agencies and leading professional bodies from the industry. BCA launched its fifth edition of CONQUAS in June 1998, known as CONQUAS 21. In the new edition, new features, including the assessment of mechanical and electrical works, had been introduced to make CONQUAS scoring even more accurate and customer-oriented (Table 9.13).

CONQUAS 21 was launched together with the Bonus Scheme for Construction Quality (BSCQ) for public projects. The building is assessed based primarily on workmanship standards through site inspection. The scoring will be done only once on three main components (structural, architectural, mechanical and electrical works) on site. Rectification and correction carried out afterwards will not be re-assessed. The objective of this practice is to encourage *doing things right the first time*.

Until the time of writing, more than 1,800 public and private building projects have been assessed by the BCA (Figure 9.7). The industry average CONQUAS score had improved steadily from 67.9 in 1989 to 76.5 in 2001. A target median CONQUAS score was set, beginning with 79 by 2005 and reaching 82 by 2010.

Buildability score

One of the areas that C21 identified as a means to improve industry practices and techniques is enhancing buildability. The Code of Practice (COP) on buildable design which covers information such as the legislation requirements, submission procedures, method of computing buildability score and submission forms was gazetted on 12 December 2000. The legislation of buildable design officially took effect from 1 January 2001 (Table 9.14).

The three main principles of standardisation, simplicity as well as single-integrated elements can be applied to achieve buildability. The introduction of government

Table 9.13 Yearly average CONQUAS score

Financial year	*1992*	*1993*	*1994*	*1995*	*1996*	*1997*	*1998*	*1999*	*2000*	*2001*
CONQUAS score (yearly average)	71.3	72.5	71.6	72.2	73.9	74.3	75.8	76.5	77.8	76.5

Source: Building and Construction Authority.

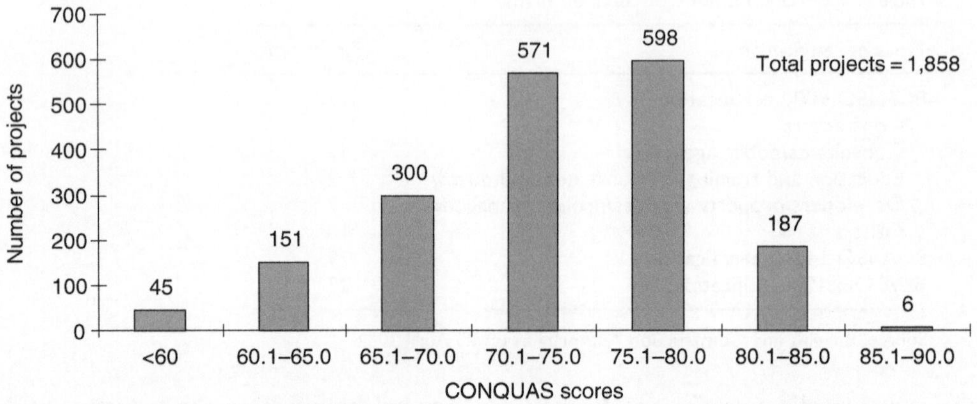

Figure 9.7 Distribution of CONQUAS scores.

Source: Building and Construction Authority.

Table 9.14 Average buildability scores

Year	*1997*	*1998*	*1999*	*2000*	*2001*[2]
Public sector					
HDB[1]	77	78	79	79	81
Private sector					
Residential					
Landed	–	47	46	44	48
Non-landed	–	55	54	56	56
Commercial	64	62	62	63	63
Industrial	65	65	64	62	64
Institutional	64	63	60	61	63

Source: Building and Construction Authority.

Notes

1 The average buildability scores comprise projects from public and private sectors. However, a separate category is still maintained for HDB's projects.

2 Although buildability legislation started in 2001, the average buildability scores for 2001 were based on non-legislated projects completed in 2001. This was because there were not enough legislated projects submitted in 2001 for computing the average figures.

regulations under the Building Control Act will require building designs to have a minimum buildability score. To ensure that the industry is prepared for the legislation, training courses are also made available to the industry.

9.3.4 *Construction safety record*

The construction industry in Singapore had been pre-occupied with increasing construction output for the last few decades, until 2002, with limited safety concerns. The notion that stringent rules on maintaining and enhancing safety were by and large argued to be against the development norms. Owing to the big increase in population in the 1960s and 1970s, the quantity and speed of construction was the major concern of public housing development, not the quality. However, this trend discontinued in the 1990s and especially after the financial crisis, quality became an important issue.

From June 2001, the Factories (Safety Training Courses) Order 2001 passed down by the Ministry of Manpower (MOM) requires mandatory training for certain categories of persons employed in construction sites and in certain industrial premises. This is an integral part of the Ministry's strategy to curb accidents through education. All workers, including Singaporeans, must undergo a Construction Safety Orientation Course (CSOC) before they are allowed to work on construction worksites. The course aims to introduce workers to the hazards associated with construction work, and what they can do to protect themselves. The Order also makes it mandatory for formwork supervisors to undergo safety training. Project managers of sites with contract sum exceeding S$10 million must also undergo compulsory safety training. Projects of this size require the attention of senior managers to manage safety issues.

It was observed that the number of accidents had been increasing in tandem with the volume of work over the past decade, until 2002. However, it peaked in 1997 and 1998, a period when construction activities had been brisk. At the same time, there appears to be a rise in the fatality rate in tandem with the rise in number of accidents in the same period when compared with the early years of the 1990s. The trend in both accidents and fatalities, however, seem broadly to be lowering in the construction industry. In 2001, only 1,455 accidents occurred, including 28 fatalities, were recorded for the local construction industry. The frequency rate in 1997 also peaked at 3.0 accidents per million man-hours and so did the severity rate to a level of 968, and both statistics dropped to 2.8 and 405 respectively in 2001 (Table 9.15). Accidents frequency rate is targeted to be lower than 2.0 accidents per million man-hours by 2005 and 1.0 accidents per million man-hours by 2010.

9.3.5 *Skill level of workforce and supervision level*

Constrained by limited human resources, the construction industry in Singapore is largely dependent on foreign workers. While it is accepted that foreign workers will

Table 9.15 Accident statistics in construction industry

Year	*1994*	*1995*	*1996*	*1997*	*1998*	*1999*	*2000*	*2001*
No. of accidents	856	887	1,243	1,538	1,532	1,504	1,395	1,455
No. of fatal cases	32	43	51	72	67	48	49	27
Fatalities	32	43	54	72	73	51	52	28
Accident frequency rate[1]	2.5	2.2	2.9	3.0	2.7	2.8	2.6	2.8
Accident severity rate[2]	664	736	828	968	846	647	654	405

Source: Ministry of Manpower.

Notes
1 Accident frequency rate is the number of accidents per million man-hours worked.
2 Accident severity rate is the number of man-days lost per million man-hours worked.

Table 9.16 Skill level of construction industry

Year	*1995*	*1996*	*1997*	*1998*	*1999*	*2000*	*2001*
Skill level of construction industry (%)	4	7	8	19	43	57	63.6

Source: Building and Construction Authority.

still dominate the construction workforce, it is not desirable that they be largely unskilled. The C21 Committee recommended the industry to achieve a workforce profile where, by 2010, a minimum of 60 per cent of its workforce is certified skilled.

The overall skill level of the construction industry itself took an enormous leap from 4 per cent in 1995 to 63.6 per cent in 2001, largely because of the concerted effort by the government (Table 9.16). Skilled workers are those who have obtained the Skill Evaluation Certificate (SEC) awarded by the Construction Industry Training Institute (CITI) or equivalent certification. The CITI is the training arm of the BCA, specialising in the training and skill certification of construction trades.

Besides imposing the SEC requirement, the Basic Skills Certificate (BSC) requirement that serves as a basic screening mechanism was also tightened. Incentives are given to promote skill upgrading. One of the successful incentives is the two-tier foreign worker levy system with a lower levy for skilled foreign workers than their unskilled counterparts. The BCA has also continuously put in place various programmes to draw more locals into the industry.

The CITI offers supervisory courses for construction personnel to upgrade themselves, e.g. Construction Trade Foreman Certificate and National Certificate in Construction Supervision (NCCS). The full-time NCCS course started in 1998, which allowed secondary school leavers to take up the course instead of only supervisors who were already working in the part-time courses. The CITI also offered a full-time Builder Certificate in plumbing and pipefitting in 1999 to school-leavers. The aim has been consistently to increase the number of supervisors in the industry.

The supervisor to worker ratio was about 1:30 in 1999, which even though higher than the figures in Australia and Japan, was considered inadequate for proper supervision of workers (Table 9.17). Noting the importance of qualified supervisors on

Table 9.17 Certifications awarded for supervisors

Year	*1996*	*1997*	*1998*	*1999*	*2000*	*2001*
Construction Trade Foremen Certificate	70	101	130	128	80	80
National Certificate in Construction Supervision (NCCS)	167	128	102	86	63	92
Total[1]	237	229	232	214	143	172

Source: Building and Construction Authority.

Note

1 Excludes other certifications for supervision in precast concrete construction, water proofing and formwork and falsework.

construction sites in ensuring high productivity and safety levels, the industry targets to increase the supervisor to worker ratio to 1:15 by 2010.

9.3.6 Research and development

Research and development (R&D) expenditure in the Singapore construction industry amounted to less than S$47 million or about 0.47 per cent of total construction volume in 1997. While the figure rose close to S$47 million in 1998, the proportion dropped to 0.45 per cent of total construction volume. In 1998, the investment further dropped to 0.4 per cent of the total construction volume. For 2000, the government set aside S$625 million to fund R&D activities (Table 9.18). The government was targeting to achieve 0.70 per cent of total construction GDP as the R&D expenditure by 2010.

The BCA, by 2002, manages the R&D fund that was meant solely for the construction sector and was actively sourcing for projects. The BCA was working closely with public sector agencies, tertiary institutions and industry on construction-related R&D. In 2001, several projects were awarded a total of S$6.8 million in R&D grants by BCA and have already commenced their R&D work.

The Centre for Total Building Performance (CTBP) was established as a joint research centre for the BCA and the National University of Singapore (NUS) in November 2000, marking a milestone in Singapore's building and construction history. CTBP, being the first building research centre in the world located in the tropics, aims to develop Singapore as the construction R&D hub in the region.

The CTBP will perform a pivotal role in promoting industry-based research collaboration with the focus on developing and implementing an approved R&D framework for achieving the best standards and best practice in the design, maintenance and management of buildings. The centre will look into three key areas of R&D, namely indoor air environment, maintainability and energy efficiency.

One of the proposals was the formation of a central body to oversee R&D in the construction industry so that R&D efforts would not be duplicated. This is to

Table 9.18 Research and development expenditure for construction (S$ m)

Year	*1996*	*1997*	*1998*	*1999*	*2000*[1]	*2001*
National expenditure on R&D	1,790	2,100	2,500	2,656	n.a.	n.a.
R&D expenditure by construction industry[2]	n.a.	4.4	2.3	4.2	2.2	6.76
R&D expenditure by civil engineering and architecture[3]	43.3	46.8	46.0	39.2	27.88	n.a.

Sources: (1) National Survey of R&D in Singapore, National Science and Technology Board (2) Building and Construction Authority.

Notes
1 R&D grants by Building and Construction Authority only.
2 R&D classified under Singapore Standard Industrial Classification (SSIC) 1996 for construction only.
3 Expenditure includes private, higher education and government sectors as well as public research institutes and centres.

facilitate dissemination of research findings and new technology breakthrough to the industry. Hence, the National Construction Research Institute (NCRI) was set up.

9.3.7 *New technology development in the construction field*

The innovation development schemes (IDS) aims to assist and encourage Singapore-registered companies to engage in innovation products, processes and applications and further enhance innovation capabilities (refer Table 9.19). It provides assistance to cover up to 50 per cent of qualifying costs.

Policy and initiative on the use of IT in construction

In October 2000, BCA awarded two contracts worth a total of S$6.6 million to SGX-ST mainboard aspirant, novaSPRINT Pte Ltd, a leading developer and implementer of proprietary industry-focused e-business solutions. This aims to develop a world's first automated plan submission cum checking system.

The first contract for S$3.9 million was to build the Integrated Plans Checking Systems for Building Plans and Building Services (IBP and IBS Systems). The development of the two systems was expected to be completed by 2003. Another was a tender award, worth S$2.7 million, to build the One-Stop Submission Centre (OSSC) which would facilitate electronic submissions, processing and approval of building project documents over the internet.

The automated plan submission cum checking systems are part of the Construction and Real Estate Network (CORENET), a major IT initiative spearheaded by the BCA in collaboration with several public and private organizations. It aims to re-engineer the business processes of the construction industry to achieve a quantum leap in turnaround time, productivity and quality. At the same time, it will help to speed up the adoption of the latest IT in the industry and revolutionise the ways building professionals carry out their design works and business processes.

To assist those who want to look into alternative approaches, the NUS, in collaboration with BCA, has developed a software package called Building Energy Standards (BEST). The software can be used by engineers, architects and building services professionals to demonstrate compliance with the prescribed energy standards.

Table 9.19 Major new technology developments in the construction field approved by IDS

Technology and brief description	*Developer*
1 Development of a field water leakage testing device for building envelopes	Winwall technology Pvt. Ltd
2 Development of an information technology-based Integrated Operation, Support and Project Management System for enterprise wide utilisation	Econ Piling Pvt. Ltd
3 Development of an information technology infrastructure network – the Enterprise One	Beca Carter Holdings (SEA) Pvt. Ltd

9.4 Buildability

9.4.1 Extent of design and build projects

The increased development of more complex intelligent buildings, the demand for greater responsiveness, guaranteed cost and delivery schedules of projects as well as the push for greater efficiency, have augmented the need to integrate the processes and roles of various players.

The BCA has identified various forms of D&B, from traditional D&B to the Design, Develop and Construct or Consultant Novation models (Table 9.20). Whatever the model chosen, D&B is a proven form of project delivery system that allows the roles of various players to be integrated up-front at the design stage, offering many advantages.

9.4.2 Prefabrication technology

Prefabrication is a key initiative towards industrialising the construction industry. It enhances buildability. Today, the architectural precast concrete offers maximum freedom of architectural expression with the enhanced economies of production, another step in the development of precast concrete.

The industry aims to achieve at least a 20 per cent precast level by 2010 (Table 9.21). However, use of precast products is only one of the ways to achieve higher buildability. The use of structural steel and flat plate design using system formwork are also gaining popularity.

9.4.3 Environmental conservation

Energy efficiency is one of the key strategies to make Singapore's economy more cost competitive. It also addresses the environmental concern with greenhouse gas

Table 9.20 Extent of design and build projects

Building works value (S$ m)	*1995*	*1996*	*1997*	*1998*	*1999*	*2000*	*2001*
Total building works[1]	14,167	16,287	19,364	11,904	9,425	9,629	9,537
Total D&B projects	1,105	1,987	2,924	1,607	1,819	1,454	1,278
% of total building works	7.8	12.2	15.1	13.5	19.3	15.1	13.4
Public sector D&B projects	837	1,327	1,466	1,181	1,001	629	419
% of total D&B projects	75.7	66.8	50.1	73.5	55.0	43.3	32.8
Private sector D&B projects	268	659	1,458	426	819	825	859
% of total D&B projects	24.3	33.2	49.9	26.5	45.0	56.7	67.2

Source: Building and Construction Authority.

Note

1 Excluding Engineering Procurement and Construction Projects.

Table 9.21 Precast level

Year	*1993*	*1994*	*1995*	*1996*	*1997*	*1998*	*1999*	*2000*	*2001*
Precast level (%)[1]	6.9	5.8	7.1	6.4	8.2	8.0	7.0	7.4	7.5

Source: Building and Construction Authority.

Note
1 Indicate per cent of cement used in precast concrete.

emissions that cause global warming. Furthermore, using energy efficiently will help to conserve fossil fuels, a resource that is rapidly depleting.

In Singapore, the residential and commercial building sectors are the major consumers of energy. Together with the other two sectors, namely, transport and industries, they each account for roughly one-third of the total energy consumed in Singapore. There is therefore potential for energy savings in buildings.

Over the years, BCA had played a part in enhancing energy efficiency in buildings by ensuring that minimum standards in energy efficiency were met. Since 1979, energy conservation standards for design of buildings had been incorporated in the local building code. The energy conservation standards prescribe an Overall Thermal Transfer Value (OTTV) of not more than 45 watts per unit area (W/m^2).

The BCA was expected to replace the OTTV with a new standard called the Envelope Thermal Transfer Value (ETTV) – a new and more accurate formulation of heat transfer into a building under local weather conditions. At the same time, BCA would also be setting a stricter compliance value of 35 watts per unit area of heat gain by a building.

9.5 Construction outlook 2002–2010

9.5.1 *Business opportunities and major projects in pipeline*

Singapore was ranked the fourth most competitive economy by the World Economic Forum (WEF) in its Global Competitiveness Report 2001–2002. The growth competitiveness ranking analyses the extent to which individual national economies have the structures, institutions, and policies in place for economic growth over the medium term. In terms of current competitiveness, until 2002, which measures the conditions that determine a nation's sustainable level of productivity, Singapore ranked tenth. The competitiveness ranking was based on two components that measure the sophistication of company operations and strategy and the quality of business environment within an economy. Fiscal policy in Singapore is guided by the principle that it should support the private sector as the engine of growth and ensures that the macro-environment is stable. The Singapore government has been prudent and conservative in its budgetary policy. It has balanced its budget in nearly every year for the last 3 decades, until 2002. The government also sets clear and transparent ground-rules and ensures that markets were competitive.

Singapore had thrived by being plugged into the global economic network. In fact, it had just been ranked the most globalised economy in the world and the world's second most liberalised economy, according to the index of Economic Freedom by the Heritage Foundation in 1997. Also, Singapore had one of the most open construction markets in the world.

It is always Singapore's intention to continue to strengthen its economic linkages with the rest of the world. While Singapore remains committed to the launch of a new multilateral trade round at WTO, it is also pursuing deeper trade liberalisation initiatives with like-minded countries through bi-lateral Free Trade Agreements (FTAs). In September 1996, Singapore's application to join the Agreement on Government Procurement (GPA) was approved by the World Trade Organisation (WTO). The GPA requires that signatory members shall not discriminate against products, services and suppliers of another member. Its coverage includes the procurement of construction goods and services.

Closer to home, Singapore continues to work towards greater economic integration with its neighbours through the ASEAN Free Trade Area (AFTA), ASEAN Investment Area (AIA) and e-ASEAN initiatives. Deeper economic integration will enhance ASEAN's attractiveness as a destination for global foreign direct investments (FDIs). Until 2002, there were no limitations on market access and national treatment under cross-border supply, consumption abroad and commercial presence. There was no restriction on ownership. Foreign construction firms like local firms need only to register with the Registry of Companies and Businesses. There was no specific requirement which construction firms need to further meet for private sector projects. Firms intending to undertake public sector projects must, however, be registered with the BCA. This applies to both local and foreign firms. The perennial concern of many countries in liberalising professional services (by architects and engineers) was the fear that they may inadvertently undermine the quality of such services. Table 9.22 shows the major projects in the pipeline.

9.5.2 *Areas where foreign investments are needed*

Infrastructure, energy and power along with telecommunication and IT were the key sectors that require foreign investments in Singapore. In 2000, the government announced a framework for a competitive energy market for the electricity and gas industries. Among other sectors, treatment plants for recycling of water had been put into focus with the announcement of PUB and Ministry of the Environment (ENV) piloted a project at the Bedok Sewerage Treatment Plant to recycle used water. The water, named NEWater will be used for non-potable industrial purposes.

A long-term initiative was also launched to develop life sciences as the fourth pillar of Singapore's industries. In addition, the government launched a $7 billion Science and Technology plan for 2001–2005 with a view to enhancing technological competence of the country, particularly in the context of increased competition among the East Asian countries. The project was a close follow up to the US$1 billion Technopreneurship Investment Fund that was launched in 1999 to spur development of the venture capital industry in Singapore.

The construction industry in Singapore was one of the best regulated in Asia in terms of quality and skill enhancement, R&D. The construction industry was more mature than those of many other Asian countries. The introduction of bench marking and buildability scores was seen only in Singapore. In September 1996, Singapore's

Table 9.22 Major projects in the pipeline

Name of project	*Brief description of project*	*Period*	*Estimated project cost (S$)*
Residential			
Public housing	Upgrading of HDB flats under Main Upgrading Programme (MUP) and Interim Upgrading Programme (IUP)	To complete in the next 10 years	4.6 billion
	Retrofitting of 4000 HDB blocks built before 1990 to enable lifts stopping at every floor under Lift Upgrading Programme	To complete in the next 10 years	4.8 billion
Commercial			
Small Office and Home Office (SOHO)	Mixed development of 250 Small Office and Home Office (SOHO) units on top of the upcoming Singapore River MRT Station	Expected to be completed by 2005	500 million
Industrial			
One-North	Development of a 200 ha self-contained life-sciences and IT hub over 4 phases. The project will be completed with housing, labs, retail outlets and a modern transport network	To complete in 15–20 years	15 billion
Desalination plant	A Build-Own-Operate (BOO) project to supply 30 million gallons of desalinated water per day for industrial and public use	2003–2005	400 million
Institutional			
Singapore Management University (SMU)	First phase development of a university with 3 schools buildings, a campus centre, a library and an air-conditional underground pedestrian tunnel linking 5 buildings at Bras Basah	2003–2005	400 million
Jurong General Hospital	Construction of the Jurong General Hospital to replace the old Alexandra Hospital at Jurong Town Hall Road	2003–2006	400 million
Supreme Court	Development of a new supreme court with 4 basement levels at High Street/North Bridge Road	2002–2005	200 million
Infrastructure			
MRT Lines	Development of 3 new rail lines – Bukit Timah MRT Line, Eastern Region MRT Line and Jurong Region LRT Line	2005–2012	11 billion

Source: Building and Construction Authority.

application to join the GPA was approved by the WTO. With effect from July 1999, all BCA registered G6-G8 contractors were certified to ISO 9000 standards. All projects in Singapore were subjected to review under the country's CONQUAS.

Note

1 These include unconfirmed projects under the BTO scheme.

Chapter 10

South Korea

Korea Research Institute for Human Settlements,
J. Raftery and M. Anson

10.1 The country

Korea lies mid-way in between China and Japan. The Korean Peninsula borders the Sea of Japan and the Yellow Sea. According to the Korean government homepage (http://www.korea.net/), the shortest distances between Korean and Chinese coasts, and between Korean and Japanese coasts are both 200 kilometres. Chinese culture thus filtered into Japan through Korea due to such geographical location. The Korean peninsula was "affected far more by the civilizations on the Asian continent than by those in Japan" in ancient times. Consequently, there is the "common culture sphere of Buddhism and Confucianism" between the three countries.

The Korean Peninsula is protruding southwards from the northeastern part of the Asian Continent. It is divided into three distinct regions comprising Central, South and North. Each region is further divided into three separate "geographical spheres". Each "sphere" has its economic, cultural and physical distinctiveness. Inside the Central region is Seoul. All the areas north of the Demilitarized Zone are referred to as the North (http://www.korea.net/).

The land area of South Korea, excluding the portion of North Korea, amounts to 99,373 square kilometres, and includes approximately 3,200 islands of various sizes. South Korea is about the size of Hungary (93,000 square kilometres) and Jordan (97,700 square kilometres). Habitable land, however, is only a quarter of the total land area where 47 million people inhabit as of 2002 thus making South Korea one of the most densely populated nations on the earth (KRIHS website: http://www.krihs.re.kr/eng/about/foundation.asp).

10.1.1 History

The Korean government website (http://www.korea.net) described that people have been living on the Korean peninsula for about half a million years. Artefacts were found in archeological digs in Central Korea. Farming and planting of rice are believed to have started in the Bronze Age in the warmer southern regions, some 3,500 years ago. The origin of the Korean people was traced to the founding of the state of Choson in 2333 B.C. Its literal meaning is the "Land of the Morning Calm". It was established by Tan-gun, a legendary figure born of the son of Heaven, and a woman from a bear-totem tribe. Since then, there were rises and falls of small town-states that clan communities combined to form. By the first century B.C., there were the Three Kingdoms that emerged on the Korean Peninsula: Koguryo (37 B.C. to A.D. 668),

Paekche (18 B.C. to A.D. 660) and Shilla (57 B.C. to A.D. 935). Further, it was the Shilla kingdom that unified Korean Peninsula in 676, and since then, Korea has been ruled by a single government. Both the Koryo (918–1392) and the Choson (1392–1910) Dynasties consolidated their power and flourished culturally, whilst resisting invasions from the Khitans, Mongols, Manchus and Japanese. It was during the Choson period that Korea isolated itself from the rest of the world and became known as the "Hermit Kingdom". Finally in 1910, Korea was annexed by Japan and was instituted colonial rule. National liberation came on the 15 August, 1945, only to be followed by territorial division.

The end of Second World War saw the division of the Korean peninsula into the northern and southern zones. The former was occupied by Soviet forces and the latter by US ones. The thirty-eighth parallel was the boundary between the two zones. The Republic of Korea was inaugurated on 12 August, 1948 in the south, whilst the Democratic People's Republic of Korea was proclaimed on 9 September, 1949 in the north. Civil war broke out on 25 June, 1950. It lasted for 3 years, during which the South Korean military was assisted by the United Nations that sent armed forces and other assistance. Over four million lost their lives before the war ended in a cease-fire on 25 July, 1953, when a new border was fixed at the Demilitarized Zone (DMZ). The Zone is a 4 kilometre-wide strip of land, running along the lines of cease-fire from the east to the west coast for a distance of about 241 kilometres (http://www.korea.net/ and http://www.korea-dpr.com/).

10.1.2 Population

In July of 2004, the population was estimated to have been almost 48.6 million, of which more than 71 per cent was of the age 15–64. People aged 65 and over represent about 8 per cent of the population, whilst the remaining 20 per cent are the young generation aged 14 and under. The population growth rate is 0.62 per cent and there is zero migrant per 1000 population, both being 2004 estimates. Like Japan, South Korea has a homogeneous ethnic group, except for about 20,000 Chinese, and again like Japan, a great majority of the population (more than 98 per cent) is literate (CIA World Factbook).

10.1.3 Government

South Korea, or Republic of Korea as the conventional long form country name, has a republic government. The Korean constitution, promulgated on 17 July, 1948, guarantees individual rights and stipulates the three independent administrative bodies: the National Assembly, the Government and the Judicial System (http://www.koreainfogate.com/).

The legislative branch is the unicameral National Assembly or Kukhoe. Members are elected by direct popular vote for a 4-year term. There are currently 227 elected legislators. In addition, there are the remaining 46 seats distributed proportionately among parties winning five seats or more in the direct election. This proportional representation system encourages "legislative participation by leading technocrats through the political parties". Among the major powers of the National Assembly are "the power to propose, deliberate and approve or reject legislative bills, to finalize and inspect closing accounts of the national budget, to consent to the conclusion and ratification of treaties and to concur in the declaration of war or the conclusion of peace", and the "right to impeach the President and to approve his emergency orders" (http://www.koreainfogate.com). Principal political parties include Democratic Labour Party (DLP), Democratic People's Party (DPP), Grand National Party (GNP), Millennium Democratic Party (MDP), United Liberal Democrats (ULD) and Uri Party (http://www.korea.net/).

The executive branch is the government. The president is the chief of state and the commander-in-chief of the armed forces. The president is elected by popular vote for a single 5-year term and cannot be re-elected. There is universal suffrage for those aged 20 and over. The State Council (cabinet) is chaired and appointed by the president upon the recommendation of the prime minister, who is the head of government and is appointed also by the president with the approval of the National Assembly. As the principal executive assistant to the president, the prime minister supervises the executive ministers.

There are two administrative tiers: the metropolitan cities and the provinces (*do*). The seven metropolitan cities, defined as urban areas with a population of 1 million or over, collectively form the first tier. Seoul is the capital and the largest metropolitan city with a population of 10 million. The country had been considering building a new administrative capital. There have been concerns of excessive costs caused by the overcrowding of Seoul. On 11 August, 2004, the Yeongi-Gongju region was chosen as the final site of the new administrative capital by the Presidential Committee on Administrative Capital Relocation. The site covers 21.6 million pyeong (one pyeong equals 3.3 square metres) including the townships of Nammyeon, Geumnammyeon and Dongmyeon in the county of Yeongi-geun, and the township of Janggimyeon in Gongju City in Chungcheongnam-do (province). The Yeongi-Gongju region is about 10 kilometres from both Daejeon and Cheongju, the region is adjacent to the Seoul-Busan high-speed railway and Cheongju airport. It is also near the Dangjin-Sangju expressway, Seoul-Busan expressway and Seoul-Busan railway (http://www.korea.net).

After Seoul and in descending order of population size are the other six metropolitan cities: Busan, Daegu, Incheon, Gwangju, Daejeon and Ulsan. The nine provinces (*do*) form the second administrative tier. Each province is divided into cities (*si*) and counties (*gun*). Each city has a population of 50,000 or more, and a *gun* comprises one town (*eup*) plus five to ten *myeon*. There are altogether 72 cities (*si*) and 91 counties (*gun*).

At the top of the judiciary is the Supreme Court, of which the Chief Justice is appointed by the president to a single 6-year term with the consent of the National Assembly. The Chief Justice recommends justices to the president for their appointments. Under the Supreme Court are the High Court and District Court. Judges at all the lower courts are appointed by the Chief Justice with the consent of the Conference of Supreme Court Justices. In addition, there is also the Constitutional Court (http://www.korea.net/).

10.1.4 Economy

As Korea underwent rapid modernization and industrialization since the 1960s, the First Five-Year Economic Development Plan (1962–1966) focused on laying a foundation for industrialization, which caused urban population to soar up to almost 87.2 per cent in 1998. Since then, manufacturing has grown rapidly right through to the early 1990s (Tables 10.1, 10.2, 10.3 and 10.4). Such a rapid change consequently placed enormous pressure on the country's limited land resources. Partly because of efficiency-oriented economic development policies, the nation's socio-economic activities have become spatially polarized as represented by the excessive concentration in the Seoul Capital Region and serious depopulation in rural areas. The resulting regional disparities and imbalanced settlement patterns impede to realizing territorially balanced development and equitable distribution of welfare.

The Korean government was instrumental in achieving the economic success through the late 1980s. There had been systems of "close government/business ties, including directed credit, import restrictions, sponsorship of specific industries, and a strong labor effort" (CIA World Factbook). There had also been conglomerate of businesses units closely knitted with each other, more commonly known as "chaebol", similar to the "keiretsu", the Japanese equivalent. The government also promoted

Table 10.1 Key economic indicators of the 1960s

	1962	*1969*	*Annual change %* 1962–1969
Per capita GNP (US$)	87	210	12.5
Real GDP growth rate (%)	2.1	13.8	8.7
Current account balance (million US$)	−55.5	−548.6	–
Trade balance	−335.3	−991.7	–
Exports	54.8	658.3	41.5
Imports	390.1	1,650.00	24.7
Gross domestic investment ratio (%)	11.8	27.9	–
Gross savings ratio (%)	11	21.4	–
Consumer price inflation (%)	–	11.8	11.2*

Source: Bank of Korea.

Note
* Annual change from 1966 to 1969.

Table 10.2 Key economic indicators of the 1970s

	1970	*1979*	*Annual change %* 1970–1979
Per capita GNP (US$)	253	1,647	22.9
Real GDP growth rate (%)	8.8	7.1	8.8
Current account balance (million US$)	−622.5	−4,151.1	–
Trade balance	−922.0	−4,395.5	–
Exports	882.2	14,704.5	36.4
Imports	1,804.2	19,000.0	22.7
Gross domestic investment ratio (%)	24.3	35.8	–
Gross savings ratio (%)	18.1	28.5	–
Consumer price inflation (%)	16.3	18.2	15.1

Source: Bank of Korea.

Table 10.3 Key economic indicators of the 1980s

	1980	*1986*	*1989*	*Annual change % 1980–1989*
Per capita GNP (US$)	1,597	2,568	5,210	12.2
Real GDP growth rate (%)	−2.7	11.6	6.4	7.8
Current account balance (billion US$)	−5.3	4.7	5.4	–
Gross domestic investment ratio (%)	31.9	29.2	33.8	–
Gross saving ratio (%)	23.2	33.7	36.2	–
Consumer price inflation (%)	29	2.8	5.7	8.4

Source: Bank of Korea.

Table 10.4 Key economic indicators (1990–1996)

	1990	*1993*	*1996*	*Annual change % 1990–1996*
Per capita GNP (US$)	5,866	7,811	11,380	11.9
Real GDP growth rate (%)	9.5	5.5	6.8	7.7
Current account balance (billion US$)	−2.0	1.0	−23.0	–
Gross domestic investment ratio (%)	37.1	35.2	38.8	–
Gross saving ratio (%)	35.9	35.2	34.8	–
Consumer price inflation (%)	8.5	4.8	4.9	6.4

Source: Bank of Korea.

the raw materials imports and facilitate transfer of technology. Such policies were implemented at the expense of consumer goods.

Construction-related investment has been increasing with both public and private sector demand. The government has initiated infrastructure expansion projects including railways, roadways, marine ports and airport facilities. There has been in addition demand for construction of new housing, office buildings and factories from the private sector.

Savings and investment were encouraged at the expense of consumption. The "formula" of success appeared to have worked well until the outbreak of the Asian financial crisis in 1997. The crisis "exposed longstanding weaknesses in South Korea's development model, including high debt/equity ratios, massive foreign borrowing, and an undisciplined financial sector" (CIA World Factbook). The country turned to the IMF for help, and economic performance has been erratic since then. Growth plunged in 1998, recovered again in 1999 and 2000, but only to fall back again in 2001. There were however growth again, spectacularly in 2002 and moderately in 2003, mostly attributable to consumer spending (*The Economist*, 11 April, 2002).

South Korea attains a GDP per capita of US$17,700 (2003 est.) at purchasing power parity. In 2002, more than half, 58 per cent, of the GDP was contributed by services sector. The industry and agriculture contributed the remaining 38 per cent and 4 per cent of the GDP respectively. Main natural resources include coal, tungsten, graphite, molybdenum, lead, hydropower potential. Major industries include electronics, telecommunications, automobile production, chemicals, shipbuilding and steel. In 2002, major exports partners included the three economies of China, Hong Kong and Taiwan (25 per cent), the US (20 per cent) and Japan (9 per cent). Major exports included Semiconductors, wireless telecommunications equipment, motor vehicles, computers, steel, ships, and petrochemicals. In the same year of 2002, major imports partners include Japan (20 per cent), the US (15 per cent), China (11 per cent) and Saudi Arabia (5 per cent). Major imports included machinery, electronics and electronic equipment, oil, steel, transport equipment, organic chemicals and plastics. In 2003, South Korea imported more than exported: US$201.3 billion f.o.b.versus US$175.6 billion f.o.b. (CIA World Factbook).

Agriculture

Population living in rural areas has quickly shrunk from more than 60 per cent at the start of the First Five-Year Economic Development Plan in the early 1960s to about 20 per cent now. The rapid urbanization has led to rapid decrease of available farmland. Demand soon outstripped supply, and by the late 1980s about half the country's needs, mainly wheat and animal feed corn, was imported. There was a major land reform in the late 1940s whereby land ownership was spread to the rural peasantry. However, individual holdings were either too small or too spread out for families to depend their livelihood on (http://countrystudies.us/south-korea).

With the government's attention focused on manufacturing and services, agriculture has remained the "the most sluggish sector of the economy", according to the Federal Research Division of the Library of Congress (http://countrystudies.us/south-korea). Agriculture, forestry and fishery together contributed less than 5 per cent to the overall GDP in 2001 (Table 10.5). Major crops include rice and barley.

Due to rapid industrialization and urbanization, there had been shortage of young labour. Wages have been rising all the time causing farming more and more expensive. Yet, the gap between rural and urban incomes has continued to widen since the 1980s. In the late 1980s, the average farming household income was about 80 per cent of the urban one. The prospects on the farm have not been helped by such government's programmes as the New Community Movement (known as the Saemaul Movement) that in the late 1980s eventually failed due to corruption. Much of the Saemaul Movement effort was directed at the construction of roads, bridges, irrigation ditches and common compost plots to help the villages as a whole (http://countrystudies.us/south-korea).

Transportation

In 1999, highways in South Korea totalled 86,990 kilometres, of which 64,808 kilometres were paved including 1,996 kilometres of expressway. There were a total of 3,125 kilometres of railway in 2002 (CIA World Factbook). With its seven lines extending a total length of 217.3 kilometres in 1998, Seoul's underground system was the eighth-largest in the world. It carried 4.4 million people each day, some 32 per cent of the city's daily commuter load. Railway, on the other hand, plays an important role on inter-city passenger and freight transport. There are over 3,000 operational kilometres, not to include the Seoul–Pusan high-speed railway that extends some 412 kilometres. When finally completed in

Table 10.5 Agriculture in the national economy

		1970	*1980*	*1990*	*1995*	*1996*	*1997*	*1998*	*1999*	*2000*
Population										
Total	Thousand	32,241	38,124	42,869	44,609	45,093	46,165	46,287	46,617	47,008
Agriculture	Thousand	14,422	10,827	6,661	4,851	4,692	4,468	4,400	4,210	4,031
% of total		44.7	28.4	15.5	10.8	10.3	9.7	9.5	9.0	8.6
Employment										
Total	1,000 won	9,167	13,683	18,085	20,377	20,764	21,048	19,994	20,281	21,061
Agriculture	1,000 won	4,846	4,654	3,237	2,541	2,405	2,324	2,480	2,349	2,288
% of total		52.9	34	17.9	12.5	11.6	11	12.4	11.6	10.9
GDP										
Total	Billion won	2,771	36,857	178,262	348,979	386,438	416,018	376,717	425,345	465,183
Agriculture	Billion won	645	5,612	15,592	23,012	24,545	24,056	22,632	24,015	24,332
% of total		23.3	15.2	8.7	6.6	6.4	5.8	6.0	5.6	5.2

Sources: Statistical Yearbook of Agriculture and Forestry, Ministry of Agriculture and Forestry (MAF), Korea, 1998; Korea Statistical Yearbook, Korea National Statistical Office, 2001.

2010, the journey takes slightly less than two hours, thus brining most areas of the country within a half day's travel distance (http://www.koreainfogate.com/ and http://www.newsworld.co.kr/cont/0106/24.html).

Labour

The total labour force was 23 million in 2003 with an unemployment rate of 3.4 per cent. In 2001, 69 per cent of it worked in the services sector. Labour force in the industry and agriculture sectors made up the remaining 22 per cent and 10 per cent respectively. Since 2003, the work week has been shortened from six days to five, following a legislation approved by the National Assembly (CIA World Factbook). The reduction of working day is part of the "new vision for industrial peace", labour policy formulated and implemented so that there is "more time for families" (www.korea.net). Wages increased substantially in the late 1980s, partly attributable to large numbers of workers moving to the Middle East to work on construction projects (http://countrystudies.us/south-korea). In 2003, the average urban household managed savings amounting to more than 22 per cent of the income. Monthly total income and monthly expenditure were 2,939,000 won and 2,280,000 won respectively, leaving a net saving of 659,000 won.

The Korea Labor Institute (KLI) was founded as a government-sponsored research organization in May 1988 to focus on policy-oriented research on labour issues. According to the KLI (http://ns.kli.re.kr/) in 2003, the construction sector contributed 50,309 billion to, that is about 7.6 per cent of, the GDP of 662,474 billion won at 2000 constant price. In the same year, the construction sector employed 1,816,000 persons, or 8.2 per cent, out of the total of 22,139,000 persons employed. The construction sector employed 8.2 per cent of the total working population but contributed 7.6 per cent of the GDP only. That is, construction labour productivity is slightly less than the overall average. Total wages earned by the average construction worker was also 9.2 per cent lower than the overall average. In 2003, wages including fringe benefits amounted to 2,350,000 won per month for an average construction worker. The overall average was 2,587,000. Construction workers, however, worked about 7.6 less hours than the average. In 2003, work hours per month in the construction and "all" industries totalled 199.3 and 215.8 respectively. We could derive from the figures that the Korean construction sector is still relatively labour intensive than the average. Construction workers work less hours and also earn less money than the national average.

10.2 The construction sector since 1997: an executive summary

South Korea, like many other East Asian countries, had experienced severe economic depression since the foreign currency crisis in 1997. Bail-out financing from the IMF helped South Korea to overcome the immediate financial panic but the difficulties facing South Korea were far from being over. Since 2002, the South Korean government's efforts and low interest rate policy had shown some effects and the economic situation began to improve. The forecast for 2002s economic performance was being adjusted upwards continually. As of 2002, economic growth was around 7.5 per cent and an unemployment rate of 6.4 per cent was expected. Also the consumer price index change was expected to be 0.9 per cent compared to 7.5 per cent in 1998. Consumption and investment were decreasing dramatically with construction investments leading the trend. The foreign trade sector fared better due to the low won, and in 2002, we hoped for a 25 billion US$ trade surplus which was expected to help build up the foreign currency reserve.

Due to its demand-driven nature, the construction sector felt the impact of the economic crisis more than any other industry. The industry enjoyed healthy growth during the 1990s with the increase in housing demand and then infrastructure demand. The situation completely changed in 1998. The already oversized industry expected its total contract value to drop by more than 30 per cent due to the sharp decrease in construction investment. Moreover, faced with interest rates higher than 20 per cent, many contractors were going bankrupt and the rest were hovering for survival. The government had tried its best to boost expenditure on construction so as to support employment and protect consumers. In 1999, the construction market had shown some symptoms of recovery due to the low interest rate policy. It was expected that the construction investment growth rate would achieve 0.9 per cent at the end of 1999.

The property market had been struck by the financial crisis and the ensuing structural reforms. Land price had decreased dramatically since the economic crisis. In 1998, overall land prices decreased 13.06 per cent, in particular, residential land prices decreased by 14.68 per cent and industrial land prices decreased by 14.03 per cent. Comparatively, commercial land prices exhibited less fluctuation. Housing prices decreased 18.44 per cent in 1998. Apartment prices, in particular, decreased by 20.18 per cent. Average rents for office space showed the same behaviour. In comparison, the rents for office space in bigger cities decreased more drastically. However, currently housing prices and rents recovered more than before the IMF regime, especially around the Seoul area.

Foreign participation in the South Korean construction market was minimal. But foreign companies applying for license were slowly increasing and many specialty contractors were already active in government procurement projects. In the architectural and engineering fields, cross-border supply of services had been on going for quite some time and several joint ventures had been formed. In the supervision and consulting fields, 13 companies were participating in projects such as Seoul–Pusan High Speed Railway, Inchon New Airport, West Sea Highway, Gayang Bridge, etc. Tendering and contracting systems were completely revamped and many deregulation and liberalization measures had been taken. Tendering methods regulated by region and by the size of the contractors were replaced by more competitive modes.

Liberalization and globalization of business activities is a worldwide phenomenon with no exception in the construction industry after establishment of WTO. Korea accepted the trend by participating in many related international treaties and by opening up of its construction market. Overseas construction had been very active and perhaps this sector was expected to prove to be a beneficiary of the liberalization measures. South Korea had prepared to welcome a foreign presence in the domestic market so as to utilize the comparative advantage fully, to learn advanced technology, procure better materials, augment investments and improve production systems with the help of foreign companies. Based on the cooperation in the domestic market, Korea also wishes to assist other nations in building up their construction stock. The Government was adopting a more practical and balanced approach to overseas construction, rather than one concerned also with prestige and diplomacy.

As of 2002, the situation of the construction industry and Government policies towards it were centred on two factors: the financial crises and globalization. Due to the economic crisis, the construction industry was in great difficulty. This was expected to last only for the short term and that the economy would recover within 2 years. The long-term trends were those of globalization, localization and information proliferation, which meant increased competition for local contractors, which in turn was expected to create a buyers' market and thus elevate the importance of R&D investment in the construction industry. There was no conflict in the short- and long-term factors, which pressurizes the industry to restructure.

The major tasks envisioned in the 'Basic Plan for Promotion of Construction Industry' are as follows:

- to ensure fair competition;
- to raise productivity;
- to assure environmental friendliness and construction quality;
- to set up a new harmonious role between general and specialty contractors;
- to promote relations between general and specialty contractors;
- to strengthen production factors (such as technology, manpower, materials) and finance capacity; and
- to encourage both inbound and outbound overseas construction.

Through these and other measures, it was hoped that the construction industry could compete head to head with top-level construction companies in the twenty-first century.

Most economies in Asia had recovered from the financial crisis faster than expected. In this regard, the Korean government and international financial institutions (World Bank, IMF) had played an indispensable role in stopping the spread of the financial crisis.

The Korean economy had improved impressively since 1997, when the Kim Dae Jung administration came into office in the depths of the financial crisis. The fundamental causes of Korea's economic troubles could be traced to structural weaknesses and pervasive moral hazard from lack of market discipline. The government immediately tackled the situation and overcame the crisis far more quickly than expected.

Korea's economic recovery has been remarkable. GDP growth recorded 10.7 and 8.8 per cent in 1999 and 2000 respectively, and was expected to be around 3.7 per cent for the first quarter of 2002. Other major macroeconomic indicators show equally favourable developments, including a historically low inflation rate of 0.8 per cent in 1999 and 2.3 per cent in 2000 and 5.2 per cent during the first quarter in 2001.

It was expected that Korea would continue to address the issues head-on and strengthen its economy through openness and reform. The reform based on market principles rather than political interests would continue to be promoted. Based on these efforts, 5 per cent economic growth, a drop to a near 3 per cent in unemployment rate, and 6 per cent increase in exports were the forecast for 2002.

Moreover, it was expected that hosting the 2002 World Cup and Pusan Asian Games would stimulate the Korean economy and would serve as a great opportunity to introduce the new Korea to the world. In addition, the construction of Incheon International Airport, Pusan New Port, Seoul–Pusan High-Speed Railway among other infrastructure facilities, alleviation of certain regulations, and incorporation of international standards would place Korea as the hub of Asia.

The construction industry played a major role in the rapid economic growth of Korea. To support economic development, the construction market grew faster than the average, increasing its share in GDP. From the late 1980s more than 20 per cent of GDP was invested in the construction industry, a very high figure by any standard. We believe that the high construction investment ratio was one of the main reasons for the vigorous economic growth in Korea.

The growth of the construction sector plunged down to 10.23 per cent in 1998 (2.3 per cent in 1997) after recording strong growth for consecutively 8 years since 1989. However, the economic recovery action plan implemented by the Government had shown an improvement in the performance of the construction sector in 1999. Since 1999, the construction market showed symptoms of little recovery based on the recovery action plan implemented by the government but showed a slow recovery in the long term. Construction investment growth rate recorded 0.9 per cent in 1999, although it recorded −4.1 per cent in 2000 and 5.4 per cent in 2001. A 6.7 per cent growth rate was recorded for the first quarter of 2002.

The construction sector, by its nature, was affected by the nation's economic situation more than any other industry. As evident in the past trend, the construction industry enjoyed healthy growth during the 1990s with the increase in housing demand and then infrastructure demand.

But activities in all construction industry sectors were more adversely affected by the financial crisis in 1998. There was a marked decline in construction activities caused mainly by deferment of selected new infrastructure projects and the slow implementation of existing infrastructure projects as well as postponement of non-residential projects.

In response to increased demands of government expenditure for the welfare system and future investment needs in new knowledge industries, there was the need to relieve the burden on government by actively inducing private investment in infrastructure. The fund reserves acquired through private investment were expected to be utilized in order to achieve a balanced development across the regions. Based on Article 7 of the Act on Private Participation in Infrastructure (PPI) and the 2001 Annual Plan for PPI, this Ten-Year Plan for PPI (The Plan) was formulated in line with the 2002 investment policy direction and by each sector's priority level. The total project cost of 179 selected private investment projects was 63.2 trillion won. Of these, the net private investment amount excluding government subsidy was estimated to be around 46.7 trillion won.

Government policy environment and the present situation had two dimensions. Due to the IMF crisis, the construction industry was in great difficulty but this was expected to be of short-term effect which could be overcome in future. The long-term trends were those of globalization, localization and 'informatization', which meant increased competition, which in turn was expected to create a buyers' market and thus elevate the importance of R&D investment in the construction industry. There was no conflict in the short- and long-term factors, both of which pressurized the industry to restructure.

Many analysts believed that the Korean economy had already bottomed out in the fourth quarter of 1998, or for some, in the first quarter of 1999. Either way the Korean economy had achieved a full recovery in 2001, even though Korea had finished paying off the IMF debt earlier than expected.

Activity in the construction industry is highly cyclical. Construction activities have recently begun to emerge from the cyclical downturn which started at the beginning of 1999. It is believed that the construction economy had already bottomed out in the fourth quarter of 1998. Accordingly, construction investment had slowly achieved a recovery in 2001.

10.3 Macroeconomic review and outlook

In order to ride out the economic turmoil, South Korea had gone through a complete structural change. The financial system and business sector were being reformed such that less competitive firms were led out of the market. Restructuring towards leaner organizations was also under way in the government policies. The most difficult task might be changing the peoples' attitude towards a more open and free economy. It was believed that through this painful process, South Korea would regain confidence from the world and thus induce more foreign investment. If the reform is successfully carried out as planned, the economy will recover significantly from 2003. Otherwise, many experts forecast 3 or 4 more years of economic hardship. Low interest rate policy, which was enforced at the end of 2001, had great effects on the recovery process. Tables 10.6 and 10.7 show the main macroeconomic indicators and the construction share of the GDP respectively.

10.3.1 Main macroeconomic indicators

South Korea, like many other East Asian countries, had experienced a severe economic depression since the foreign currency crisis in 1997. A financial bail out from the IMF helped South Korea overcome the immediate financial problem but the difficulties facing the South Korean economy persist. Since 2002, South Korean government efforts and a low interest rate policy have shown some effects and the economic situation have begun to recover. As of summer 1999, economic growth was around 7.5 per cent and an unemployment rate of 6.4 per cent was expected in 2002. Also, the consumer price index change was expected to be 0.9 per cent, compared with 7.5 per cent in 1998. Consumption and investment were decreasing dramatically as construction investment led the trend. The foreign trade sector fared better due to the low won, and South Korea was expecting a US$25 billion trade surplus in 1999 which would help build up foreign currency reserve.

Gross domestic product indices

Korea's GDP had resumed its growth mode after the setback of 1997. In 2001, GDP attained US$422.2 billion, much more than the 1998 figure of US$317.7 billion.

Table 10.6 Main macroeconomic indicators (base year: 1995)

	1995	*1996*	*1997*	*1998*	*1999*	*2000*	*2001*	*2002.6**
GDP and components								
GDP at real prices* (bn won)	377,349.8	402,821.2	423,006.7	394,710.4	437,709.4	478,532.9	493,025.5	250,006.9
GDP at current market prices (bn won)	377,349.8	418,479.0	453,276.4	444,366.5	482,744.2	521,959.2	545,013.3	143,422.0
GDP growth (%)	8.9	6.8	5.0	−6.7	10.9	9.3	3.0	6.0
Primary sector (mining, agriculture, etc.)	25,129.8	25,895.3	26,993.2	24,907.0	26,242.2	26,299.2	27,116.3	8,342.9
% growth	6.1	3.0	−4.1	8.3	−5.1	−0.2	3.0	2.1
Agriculture, forestry and fishing	–	–	–	23,569.4	24,833.4	25,318.3	25,664.7	2,221.6 (2002.3)
% growth	–	–	–	−6.6	5.4	2.0	1.4	1.3 (2002.3)
Mining and quarrying	–	–	–	1,337.6	1,408.8	1,443.9	1,451.6	805.3
Manufacturing sector	110,826.9	118,342.7	126,117.2	116,743.8	141,295.1	163,733.1	166,506.2	86,056.0
% growth	11.3	6.8	6.6	−7.4	21.0	15.9	1.7	6.6
Services sector				106,638.4	112,659.0	117,721.7	123,255.8	6,459.6
% growth	10.3	5.3	5.5	−7.2	11.9	9.5	4.1	1.1
Construction sector	42,564.1	45,513.3	46,137.4	42,161.3	38,305.8	37,125.3	39,220.1	17,597.6
% growth	8.8	6.9	1.4	−8.6	−9.1	−3.1	5.6	5.5
Demographic indicators								
Population (1000)	45,093	45,545	45,991	46,286	46,616	47,008	47,342	47,639
Population growth rate (%)	1.01	1.00	0.98	0.72	0.71	0.84	0.71	0.63
Total labour force (1000)	20,853	21,243	21,662	21,456	21,634	21,950	22,181	22,274
Labour force growth rate (%)	2.3	1.8	1.9	−1.0	0.8	1.4	1.0	0.4
Unemployment rate	2.0	2.0	2.6	6.8	6.3	4.1	3.7	2.7
Financial indicators								
Short-term interest rate	11.73	13.53	18.55	15.22	6.81	7.08	5.32	4.74
Long-term interest rate	9.24	11.44	15.32	12.80	8.72	8.50	6.66	6.94
Changes in consumer price index		86.4	90.2	97.0	97.8	100.0	104.1	106.3
Base lending rate (Commercial banks)				15.25	9.29	8.53	7.77	7.01
Base lending rate (Finance companies)				16.82	13.21	11.58	9.99	4.11
Change against US$	774.7	844.2	1,415.2	1,207.8	1,145.4	1,259.7	1,326.1	1,326.4

Note
* 2002.6 exchange rate to US$.

Table 10.7 The construction share of the GDP (current prices)

Year	*1995*	*1996*	*1997*	*1998*
%	11.3	11.6	11.6	10.1

The growth rate of GDP since the 1997 financial crisis also indicated a return to economic health. Initial restructuring efforts achieved quick results much faster than most analysts had forecast. From its 6.7 per cent contraction in 1998, the GDP growth rate soared to 10.9 per cent in 1999, 9.3 per cent in 2000 and 3.0 per cent in 2001 (Table 10.8 and Figures 10.1 and 10.2).

Table 10.8 Gross domestic product indices (unit: KRW tril., US$ billion per cent)

	1995	*1996*	*1997*	*1998*	*1999*	*2000*	*2001*	*2002.6**
GDP in KRW	377.3	418.5	453.3	444.4	482.7	522.0	545.0	143.4
GDP in US$	489.4	420.0	476.6	317.7	405.8	461.7	422.2	119.3
Annual average increase rate	8.9	6.8	5.0	−6.7	10.9	9.3	3.0	6.3

Note
* 2002.6 exchange rate to US$1201.8 won/$.

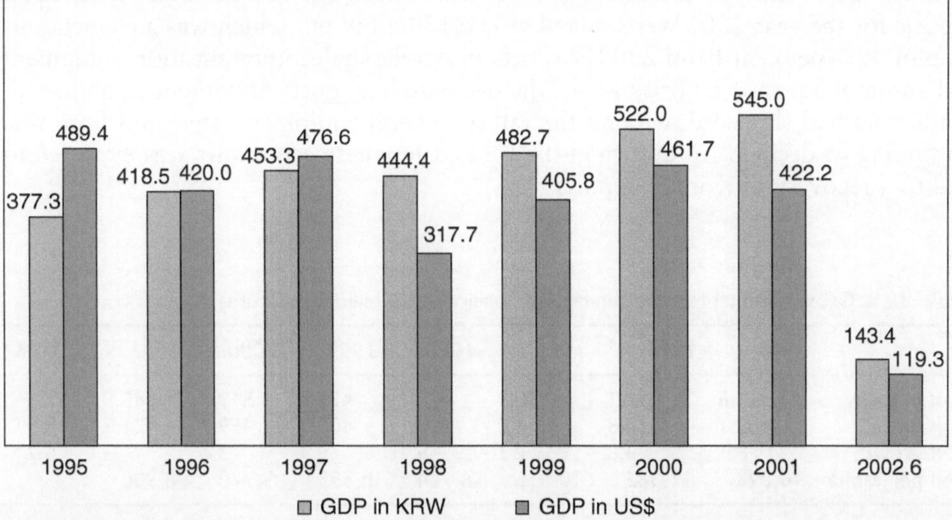

Figure 10.1 Gross domestic product in KRW, US$.

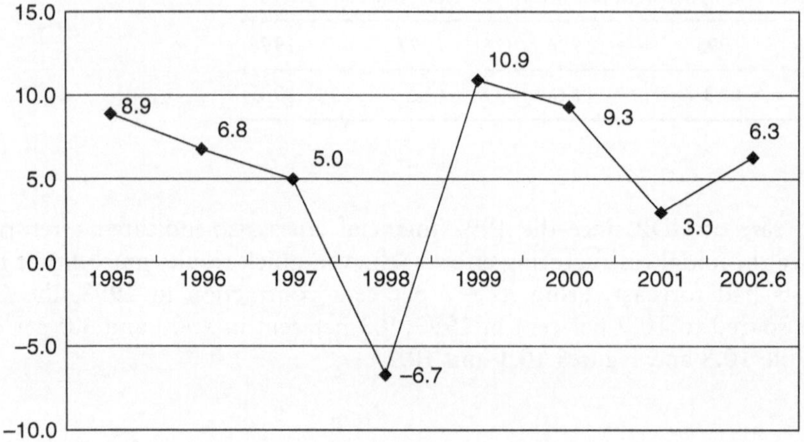

Figure 10.2 Gross domestic product: average increase rate.

Gross national income

In much a mirror reflection of trends for the GDP, the Gross National Income (GNI) statistics had registered great strides since the financial crisis, but have moderated in recent quarters (Table 10.9). GNI per capita vaulted back to US$8,900 in the year 2001, jumping from US$6,744 in 1998 (Figures 10.3 and 10.4).

Exports and imports

The global synchronization of Korea's economy, particularly in concert with economic uncertainty in Japan and the US, was reflected in the latest statistics. Total exports for the year 2001 were valued at US$150.4 billion, which was a tremendous drop of 12.7 per cent from 2000. Exports in wireless telecommunication equipment and automobiles showed brisk gains, the decrease in exports of computers and semiconductors had slowed down and the fall of general equipment, steel and fibre was continuing its decline. An upturn in the US and Japanese economies was expected to lead to a recovery of Korean exports.

Table 10.9 Gross national income (unit: KRW trillion, US$ billion, per cent)

	1995	*1996*	*1997*	*1998*	*1999*	*2000*	*2001*	*2002.6**
GNI in KRW	37.6	417.1	450.9	436.6	476.6	519.2	543.9	142.5
GNI in US$	488.1	518.3	474.0	312.1	400.7	459.2	421.3	118.5
GNI growth	8.1	4.8	2.1	−9.1	9.4	3.6	1.3	6.2
GNI per capita	10,823	11,385	10,315	6,744	8,595	9,770	8,900	–

Note
* 2002.6 exchange rate to US$1201.8 won/$.

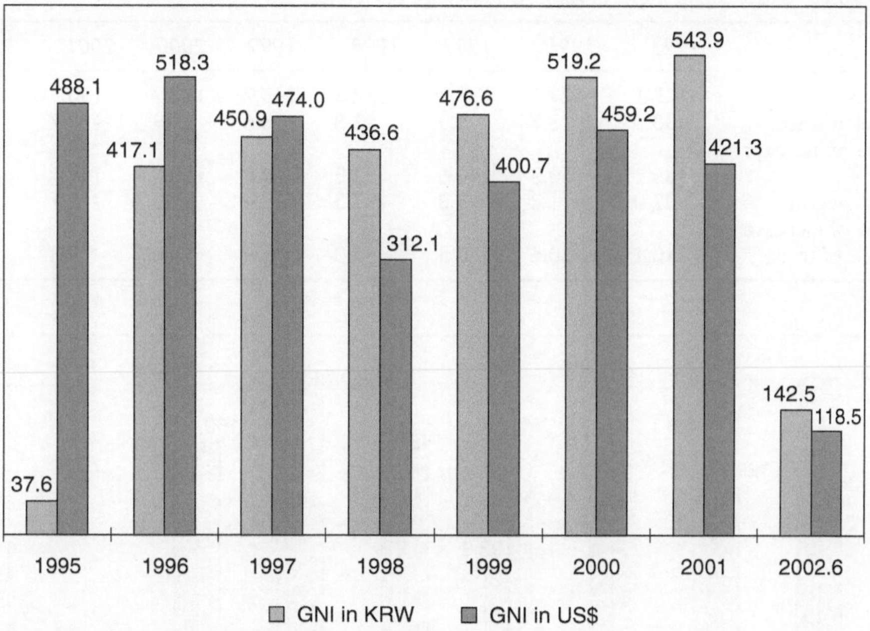

Figure 10.3 Gross national income in KRW, US$.

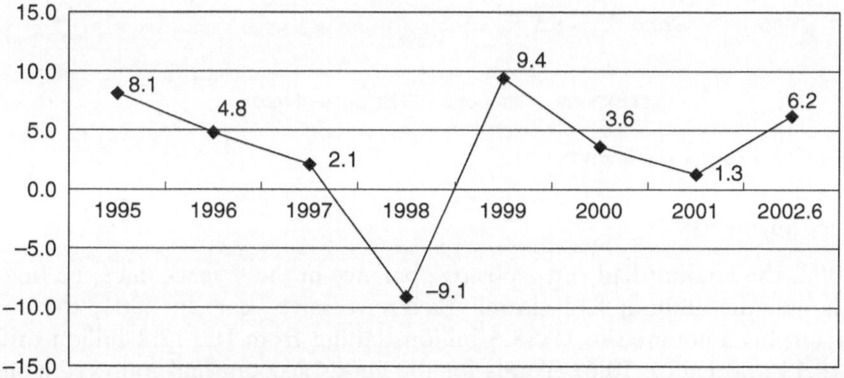

Figure 10.4 Gross national income: average increase rate.

Import activity displays split movements. Consumer goods imports had risen, but capital goods and raw materials imports had suffered much greater declines. The decrease in domestic facility investment had led to a drop in demand for the latter. In the year 2001, Korean imports were valued at US$141.1 billion, which was a 12.1 per cent decline from the previous year (Table 10.10 and Figure 10.5).

Table 10.10 Exports and imports (unit: US$ billion, per cent)

	1995	*1996*	*1997*	*1998*	*1999*	*2000*	*2001*	*2002.8*
Exports	125.1	129.7	136.2	132.3	143.7	172.3	150.4	14.2
Annual average rate of increase	30.3	3.7	5.0	−2.8	8.6	19.9	−12.7	20.4
Imports	135.1	150.3	144.6	93.3	119.8	160.5	141.1	12.9
Annual average rate of increase	32.0	11.3	−3.8	−35.5	28.4	34.0	−12.1	13.8
Balance of trade	−10.1	−20.6	−8.5	39.0	23.9	11.8	9.3	−27.1

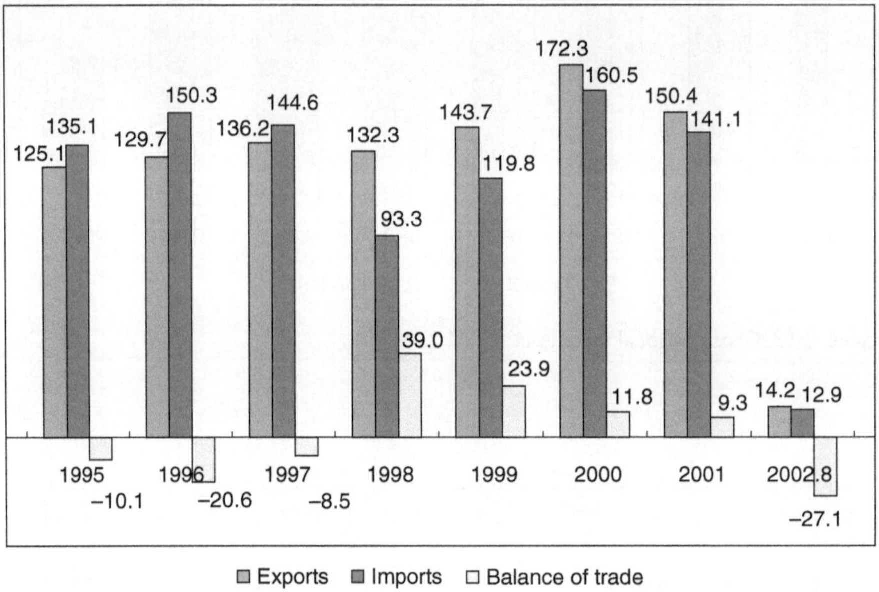

Figure 10.5 Exports and imports.

Balance of payments

As of 2002, the account had run a positive balance in the 4 years since the financial crisis, though the amount had shrunk each successive year. In 2001, the current account surplus amounted to US$8.6 billion, falling from US$12.2 billion in 2000 (Table 10.11 and Figure 10.6). Totals for the goods account had followed a similar pattern.

Table 10.11 Balance of payments (unit: US$ billion)

	1995	*1996*	*1997*	*1998*	*1999*	*2000*	*2001*	*2002.7*
Current account	−8.5	−23.0	−8.2	40.4	24.5	12.2	8.6	3.6
Goods account	−4.4	−15.0	−3.2	41.6	28.4	16.9	13.4	8.6

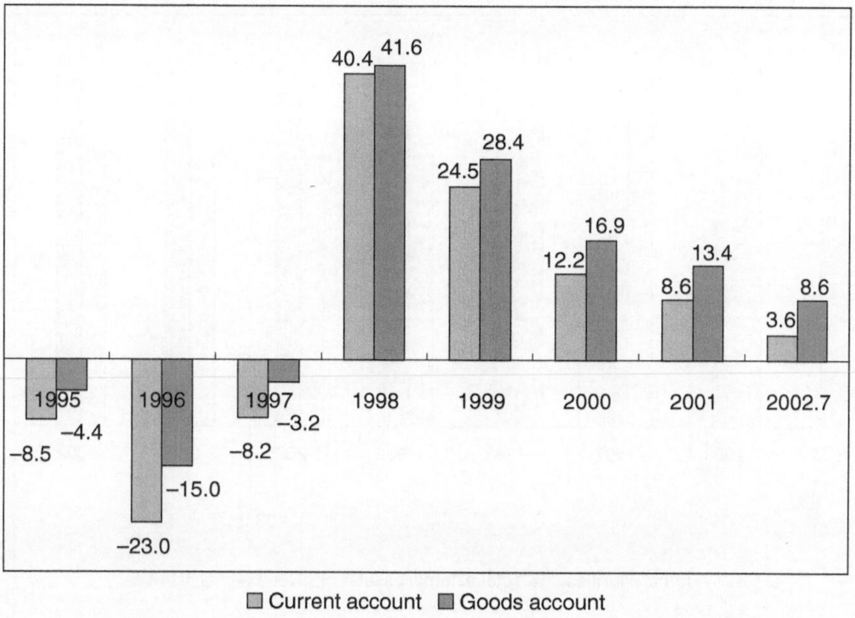

Figure 10.6 Balance of payments.

Total external liabilities and assets

The nation's total external liabilities amounted to US$119.9 billion in 2001, which was a steady decline since 1997. Total external assets had sustained a growing trend, but it has slightly declined in 2001. External assets rose up to US$161.9 billion in 2001, from US$105.2 billion at the end of 1997 (Table 10.12 and Figure 10.7).

Foreign reserves

The foreign reserves have surged dramatically since 1997 even as the nation continues with early repayment of IMF loans. Reserves leaped to US$106.1 billion by the end of March 2002, an increase of more than tenfold compared to the end of 1997 reserves of just US$8.9 billion (Table 10.13 and Figure 10.8).

Table 10.12 Total external liabilities and assets (unit: US$ billion)

	1995	*1996*	*1997*	*1998*	*1999*	*2000*	*2001*	*2002.3*
Total external liabilities	127.5	163.5	159.2	148.7	137.1	131.7	119.9	123.6
Total external assets	91.9	109.7	105.2	128.5	145.4	164.7	161.9	169.6
Net external liabilities	35.5	53.8	54.1	20.2	−8.3	−33.0	−42.0	46.1

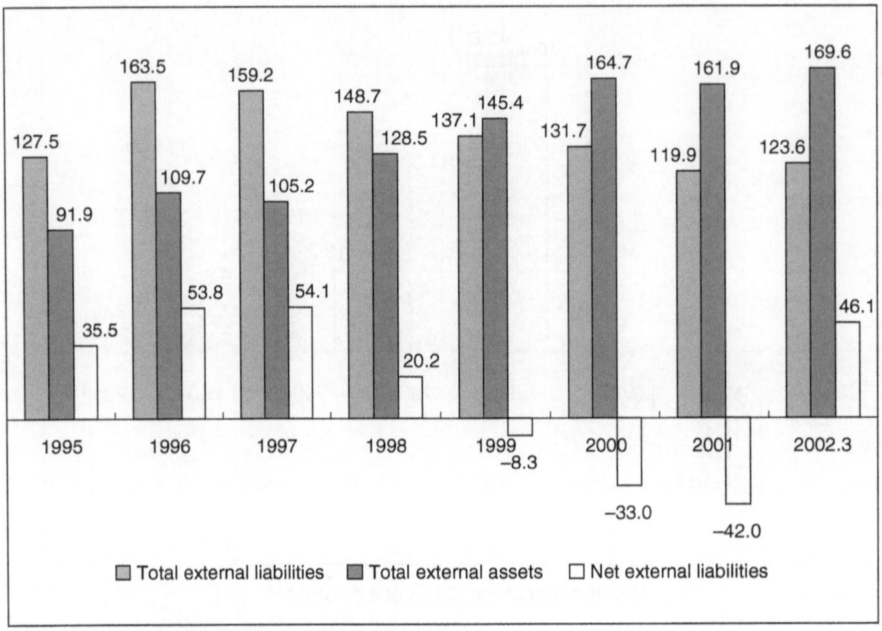

Figure 10.7 Total external liabilities and assets.

Table 10.13 Foreign reserves (unit: US$ billion)

	1995	*1996*	*1997*	*1998*	*1999*	*2000*	*2001*	*2002.6*
Foreign Reserves	29.4	29.4	8.9	48.5	74.1	96.2	102.8	112.4

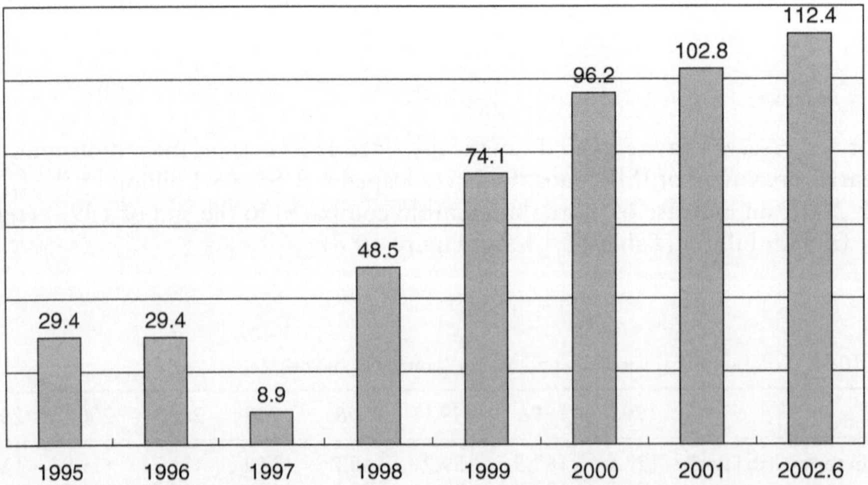

Figure 10.8 Foreign reserves.

Foreign direct investment

International investors have quickly responded to the Kim Dae Jung administration's campaign to eradicate barriers to foreign direct investment (FDI). In just the 4 years from 1998 through 2001, Korea attracted in excess of US$52 billion of FDI, which was more than for all the years of 1962 through 1997 combined.

In 2001, FDI totalled US$15.7 billion, a stunning jump from the US$3.2 billion of FDI that Korea attracted in 1996. The main characteristics of FDI display continued investments from the EU, US and Japan: an increased number of small-scale investments of under US$5 million; increased investments into the service sector and a decline into the manufacturing sector; and increased investments via acquisition of existing shares in domestic corporations. This last trend was exemplified in the Korea Stock Exchange, as foreign shareholdings account for 36.6 per cent of total market share in 2002 (Table 10.14 and Figure 10.9).

Table 10.14 Foreign direct investment (FDI) (unit: US$ million, per cent)

	1995	*1996*	*1997*	*1998*	*1999*	*2000*	*2001*	*2002.6*
Foreign direct investment (US$ million)	1,941	3,203	6,971	8,852	15,541	15,697	11,870	1,323
Market share of foreign investment (%)	11.9	13.0	14.6	18.6	21.9	30.1	36.6	

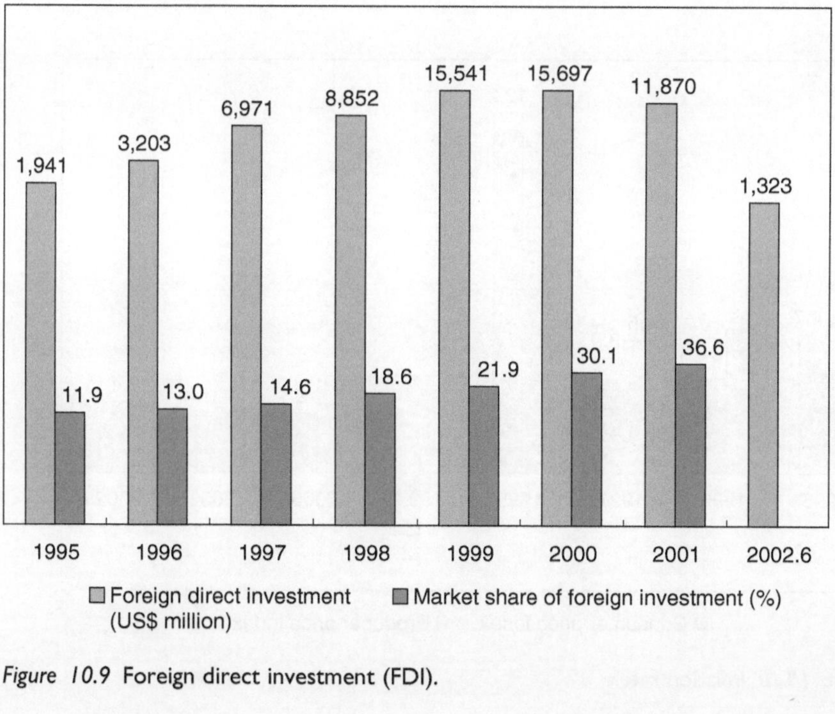

Figure 10.9 Foreign direct investment (FDI).

Inflation rate

The Consumer Price Index (CPI) was held in check through the years 2000 and 2001, recording just 2.3 and 4.1 per cent increases respectively. From 2002, however, prices had spiked, as a result of increased energy costs, a severe winter, and an upward trend in exchange rates. The Producer Price Index (PPI) had also displayed a rising trend, but at a more moderate rate. The index rose only 1.9 per cent in 2001 and 2.0 per cent in 2000. In 1998, the crisis-affected PPI leaped a startling 12.2 per cent (refer Table 10.15 and Figure 10.10).

Unemployment

Following an average unemployment rate of 3.7 for the year 2001, the government had set a goal to constrain the jobless rate to less than 3.5 per cent for 2002. The unemployment rate was immensely better than the average unemployment rates of 6.3 per cent in 1999 and 6.8 per cent in 1998.

In total, there were some 802,000 jobless workers in Korea as of February 2002 (Table 10.16 and Figure 10.11). Since 2001, the number of participants in economic

Table 10.15 Inflation rate (unit: per cent)

	1995	*1996*	*1997*	*1998*	*1999*	*2000*	*2001*	*2002.8*
Consumer price index	4.5	4.9	4.5	7.5	0.8	2.3	4.1	2.4
Producer price index	4.7	3.2	3.9	12.2	−2.1	2	1.9	1.6

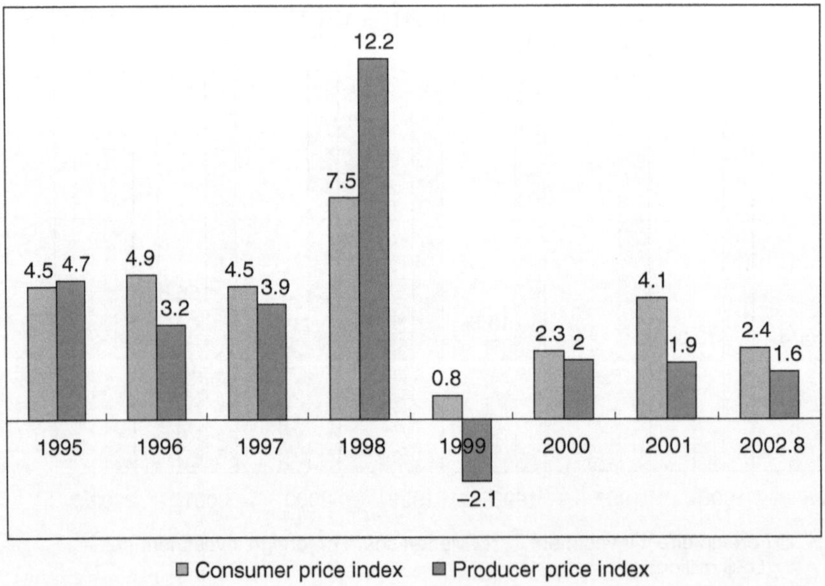

Figure 10.10 Inflation rate.

Table 10.16 Unemployment (unit: thousand persons, per cent)

	1995	*1996*	*1997*	*1998*	*1999*	*2000*	*2001*	*2002.8*
Unemployed	420	426	556	1,462	1,353	889	819	658
Unemployment rates	2.0	2.0	2.6	6.8	6.3	4.1	3.7	2.9

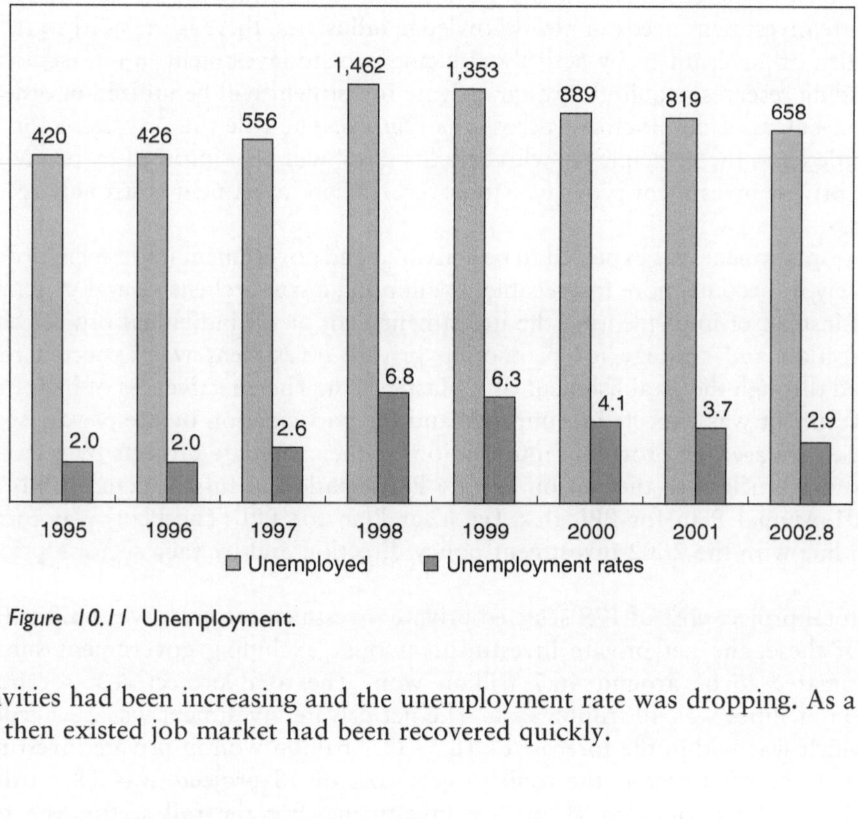

Figure 10.11 Unemployment.

activities had been increasing and the unemployment rate was dropping. As a result, the then existed job market had been recovered quickly.

10.4 The construction industry

10.4.1 Construction and economic development

The construction industry played a major role in the rapid economic growth of South Korea. To support economic development, the construction market grew faster than the average, increasing its share of GDP. From the late 1980s, more than 20 per cent of GDP was invested in the sector, a very high figure by any standard. This high construction investment ratio was one of the reasons for the compact economic growth in South Korea.

The construction industry also reaped the rewards of economic growth. Entry into the market was regulated by a licensing system in which existing contractors could increase their size and efficiency. The wages of construction workers and the prices of construction materials were also held low by government guidelines. These measures helped the industry maintain its competitiveness in both domestic and overseas construction markets.

However, with political reforms within Korea, the talks at the Uruguay Round, and the joining of OECD in the 1990s brought about a completely new business environment. The construction market opened up gradually and rules and regulations related to the industry were revised to conform to international business standards.

In response to increased demands of government expenditure for the welfare system and future investment needs in new knowledge industries, there is the need to relieve the burden of government by actively inducing private investment in infrastructure. The funding reserves acquired through private investment will be utilized in order to achieve a balanced development across the regions. The role the private sector will be extended and the principle of 'who benefits pays' will be reinforced by promoting feasible private investment projects. Moreover, foreign investment is strongly recommended.

Private investment was expected to be activated and government investment policies were likely to become more foreseeable by adopting a comprehensive and systematic method instead of implementing the investment plan on an individual project basis. The direction and strategy for promoting private investment was expected to be proposed through the establishment of a Master Plan. The effectiveness of infrastructure investment was expected to improve and the participation by the private sector would be increased by providing information on the candidate projects by sectors.

Based on Article 7 of the Act on Private Participation in Infrastructure (PPI) and the 2001 Annual Plan for PPI, this Ten-Year Plan for PPI (The Plan) was formulated in line with the 2002 investment policy direction and by each sector's priority level.

The total project cost of 179 selected private investment projects was 63.2 trillion won. Of these, the net private investment amount excluding government subsidy was estimated to be around 46.7 trillion won. The total project cost for transportation facilities was 40 trillion won. The net private investment was 26.1 trillion won, which was within the forecast of 18.5–39.7 trillion won in private investment needs. For the road sector, the total project cost of 18 projects was 18.9 trillion won with 13.2 trillion won as private investment. For the rail sector, the total project cost of 23 projects was 13.3 trillion won with 8.0 trillion won as private of investment. For the port sector, the total project cost of 29 projects was 7.0 trillion won with 4.2 trillion won as private investment. For the logistics sector, the total project cost of 2 projects was 0.9 trillion won with 0.7 trillion won as private investment.

The total project cost of environmental, tourist and energy facilities was around 23.2 trillion won with 20.6 trillion won as net private investment. But, when excluding the energy sector that was expected to be privatized, the scale of pure private investment was 6.2 trillion won. In the case of the environmental sector, the total project cost of 89 projects was 3.9 trillion won with private investment of 2.2 trillion won. The scale of private investment for other facilities excluding environmental sector was 4.1 trillion won for 5 tourist projects and 14.4 trillion won for 13 energy projects (Table 10.17).

Table 10.17 Private investment project scale (unit: billion won)

Classification	*Projects*		*Government subsidy*	*Private investment*
	No. of projects	*Total project cost*		
Road				
Expressway	10	11,526.6	3,458.0	8,068.6
National roads	4	1,619.4	458.8	1,133.6
Local roads	4	5,727.9	1,718.4	4,009.5
Subtotal	18	18,873.9	5,662.2	13,211.7
Rail				
Network/metropolitan	5	5,200.1	2,080.0	3,120.1
Urban rail	3	1,329.6	531.9	797.7
LRT	15	6,774.5	2,709.8	4,064.7
Subtotal	23	13,304.2	5,321.7	7,982.5
Port				
Trading port	23	6,001.2	2,400.5	3,600.7
Fishing port	2	140.4	56.1	84.3
Waterfront	3	837.9	335.2	502.7
Passenger facility	1	22.3	8.9	13.4
Subtotal	29	7,001.8	2,800.7	4,201.1
Logistics	2	869.7	173.9	695.8
Transportation facilities total	72	40,049.6	13,958.5	26,091.1
Environmental	89	3,942.5	1,769.3	2,173.2
Energy	13	14,419.3	–	14,419.3
Tourism	5	4,764.7	714.7	4,050.0
Other facilities total	107	23,126.5	2,484.0	20,642.5
Total	179	63,176.1	16,442.5	46,733.6

10.4.2 Productivity

Value-added products

Construction production amounted to 41,964.7 billion won in 1998, a 13.16 per cent decrease compared with 1997. This growth rate was lower than for other industries such as manufacturing (−11.74) and services (−12.86).

The growth rate of construction products recorded at 37,125 in 2000 showed a decrease of 3.1 per cent over 1999. This was remarkable, since other industries showed positive growth rates such as manufacturing (13.7 per cent) and services (9.4 per cent). Construction products increased 5.3 per cent in 2001, however, and 8.2 per cent in the first half of 2002. Compared with manufacturing growth rates of 1.6 and 3.4 per cent in the same period, these rates were remarkable (see Tables 10.18 and 10.19).

Physical measurement of construction productivity

Physical measurement of construction productivity records 47,856 square metre of residential, 25,888 square metre of non-residential, 13,538 square metre of industrial

Table 10.18 Value-added product (at 1995 price: billion won)

Industries	*1995*	*1996*	*1997*	*1998*	*1999*	*2000*	*2001*	*2002.3*
Construction	42,564.1	45,513.3	46,137.4	42,161.3	38,305.8	37,125.3	39,220.2	7,555.5
% change	8.8	6.9	1.4	−8.6	−9.1	−3.1	5.3	8.2
Manufacturing	110,826.9	118,342.7	126,117.2	116,734.8	141,295.1	163,733.1	166,506.2	41,330.3
% change	11.3	6.8	6.6	−7.4	21.0	13.7	1.6	3.4
Services	157,564.2	169,337.1	180,316.6	171,776.3	189,005.3	205,535.9	216,359.9	57,345
% change								
Wholesale and retail trade, restaurants and hotels	8.4	7.8	4.0	−10.9	14.1	9.4	3.88	5.50
Transport, storage and communication	11.1	10.6	13.4	−0.8	14.5	16.7	6.84	7.95
Finance, insurance, real estate and business service	10.4	7.2	5.9	−1.9	5.5	4.7	4.38	8.46
Community, social and personal services	10.3	5.3	5.5	−5.9	11.1	4.5	6.80	10.41
Primary sector	25,129.8	25,895.3	26,993.2	24,907.0	26,242.3	26,299.2	26,762.2	2,610.9
% change								
Agriculture, forestry and fishing	6.6	3.3	4.6	−6.6	5.4	0.1	1.4	6.7
Mining and quarrying	−0.6	−0.1	−0.9	−24.0	5.3	2.2	0.5	16.5

Note

Primary sector includes agriculture, mining, etc.

Table 10.19 Value-added per employee (unit: million won per person at 1995 price)

	1998	*1999*	*2000*	*2001*	*2002.3*
Construction ($)	26.72	25.96	23.32	24.90	4.56
% change	13.85	−2.93	−11.32	6.35	−2.85
Manufacturing ($)	29.95	35.27	38.59	39.65	9.91
% change	6.04	15.08	8.60	2.67	3.43
Services ($)	12.64	13.14	13.79	13.94	3.73
% change	−0.3	3.81	4.71	1.08	4.85
Primary sector ($)	9.96	11.08	11.40	12.10	1.31
% change	−12.45	10.11	2.81	5.79	7.63

Note
Primary sector includes agriculture, mining, etc.

construction in 2001. In 2002, 13,583 and 11,410 square metre were recorded in residential and non-residential construction respectively in the first quarter (see Table 10.20 and 10.21).

10.4.3 Average construction material price

During the early 1990s there was a severe shortage of construction materials due to the overheated construction business cycle. The construction sector was singled out as the culprit behind inflation and the growing trade deficit. Since then, the construction boom has cooled down and the demand and supply of construction materials have been more or less at equilibrium domestically. There were, however, seasonal shortages, especially in the spring season when project orders intensify.

As of 2002, around 62 million tons of cement and 11 million tons of steel beams were produced and consumed domestically. In 1999, the problem was excess capacity rather than shortage. The official prices of major construction materials were influenced by government guidelines but actual transaction values change according to market conditions. During the mid-1990s, the price of construction materials remained stable as shown in Table 10.22.

Table 10.20 Physical measurement of construction productivity (unit: sq. m of built space)

	1998	*1999*	*2000*	*2001*	*2002.3*
Residential	31,156	44,606	41,283	47,856	13,583
Commercial	8,751	11,746	18,882	25,888	11,410
Industrial	2,797	7,824	11,646	13,538	3,507
Others	8,259	8,359	9,247	10,435	3,099
Total	50,965	72,534	81,058	97,717	10,259

Note
Area (sq. m) refers to total built-up area.

Table 10.21 Unit construction cost (unit: million won/sq. m of gross floor area)

	Residential		*Residential (row house)*	*Residential (apartment)*				*Commercial office*			*Industrial (small)*		*Hotel*	
	Standard	*Luxurious*		*RC* (Less than 10 F)	*RC* (11 ~ 15 F)	*RC* (More than 16 F)	*SRC* (11 ~ 15 F)	*RC* (Less than 5 F)	*RC* (6 ~ 15 F)	*SRC* (More than 16 F)	*RC*	*S*	*RC*	*SRC*
1996	0.61	1.53	0.98	0.48	0.49	0.49	–	0.58	0.65	0.79	0.47	0.60	0.74	0.96
1998	0.86	1.89	1.18	0.55	0.55	0.57	–	0.67	0.77	0.95	0.55	0.70	0.87	1.13
2000	0.83	1.83	1.11	0.52	0.52	–	0.90	0.64	0.74	0.91	0.52	0.67	0.95	1.19

Note
Unit prices exclude sub-structural works.
RC: reinforced concrete structure, M: masonry structure, S: steel frame construction and SRC: steel framed reinforced concrete construction.

Table 10.22 Price of construction materials

	Cement in bulk (won per 40 kg)	*Steel bars* (won per tone) #16	*20 mm aggregates* (won per m^3)	*Concreting sand* (won per m^3)	*Common bricks* (won per thousand pieces)	*RMC 210 kg/cm^3* (won per m^3)
1995	2,068	315,250	9,208	9,417	37,000	35,015
1996	2,168	302,333	9,666	9,841	40,000	39,857
1997	2,200	295,333	10,000	10,333	43,000	43,974
1998	2,415	316,100	10,000	10,458	49,000	46,555
1999	2,600	297,916	10,000	9,333	43,000	45,532
2000	2,728	301,666	10,000	9,000	42,000	45,560
2001	3,000	313,333	10,916	9,125	43,000	47,768
2002.6	3,300	335,000	12,000	10,000	45,000	50,930

Note
RMC: Ready-mix concrete.

10.4.4 Construction industry wages

The average wage had increased, but in an erratic fashion. Up to the Seoul Olympics in 1988, wages rose slowly due government wage guidelines. After the Olympics, however, the changed political atmosphere and implementation of a 2 million housing construction plan caused the wages to go up very rapidly. In 1987, the average (daily) wage was 12,903 won. Five years later, in 1992, the wage had more than tripled. For the next 5 years until 1997 the trend in rising wages levelled off. Wages were expected to drop further in 1999 due to decreased construction demand, which was same as in 1998. But, in 2000, salaries and wages increased again (Table 10.23). Then average salaries and wages were increasing steadily until the time of writing.

10.4.5 Average sectoral wages

Construction and manufacturing sector wages went down from 1997 to 1998. However, average sectoral wages increased in both construction and manufacturing in 1999, 2000 and 2001 (Table 10.24).

10.4.6 Construction quality

The number of firms with ISO 9000 certification increased until 1999, but after 2000, the number was decreasing, until 2002. In 1999, the number of companies with ISO 9000 certification increased to 2,639, while the number decreased in 2000 and after that too (Table 10.25).

Table 10.23 Construction industry salaries and wages (daily)

	1995	*1996*	*1997*	*1998*	*1999*	*2000*	*2001*	*2002.5*
Chief worker	53,714.5	59,935	60,544.5	55,777.5	56,294.5	56,598.5	61,280.5	64,905
Special daily wage	46,532.5	54,451	54,713.5	49,166.5	49,578	51,861.0	54,379.0	62,902
Normal daily wage	30,899.5	34,476	36,845.5	33,926.5	33,841.5	37,267.5	39,927.0	45,031

Table 10.24 Average sectoral wages per month (won at 1995 constant price)

	1995	*1996*	*1997*	*1998*	*1999*	*2000*	*2001*	*2002.3*
Construction	1,384,194	1,501,257	1,623,638	1,502,275	1,690,873	1,839,844	1,834,683	1,915,788
Manufacturing	1,123,895	1,261,168	1,326,241	1,284,477	1,475,500	1,601,468	1,702,350	1,693,183
Services	1,409,529	1,549,282	1,667,299	1,631,334	1,726,280	1,860,788	1,893,929	1,783,995
Primary sector	1,195,996	1,379,913	1,534,912	1,524,520	1,581,067	1,702,280	1,779,553	2,059,925

Table 10.25 Number of ISO 9000 certified construction companies

1995	*1996*	*1997*	*1998*	*1999*	*2000*	*2001*	*2002.6*	*Total*
17	155	640	1,468	2,639	889	370	274	6,442

10.4.7 Safety record

To secure quality in construction works, contractors, sub-contractors and relevant technical personnel responsible for minor faults in construction works, other than suspension of business and other administrative measures pursuant to construction related regulations, would be penalized. Further, those penalized would be restricted from participation in tendering or receive lower scores on PQ inspections.

All efforts on construction works were made towards securing safety of facilities and labourers. According to Table 10.26, the accident frequency rate had gone down. In 1997, the rate of accidents per million man-hours worked decreased to 4.33, while the rate was 5.40 in 1995. In 2000 and 2001, the accident rate decreased to 2.56 and 0.69 respectively.

10.4.8 Skill level of workforce and supervision level

Table 10.27 shows the skill level of workforce and supervision level.

10.4.9 Research and development

Research and development expenditure had never been a priority issue. In 1996, the construction industry as a whole invested 0.74 per cent of its sales. The MOCT invested 0.54 per cent of its budget, and the four public corporations (Highway,

Table 10.26 Accident frequency rate

	Accident frequency rate (accidents/million man-hours worked)
1995	5.40
1996	4.69
1997	4.33
1998	3.32
1999	3.17
2000	2.56
2001	0.69

Table 10.27 Skill level of workforce and supervision level

	Total construction workforce (nos.) <A>	*No. of skilled workers <B>*	*Skill level (%) <B/A>*	*Ratio of supervisor to workers (1:XX)*
1995	1,550,017	n.a.	–	–
1996	1,610,910	n.a.	–	–
1997	1,644,000	37,376	2.3	–
1998	1,578,000	44,846	2.85	–
1999	1,476,000	50,025	3.39	–
2000	1,581,000	45,162	2.86	–
2001	1,575,000	47,989	3.05	–
2002.3	1,656,000	–	–	–

Note
Skilled Workers refer to sum of middle, high, special level of workers.

Table 10.28 Research and development expenditure

	National expenditure on R&D	*R&D expenditure by construction industry* (billion won)	*% of annual construction volume*
1995	9,440,606	467.6	0.97
1996	10,878,051	437.6	0.72
1997	12,185,807	433.5	0.65
1998	11,366,602	n.a.	n.a.
1999	11,921,893	n.a.	n.a.
2000	13,848,893	509.1	1.26
2001	14,236,451	701.6	–
2002.3 (expected)	15,291,593	752.4	–

Water, Housing, Land Corporation) 2.0 per cent of their total sales on R&D. The figures were meagre compared to the domestic industry average of 6.8 per cent, not to mention foreign firms (Table 10.28). In order to increase R&D investment, the MOCT had plans to double R&D investment by the year 2002. R&D expenditure for the construction industry recorded 701.6 billion won in 2001 and was expected to be 752.4 billion won in 2002.

New construction technology development

It is not easy to document new developments in technology. It can get either too technical or too broad to convey any meaning. We will opt for the second option rather than risk getting lost in the maze of technical details. At the time of writing, we were in the last year of Five-Year (1998–2002) Construction Technology Development Master Plan carried out by Ministry of Construction and Transportation (Table 10.29). It delineated major technology fields that were to be focused in the coming years as follows:

- increase productivity, reduce construction costs – mechanization and automation;
- promote efficient land use – super high-rise building, utilize mountainous terrain and underground space;
- introduce environment friendly construction – energy saving and recycle technology;
- adopt construction management and 'life cycle cost' approach for mega-projects;
- further develop GIS (geographical information system), ITS (intelligent transport system) and CALS (continuous acquisition and life cycle support);
- water supply and treatment, management of ground water; and
- upgrade engineering and architectural technology – standardization of materials, rehabilitation and redevelopment of existing structures, repair and maintenance.

10.4.10 Environmental protection

The number of firms with ISO 14000 certification increased continuously (Table 10.30). The number in 2000 was 68, making the total 186. In 2002, up to June, 43 construction companies acquired ISO 14000 certification, making the total number 272.

Table 10.29 New technology in automation, mechanization and prefabrication

Technology and brief description	*Developer*
Semi-automatization of bulldozer	2000
1 a computer-controlled semi-auto equipment for bulldozer 2 to manipulate blade to control the digging resistance 3 unskilled worker is able to operate the equipment in difficult digging condition.	Korea Institute of Construction Technology
The development of high-performance robot for concrete floor finishing	1998
1 small-sized advanced omni-directional robot for concrete floor finishing 2 to enable the easy cooperation of the operator and the robot for maximizing the floor finishing capabilities 3 it can be used at the apartment and the office buildings with small floors.	Korea Institute of Industrial Technology
Automation of the steel frame construction of buildings	1997
1 the automation of the steel frame construction of buildings 2 the control system monitors the states of the mechanism and displays them for the operator and drives the mechanism safely and efficiently according to the commands from the operator.	Hanyang University, Halla Engineering and Construction Corp.
Semi-automatization of tower crane	1997
1 develop a distance measuring device to the target object 2 develop the scope measuring device of the site at the driver's seat of tower crane 3 develop the controller which manages the apparatus and controls the operation of motor automatically to improve the operating speed of tower crane.	Korea Institute of Construction Technology

Table 10.30 Number of ISO 14000 certified construction companies

1996	*1997*	*1998*	*1999*	*2000*	*2001*	*2002.6*	*Total*
3	11	12	39	68	96	43	272

10.5 The Asian crisis and its impact on the construction sector

10.5.1 The construction and property markets

The construction market

The construction sector, due to its derived-demand nature, acutely felt the impact of the Asian economic crisis. The industry enjoyed healthy growth with the increase in housing and then infrastructure demand during the 1990s. The situation completely changed in 1998. In 2002, the already oversized industry expected its total contract value to drop by more than 30 per cent due to the sharp decrease in construction investment. Moreover, faced with interest rates higher than 20 per cent, many contractors were declared bankrupt and most others were struggling for survival. The government had tried to boost expenditure on construction and public works in order to create employment and protect consumers. During 1999, the construction sector began to show symptoms of recovery due to low interest rates.

Private sector demand had stayed ahead of public sector demand for quite some time at a ratio of 6:4. The consistent increase in private building construction was the main reason for this reversal. Similarly, the building and civil construction ratio was about 7:3. The amount for repair and maintenance was not announced separately but it was estimated to be around 9 per cent of the total construction output and was expected to grow quickly as the construction stock accumulates. Total contract amounts, however, were decreasing rapidly due to the economic crisis. Public and civil construction would rise, though, as the government spends more on construction to pump-up the economy.

The construction market situation completely changed from 1998. The growth of the construction sector plunged down to −10.23 per cent in 1998 (2.3 per cent in 1997), after recording a drastic growth for consecutively 8 years since 1989. However, the economic recovery action plan implemented by the government was showing an improvement in the performance of the construction sector in 1999. From 1999, construction market showed few symptoms of recovery due to the recovery action plan implemented by the government but it also showed a slow recovery in the long term. Construction investment growth rate recorded 0.9 per cent in 1999, although it recorded −4.1 per cent in 2000 and 5.4 per cent in 2001. At the time of writing, 6.7 per cent growth rate was recorded for the half of 2002.

Figure 10.12 and Table 10.31 show the sectoral trends of construction investment. According to this, the growth rate of residential building investments had improved from −11.1 per cent for 2000 to 11.7 per cent for 2001 and 28.4 per cent for the first half of 2002. But non-residential building construction was more adversely affected by the financial crisis in 1998. There was a marked decline in non-residential building construction activities caused mainly by deferment of selected new projects and the slow implementation of existing projects. Investment continued to decline until 1999. However, it was recorded that the growth rate of the non-residential building investment had improved to 15.2 per cent in 2000 and 4.6 per cent in 2001. Until 2002, when an 8.2 per cent growth rate was recorded in the first half of the year.

The economic recovery action plan of the government showed results with an improvement in the performance of the civil construction sector in 1999. The growth rate of civil construction had improved to 9.5 per cent for 1999. Although it was recorded at −7.6 per cent for 2000, 2.6 per cent growth rate was recorded for 2001 and 2.7 per cent growth rate was expected for 2002. The government continued to encourage private participation into public infrastructure. Table 10.32 shows the construction volume.

The property market

The property market keenly felt the effects of the financial crisis and ensuing structural reforms. During the developmental era, property prices rose sharply, outpacing even

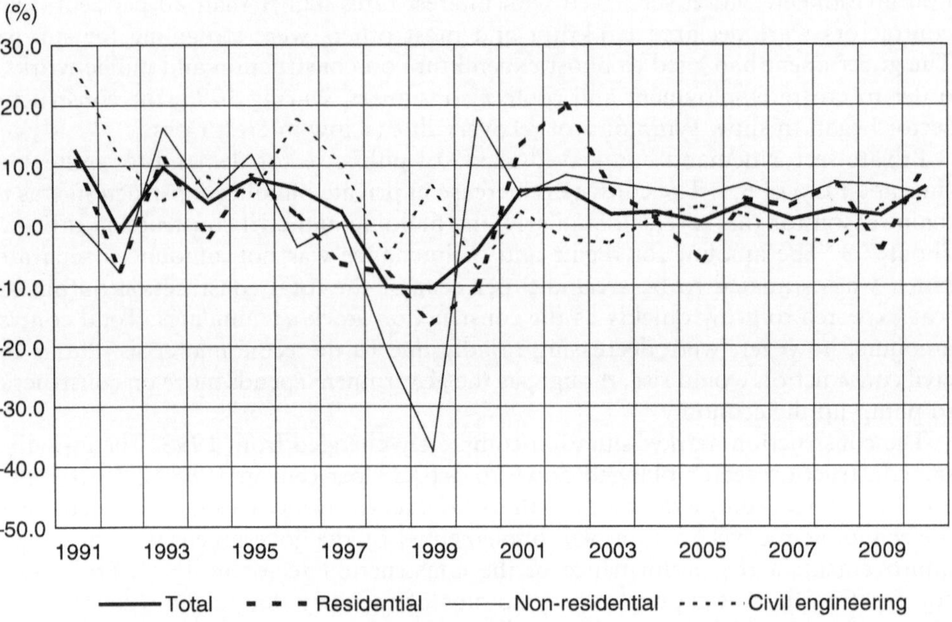

Figure 10.12 Changes in construction investment.

Table 10.31 Construction investment (unit: billion won at 1995 year fixed price)

		Total	*Residential*	*Non-residential*	*Civil (SOC)*
1995	Actual value	82,197.6	28,026.9	23,262.5	30,908.1
	(change ratio (%))	7.96	8.26	11.80	4.97
1996	Actual value	87,247.2	28,448.8	22,441.3	36,357.1
	(change ratio (%))	6.14	1.51	−3.53	17.63
1997	Actual value	89,284.8	26,667.8	22,544.5	40,072.6
	(change ratio (%))	2.34	−6.26	0.46	10.22
1998	Actual value	80,295.3	24,548.2	17,793.7	37,953.3
	(change ratio (%))	−10.23	−7.64	−21.22	−5.78
1999	Actual value	72,059.5	20,506.3	11,125.5	40,427.7
	(change ratio (%))	0.89	1.13	−17.68	9.47
2000	Actual value	69,140.1	18,454.2	13,117.2	37,568.7
	(change ratio (%))	−4.2	−11.1	15.2	−7.6
2001	Actual value	73,123.5	20,910.6	13,747.6	38,465.3
	(change ratio (%))	5.4	11.7	4.6	2.3
2002	Actual value	33,894.4	12,344.9	7,285.9	14,263.6
	(change ratio (%))	6.7	28.4	8.2	−7.5
2002	Forecasted value	78,310.0	23,850.0	14,820.0	39,530.0
	(change ratio (%))	6.6	12.3	7.2	2.7

Table 10.32 Construction volume (Gross built-up area: thousand sq. m)

Type of development	*1997*	*1998*	*1999*	*2000*	*2001*	*2002 (forecast)*
Total	113,372	50,965	72,534	81,058	97,717	
Residential	62,677	31,156	44,606	41,283	47,856	
Commercial	27,918	8,751	11,746	18,882	25,888	
Industrial	9,289	2,797	7,824	11,646	13,538	
Educational	6,600	3,982	4,478	5,126	5,534	
Others	7,377	4,279	3,880	4,121	4,901	

the economic growth rate. The property market calmed down somewhat with the introduction of drastic 'public concept of land' measures, but investors expected prices to go up sooner or later. The expectation of rising prices led many companies to stock up their portfolios with property shares. As of 2002, as a result of the financial crisis, the market was flooded with property as investors and financial institutions were trying to convert them into cash. Thus, the 'myth of property speculation' collapsed along with the bubble in the property value.

In 1998, the average property price fell by 30–40 per cent from 2001 and the market function of matching demand and supply was not working properly. The situation was much worse for the land sector than the built sector. The recovery of the South Korean economy was held back by the property market, impeding the economic restructuring process. This created a vicious cycle of a depressed economy causing a depressed property market and vice versa. However, the trend changed completely from 1999, by showing positive growth rates in land prices and rents.

10.5.2 Land prices in major cities

Land prices dramatically decreased after the economic crisis. In 1998, overall land price decreased 13.06 per cent, especially residential land decreased 14.68 per cent and industrial land 14.03 per cent. Comparatively, commercial land prices changed less. But this situation changed in 1999 and 2000. As of 2002, most prices related with land showed positive growth rates as shown in the Table 10.33.

10.5.3 Average rents in major cities

Housing price decreased 18.44 per cent in 1998. Apartment prices outstandingly decreased by 20.18 per cent. Average rents for office space showed the same

Table 10.33 Land price change (per cent change from previous year)

	1995	*1996*	*1997*	*1998*	*1999*	*2000*
Commercial land	0.37	0.80	−0.07	−17.07	1.56	−0.26
Residential land	0.30	0.71	0.15	−14.68	2.07	0.03
Industrial land	0.76	0.77	−0.17	−14.03	2.53	0.43
Average	0.55	0.96	0.31	−13.06	2.94	0.67

Table 10.34 Average rent change (per cent change from previous year)

	1995	*1996*	*1997*	*1998*	*1999*	*2000*
Total	3.62	6.5	0.85	−18.44	14.37	11.14
Detached dwelling	2.67	3.2	−0.10	−16.59	3.37	8.2
Row house	4.38	5.4	0.66	−18.28	13.30	13.4
Apartment	4.71	10.2	1.63	−20.18	21.09	12.18

Table 10.35 Average rents for office space in major cities (per cent change from previous year)

City	*1995*	*1996*	*1997*	*1998*	*1999*
Seoul	0.18	0.94	0.29	−16.25	2.66
Pusan	0.14	0.40	−0.90	−16.52	1.24
Daegu	1.27	1.09	−0.71	−15.24	1.93

behaviour. Relatively, the rents for office space in bigger cities decreased more. However, as of 2002, housing prices and rents had recovered to more than before the IMF regime, especially around the Seoul area (refer Tables 10.34 and 10.35).

10.5.4 Consultants, contractors, material and machinery suppliers, professional and technical personnel, and workers

Number of construction companies

As the government's control over the industry loosened after 1989, licenses were issued to eligible contractors, first at regular intervals and later upon request. During the last decade, the number of general contractors increased more than sevenfold as a result, numbering up to 4,000 firms; likewise, specialty contractors increased almost five times to around 20,000 firms. The overall number of construction firms (including contractors in electrical works, telecommunication, etc.) also more than doubled. Competition within the industry has recently intensified. Some newcomers stayed afloat, and some large firms got larger in general.

In 1996, 196 general contractors and 595 specialty contractors went bankrupt, resulting in bankruptcy rates of 5.6 and 2.8 per cent respectively (Table 10.36). In 1997, the bankruptcy figure increased to 291 general contractors and 1,058 specialty contractors. In 1998, the bankruptcy rate more than doubled from 1997. Altogether 524 general contractors and 1,321 specialist contractors went bankrupt as a result of restructuring. By 2002 the situation was improving.

Increased participation of foreign companies was also a notable fact. Even before the Uruguay Round of the GATT talks, there already existed a few small specialist contractors involved in joint ventures. They came into the market as partners of the South Korean companies rather than actively taking part in the market. However, 100 foreign entities had applied for general construction licenses. They had shown interest in such major projects as the Inchon Airport and rapid railway projects.

Table 10.36 The number of bankrupt construction companies

	1996	*1997*	*1998*	*1999*	*2000*	*2001*	*2002.6*
General contractor	196	291	524	112	132	78	20
Constructor	595	1,058	1,579	354	449	430	145

There had been more foreign participation in the fields of engineering, architecture and construction management where South Korean competitiveness was rather weak.

Number of employees and construction labour

EMPLOYEES AND CONSTRUCTION LABOUR

In 1970, the construction industry employed 280,000 workers, 2.9 per cent of total employment. The number of workers increased steadily and in 1997, more than 2 million employees were working in the construction field, 9.6 per cent of total employment. If we compare the trends in the construction industry's share of employment to its share of value-added in GDP, we can safely conclude that the industry's productivity had improved consistently. For the first quarter of 2002, more than 1.6 million employees were working in the construction field, 7.4 per cent of total employment.

The main indices in construction, the distribution of contractors by employment size and the turnover for top-10 local construction companies are shown in Tables 10.37, 10.38 and 10.39 respectively.

Table 10.37 Main indices in construction

Classification \ *Year*	*1997*	*1998*	*1999*	*2000*	*2001*	*2002.3*
Value of construction orders (trillion won)	79.9	47.9	51.1	60.1	67.8	18.3
Number of general contractor	3,900	4,208	5,155	7,978	11,961	11,971
Average contractor amount (100 million won)	202	114	99	75	57	15
Number of employee in construction (thousand person)	2,004	1,578	1,476	1,581	1,575	1,656
Construction permission square (thousand m^2) (growth rate, unit: %)	113,372 (−0.39)	50,965 (−55.05)	72,534 (42.32)	81,058 (11.75)	97,717 (17.04)	31,599 (50.53)

Source: Korea construction association.

Table 10.38 Distribution of contractors by employment size

*2000**	*Enterprise*	
	Number	*%*
Less than 5 persons	18,881	36.87
5–49 persons	26,752	52.24
50–200 persons	4,778	9.33
200–299 persons	339	0.66
300–499 persons	250	0.48
500–999 persons	116	0.22
1000 persons and above	75	0.14
Total	51,203	100

Note
*in 10 billion won.

Table 10.39 Turnover and number of employees for top-ten local construction companies (in 10 billion won)

Company	*Revenue*					
	Total	*Building*	*Civil engineering*	*Plant*	*Others (specify)*	*Overseas*
Hyundai	57,333	18,803	13,004	2,132	3,834	19,560
Daewoo	30,121	14,666	8,056	2,704	974	3,721
Samsung	38,510	22,250	5,800	2,300	–	8,160
Dongah	22,599	11,135	4,803	1,635	1,517	3,509
Daelim	24,466	8,762	4,916	2,980	428	7,380
Hyundai industries	20,695	16,368	3,730	222	375	–
LG	22,804	11,741	5,363	2,200	52	3,448
Ssangyong	27,076	1,963	3,262	21	11,516	19,296
SK	8,094	464	758	6,313	11	548

10.5.5 Policies and strategies adopted by the public and private in response to the financial crisis

The South Korean government had deregulated construction, housing and land to encourage construction investment. It had also reduced capital gains in trading apartments for a restricted period. The government also abolished the limitation of foreign ownership of land. The steps taken were as follows:

- abolish regulation on trading new apartment after acquisition;
- deregulate construction requirements for new apartment – space, kindergarten, pharmacy, facilities for normal life, etc.;
- abolish maximum limitation of owning residential land;
- permit foreign investment firm's founding factory; and
- exempt acquisition tax and registration tax in case of acquisition of enterprise's land.

The MOCT had tried to manage large infrastructure projects in order to use its constrained budget more efficiently. At the beginning of 1999, the MOCT announced 'Policies for Efficient Public Construction Projects' whose goal was to reserve its construction budget for dealing with the worst construction conditions caused by economic crisis.

10.6 Prospects for the construction sector

In 1996, the 'Framework Act on the Construction Industry' was enacted in place of the 'Construction Business Act'. The substitution was aimed at deregulating and simplifying construction-related Acts and also to promote the construction industry. The Act states that the MOCT must draft and execute a basic plan for the promotion of the construction industry every 5 years. The plan must include fundamental policy direction, measures regarding technology and manpower, ways to secure the quality of construction works, and policies to encourage small and medium construction businesses. The first Plan was drawn up by the MOCT with the assistance of KRIHS.

The economic crisis coupled with the phenomenon of globalization had pressurized the construction industry to change. Major policies that dealt with these situations were as follows:

- establish a fair competition rule;
- raise productivity, raise the environmental friendliness and quality of construction work;
- set up a new harmonious role between general and specialist contractors;
- strengthen production factors such as technology, manpower, material and finance capacity; and
- encourage both inbound and outbound overseas construction.

10.6.1 Business opportunities and upcoming major projects

Investment in infrastructure (roads, railways, airports, harbours, freight logistics, water supply, etc.) had been kept at a high level of 8–9 per cent of GDP during the development stage of the 1970s and early 1980s. By the late 1980s, however, infrastructure investment began to slow down, which subsequently caused excessive congestion costs in the 1990s. In particular, during the 1990s when international competition intensified, many identified the insufficiency of infrastructure as a major obstacle to improving the international competitiveness of South Korean industries.

Infrastructure became the top policy issue during the mid-1990s, which was reflected in the over 20 per cent growth rate of infrastructure-related expenditure. In 1997, almost 40 trillion won (approximately 9 per cent of GDP) was invested in infrastructure. The burden was shared by the central government (25 per cent), local governments (30 per cent), public corporations (15 per cent) and the private sector (30 per cent). Even during the economic crisis, consensus was to keep infrastructure expenditure cuts to a minimum. Thus from 1999 to 2003 the government planned to invest approximately 100 trillion won.

Difficulties associated with infrastructure investment had always been the raising of enough financial resources. In 1994, the Private Sector Investment Inducement

Act was introduced to promote private participation in infrastructure projects. As of July 1998, 40 projects had been selected as candidates for private sector participation and only five of their projects had started construction. The result was less than what was hoped for. This outcome coupled with the deteriorating financial situation necessitated a complete overhauling of the Private Sector Investment Inducement Act. The revised act would be simplified and more incentives would be offered to induce private funds.

10.6.2 Intermediate and long-term plans for infrastructure projects

Table 10.40 shows the main intermediate and long-term plans for infrastructure projects.

10.6.3 Implementation body and burden sharing for infrastructure projects

Table 10.41 shows the implementation body and burden sharing for infrastructure projects.

10.6.4 Financial resources for infrastructure projects

The fiscal budget provided funds for infrastructure projects from four financial sources: the general account, the transportation special account, the land special account and the railway special account. The supplementary budget for 1998 allocated 8,826 billion won for infrastructure projects. An amount of 1,287 billion won (14.6 per cent) came from the general account, 5,103 billion won (57.8 per cent) from the transport special account, 405 billion won (4.6 per cent) from the land special account and 2,031 billion won (23.0 per cent) from the railway special account.

The South Korean government substantially revised the enforcement ordinance of the Private Social Overhead Capital (SOC) Investment Act on 1 April 1999 to shift its policy from one that focuses on regulating private participation to one that focuses on actively supporting it. In particular, the government abolished discrimination between Korean and foreign investors in the process of promoting private investment, thus opening the door for introduction of advanced management techniques and efficient operational system from overseas along with foreign capital.

To this end, the government improved the domestic system to make it meet international standards and introduced a way to evaluate business plans and designate managing agents according to those standards (LHP://www.kisc.org/in part).

Type of projects available for private investors

1 Public projects: For projects assessed to be profitable for private participation according to the result of a feasibility study provided by Paragraph 1, Section 2 of the Act (see the Table 10.42).
2 Private projects: Private investors may select from among those facilities in Table 10.42 ones from not yet announced as public projects, but which can be promoted through private investment, and submit proposals for them to the competent authorities.

Table 10.40 Main intermediate and long-term plans for infrastructure projects

Name of plan	*Purpose, content*	*Size (trillion won)*	*Period*
Highway	• Increase distribution efficiency • Improve road condition • To build a main road net with seven South–North axes and nine East–West axes • To increase total length of highway up to 2,800 km (1.5 times) and the ratio of four-lane national road to 35%	Est. 54 (341 routes, 5,503.5 km of total work length)	To 2005
Airport	• To make an aerial centre in Asia • To have Inchon airport serve as a hub airport and an Asian transit depot • To construct new airports and to expand five existing airports	Est. 15.0	To 2011
Railway	• To construct electrified railroads net by regions • To establish railroad nets to connect Korea to the continent • To increase total length of double-lane railroads up to 1,251 km, raising a double-lane ratio from 29 to 46% • To increase total length of electrified railroad up to 1,460 km, raising an electrified railroad ratio from 20 to 46% • To expand total length of subway from 262 to 499 km by 2002	Est. 21.0	To 2002
Water	• To secure enough water supply • To construct dams.	Est. 10.9	To 2002
Logistics	• To expand logistics and to improve distributing facilities • To build four new inland container depots	Est. 3.1	To 2002
Harbour construction	• Extending loading and unloading facilities • To construct or expand 15 harbours	Est. 21.0	To 2011

Different forms of investment in private projects

1 BTO (Build-Transfer-Operate): Upon the completion of the SOC facility ownership reverts to the central or local government, but the rights to manage and operate are given to its managing agent for a certain period of time.
2 BOT (Build-Operate-Transfer): The ownership of the SOC facility is given to its managing agent for a certain period of time following the completion and reverts to the central or local government later.
3 BOO (Build-Own-Operate): Immediately upon the completion of the facility the ownership is given to its managing agent.

Table 10.41 Implementation body and burden sharing for infrastructure projects

Authority	*Burden sharing*	*Major projects*
National	National only	National roads, general railway construction, logistics site construction, airports, harbours, water conservation system, public housing
	National and regional public bodies	Metropolitan railroad nets, metropolitan roads
	National, regional public bodies and beneficiaries	Public housing
Regional public bodies	National and regional public bodies	National support local roads, detours, subway
	National, regional public bodies and beneficiaries	
Public corps, etc.	National only	Dam, industrial water supply system, local industrial site building
	National and public corps, etc.	Expressway, high-speed railway
	National, public corps, etc., regional public bodies and beneficiaries	Public rent housing
Beneficiary groups	National, regional public bodies and beneficiaries	

In addition, the competent authorities may suggest alternative ways in the basic plans for private investment facilities made under the rules in Section 10 of the Act.

As of 2002, a number of SOC construction projects were under way, which involved equity participation by foreign investors and private domestic companies. To attract more private investment in the field of SOC the government amended the Private Capital Inducement Promotion Act for the Expansion of Overhead Capital in 1999 to create the new Act on Private Participation (http://www.kisc.org/weekly/2000_04/region.htm).

Under the new act, the methodology for project investment and participation was diversified with the introduction of methods such as BOT, while the return on investments increased to 18 per cent from 13 per cent. The government had also raised the guaranteed level of operating revenue from 80 per cent of estimates to 90 per cent.

Moreover, the act provides for solutions to mitigate foreign exchange risk, and an SOC investment company would be established to extend financial support. The Private Infrastructure Investment Centre of Korea (PICKO) has been established to support private investment projects by providing necessary information and consultation.

In response to this active level of support, companies from Singapore, Japan, France, Canada, and the United States have already finalized their investments in Korean SOC projects, and in some cases feasibility analyses are already under way.

Projects include a contained pier at Inchon, the Kwangju Circular road, the Pusan's north port bridge, rail connection to Seoul from Inchon airport, a second bridge connecting the past and the new airport at Inchon, LRT projects, and various toll road projects.

Table 10.42 Types of SOC facilities (34 in total)

Area	*Competent agency*	*Types of facilities*
Road and Highway (3)	Ministry of Construction and Transportation	Roads and auxiliary facilities, outdoor parking lots, intelligence transportation network
Railroad (2)	Korean National Railroad	National railroads
	Ministry of Construction and Transportation	City railroads
Harbour (3)	Ministry of Maritime and Fisheries	Ports, harbours, fishing ports, passenger terminals
Airport (1)	Ministry of Construction and Transportation	Airport facilities
Water Resource (4)	Ministry of Construction and Transportation	Multipurpose dams, river facilities
	Ministry of Environment	Sewage system, water supply
Communication (2)	Ministry of Information and Communication	Telecommunication, information network
Energy (3)	Ministry of Industry and Energy	Power plants, gas supply, regional energy supply facilities
Environment (6)	Ministry of Environment	Waste and sewage disposal, livestock waste disposal, water purification and recycling facilities
Distribution (3)	Ministry of Construction and Transportation	Distribution centre, cargo terminals, and warehouses, passenger vehicle terminals
Culture and Tourism (7)	Ministry of Culture and Tourism	Vacation and recreation resorts, youth camps and training facilities, sports facilities, libraries, art galleries and museums, international conference facilities
	Ministry of Construction and Transportation	City parks

SOC (Social Overhead Capital) investment ratio and main sector in national budget after 1990

Table 10.43 shows that SOC investment in the national budget is continually increasing. Table 10.44 shows the main sectoral SOC investment during 1996–1999. The increasing ratio of Road, Port, Airport sector is comparatively high.

10.6.5 Major cases of private sector capital participation

The major cases of private sector capital participation during the period 1995–2001 are shown in Table 10.45.

10.6.6 Areas in need of foreign investment

Foreign participation in the Korean construction market was still in its initial stage until 2002. Companies applying for a license were slowly increasing in number and many specialist contractors were already active in government procurement projects as well. In the architectural and engineering fields, cross-border supply of services

Table 10.43 SOC investment ratio in national budget (unit: million won, per cent)

	SOC investment	*Budget size*	*SOC ratio*
1995	4,163,156	51,498,089	8.1
1996	5,021,080	58,480,817	8.6
1997	5,899,254	63,962,100	9.2
1998	7,204,020	73,225,982	9.8
1999	9,200,348	80,509,887	11.4
2000	10,550,363	88,736,307	11.9
2001	12,285,382	94,124,600	13.1

Source: Annual major financial indicators, Ministry of Planning and Budget (2001).

Table 10.44 Sectoral SOC investment (unit: 100 million won)

	Road	*Rail*	*Port*	*Airport*	*Water Resource*	*Subway*
1996	40,833	13,584	6,147	4,452	6,441	8,399
1997	49,901	17,590	9,292	6,099	7,686	8,137
1998	56,932	18,283	10,058	9,913	6,929	9,441
1999	62,244	18,080	9,470	10,253	7,862	10,334
1996–1999 % increase	52.44	33.10	54.06	130.30	22.06	23.04

Source: 1999 The budget of Korea, Ministry of Planning and Budget (1997, 1999).

had been on going for quite some time and now several joint ventures were being formed. In the supervision and consulting fields 13 companies were participating in projects such as the Seoul–Pusan High Speed Railway, Inchon New Airport, West Sea Highway, Gayang Bridge, etc.

Globalization of the construction services

In 1996 the 'Framework Act on the Construction Industry' was enacted in place of the 'Construction Business Act'. The substitution was aimed at deregulation and simplifying construction related Acts and also to promote the construction industry. The Act states that the MOCT must draft and execute a basic plan for the promotion of the construction industry every 5 years. The plan must include fundamental policy directions, measures regarding technology and manpower, ways to secure the quality of construction works, and policies to encourage small and medium construction businesses. The first Plan was drawn up by the MOCT with the assistance of KRIHS.

Our diagnosis of the policy environment and the present situation has two components. Due to the IMF crisis the construction industry is in great difficulty but this will be a short-term effect that can be overcome in future. A long-term underlying current is that of globalization, localization and information management which translates into increased competition, forming of a buyers' market and elevation of the importance of R&D investment for the construction industry. There is no

Table 10.45 Major cases of private sector capital participation during the period 1995–2001

Name of facility	*Type of facility*	*Operation format*	*Total costs (billion won)*	*Period*
Inchon Airport Highway	Toll highway	Built by a joint capital company involving the private sector, transferred to the government and operated by the construction companies	1,740	1995.11–2000.11
Inchon International Airport Freight Terminal B&D	Freight terminal	Same as the above	61.7	1998.6–2000.10
Inchon International Airport Freight Terminal C	Freight terminal	Same as the above	52.2	1998.7–2000.10
Kwangju Belt Highway	Highway	Same as the above	214.9	1997.6–2000.12
Daegu–Dadong Highway	Highway	Same as the above	2,189.5	Est. 1998–2001
Chonan–Nonsan Highway	Highway	Same as the above	2,032.1	1997.12–2002.12
Pusan New Harbour	Harbour	Same as the above	3,658.7	1999–2006
Kyoung–In Canal	Waterway	Same as the above	1,959.7	1998–2002
LNG	Power plant	Built, operated and owned by a private company	380	1999–2001

conflict between the short- and long-term factors. Both are pressuring the industry to restructure.

Major tasks envisioned in the Plan are as follows: establishing a fair competition rule, raising productivity, securing environmental friendliness and quality of the construction works, setting up a new harmonious role between general and specialist contractors, strengthening production factors such as technology, manpower, material, and finance capacity, and encouraging both inbound and outbound overseas construction. Refer Table 10.46 for construction volume of foreign firms.

Table 10.46 Construction volume of foreign firms ($ m)

Type of development	*1998*	*1999*	*2000*	*2001*	*2002 (forecast)*
Total imports	4,055,313	9,188,750	5,433,154	4,354,901	3,921,374

10.6.7 Efforts to open markets, relax foreign capital regulations and promote foreign investment

Table 10.47 deals with various policies on open markets, foreign capital regulations and foreign investments.

10.6.8 Business alliance opportunities

Liberalization and globalization of business activities is a worldwide phenomenon with no exception in the construction industry after the establishment of the WTO. South Korea accepted the trend by participating in many related international treaties

Table 10.47 Efforts to open markets, relax foreign capital regulations and promote foreign investment

Policy	*Content*	*Implementation date (period)*
Lifting regulations on alien ownership of land	Restrictive measures on foreign ownership of land will be lifted allowing in principle all types of foreign ownership	1998 June 26
Other market opening (all construction related markets are opened)	Building rental and marketing services	1998 April
	Land rental and land development and supply services	1998 May
Deregulation measures for effective market access	License issued upon request	1994
	Obligatory membership to construction association and bond company relaxed	1997
	Setting ceiling amount on one contractor can take through company evaluation abandoned	1997
Reform 'Private Sector Investment Inducement Act'	Accept unsolicited private sector proposal Make evaluation and selection process more transparent Ensure higher profit margin lower risk Establish specialized organization Change process to accommodate foreign investments	1998 September
Enactment of 'Foreign Investment Promotion Act'	Attitude change from restriction/management to promotion/assistance Provide free foreign investment zone Increase tax incentives Simplify investment process Establish one stop information and service centre	1998 September

and by opening up its construction market. Our overseas construction has been very active and perhaps in this sector we may prove to be a beneficiary of the liberalization trend. We are now prepared to welcome a foreign presence in the domestic market. We would like to learn advanced technology, procure better materials, augment investments, and improve production systems with the help of foreign companies. Based on cooperation in the domestic market Korea also wishes to assist other nations in the building up of their construction stock if possible. The Government is now adopting a more practical and balanced approach to overseas construction as opposed to its use for prestige/diplomatic purposes.

Asia being the most active construction market in the world, and with such mutual endeavour, we see a bright future for the Asian construction industry. With exchange and international cooperation, Asian countries will be able to improve their construction technology level. In addition, they will have a chance to develop construction management and other advanced skills in the construction area. Like developed countries, eventually, the pattern of the construction industry in Asian countries will be shifted from simple construction work to technology-intensive work.

10.6.9 Rules and regulations for participation of foreign construction and professionals

In Korea, there is no difference between the treatment of domestic and foreign companies. The same conditions apply and there is no special regulation in construction. Table 10.48 shows the most important foreign general contractor companies operating in Korea.

10.6.10 Overseas construction

Overseas construction plays a leading role in bringing in foreign currency. Since the first overseas construction project in Thailand in 1965, 4,054 projects valued at $158.3 billion have been implemented by Korean construction companies.

Overseas construction was at its height from the late 1970s to the mid-1980s recording an annual average of about $10 billion in contracts in Saudi Arabia and throughout the Middle East region.

Overseas construction works in the Middle East fell drastically after the late 1980s due to increased wages and decrease in the area's infrastructure demand, but began

Table 10.48 The present foreign companies of general contractor license in Korea

Company name	*Country*	*Acquired date of license*
Hugita	Japan	1996.10.14
Becktel international in corporate	USA	1996.10.14
HuruwarDaniel eastern ink	USA	1996.10.14
China construction process company	China	1996.10.14
Daesung construction	Japan	2000.12.11
Iconcorporation international limited	Singapore	2000.12.13
Cosmopolitan construction	USA	2000.12.19

Table 10.49 Change in overseas construction

Year	*Total contract value* ($ m)
1981	13,681
1988	1,602
1993	5,117
1995	8,508
1996	10,779
1997	14,032
1998	4,055
1999	9,189
2000 (as of Dec.)	5,412

Table 10.50 Top-five countries for construction export

Country	*Value* ($)
Iran	1,208
Nigeria	582
Libya	482
Singapore	452
China	209

to show a turn around beginning in the late 1990s leading to a record high of $14 billion contracted in 1997 (Table 10.49).

However, the 1997 Asian economic crisis caused the figure to fall again. The reasons this time were problems in financing due to the economic crisis and huge shrinkage in the Southeast Asia's construction market. Overseas construction has shown signs of recovery again in 1999 as Korea's international credit rating went up again and thanks to the restructuring efforts of the construction companies. Table 10.50 shows the top-five countries for construction export.

Chapter 11

Sri Lanka

Institute for Construction Training and Development, Sri Lanka and Lennon H.T. Choy

11.1 Introduction

11.1.1 Background

The Sri Lankan island was occupied by the Portuguese, Dutch and British in the sixteenth, seventeenth and nineteenth centuries respectively. Ceylon proclaimed its independence in 1948 and the country was then renamed to Sri Lanka in 1972. Persistent tensions among the Sinhalese majority and Tamil separatists triggered an ethnic war in the mid-1980s that cost tens of thousands of lives (MFA, 2002).

The population of Sri Lanka was 19,408,635 (July 2001 est.) which was steadily growing at the rate of 1.4 per cent (2001 est.). The sex ratio for the total population was 0.95 male per female at the time of writing. Sri Lanka is a multi racial country with a large Buddhist majority (70 per cent). Ethnic Hindu comprises 15 per cent, Christian 8 per cent and Muslim 7 per cent of the total population. The ongoing civil strife in the northern region of the country has taken its toll both on economy as well as development of this otherwise highly literate (about 90.2 per cent) country.

Geography

Sri Lanka is located in southern Asia, an island in the Indian Ocean, south of India at coordinates 7 00 N and 81 00 E. Sri Lanka has a total area of 65,610 square kilometres divided into land (64,740 sq. km) and water (870 sq. km) with zero land boundaries and a coastline of 1,340 kilometres.

Sri Lanka is a fairly uniform country in terms of climate, terrain, people and the government. Climate is mostly tropical monsoon and its topography varies from mostly low, flat to rolling plain to mountains in the south-central interior. The elevation extremes range from the lowest at Indian Ocean (0 metre) to the highest point at Pidurutalagala (2,524 metres).

Sri Lanka's staple natural resources are limestone, graphite, mineral sands, gems, phosphates, clay and hydropower including arable land (14 per cent of total lands), permanent crops (15 per cent), permanent pastures (7 per cent), forests and woodland (32 per cent) and others (32 per cent). A total of 5,500 square kilometres lands (1993 est.) constitute irrigated lands.

The capital city of Sri Lanka, Colombo, is strategically placed along the main shipping route connecting Asia with Europe and America. It is one of the few deep-water ports in the Indian subcontinent and could become an international shipping hub for South Asia with enhancement on operational efficiency (ADB, 2001a).

11.1.2 Political structure

The Democratic Socialist Republic of Sri Lanka has a republic type government, whose administrative divisions comprise 9 Provinces, 25 Districts, 256 Divisional Secretariats and 160 Electorates with a highly complex mixture of Kandyan Law, Tesawalamai Law and Customary Law and political leaders elected based on universal suffrage.

11.1.3 Development process: through the years

Since independence from the British regime, development in Sri Lanka has been based on fluctuations between the Capitalistic and Socialistic ideal. The state assumes the central role in regulating, allocation of resources and steering the economic development of the country. Since 1970 the country has been steadily opening up to the outside market. In 1977, Colombo superseded its statist economic policies by market-oriented policies, and replaced its import substitution trade policies with export-oriented trade policies. The food processing, textiles and garments, food and beverages, telecommunications, and insurance and banking are now the biggest sectors in Sri Lanka's economy. Plantation crops once dominated Sri Lanka's exports (93 per cent in 1970), now textiles and garments become the leader (63 per cent in 1996). The GDP growth rate was high (5.5 per cent) throughout the 1990s except in 1996 when the double blow of a drought and a deteriorating security situation lowered the growth rate to 3.8 per cent. With tourism and exports continued to blossom, GDP growth rate maintained at 6.0 per cent in 2000. However, the resurgence of ethnic war between Sinhalese and Tamils has started dampening the economic performance since 2001. For economic reform, the central bank of Sri Lanka has recommended Colombo to (i) expand market mechanisms in non-plantation agriculture, (ii) dismantle the government's monopoly over wheat imports, and (iii) promote competition in the financial sector (MFA, 2002).

11.2 The national economy

11.2.1 Macroeconomic review and outlook

The Sri Lankan economy has been growing and remained relatively stable over the past decade, until 2002, with yearly GDP growth at between 3 and 6 per cent. This was also attributed to growth in the share of construction (at about 5 per cent) and service sector in the national economy. Manufacturing, however, dipped in the year 1998 and 1999 following poor export due to the Asian financial crisis. In the domestic job market, although the labour force grew by 2.9 per cent towards the close of the millennium as compared to a mere half a per cent in 1995, the unemployment rate subsequently improved to 7.7 per cent in 2000 compared to 13.8 in 1993.

A significant increase in employment and a moderation of inflation in 2000 showed that the Sri Lankan economy has picked up the momentum of economic growth in the 1990s. The GDP and GNP growth rates were 4.3 and 3.8 per cent respectively in 1999 and increased to 6.0 and 5.8 per cent respectively in 2000. Following a period of slow down in the second half of 1998 and 1999, economic recovery has started in the second half of 1999 and continued through the first three quarters of

2000. The relatively low population growth together with the high growth in GDP had rendered an increase of GDP per capita from Rs. 58,077 (US$825) in 1999 to Rs. 64,855 (US$856) in 2000. As a result, Sri Lanka moved up further within the lower middle-income category in the international classification, which comprises countries with GDP per capita ranging from US$795 to US$3,125.

Owning to the surging global petroleum prices, a shortfall in government revenue and a sudden expansion of security expenditure, Sri Lanka's balance of payments and fiscal deficits widened towards the end of the year 2000. Dependence on costly thermal generation also increased with an unusually low rainfall that caused a shortfall in hydropower generation. The pressure exerted on foreign reserves has called upon a change in the exchange rate policy with greater flexibility. Meanwhile, monetary policy was tightened so as to stabilize domestic prices and protect foreign reserves. As a consequence, short-term interest rates increased substantially.

Continuing the upward trend since 1997, the ratio of gross domestic investment over GDP increased to 27.3 per cent in 2000. It was mainly due to the increasing amount of private sector investment. The increase of private sector investment is further evidenced by a rise of relative share of private investment in total capital accumulation, which has reached 89 per cent in 2000. Public sector investment, in the mean time, mainly focused on economic and social infrastructure development. Adverse effects on foreign investment activities due to the ongoing ethnic conflicts were considered insignificant, with inflows of foreign direct investment (FDI) amounting to US$175 million in 2000.

To meet fast growing demand, private investment in infrastructure facilities continued to increase in 2000. Government investment in infrastructure fell short of expectations owning to the curtailing of capital expenditure so as to accommodate the increasing demand on defence expenditure in the middle of the year. Actual government investment including infrastructure development only amounted to 6.4 per cent of GDP despite that it was budgeted for 8.1 per cent in the fiscal year (CBSL, 2000). The main macroeconomic indicators are tabulated in Table 11.1.

Economy in 2000

Economy was recovered in 2000 with a GDP growth rate of 6.0 per cent. Service sector (53.4 per cent) was the largest among the GDP compositions, followed by primary sector (22.2 per cent), manufacturing sector (17.4 per cent) and construction sector (7 per cent). The service sector had grown gradually since the 1990s, which was led by the fast growth of telecommunications services and the steady tourism industry. Though ranked second, the primary sector (including mining and agriculture, etc.) experienced a slow-down in growth rate (2 per cent only). It was attributed to the decline in paddy output due to unfavourable weather conditions. The strong growth of manufacturing sector (9.2 per cent) was largely due to the favourable external trade environment embarked from the second half of 1999. Production of textiles and garments, and rubber and plastic products increased in response to the rising international demand. The unemployment rate plunged from 8.9 per cent in 1999 to 7.7 per cent in 2000. Graduates still found difficulty in securing a job. Unemployment problem was dampened by the increasing number of migrants from poor rural areas and foreign countries.

Table 11.1 Main macroeconomic indicators

	1992	*1993*	*1994*	*1995*	*1996*	*1997*	*1998*	*1999*	*2000*
GDP at constant (1996) prices Rs per million	425.3	499.8	579.1	598,327	695,934	739,763	774,796	808,340	857,035
GDP growth (%)	–	–	–	5.5	3.8	6.3	4.7	4.3	6.0
Primary sector (mining, agriculture, etc.) Rs million	–	–	–	37,707	170,035	175,213	178,481	186,476	190,238
% growth	–	–	–	3.35	−4.2	3.04	1.9	4.5	2.0
Manufacturing sector Rs million	–	–	–	34,294	112,724	122,929	130,733	136,498	149,115
% growth	–	–	–	9.2	6.5	9.1	6.3	4.4	9.2
Services sector Rs million	–	–	–	84,388	364,941	390,799	411,121	428,291	457,867
% growth	–	–	–	5.8	6.2	7.0	5.2	4.2	6.9
Construction sector Rs million	–	–	–	11,564	48,234	50,842	54,461	57,075	59,815
% growth				4.9	3.4	5.4	7.1	4.8	4.8
Population million person	17.4	17.6	17.9	18,100	18,315	18,552	18,774	19,043	19,359
Population growth rate (%)				1.5	1.1	1.3	1.2	1.4	1.6
Total labour force million per person				6,106.1	6,242	6,266	6,661	6,673	6,867
Labour force growth rate (%)				0.5	2.2	0.4	6.3	0.2	2.9
Unemployment rate	13.3	13.8	12.1	12.7	11.3	10.5	9.2	8.9	7.7
Changes in consumer price index (%)	9.4	8.4	4.7	7.7	15.9	8.7	8.4	4.4	6.2
Short-term interest rate (Commercial Banks, average lending interest rates)	17.69	18.09	18.73	19.26	20.2	15.2	14.9	15.9	21.5
Long-term interest rate (Treasury Bill rate)	18.99	19.38	19.43	18.97	17.5	10	12	11.8	17.8
Annual average exchange rate with US$				51.25	55.27	58.99	64.59	70.39	75.78

Due to the unfavourable weather, food prices increased significantly in 2000. Together with the increase of oil prices, inflation rate increased from 4.7 per cent in 1999 to 6.2 per cent in 2000.

Meanwhile, Sri Lanka increased its defence expenditure from 4.4 per cent in 1999 to 5.6 per cent in 2000. To accommodate the increased expenditure, the government raised the national security levy and excise taxes on alcohol and tobacco products. Non-essential capital expenditure was also reduced by 10 per cent to alleviate budget deficit. Notwithstanding of these efforts, budget deficit increased from 7.5 per cent in 1999 to 9.8 per cent in 2000.

Both exports and imports swelled in 2000. The exports growth rate was high at 17.8 per cent, mainly due to the strong economic growth in the US, Europe and Asia. Exports of textiles and garments goods had increased by 20.5 per cent in a single year. However, aggregate imports also soared 22.9 per cent in 2000. Though imports of investment goods including transport equipment and machinery decreased by 13.4 per cent, imports of intermediate goods increased by 22.9 per cent as a consequence of the sharp surge of petroleum price. As a result, the current account deficit increased from 3.7 per of GDP in 1999 to 7.3 per cent in 2000. Balance of payments further deteriorated in 2000 as a consequence of the depreciation of rupee by an average of 9.9 per cent in 2000 (ADB, 2003).

Role of construction/property market in national economy

The role of the construction/property market in Sri Lanka had grown significantly, as indicated by the steady increases in construction investment and land prices/rents in Tables 11.2, 11.3 and 11.4.

Table 11.2 Construction investment (or market) by type (local currency at current price)

	1996	*1997*	*1998*	*1999*	*2000*
Residential construction					
Public	1,138.5	1,397.2	978.4	972.8	–*
Private	31,319.0	41,796.3	46,408.5	49,339.5	52,662.9
Total residential (A)	32,457.5	43,193.5	47,386.9	50,312.3	–
Non-residential construction					
Public	3,305.1	4,375.1	6,602.1	10,122.1	–*
Private	14,913.8	21,515.9	24,212.9	26,603.1	25,077.6
Total non-residential (B)	16,218.9	25,891.0	30,815.0	36,725.2	–
Civil engineering construction					
Public	7,432.4	15,535.0	11,761.0	13,429.9	–*
Private	28,336.3	37,811.1	42,372.9	45,205.6	47,547.4
Total civil engineering (C)	35,768.7	53,346.1	54,133.9	58,635.5	–
Total construction investment (A+B+C)	86,445.1	122,430.6	132,342.8	145,673.0	143,881.9

Note

* Public (Residential + Non-residential + Civil engineering) = 18,494.0. This is included in the grand total. Sub-breakdown is not available.

Table 11.3 Land prices in major cities (per sq. m)

	1995 (SLR 0000)	*1996* (SLR 0000)	*1997* (SLR 0000)	*1998* (SLR 0000)	*1999* (SLR 0000)	*2000* (SLR 0000)
Commercial land	3.25	3.50	3.50	3.50	3.50	3.60
Residential land	2.25	2.50	2.50	2.50	2.50	2.55

Table 11.4 Average rents in major cities (in Sri Lanka Rs/sq. m per month)

	1995	*1996*	*1997*	*1998*	*1999*	*2000*
Commercial (office)	750	750	750	750	750	760
Residential	250	250	250	250	250	255
Industrial	140	130	130	130	130	130

11.2.2 Changes of economic policies due to the financial turmoil

As a part of the ongoing reform programme, the government broadened the tax base and rationalized the tax system. Two major actions were done to widen the tax base, namely, (i) amendment of the automatic waiver for excise, and (ii) extension of the National Security Levy to wider range of services. To rationalize the tax system, a Revenue Protection Unit was established to strengthen the enforcement regulations as well as to prevent abuses of the system. In order to ease hardship cases and to ensure smooth operations, the Goods and Services Tax Law was amended accordingly. For public enterprises and statutory agencies, a number of audit and management committees were set up to enhance financial prudence. In addition, a separate unit under the Ministry of Finance and Planning was established to evaluate government expenditure. Foreign exchange trading was widened by the central bank by allowing rupee to depreciate against the US dollar in 2000. As a monetary tool, overnight re-purchase rate surged from 9.2 per cent in 1999 to 20 per cent in 2000 so as to maintain stability of the foreign exchange market (ADB, 2003).

Policies and strategies adopted by public and private sector to overcome the economic slow down in the construction sector

Sharp escalation of World Petroleum prices, shortfall of government revenue and the domestic crisis by the terrorists had a direct bearing on the prospective growth rate of the construction sector. As a result, domestic construction industry stake holders such as contractors, consultants, materials manufacturers and suppliers were deprived of their potential growth opportunities. This was aggravated due to heavy dependence on foreign funding, as a result of some of the conditions laid down by the funding agencies, which were none, conducive for the domestic construction industry.

In order to meet the challenges due to economic slow down, the government of Sri Lanka and its organizations along with the private sector had resorted to various policies and strategies. Government continued to adopt liberal policies and a free market system.

Government has given precedence to infrastructure development to meet the development needs of the country while focusing attention on the housing needs of low and middle income dwellers.

Government continued to encourage private sector investment in developing commercial infrastructure facilities such as port services, electricity and telecommunication projects, selected highways and industrial towns on a BOO/BOT basis. The investment policy allows for:

a total foreign ownership in virtually all areas of the economy;
b safety of investments guaranteed by the constitution; and
c equal treatment for foreign and local investors under the general laws of the country.

The Public Sector Infrastructure Development Company (PSIDC) established in 1996 is basically a means of channelling concessionary foreign funds, as long-term loans, to those private enterprises for investment in medium- and large-scale projects.

The Board of Investment (BOI) promotes private investment by providing alternative incentives and intensified promotional activities. The law empowers the BOI to grant special packages of incentives to enterprises, which meet BOI criteria. The incentive package includes tax and other fiscal concessions, where projects are based on bilateral aid and tenders are to be restricted to contractors of the aid giving countries. Sri Lankan contractors would be permitted to bid in collaboration with foreign construction contractors or construction management organizations from the aid giving countries. The National Construction Contractors Association of Sri Lanka (NCCASL) would provide the necessary assistance to find local collaboration for foreign firms seeking to work in this country.

Foreign investment in Sri Lanka is safeguarded in various ways. Sri Lanka has entered into separate investment protection agreements with many countries. Sri Lanka's constitution guarantees the safety of foreign investment through these agreements and Sri Lanka is a founder member of the Multilateral Investment Guarantees Agency (MIGA), which provides foreign investors with further safeguards against expropriation, discriminatory measures and other non-commercial risks. Investors in priority sectors are also eligible for freedom to repatriate profits on capital proceeds and also free transferability of shares.

Presidential task force recommendations made in 1998 for the improvement of the construction industry are in force. On the request of the NCCASL, a construction task force had been established to look in to the construction policy reformulations needed to suit the current set backs in the construction industry, which includes the following elements:

1 Continuity of work
2 Standardization
3 Minimization of financial risks
4 Dialogue and share of information and data between the government and the domestic stakeholders of the construction industry

5 Material development
6 Technology development
7 Capacity building
8 Quality and productivity improvements.

The NCCASL is the only association of contractors recognized by the state. At the time of writing, the NCCASL had a membership of over 2000 and had branches operating in all the nine provinces of the country. The association had brought forward proposals for the following for the attention of the government as a strategic move.

1 Funding package for construction contractors – It was observed that the existing domestic construction capacity at the time of writing was insufficient to meet the future-projected demand on the construction sector, mainly due to relatively low capital base deployed at present by the construction companies. This had resulted in a vast influx of multi-national construction companies into the country with almost unlimited resources. In order to rectify this situation, a funding package was proposed to develop a selected group representing a fair cross-section of the domestic contractors while focusing on skills development, technology development, acquisition of plant and equipment, etc. on a pre-determined Co-operative Plan.
2 To convert the existing Construction Guarantee fund to a Construction Development Bank – The construction contracting companies required several finance-related inputs such as bonds and guarantees, working capital, bridging finance, equipment finance, etc. for their existence. As of 2002, most of the above requirements were met through commercial banking institutions at a substantial cost, while requiring almost hundred per cent co-laterals. In order to ease the above burden to some extent, the Construction Guarantee Fund was established in 2000 with government patronage with a view to issuing bonds and guarantees. Since the above institution had met only a part of the needs of contractors, it was proposed that the above Fund be restructured to establish a 'Construction Development Bank' with much wider scope to cover all finance-related requirements of construction contractors and also with future prospects of funding large construction projects.
3 Local LC for claims and bills – This is a bridging financing mechanism to cater for delays in due payments to construction industry service providers.
4 Establishment of material parks – The purpose of this is to create competitiveness in a positive sense while promoting the concept of value of money.
5 Continuity of work.
6 Improvements to procurement system.

At the time of writing, after the scrutinization process, the necessary regulatory framework was expected to be in place once the proposed Construction Industry Bill having the following objectives comes into effect.

The objectives and functions of the Authority should be to:

a Administer the provisions of this Act and the regulations made there under;
b Formulate strategies for the development of the construction industry;

c Promote and upgrade professionalism of constituents in the construction industry by liaising the activities of professional bodies and encouraging formation of bodies of specialists;
d Promote standards for construction works and maintenance, including materials and regulate the adoption of overseas technologies;
e Standardize conditions of contract, procedures and practices relating to the construction industry;
f Foster and promote the advancement of the construction contractors of Sri Lanka;
g Determine the registration and grading of construction contractors;
h Promote and foster the advancement of domestic consultancy services in specialist disciplines pertaining to the construction industry;
i Promote and grant assistance to the export of consultancy services in specialist disciplines pertaining to the construction industry;
j Provide information services to the public on matters relating to the construction industry;
k Develop and publish documentation pertaining to construction industry;
l Regulate and monitor technical auditing;
m Review and monitor the resource requirements of the construction industry;
n Set standards for training and upgrading of the skills and expertise of persons and bodies engaged in the construction industry;
o Facilitate the training of construction industry personnel;
p Regulate the engagement of workers in the construction industry;
q Promote technology transfer and technology development;
r Promote appropriate and affordable technology innovations;
s Promote the use and development of local materials in construction works;
t Make recommendations for the grant of concessionary terms on the import of construction machinery, plant, equipment spares and materials for the use or incorporation of construction works;
u Promote research in respect of any matter relating to the construction industry;
v Promote and grant assistance for the development of organisations related to the construction industry;
w Undertake or assist any other activity for promotion of the construction industry.

Broader policy and development issues

At the time of writing, the Government had prepared its Framework for Poverty Reduction (Framework thereafter) after a two-year consultative process which involved many government agencies, non-government organizations and research institutes. The Framework promotes three major targets, namely (i) reduce poverty by creating pro-poor growth opportunities, (ii) strengthen the social protection system, and (iii) empower the poor by strengthening governance. To achieve the first target, the Government shifts the role in poverty reduction from direct public spending to a facilitator of wealth creation. An array of measures had been carried out to encourage the poor to participate in the growth process. To achieve the second target, the Framework reforms social safety net so as to remove the adverse incentives created in the past. The reforms include a shift from cash allowance to social insurance and better protection mechanism to those displaced by emerging conflicts. To achieve the third target, institutional frameworks of governance are strengthened by reviewing management practices on public services. Financial accountability of regional fiscal policies is tightened while motivations of decentralization are promoted. With bigger share of costs by the participants, community-based initiatives are encouraged to assist the poor. Asset re-distribution is also proposed by the Framework to help the hard-core socially excluded groups to rejoin mainstream economic activities (ADB, 2001b).

Also a key feature of the structural reform is the comprehensive medium-term vision for enhanced private sector participation. This is to be implemented between 2000 and 2002. A multi-sectional approach is adopted. Measures include (i) accelerate restructuring and divestment of public enterprises, (ii) promote competition policy and consumer protection, (iii) strengthen domestic market competition, (iv) reform the labour market to promote mobility, and (v) reform financial market to facilitate private access to financing (ADB, 2000).

11.2.3 Economic outlook in the next decade

At the time of writing, a slow down of GDP growth rates at 4.5 and 5.0 per cent were forecasted for the year 2001 and 2002 respectively. Due to the droughts that happened in the previous year, improvement on agricultural outputs was expected to be limited in 2001. The outputs, however, should resume to the normal level in 2002. As a consequence of currency devaluation that exerted an upward pressure on energy prices, high inflation rates were expected for 2001 and 2002, at 8 and 6 per cent respectively. Investment incentive was expected to be affected by high interest rate and economic uncertainties. Rate of investment was likely to fall slightly in 2001. However, some improvement on investment incentive was expected to be seen in 2002.

Domestic saving was likely to decline slightly in 2001. To reduce budget deficit, the government was expected to take strong action to cut public expenditure. A slight improvement on budget deficit was forecasted in 2001–2002. Exports were expected to fall significantly to just over 3 per cent in 2001. Imports were also expected to fall to below 3 per cent. The decrease in exports and imports was mainly attributed to the slower growth of international demand on Sri Lankan textile and garments outputs as well as the downturn of domestic demand on capital goods. The account of balance of payments was expected to improve in 2001–2002 because the volumes of imports were likely to shrink substantially. More earnings on services sector and steady increase of private remittances were expected to improve the balance of payments account further (ADB, 2001b).

According to the Sri Lankan government (2002), GDP was targeted to grow between 7 and 8 per cent during the decade 2000–2010. The growth would be led by the manufacturing and services-related activities such as information technology, electronics, communications, construction, trans-shipment, and financial and business services.

Major development targets in the decade 2000–2010 are as follows (Sri Lankan Government, 2002):

- A 9.0 per cent GDP growth rate in 2010.
- The reduction of the budget deficit from 9.9 per cent of GDP (2000) to 3.0 percent of GDP (2010).

- The reduction of the current account deficit in the balance of payments from 6.6 per cent of GDP (2000) to 0.5 per cent of GDP (2010).
- The reduction of inflation from 6.2 per cent (2000) to 2.4 per cent (2010).
- An increase in national savings from 21.4 per cent of GDP (2000) to 32.9 per cent (2010).
- An increase in investment from 28.0 per cent of GDP (2000) to 39.7 per cent (2010) of which private investment will be 29.7 per cent of GDP in 2010.
- An increase in exports from 39.1 per cent of GDP (tentative figure) (2000) to 50.8 per cent of GDP (2010).

11.3 The construction sector

11.3.1 Value-added per employee

Planning and productivity are two keys promoting the enhancement and development of the construction industry. Careful planning ranging from macrolevel to microlevels helps to a greater extent towards achieving productivity. Although there is adequate macrolevel planning sufficient planning is lacking at microlevels.

In Sri Lanka, very limited work had been done to study construction productivity in respect of labour and plant. It had been identified that the disruption factors such as lack of continuity and information were the main causes of loss of productivity in construction projects in Sri Lanka.

In general, labour productivity depends on the following factors:

a quality, size and composition of labour force
b motivation of labour force
c degree of mechanization
d continuity of work as affected by

- supply of materials
- performance of other contracts or sub-contracts
- availability and adequacy of technical information
- varieties

e completion of the projects
f required quality of finished works
g methods of construction
h type of contract
i quality and number of managers
j weather conditions.

The main factors that are applicable to Sri Lanka are quality and motivation of the labour force, supply of materials, required quality of finished work, quality of managers and weather conditions.

Sri Lanka has a highly skilled labour force. Their availability for the local construction industry, however, is very limited due to heightened overseas demand particularly by the Middle East countries coupled with reluctance of the youth to enter the industry. This too has had a negative impact on quality and productivity in the local industry.

The government of Sri Lanka has declared the years from 1997 to 2006 as the National Productivity decade. It had been identified that two major causes of low productivity were the lack of understanding of the concept of productivity and the inadequacy of knowledge of the many techniques available for improving organizational and personal productivity. The productivity concept in construction industry is directly coupled with quality and economy. As of 2002, training and awareness programmes were used for productivity improvements in the industry.

11.3.2 Construction cost

Construction costs in Sri Lanka had been increasing continuously (refer Table 11.5). The average construction material prices and construction industry salaries and wages had been on the rise (refer Table 11.6). Financing mechanisms (e.g. commercial banks) available in Sri Lanka were not supportive of construction activities and maintain high lending rates. This factor coupled with inflation had a debilitating effect on construction activities. This in turn affected the industry ranging from individual house builders to infrastructure projects. In this context contractors with unlimited resources (e.g. foreign contractors) had an incomparable competitive edge over the domestic construction contractors thereby denying growth opportunities for the latter. The increase in the construction industry salaries and wages are shown in Table 11.7.

Table 11.5 Unit construction cost (SLR/sq. m of gross floor area)

	Residential (high rise)		*Commercial office*		*Industrial*	*Hotel*
	Standard	*Luxurious*	*Standard*	*Prestige*	*Standard*	*Luxurious*
1995	9,415	15,925	18,173	47,021	9,415	15,925
1996	10,296	16,947	19,476	49,926	10,296	16,947
1997	10,760	17,862	19,755	51,853	10,760	17,862
1998	13,450	20,830	20,358	54,876	13,450	20,830
1999	13,800	22,500	22,000	56,500	13,800	22,500

Note
Unit price excludes sub-structural works.

Table 11.6 Average construction material prices

Material Type	*1995*	*1996*	*1997*	*1998*	*1999*
Cement (50 kg)	301.74	314.99	292	287	270
Sand (cube)	1,089	1,090	1,213	1,290	1,340
Metal 3/4" (cube)			2,831	2,907	2,951
Bricks (1000 nos)	1,470	1,395	1,468	1,704	1,840
Steel (r/f-12 mm) (kg)	31,326	33,200	33.87	34.22	31.71
Asbestos sheets (10'0" long roofing sheet)	–	–	763	754	743
Timber (cu. ft)	–	–	941	958	957
Aluminium (sq. ft)	–	–	739	784	796
Calicut tiles (1000 nos)	–	–	8,869	9,521	10,280

Table 11.7 Construction industry salaries and wages

Cost (Rs.)

Labour	*1995*	*1996*	*1997*	*1998*	*1999*	*2000*	*2001*
Unskilled (day)	93.9	146	175	200	200	225	250
Semi-skilled (day)	–	–	240	285	285	310	330
Skilled (day)	140	145	250	300	300	330	350

11.3.3 Construction quality

The Institute for Construction Training and Development (ICTAD) which is playing a very positive role as the lead agency for acting as a facilitator, co-coordinator and promoter of the development of construction industry in Sri Lanka had identified parameters affecting quality such as standards, processes, procedures, documentation, practices, etc. had several mechanisms in place to promote quality of construction. Some of the main mechanisms are:

1 Committee on standards for construction materials to:

   i Identify problems related to quality in construction materials and find solutions
   ii Formulate standards for new materials
   iii Recommend mandatory standards
   iv Implementation of standards.

   The committee had resolved many problems related to the quality of materials and made the standards mandatory for Portland cement, structural steel, corrugated asbestos roofing sheets and flat asbestos sheets.

2 Registration and grading of construction contractors: Registration is done according to the contractors' ability and fields of specialty in order to make sure that they do not undertake jobs beyond their capabilities.

   The details of ICTAD grading system are appended below:

   i ICTAD grading system for main contractors:

Grade	*Financial Limits* (Rupees million)
M1	300 ~ above
M2	150 ~ 300
M3	50 ~ 150
M4	20 ~ 50
M5	10 ~ 20
M6	5 ~ 10
M7	2 ~ 5
M8	1 ~ 2
M9	0 ~ 1
M10	0.5 ~ 1

Source of Information – ICTAD

Field of Specialization

- Buildings
- Highways
- Bridges
- Water supply and drainage
- Irrigation and land drainage
- Dredging and reclamation

ii Grading system for electrical and mechanical services:

Grade	*Limits* (SLR M)
EM1	$x \geq 10$
EM2	$10 > x \geq 5$
EM3	$5 > x \geq 3$
EM4	$3 > x \geq 1$
EM5	$1 > x$

Source of Information – ICTAD

Specialties

- Steel fabrication
- Air conditioning and refrigeration
- Security system/fire engineering
- Electrical engineering (low tension)
- Electrical engineering (high tension)
- Telecommunication

iii Grading system for Finishing Trades

Grade	*Limits* (SLR M)
F1	$x \geq 3.0$
F2	$3.0 > x \geq 1.0$
F3	$1.0 > x \geq 0.5$
F4	$0.5 > x$

Source of Information – ICTAD

Specialties

- Interior decoration
- Landscaping
- Floor finishing
- Carpentry and joinery
- Aluminium work
- Plumbing

3 Recognition for quality achievements – ICTAD award for construction excellence: The Annual ICTAD award for Construction Excellence symbolises its commitment towards the promotion of quality and professionalism in the construction industry in Sri Lanka.

ICTAD's mission is to develop the domestic construction industry. The objective is to encourage contractors to achieve quality and efficiency in the construction industry by encouraging innovative approaches in technology and timely completion of projects to acceptable standards and costs.

The construction industry plays a vital role in the national development efforts of Sri Lanka and the part played by construction contractors is very important. ICTAD award for Construction Excellence recognises outstanding performance in construction management and organisational skills, technical expertise and quality standards. It also recognises training of craftsmen/tradesmen as well as innovative and appropriate technology in achieving excellence in construction projects.

The building projects are assessed according to residential and non-residential buildings and according to the following three financial categories:

- i Category 1 – above Rs 150 million
- ii Category 2 – between Rs 50 million and 150 million
- iii Category 3 – between Rs 20 million and 50 million

4 Promotion of ISO Quality systems and recognition of stakeholders adhering to ISO Quality Systems.
5 Preparation of guidelines for registration of selected material manufactures and suppliers.
6 Development of a standard type of sand for testing of concrete, with the participation of Sri Lanka Standards Institution and the University of Moratuwa.
7 Development and review of publications for the construction industry.

Sri Lanka Standards Institution too promotes the standards and quality of construction and has the following schemes in operation:

1. Product certification
2. System certification
3. National quality award
4. Import inspection
5. Development and review of standards.

11.3.4 Skill level of workforce and supervision level

The ICTAD carried out a comprehensive manpower survey with the participation of the University of Moratuwa. Based on the manpower requirements derived from this survey, a vocational and educational training plan had been worked out with the assistance of the Tertiary and Vocational Education Commission (TVEC).

For the construction workers already in employment, upgrading of skills is provided in the following areas with facilitation by ICTAD:

- Plastering
- Plumbing
- Tiling
- Painting
- Concreting

- Scaffolding
- Steel fixing and bar bending.

Training modules were developed for safety, sanitary appliance installation and formwork.

11.3.5 New technology development in the construction field

Apart from major infrastructure projects traditional systems of construction were prevalent in the Sri Lanka construction industry until the time of writing. Need for new technology development was a priority due to the quality and productivity demand and high cost of construction using traditional techniques. In this context the following proposals had been made to the government:

1 to establish a research and development fund to carry out applied research work; and
2 to establish a database for retaining technology.

Sri Lanka had developed a satisfactory precast base for several years until 2002, but there was a large scope for expansion. Precast and precast-prestressed technologies might be the answer to high construction costs and a cost-effective alternative to depleting natural resources such as timber. The State Engineering Corporation (SEC) and a few private firms had built up the required expertise in this field and had shown remarkable results in large-scale projects.

11.3.6 Buildability

Prefabrication technology

Prefabricated concrete items were introduced to Sri Lanka by the SEC and State Development and Construction Corporation (SD&CC).

The demand for prefabrication arises when speed and high-quality construction are required. This is particularly true for repetitive type of works. Prefabrication is popular in the following areas:

1 Bridge beams, road kerbs
2 Hume pipes, septic tanks
3 RC purlings, thin RC beams, roof tiles
4 Decorative concrete items
5 Doors, windows
6 Suspended ceilings, floorings, tiles, etc.

At the time of writing, most of the major concrete precasting works belonged in public organizations, but increasingly private sector organizations were getting involved in the business as well. However, precasting of concrete items was a common feature in all large construction sites in order to achieve the required progress. The precast concrete works industry was not popular among the normal house builder due to the requirement for cranes, etc. for lifting and placing.

Environmental conservation

The Central Environmental Authority (CEA) has several regulations in place to prevent environmental degradation. In doing so, CEA has made environmental impact assessments mandatory for prescribed projects. National environmental action plan (1998–2001) has the following action framework:

1 actions promoting policy initiatives or recommendations;
2 actions to improve technical and management tools and the skills needed to use them;
3 actions to facilitate research, extension and education on environmental issues; and
4 action to increase effectiveness and efficiency of the institutional and legal structures to manage environmental issues.

Major construction materials prone for potential environmental degradation are:

1 River sand
2 Coral base lime
3 Timber
4 Clay products.

Items (1) and (2) were included in the Public Investment Programme under high-priority projects. Promotional activities were on for the use of plasticizers in place of lime, identification of sources of sand other than river beds and sea beaches, and use of sand substitutes by the construction industry.

As far as timber is concerned use of aluminium, precast concrete components, fibre glass components, structural steel, etc. were promoted. The use of micro-concrete roof tiles and cement sand blocks helped to minimize environmental degradation due to the use of clay products.

In general the private sector is promoted through self-regulation and motivation and through the use of measures such as environmental audits, ISO 14000 environmental management systems, and the private sector is encouraged to use clean technology.

11.4 Construction outlook 2002–2010

11.4.1 *Business opportunities and major projects in pipeline*

The government is continuing with an active policy of economic liberalization. There are no restrictions to foreign participation on domestic competitive bidding. This policy has several impacts for the development of the domestic construction services because:

1 foreign participants had almost unlimited resources relative to domestic participants;
2 the conditions laid down by the donor agencies favoured foreign participation; and
3 Foreign participation provided no significant yield of technology transfer.

Though foreign companies were allowed to bid also for projects of a local nature, the foreign construction volume was structured to be relatively small, in order to protect the domestic construction services. Therefore, foreign competition was minimized in this respect.

Sri Lanka is an active member of the SAARC Construction Industry Council (SCIC) and has the following regional agenda for action as strategies towards cooperation in the Asian Construction Market:

1 WTO/GATS-related issues – Washington accord for engineers mobility and ASEAN accord for workers' mobility.
2 Disaster management plans and formation of a work cell.
3 Human resource development in construction industry – workers/contractors/supervisors training, harmonization of contents in SAARC nations, initiating a dialogue with WTO secretariat, placing bilateral/multilateral accords, among and within SAARC Nations.
4 Drawing up of the guidelines for trade in construction materials/equipment/LE goods/services, etc., and requesting SAARC secretariat to push these for implementation with all governments of SAARC nations.
5 Drawing up to the guidelines for a preferential treatment for the joint ventures/collaborations/associations from amongst the organizations of SAARC nation origin, for award of the construction/consultancy contracts in multilaterally funded contracts, being executed in SAARC nations. Simultaneously press the Governments of SAARC nations, through SAARC Secretariat for insisting on introducing such clauses in the loan agreements, being signed by them.
6 To prepare a SCIC database containing all relevant data/details, which should be used for substantiating such claims, and also to the used by the business entities to prepare their business plans and to facilitate information flow. (This could be a used charge-based service.)
7 Construction financing, insurance and help in accordance of industry status to various National Construction Sectors (those, who had not been able to get it till then).
8 Development of Risk Identification, profiling and products for the construction industry.

11.4.2 Areas where foreign investments were needed

Despite the slow down in global and local markets, the signing of BOI investment agreements had maintained a consistent flow as September 2001 approvals reach Rs 1.2 billion. Although, to put this in context, 'all approvals granted did not translate into investment'. Typically, about 80–85 per cent of approvals moved on to signed agreements and approximately 60 per cent of these translated into actual investment. The BOI signed 10 agreements during the month of September 2001, amounting to an estimated total investment of Rs 1.3 billion of which Rs 252 million consisted of foreign investment and the projected employment was about 1057. An interesting feature of the September 2001 agreements signed was the high concentration of investment committed to Infrastructure projects amounting to as much as 80 per cent of the value of total agreements signed. Three projects in this sector totalling one billion rupees included Expo Property Developers (Pvt.) Ltd, investing Rs 523 million. This project was to construct a mixed development complex that comprised residential apartments, shops and offices at Messenger Street, Colombo 12. Further, Jia Mai General Development Construction (Pvt.) Ltd had obtained approval to construct 150 housing units at Mattakkuliya. This Chinese and local investment was expected to invest Rs 450 million and create 25 direct jobs. In the power sector Hydro Jet (Pvt.) Ltd was expected to establish a project to start up a mini Hydro Power Plant at Deiyannewela, Aranayake. At the time of writing, this project envisages establishing a mini Hydro Power Plant of 1.2 megawatt under BOO basis with an investment of Rs 75 million generating 51 job opportunities. The Swadeshi Herbals (Pvt.) Ltd had signed to set up a project to grow Herbal plants and manufacture herbal products for the domestic market. This project was expected to be located at Handapanagala Estate, Giriulle investing Rs 68 million with 93 new job opportunities. Orient Building Systems (Pvt.) Ltd had signed an agreement to construct 175 low-cost housing units at Horana with an investment of Rs 50 million that generated 77 direct jobs. The main feature of this project was that a special brick called 'Inter Locking bricks' manufactured by the enterprise itself would be used for the construction of houses. For this purpose, a special brick-making machine at a cost of Rs 2.2 million was to be imported from India. The company had planned to construct 3 types of single unit houses within a period of two and half years.

Private and public investment programmes had been implemented to improve the country's basic infrastructure facilities. As of 2002, Sri Lanka's infrastructure ranked as the best available in South Asia. The government was actively encouraging private sector investment in infrastructure on the basis of BOO (Build, Operate and Own) and BOT (Build, Operate and Transfer). As of 2002, there were private investment projects in power, ports, water sanitation, transportation (highways and rail) and IT. Among the broad infrastructure sectors where foreign investments were most needed are the followings:

Telecommunications

Development of Sri Lanka's telecommunications sector was based on private sector investments. Significant aspects of this sector were operations by multiple public and private telephone service providers, including three basic, four cellular and six payphone services, making the industry one of the most competitive in Asia. Deregulation had resulted in the country's largest ever privatization initiative, with Nippon Telegraph and Telephone Corporation (NTT) of Japan purchasing 35 per cent of Sri Lanka Telecom for US$225 million.

Power

The Ceylon Electricity Board (CEB) is the main supplier of electricity to the National Grid. Independent Power Producers (IPP) share of total power generation was 8 per cent in 1999. The total installed capacity of electricity under both CEB and IPP stood at 1,691 megawatt in 1999.

Roads

The construction of 4 major expressways was expected to improve the existing road network. These included the Colombo/Matara, Colombo/Katunayake and the Colombo/Kandy highways as well as the Colombo outer circular road. At the time of writing, construction work of Colombo/Katunayake expressway had already commenced.

References

ADB (2000) "Sri Lanka Opens up to Private Sector in Bid to Grow Faster", News release of Asian Development Bank, 12 December 2000.

ADB (2001a) "Turning Colombo Port into Regional Shipping Hub", Asian Development Bank.

ADB (2001b) "Asian Development Outlook 2001", Asian Development Bank.

ADB (2003) "Regional and Country Highlights – Sri Lanka", Asian Development Bank.

CBSL (2000) "Annual Report - 2000", Central Bank of Sri Lanka, ICTD, "Sri Lanka – Country Report", *Proceedings of Asiaconstruct Conference*, various issues.

MFA (2002) "Sri Lanka", Ministry of Foreign Affairs, Thailand, www.mfa.go.th/internet/document/886.pdf.

Sri Lankan Government (2002) "Vision 2010 – The Policy Framework", The Government of Sri Lanka.

Chapter 12

Vietnam

National Institute for Urban and Rural Planning, Vietnam and B.S. Tang

12.1 Introduction

12.1.1 Background

Vietnam is located in southeastern Asia. It is bordering the Gulf of Thailand, the Gulf of Tonkin, and the South China Sea. Its adjoining neighbouring countries include China, Laos and Cambodia. The country has a total area of about 329,560 square kilometres covering land (325,360 square kilometres) and water (4,200 square kilometres). It has a long coastline of about 3,500 kilometres (excluding islands) (CIA World Factbook, 2001).

Vietnam has tropical climate in the south, and monsoonal climate in the north. Its hot, rainy season covers the period between mid-May to mid-September. Warm, dry weather is found mainly between mid-October and mid-March. The topography of Vietnam is characterized by low, flat deltas in the south and north. Central highlands and mountainous areas are located in the far north and northwest regions. The elevation ranges from the lowest at the South China Sea to Phanxipan peak (3,143 metres). Vietnam is rich in the natural resources such as phosphates, coal, manganese, bauxite, chromate, offshore oil and gas deposits, forests and hydropower. Land uses include arable land (17 per cent of total land), permanent crops (4 per cent) permanent pastures (1 per cent) forests and woodland (30 per cent) and others (48 per cent). A total of 18,600 square kilometres (1993 estimates) was classified as irrigated land (CIA World Factbook, 2001).

Vietnam has a population of about 80 million. About 90 per cent of the population are ethnic Vietnamese. They originated in southern China and northern Vietnam and moved southwards during the past centuries to occupy the eastern seaward part of the Indochinese Peninsula. The remaining 10 per cent of the population comprises many other ethnic groups. For example, approximately 1.2 million Chinese are concentrated mainly in southern Vietnam. The central highland people constitute another large group, commonly known as the Montagnards (mountain people), which comprise two main ethnolinguistic groups: Malayo-Polynesian and Mon-Khmer. In the highland territory, there are over 30 groups of various local cultures and dialects. Furthermore, there are about 600,000 of Khmer Krom (Cambodians) concentrating in southern provinces near Cambodia and at the mouth of the Mekong River. Other minority groups in the north include Chams, Hmong and Thai (Bureau of East Asia and Pacific Affairs, 1998).

Vietnam today is a socialist republican state that operates on the basis of the communist ideology and French civil law system. There are 58 provinces and 3 municipalities.

12.1.2 Economic development profile: an overview

Vietnam is a poor, densely populated country that is recovering from wars, a legacy of international politics, and the rigidities of a centrally planned economy. From the late 1970s to the early 1990s, Vietnam was heavily reliant on the Soviet Union and its allies for trade and economic assistance. Throughout the 1980s, the Soviet Union provided about $3 billion of economic and military aids to Vietnam annually. At that time, most of Vietnamese external trade was conducted with the Soviet bloc and with the Council for Mutual Economic Assistance (CEMA) countries (Bureau of East Asia and Pacific Affairs, 1998). However, by the mid-1980s, Vietnam began to re-orientate its economic and foreign policies.

Since the Sixth Party Congress in December 1986, Vietnam has prepared to open its economy and rationalize its international relations in Southeast Asia. Vietnam began to step up its efforts in regularizing its relations with the Western non-communist countries, attracting foreign capital and restoring its links with the world financial and political organizations. After the lifting of the US ban on multilateral loans to the country during the 1990s, Vietnam secured membership of the World Bank, the International Monetary Fund (IMF) and the Asian Development Bank (ADB). Full diplomatic ties between Vietnam and China were restored in 1991. Vietnam became a member of ASEAN in July 1995 and joined APEC in November 1998. With its increasing integration into the global community, Vietnam begins to receive aids from the West again. At present, Vietnam has established diplomatic relationship with over 160 countries. Furthermore, the country has expanded trading with its Asian neighbours as well as with countries in Western Europe and North America. Vietnam has been participating in the ASEAN Free Trade Area (AFTA) and will probably join the World Trade Organization (WTO) in the near future (Bureau of East Asia and Pacific Affairs, 1998).

The centrally planned economic model and poor economic management were the major internal obstacles in constraining economic growth of Vietnam. After the reunification of the country, economic stagnation took place in the period between 1975 and 1985. During this period, the Vietnamese government made little improvement in increasing its economic outputs and raising the general living standards. Economic crisis broke out by the mid-1980s with inflation running at 700 per cent in 1986. The need for economic reforms was agreed by the Vietnamese authority. In that year, the Sixth Party Congress approved strategies for substantial reforms in key economic areas such as exchange rate, foreign investment and government budget management. The Vietnamese generally refers this as the beginning of a major economic reform known as 'Doi Moi' or Renovation (Bureau of East Asia and Pacific Affairs, 1998).

The Vietnamese party-state gave strong commitment and high priority in the implementation of economic reforms. Significant economic progress was achieved in the areas of industrial production, external trade and foreign investment in the country. As a result, Vietnam became one of the fastest growing economies in the world. Its average GDP growth reached 8 per cent from 1990 to 1997 and its inflation rates fell sharply from over 300 per cent in 1987 to less than 4 per cent in 1997. Furthermore, investment increased nearly threefold and domestic savings recorded a fivefold increase. Per capita income rose from $220 in 1994 to $320 by 1997. The proportion of the population in acute poverty also reduced. All these reflected a significant improvement in the economic well-being of the Vietnamese population (Bureau of East Asia and Pacific Affairs, 1998).

Economic reforms had not only doubled the agriculture production in Vietnam, but it had also transformed the country from a net food importer to a major rice exporter. The recovery of the agricultural sector was fuelled by the increasing procurement prices and private production since 1980. Land reform and the decollectivization further improved the agricultural productivity in the 1990s. Vietnams economic structure is gradually being transformed. The country now depends heavily on its primary product exports such as rice, coffee, tea, cashews and rubber to fund its industrialization. The GDP share of agriculture declined from 42 per cent in 1989 to 26 per cent in 1997. Industrial production accounted for nearly 32 per cent of GDP in 1997. However, state-owned enterprises that comprise the backbone of its modern industrial sector were characterized by low productivity and inefficiency. Comparatively speaking, southern industry was more efficient and it is largely composed of textiles, food processing and light manufacturing. At present, Vietnam's key industrial exports comprises labour-intensive, light industrial products. While the government had removed subsidies to many inefficient state enterprises and repeatedly announced its intention to privatize the state enterprises, few state-owned enterprises companies had been privatized to date. Poor competitiveness in its industrial production remains problematic. Hence, low quality of industrial capacity will continue to constrain Vietnam from becoming a world producer of advanced manufactured goods (Bureau of East Asia and Pacific Affairs, 1998).

On the other hand, economic reforms also resulted in a dramatic increase in foreign trade, which now represents about 80 per cent of GDP. Exports recorded a growth of 20–30 per cent per year and made up some 35 per cent of GDP by 1997. Agriculture and forestry products were Vietnam's largest exports (36 per cent in 1994). Light industrial exports, such as textiles, were also rising (from 14 per cent in 1991 to 19 per cent in 1994). The country is now more closely integrated than previously with the regional economies. By 1994, Vietnam's top five trading partners were all Asian countries – Japan, Singapore, Hong Kong, South Korea and China. However, imports also rose considerably during the same period leading to persistent trade and current account deficits. In 1997, the trade deficit was recorded at $2.4 billion. The economic prospects for Vietnamese exports and current account performance are expected to improve when the country has completed its accession to the World Trade Organization (Bureau of East Asia and Pacific Affairs, 1998).

Foreign investment in Vietnam rose rapidly in recent years. In 1997, foreign direct investment inflows accounted for 8 per cent of GDP and achieved an all-time high at $2.6 billion. This figure exceeded the $2 billion current account deficits of the country at the time. Vietnam became one of the emerging markets in southeast Asia for foreign investors. As of end 1997, the Vietnamese government issued more than 2,200 investment licenses valued at $31.4 billion to foreign firms coming mostly from the East Asia region. For example, Singaporean firms and the United States had received licenses to invest $4.9 billion and over $1.1 billion respectively in Vietnam. Other key Asian investment sources included Taiwan, Hong Kong, Japan and South Korea (Bureau of East Asia and Pacific Affairs, 1998).

12.2 The national economy

12.2.1 Macroeconomic review

In 1986, Vietnam took a step forward in launching an extensive economic programme called 'Doi Moi' (Renovation). This signified the move from a command economy to a more market-oriented one. According to information provided by the Bureau of East Asia and Pacific Affairs (1998), the major reform measures included:

- decollectivization of agriculture;
- land reform that provided a greater security of land tenure;
- shift of investment focus from heavy industry to agriculture, light industry and exports;
- elimination of virtually all administered prices;
- liberalization of foreign trade and investment;
- interest rate liberalization;
- reform of exchange rate; and
- promotion of private-sector led growth through establishment of a legal framework.

The programme had enabled the Vietnamese economy to grow at an average annual rate of 3.7 per cent during the period between 1987 and 1991. Total foreign trade made up about US$6.3 billion in 1993. The country emerged to become a major rice exporting country. In December 1990, the ruling party came up with a new plan that was intended to give Vietnam an economic push whilst maintaining the state control of the society. This plan, known as the Draft Strategy of Social-Economic Stabilisation and Development Up to the Year 2000, comprised three main objectives (UNIDO, 2004):

1 doubling the country's per capita income;
2 establishing an open economic system; and
3 creating a multi-sectoral economy.

The plan also targeted at reducing the inflation rate from 85 per cent in 1990 to less than 10 per cent by the year 1995, and also expanding exports by five times before the year 2000. As an outcome of this plan, the Vietnamese GDP growth rate recorded an average of 8.2 per cent between the period 1991 and 1995, more than double that of the period between 1986 and 1990 (Table 12.1).

According to UNIDO (2004), the social and economic policy reforms of Doi Moi (Renovation) had achieved significant outcomes especially during the late 1990s. GDP was doubled between 1991 and 2000. Domestic savings reached 25 per cent of GDP in 2000. The living standards of the people had improved enormously. According to the official statistics, the portion of poor households (based upon Vietnam's poverty line) had declined from over 30 per cent to 11 per cent. According to internationally comparable poverty criteria, the poverty level dropped from 70 per cent in the mid-1980s to 37 per cent in 2000 (UNIDO, 2004).

Agriculture continues to dominate the Vietnamese economy. In 1991, the agricultural sector contributed significantly to the country's Produced National

Table 12.1 GDP growth rate 1976–1995

1976–1980	GDP growth rate	1981–1985	GDP growth rate	1986–1990	GDP growth rate	1991–1995	GDP growth rate
1976	–	1981	2.3	1986	6.5	1991	6
1977	2.8	1982	8.8	1987	3.4	1992	8.6
1978	2.3	1983	7.2	1988	4.6	1993	8.1
1979	2	1984	8.3	1989	2.7	1994	9
1980	1.4	1985	5.7	1990	2.3	1995	9.5
Average	0.4	Average	6.4	Average	3.9	Average	8.2

Income (PNI). It accounted for 51.7 per cent of the PNI. The industrial sector was the second largest contributor and accounted for 22.8 per cent. This was followed by commerce (18.0 per cent) and construction (4.4 per cent). Following the drive towards industrialization, the share of agriculture in GDP declined by 11.5 per cent by 1995, while the share of industry and construction sectors increased by 7.6 per cent, reflecting a major structural change in the economy (Table 12.2).

The gross domestic product (GDP) in Vietnam expanded at an annual rate of 9 per cent during 1992–1997 and such growth rate was expected to continue. Since 1993, the construction industry expanded by 18 per cent annually and at much higher rates in urban centres. The annual increase of industrial output recorded at 12–13 per cent, the service sector at about 9 per cent and the agriculture sector at about 3.5 per cent. Although economic re-structuring continued, agriculture still accounted for about one-third of GDP and for more than 70 per cent of the total employment in Vietnam. The expansion of output from state enterprises was more rapid than that of the non-state industrial sector. Its performance was better than expected, despite there was a cut of 25 per cent in employment since 1990 (ADB, 1995).

The Asian financial crisis in 1997 gave a major blow to the economy of Vietnam. Its economic growth stalled for the first time since the implementation of Doi Moi in 1986. The growth rate of Vietnam decelerated to 4.4 per cent in 1998 (the official estimate is 5.8 per cent), followed by a pickup to 4.7 per cent in 1999. Its economy began to show signs of recovery by the end of 1999. This modest growth recovery in 1999 was mainly led by net exports. It resulted from 'demand-stimulus' policy supported by favourable fiscal and monetary means. GDP growth in 2000 improved to 6.7 per cent with major contributions from industry construction (3.4 per cent) and service (2.3 per cent). Economic growth that mainly occurred in the protected industries (cement, ceramics, cars, motorcycles and sugar) suffered after the crisis. These industries had faced increasing competitions (investment, quality and price) as

Table 12.2 Change in economic structure of Vietnam

% contribution to GDP	1990	1995
Agriculture and fisheries	38.7	27.2
Industry and construction	22.7	30.3
Services	38.6	42.5

Vietnam became integrated into the regional economy. Following a declining trend from 1999, the 2000 consumer price index (CPI) fell steadily from February until the end of the year. The annual inflation rate was −0.6 per cent, the first time of deflation during the last 10 years, until 2002.

In respect of individual economic sectors, the agriculture sector performed well in 1999. In contrast, the manufacturing and services sectors that had been the main drivers of rapid growth prior to 1997, slowed down in 1998 and decelerated further in 1999. The pace and direction of policy reforms in the state enterprises and banking sectors was expected to generate the momentum necessary to transform modest economic recovery to sustainable higher growth. Moreover, the impact of reforms had been limited in the context of 2 years of economic slowdown, until the time of writing (ADB, 2000).

According to the official data, Vietnam's real GDP growth was recorded at 6.8 per cent in 2000, an increase from 4.8 per cent in 1999. Strong performance in exports and industry supported this growth. The resurgence of industry sector growth to 9.7 per cent was spurred by rising domestic and external demand. The domestic non-state sector was found to outperform the state and foreign private sectors. The agriculture sector performed well (4.0 per cent according to official statements) despite the floods in the central region and droughts in the north (Table 12.3).

The economy of major cities like Hanoi had grown faster than the national economy, at an average of 28 per cent over the last 7 years until 2002. In December 2000, the National Assembly passed a Resolution for 2001 development of which major indicators were:

- GDP growth at 7.5 per cent (5.4 per cent by World Bank forecast, 5.0 per cent by IMF);
- sector growth: agriculture 4.5 per cent, industrial 14 per cent and service 7 per cent;
- export growth at 16 per cent;
- total GDP investment 30 per cent;
- inflation rate not exceeding 5 per cent (same as forecast by ADB);
- new employment for 1.4 million;
- vocational training for 0.84 million; and
- poverty rate 16 per cent (by new poverty line).

According to some economists, these indicators seemed feasible initially. However, in order to make it sustainable, it depends on a number of factors including macro-policy performance, improvement of business environment, and other external factors (e.g. oil price, FDI flow and growth of key Vietnam partners). Some economists were of the view that Vietnam economy was behind that of Thailand, which had about the same population and land area, by about 20–25 years. Nonetheless, it was believed that appropriate policies can reduce the gap of development and quickly bring Vietnam on a par with Thailand.

12.2.2 Economy in 2001

The Vietnamese economy showed positive growth signals in the first eight months of the year 2001, with the agricultural, industrial and service sectors all improving

Table 12.3 Main macroeconomic indicators

	1996	*1997*	*1998*	*1999*	*2000****
GDP and components					
GDP at real prices* (VND billions)	213,833	231,264	244,596	256,269	270,363
GDP at current market prices (VND billions)	272,036	313,653	361,016	399,942	421,939
GDP growth (%)	9.34	8.2	5.8	4.8	5.5
Primary sector (agriculture, forestry and fishing) (VND billions)	53,577	53,690	57,866	60,883	63,318
% growth	4.4	2.1	7.8	5.2	4
Manufacturing sector (VND billions)	50,078	56,619	63,003	68,836	74,480
% growth	13.5	13.1	12.8	9.3	8.2
Services sector (VND billions)	110,178	120,955	123,727	126,550	131,232
% growth	8.9	9.8	2.3	2.3	3.7
Construction sector (VND billions)**	16,938	18,855	18,761	19,211	n.a.
Private construction	n.a.	n.a.	n.a.	n.a.	
Public construction	n.a.	n.a.	n.a.	n.a.	
Repair and maintenance	n.a.	n.a.	n.a.	n.a.	
% growth	12.7	11.3	−5.0	2.4	
Demographic indicators					
Population (million people)	75.3	75.8	76.3	76.9**	77.5
Population growth rate (%)	1.88	0.7	0.7	0.8	0.8
Total labour force (1000 people)	35,792	36,994	37,407	37,800	n.a.
Labour force growth rate (%)	3.48	1.41	1.12	0.4	n.a.
Unemployment rate of labour force in urban area	5.88	6.01	6.85	6.74	n.a.
Financial indicators					
Short-term interest rate	n.a.	1	1.2	1.1	1.1
Long-term interest rate	n.a.	1.1	1.25	0.85	0.75
Changes in consumer price index	4.5	3.6	9.2	0.1	6
Change against US$ (VND per US$ 1)	11,170	11,996	12,989	13,953	14,070

Sources: SPH 1998, 1999; CIEM, 2000.

Notes

* GDP at constant 1994 prices.

** The construction products are not divided into private, public, and repair and maintenance.

*** 2000 data is forecast.

noticeably over their last year's performance. Table 12.4 present the economic forecasts for 2001.

In agriculture, by mid-August, the total area planted with autumn–winter rice stood at 1,591,700 hectares and the harvest of summer–autumn rice crop was completed on 1,099,400 hectares in the southern provinces, representing 92.9 and 84.7 per cent of the figures of the same period in 2000. Expert views reveal that these figures were really significant, given the adverse weather conditions in early August when torrential rains inundated large areas of rice seedlings.

Industrial production value increased by 14.3 per cent in the first eight months over the same period in 2000. The state-owned sector rose by 11.6 per cent, non-state sectors by 19.1 per cent and FDI sector by 14.4 per cent. Production values of the oil

Table 12.4 Forecast of Vietnam's economy: some macroeconomic indicators for 2001

Description	*Scenario*					
	Baseline	*1*	2	3	4	5
Assumptions that there will be different from the baseline scenario			No change in foreign investment	No change ininvestment from the state budget	5% nominal exchange rate depreciation	5% GDP increase of key trading partners
Real GDP growth (%)	7.3	6.9	6.8	7.2	7.5	7.5
Inflation rate (%)	4.8	4.5	4.8	4.8	5.3	4.8
Nom. GDP growth (%)	14.1	13.0	13.6	13.7	14.9	14.7
Trade bal. (%) of GDP	−1.4	−2.3	−1.0	−1.4	−0.8	−1.0

Source: Central Institute for Economic Management (CIEM, 2000).

Note
According to the analysis of Central Institute for Economic Management (CIEM), the above economic growth performance showed that the growth indicators set by the Tenth Vietnamese National Assembly in the Resolution on the targets for 2001 were feasible.

and gas industry rose 12.9 per cent. Fairly large industrial production can be found in localities such as Hai phong, Quan gninh, Da nang, Khanh hoa, Ho Chi Minh City, Binh duong, Dong nai and Can tho. Hanoi's industrial value rose 8.4 per cent over the same period in 2000. Considerable industrial growth came from sectors including fuel, energy, service industry, seafood, food processing, automobile assembly, and electronics and information technology sectors. However, textiles and garments, chemical fertilizers, brewery, sugar and sanitary ware industries were still facing difficulties in selling their products until 2002.

The country's export revenue in the first eight months of in 2001 was estimated at US$ 10.4 billion, up by 12 per cent over the corresponding period in 2000. Domestic businesses recorded a $5.6 billion export turnover for a 13.4 per cent year-on-year rise. The FDI sector increased its export turnover (including crude oil) over the same period by 11.3 per cent reaching $4.8 billion. Significant increases in export revenues came from seafood (42.8 per cent), fruits and vegetables (84.9 per cent), coal (33 per cent) and crude oil (9.5 per cent). Export volumes of cash crops like coffee, pepper and cashew nut increased, but their values had dropped because of falling international prices. The trade deficit stood at $122 million or 1.2 per cent of export revenues. Imports of some products also showed an upward trend, with that of automobiles increasing 2.2 times in volume and 2 times in values.

By mid-August 2000, capital constructions funded by the State budget achieved 64.5 per cent of the yearly plan, up by 3.2 per cent over the same period. The plan of the year 2001 was some VND 5000 billion higher than that of 2000. Disbursement from the central budget attained 68.6 per cent of the yearly plan, but provincial and lower-level budgets lagged behind at about 50.7 per cent. As many as 281 new FDI projects were licensed with a total registered capital of $1.13 billion – an increase of 35.7 per cent in the number of projects and 14.1 per cent in capital commitment. The industrial sector took up 78.6 per cent of these projects and 90.6 per cent of their registered capital (Vietnamese News Agency, 2001).

12.2.3 Construction and property markets

Vietnam's poor infrastructure provision created a huge demand for construction activities. A national reconstruction program was half way through, with the government putting the highest priority on infrastructure projects such as roads, bridges, airports, seaports, irrigation systems, office buildings and residential units. Government spending as well as international loans and aids were expected to make the construction industry one of the fastest growing sectors in the economy. Foreign investment was also encouraged by the Vietnamese government through BOT projects. According to the United Nations Industrial Development Organization, its construction sector employs about 250,000 people. The workforce in this sector was highly professional and skilful in using modern building machinery. The construction sector was dominated mainly by state-owned enterprises. The Vietnamese building industry had not only undertaken the construction of projects at home, but also obtained building contracts in the former USSR and Western Asia and Africa (Asia Business Network, 2002).

In 2000, the construction sector completed the implementation of VND 101.82 billion worth of capital construction assigned by the state. The domestic building materials can meet the local demand. The growth of construction was at 7.5 per cent and that of consultancy 10 per cent. Fifteen million square metres of housing were built. Many companies that used to be sub-contractors had became main contractors such as the Vietnam Machinery Installation Corporation (Lilama) and Song Da Construction Corporation. In 2001, the sector set a growth target of 14.5 per cent for building materials, 8 per cent for construction, 9 per cent for consultancy, 29–30 per cent for export, and building 17 million square metres of housing and implementing 100 per cent of the State's capital construction.

Table 12.5 provides the statistics for national construction investment for the period between 1996 and 1998. The total outlays of state investment in all types of development were generally reduced. Investment in 1998 was either equal or less than that of the 1996. On the contrary, private investment showed a noticeable increase particularly in the civil engineering field. As a result, the total outlay of investment even recorded a small increase. Unfortunately, data were not available to demonstrate the longer-term effects of the Asian financial crisis. Understandably, the investments

Table 12.5 Construction investment (in Mn VND at 1994 constant prices)

Type of development	*1996*	*1997*	*1998*
Residential construction (A)	2,035	2,283	2,154
Public	727	766	698
Private	1,310	1,517	1,546
Non-residential construction (B)	54,399	62,757	60,746
Public	n.a.	n.a.	n.a.
Private	n.a.	n.a.	n.a.
Civil engineering construction (C)	11,055	14,165	12,680
Public	9,907	14,025	9,948
Private	1,148	140	2,732
Total construction investment (A + B + C)	67,489	79,205	75,580

in both private and public construction had dropped immediately following the crisis but generally picked up again in 2000.

Land prices and rents in Vietnam are not determined entirely by the market. Government Decree No. 87/CP dated 17 August 1994 stipulated a price table for each type of land for the purpose of tax calculation (with respect to land allocation and rent) and compensation for damages when the government recovers the land. The various types of land are classified as follows:

- Agricultural and forestry land to be divided into 3 categories – plain, midland and mountainous region.
- Rural residential area to be divided into 3 categories – plain, midland and mountainous region.
- Residential areas adjacent to cities, towns, and traffic junctions and major land routes, commercial centre tourist sites and industrial areas.

The land prices in rural and urban residential areas, communication centres and main roads, business areas, tourist spots and industrial zones are based on the market prices.

For land in the urban areas, the land prices are determined mainly according to the types of street. Each type of street is classified into four to six different land locations. Classification of streets inside the city is determined on the basis of their position and frontage, business potential and profitability, level of urban infrastructure and market prices.

The People's Committee at each province or city is responsible for executing the classification of land category, particularly in terms of the type of street and position/location, in determining land prices. They are also allowed to readjust land prices to suit the current market conditions, when there have been changes to investments in infrastructure, new industrial areas, commercial centre and tourist sites.

The first half of 2001 saw the steady rise of land prices in Vietnam, especially in the inner city of some major cities such as Hanoi and Ho Chi Minh City. Increasing land prices had pushed some urban residents to move out to the periphery. Some were buying new urban housing mostly located in the outskirts of the cities, thus pushing up the land prices in these areas.

Land leased by foreign organizations or individuals with foreign invested capital was subject to the relevant government regulations. In general, retail, hotel and office developments occupy relatively central locations in the cities. Their rental rates were therefore higher than those for residential and industrial uses (Table 12.6). The leases and terms of payment are dependent upon negotiations and the regulations of the government at the time.

Table 12.7 presents the average rents between the period 1996 and 1999. It showed a decreasing trend in the rents of private residential space. Data for commercial rents were not available. However, based on anecdotal evidence, the commercial rents were found to go down significantly. It was observed that occupancy levels of commercial and residential spaces had also declined, mainly due to oversupply and economic slowdown. The take up of modern office space had dropped in major cities like Ho Chi Minh City and Hanoi. Vacancy levels of buildings that came into stream before 1997 stood at about 40 per cent of total stock in Ho Chi Minh City (Business

Table 12.6 Average land rental fees of project and companies in February 2001

Type of projects and companies	*Average land rental fees* (US$/sq. m/yr)
State-owned housing projects	2.57
State-owned office projects	2.42
Private companies hotel	13.04
Private companies offices	15.11
Private companies retail	14.91
Private companies housing	11.42
Private companies factories	2.74

Sources: Investconsult group of Korea international corporation agency (February 2001).

Table 12.7 Average rent for the period between 1996 and 1999

Product	*1996*	*1997*	*1998*	*1999*
Public residential	0.1	0.1	2.2	n.a.
Private residential	0.4	0.3	0.2	0.2
Commercial	n.a.	n.a.	n.a.	n.a.
Industrial	n.a.	n.a.	n.a.	n.a.

Source: SPH, 1998, 1999.

property, 1997). The vacancy level was at 35 per cent in Hanoi. The situation worsened as a result of the overbuilding of prime office space in 1998. Accordingly, average monthly rates dropped by 20 per cent between 1997 and 1998. Vacancy dropped to 45 per cent by 1999 and rents by 7 per cent from the year 2000. During the Asian financial crisis, the rents of high-luxury homes and villas dropped by over 40 per cent in both Ho Chi Minh City and Hanoi. Industrial rents also fell over the past years.

In the major cities of Vietnam, the residents tend to be more concerned about the number of floors in the apartment buildings. This is because housing prices are usually based on the number of floors. Higher-floor apartments are generally cheaper. The housing prices cover a range between VND 200 million and 350 million per apartment (Table 12.8). At the time of writing, house prices in new urban areas continued to increase. The minimum price for an apartment in the Bac Linh Dam residential district of Hanoi rose from around VND130 million ($8,000) a few years ago to VND200 million ($13,300). Property experts believed that a lack of market information on transactions

Table 12.8 Average price, typical size and share of house in Hanoi

	Low-density housing	*Medium-density housing*	*High-density housing*	*Total*
Average price (million VND)	300	158	148	
Typical size for HNT (sq. m)	120	70	50	
% share of people who can afford house price	25	25	50	100

Source: Feasibility study on the Hanoi New town development Project (February 2001).

Table 12.9 Economic forecast for Vietnam in comparison to total Asia-Pacific region and ASEAN

Country/region	*2001*	*2000–2010*
Vietnam	5.0	6.0
Total Asia-Pacific	5.1	5.3
ASEAN	4.8	5.1

Source: Ernst & Young International, 2000 at www.eyi.com.

and purchase procedures explained the increasing housing prices in new urban areas. The lack of transparency had led the buyers to rely on brokers, which in turn sustained an artificial surge in house prices. On the other hand, land prices in suburban districts, particularly those close to new urban areas, had almost doubled in the past 2 years until 2002 (Vietnam Investment Review, 2002).

12.2.4 Economic outlook in 2001–2010

At the time of writing, some economists had made the forecast that Vietnam can develop with an average GDP growth rate of between 8 and 9 per cent per annum for the next 20 years. This would enable Vietnam to reduce the income disparity with other southeast Asia countries (Table 12.9). However, to achieve this goal, it was imperative that the government directed the country towards innovation, competition and foreign investments (Vietnamese News Agency, 2001). The Vietnamese government had announced a target annual GDP growth of at least 7 per cent over the next few years since the time of writing. To achieve this policy objective, the investment rate was expected to be raised by over 30 per cent. It was also expected that a significant proportion of this increase has to come from private sources. To finance the required magnitude of investment, it was also necessary to mobilize sufficient domestic resources, particularly private saving.

12.3 The construction sector

12.3.1 Labour productivity

By survey of MOLISA July 2000, the total workforce in Vietnam comprised 38.6 million of which 95 per cent were of working age. About 62.7 per cent (22.7 million) were working in agriculture, 13.2 per cent in industry and construction, and 24.3 per cent in service sector. The unemployment rate reached its peak in 1998, particularly at major cities, and slightly reduced in following years. To comply with a policy to reduce 15 per cent of the government staff, state-owned enterprise (SOE) reform will lay off about 400,000 workers. After a year of implementation of the Enterprise Law, new enterprises created about 300,000 jobs. As of June 2000, foreign investors provided 330,000 jobs.

Table 12.10 presents the gross product per employee among different economic sectors during the period 1996–1999. Because the statistics on the value-added per employee was not available, our estimations were based upon the GDP data and

Table 12.10 Gross product per employee (US$ at 1994 constant price)

Sector	*1996*	*1997*	*1998*	*1999*
Construction	1582	1755	1736	1772
% change	18.4	10.9	−1.1	2.4
Manufacture	1264	1409	1518	1283
% change	13.1	11.5	7.7	−15.5
Service	1318	1309	1345	1336
% change	−0.7	−0.7	2.8	−0.7
Agriculture	200	200	209	218
% change	4.7	0	4.5	4.3

employment figures of different sectors in order to compare their respective productivity levels. Our findings illustrates that the construction sector productivity had experienced a significant drop since the Asian financial crisis.

During the pre-crisis years, however, productivity as a measure of the construction output per worker ratio showed a steady rise. For example, Table 12.11 shows that the construction output per worker in the state sector rose from 50 million VND in 1994 to 62 million VND in 1996. The increasing use of equipment and machinery, coupled with the application of higher level of technology, caused such a rather quick progress in productivity.

Regarding physical measurement of construction productivity, Table 12.12 shows the available statistics on the public housing produced by the SOEs. The data on commercial and industrial buildings, however, cannot be transferred to the unit of square metres for comparison.

Table 12.11 Construction output and workers

	1994	*1995*	*1996*
Construction output (m VND)	13,500	17,000	265,000
State sector	13,500	17,000	18,200
Non-state sector	–	–	8,300
Labour-workers (000)			
State sector	265	293	292
Non-state sector	–	–	–
Construction output (m VND)/workers			
State sector	50	58	62
Non-state sector	–	–	–

Table 12.12 Construction productivity (sq. m per manday)

Product	*1996*	*1997*	*1998*	*1999*
Public residential	0.1	0.1	2.2	n.a.
Private residential	0.4	0.3	0.2	0.2
Commercial	n.a.	n.a.	n.a.	n.a.
Industrial	n.a.	n.a.	n.a.	n.a.

12.3.2 Property and construction costs

The housing market had generally been vibrant in the main cities of Vietnam. It appeared to have been less affected by the Asian financial crisis in comparison with the commercial and industrial property markets. The available data on the unit cost of housing in Hanoi had witnessed a rising trend that peaked in 1999 (Table 12.13). The unit cost of housing had come down marginally (Table 12.14). The falling prices in 2000 and 2001 can be attributed to the oversupply as well as the mismatch in the type of supply (e.g. fewer affordable dwelling units).

Table 12.15 demonstrates the changes in the cost of residential construction in Hanoi during the past 8 years. The cost of higher standard residential units was relatively stable while the costs of medium to low standard residential units escalated mainly due to stronger demand and fewer supply.

Table 12.16 shows the average prices of major construction materials in Hanoi. Overall, the prices increased steadily but dropped considerably in the aftermath of the Asian financial crisis. But they began to pick up again in 2000. In 2001, the prices of most construction materials, such as steel bars, concrete sands and bricks, already exceeded their peak levels in 1997–1998 period as a result of strong economic recovery.

Table 12.13 Construction cost in Hanoi (VND1000 per gross floor sq. m)

Product	*1997*	*1998*	*1999*	*2000*	*2001*	*2002*
Housing	1,450	1,600	1,600	1,500	1,500	1,550
Office	n.a.	n.a.	n.a.	n.a.	n.a.	n.a.
Hotel	n.a.	n.a.	n.a.	n.a.	n.a.	n.a.
Industrial	n.a.	n.a.	n.a.	n.a.	n.a.	n.a.

Table 12.14 The residential construction cost (VND1000 per gross floor sq. m)

Locality	*Villas (categories)*				*Houses (categories)*			
	I	*II*	*III*	*IV*	*I*	*II*	*III*	*IV*
Hanoi	1,100	1,200	1,400	2,000	900	800	550	400
Ho Chi Minh City	1,200	1,500	1,500	2,000	1,000	900	550	400
Secondary cities	1,100	1,200	1,300	2,000	800	700	400	300

Source: The Government Decree No. 61CP dated 5/7/1994.

Table 12.15 The residential construction cost (VND1000 per gross floor sq. m) in Hanoi

No. Year	*Villas (categories)*				*Houses (categories)*			
	I	*II*	*III*	*IV*	*I*	*II*	*III*	*IV*
1994	1,100	1,200	1,400	2,000	900	800	550	400
1995	1,180	1,290	1,500	2,000	960	850	660	435–525
1997	1,350	1,200	1,300	2,000	1,300	1,011–1,186	794	531–606
1999	1,350	1,600	1,850	2,100	1,300	1,011–1,186	794	531–606

Table 12.16 Construction material price in Hanoi (VND1000)

Products	*1996*	*1997*	*1998*	*1999*	*2000*	*2001*	*2002*
Bim Son cement PC300 (ton)	840	840	840	746	750	740	780
Steel bar (ton)	4,250	4,155	4,657	3,890	4,500	4,600	n.a.
20 mm stone aggregates (m^3)	90	105	116	101	110	220	n.a.
Concrete sand (m^3)	59.5	60	58	50	70	140	120
Bricks (1000 pcs)	330	386	406	340	350	361	370
Ready-mix concrete M300 (m^3)	720	647	666	498	550	560	n.a.

Source: HN Construction Material Price.

Table 12.17 Average income per month of labour in state sector (VND1000)

Sector	*1996*	*1997*	*1998*	*1999*	*2000*	*2001*	*2002*
Construction	572	738	807	787	780	770	785
Manufacturing	639	762	806	806	n.a.	n.a.	750
Service	642	615	646	655	720	750	900
Agriculture, fishery and forestry	422	480	514	493	n.a.	n.a.	n.a.

Source: SPH 1996, 1997, 1998, 1999 and survey 2001, 2002.

Table 12.17 shows that average income of construction labour reached a peak in 1998 and then decreased gradually in the subsequent years. In comparison, the wages of the service sector employees were steadily rising. This may illustrate that construction sector employment was vulnerable to external influence as well as the market fluctuation in the economy. After 1998, the statistics on official incomes recorded a general decline. But it was noted that official income had reflected only part of the labour earning. In practice, the workers often had to get income from other sources in order to cover their living expenses.

The Vietnamese living standard surveys show a widening gap between rural and urban salaries. The highest growth rates of salaries and wages were recorded in the Northern provinces, especially in the northern mountains and the Central North region, rather than in the South. Furthermore, the salary gap between those with university/college degrees and unskilled workers increased from 0.95 to 1.7 times for the same period. In 2000 a modest reform measure involving the adjustments of minimum salaries in the state and FDI sectors was implemented. As of 2002, changes in salaries and wages had mirrored the development of the construction labour market. Salaries, at the time of writing, reflect more closely the cost of human capital in the construction labour market.

12.3.3 Construction quality

Quality assurance for domestic projects is mainly enforced through a set of Vietnamese construction standards. Many of these standards are largely compatible with ISO 9000. However, these standards are regarded as incomplete and out-of-date, and the actual enforcement of construction laws is still weak. Therefore, the

Government Decree 87/CP 1997 allows ODA projects to have the rights to select relevant technical standards (e.g. ISO, British, American, France, Japan, etc.).

In addition, in order to protect the interest of the clients and the local communities, the government had also imposed certain standards for compulsory application in other aspects (e.g. soil and hydrology, fire, explosion, climate, safety, sanitation, etc.). Decree 52/CP July 1999 and 12/CP prescribe the government role in tackling violations of environmental and planning regulations, and handling the application of Vietnamese Codes. If the standards are still not available for these requirements, foreign standards may be adopted with the written agreement of the Ministry of Construction (MOC). However, there are still some pitfalls that have to be improved in terms of (i) validity, safety and efficiency; (ii) compatibility with the local literacy; and (iii) anthropology of local communities.

Previously, the regulations only indicated a guaranteed period of 2 years for type A projects under the supervision of MOC, and a period of 1.5 years for type B projects supervised by provincial level authorities. However, these regulations have largely been ignored until it is enforced in accordance with Article 53 of Decree 52/CP, and Article 22 of Decision 22/QD-BXD November 1999.

12.3.4 Construction safety

The Ministry of Construction (MOC) had paid increasing attention about construction safety. In every construction project, risk and safety analysis is always undertaken in the early stages of project. Based on Mr Tran Chung's assessment, about 66.14 per cent of the construction accidents were related to the working with concrete structures, 24.74 per cent with brick blocks and 9.21 per cent with metal structures. This may partly be attributed to widespread use of concrete in construction. Other causes and consequences are outlined in Table 12.18.

About one-third of the accidents occurred in the construction sites. Table 12.19 presents only the safety records of firms under MOC. Full statistics for the whole construction sector are not available.

12.3.5 Skill levels of construction workforce

The Vietnamese government administers virtually all education facilities. Literacy of the population was high and most Vietnamese had received at least primary school education. However, the government efforts to upgrade the then school facilities and

Table 12.18 Construction safety

Cause	*Consequence*	*Preventive measure*
Climate	Land slide, flooding, damp break, house collapse	Cease working
Careless operation	Machine accident	Safe condition check
Poor construction	Structure falls	Strict supervision
Low construction condition	Human accidents	Safety equipment (safe hat, ring, glove, glass, etc.)

Source: Nghi, 2000.

Table 12.19 Safety record (construction sector only)

Year	*Number of accidents (unit)*	*Death number*
1997	25	29
1998	33	41
1999*	13	13

Source: Xay Dung Newspaper, No. 31, 5–11 August 1999.

Note
* The first 6 months of the year 1999.

improve the educational infrastructure had been hampered by a high birth rate and continuing economic problems in the country. The number of parochial and private schools had grown as the Vietnamese education system had deteriorated. Education had put an emphasis on applied sciences and vocational training. In the 1980s, over 200,000 Vietnamese were sent to the Soviet Union and East European countries in a labour and training program. In addition, since 1986, Vietnam had sent both skilled and unskilled workers to Algeria and Iraq, expanding the presence of Vietnamese labour supply to countries in the Middle East and North Africa. A large number of temporary workers live in Germany and other former East Bloc countries. Growing number of Vietnamese students were studying in Western Europe, Australia and the United States.

The Vietnamese labour force in the construction industry had increased considerably in both quantity and quality in the SOEs as well as in other sectors. Training and retraining of cadres had been considered essential with a view to increasing the size of the labour force and upgrading their skills to meet the demand in construction sector. Until 2002, the total workforce had reached some 350,000, of which 230,000 were at the central level and 120,000 at the provincial one. In particular, the MOC had taken up over 140,000. Graduate cadres in the whole sector make up about 20,000 of which over 11,000 belong to MOC alone. The quality of the workforce had been constantly improved in terms of its management skills, specialist knowledge and foreign languages. In terms of education credentials, the whole sector comprised of 220 with doctoral degrees, of which 200 were engaged by MOC. There were over 200 cadres with Masters degrees, of which nearly 150 were working with the Ministry. Qualifications of the technical workers in the industry were also upgraded.

12.3.6 Project supervision

As of 2000, there were two supervisory levels for domestic projects. The MOC quality inspection department looks after type A projects (size bigger than 2 hectares of land and with a project cost of VND 500 millions or more). Provincial inspection unit looks after type B projects (size less than 2 hectares of land and with a project cost below VND 500 millions cost). For international projects, the quality control and inspection (Circular O8/BXD-CSXD of MOC) includes full decision of the investor to hire consultants and a participatory role of relevant government agencies (MOC and provincial authority) if required. For some important national projects, the main initiative is to appoint qualified independent foreign consultants to take part in management their construction process.

At the time of writing, what was being considered was to separate the technical construction laws (e.g. codes, standards, quality control, consultant, bidding) from issues concerning investment and open competition in the construction market (Vietnam enter AFTA 2003). There were also recommendations that MOC should establish an organization with qualified experts to handle issues including standard planning and composition, defining areas for compulsory, participatory and transferred standards, and tackling feasible and efficient enforcement of standards. There was also a need for one investment management system to integrate both foreign and domestic sources of investments (state revenue, state investment, ODA, credits and other sources).

12.3.7 Research and development

Research and application of technological advances in construction had undergone a positive change in Vietnam. Annual investment in new technology by SOEs of MOC kept increasing, from VND 459 billions in 1995, VND1279 billions in 1996 to VND817 billions in 1997. In 1998, there were 20 new technology lines transferred into construction and material production. Construction technology used under foreign design and joint venture construction projects contributed significantly to the development of local construction industry. At the time of writing, there were 7 research institutes and 22 centres that were primarily responsible for scientific and technology development in the industry.

In some construction categories, such as foundations, transport and irrigation, the level of mechanization was relatively high. Most of the large corporations had recognized the benefits of mechanization and had begun to develop mechanized teams so as to increase their competitive advantage.

However, there had been a common misunderstanding that automation and mechanization were restricted to the use of machines in the construction sites. In other words, there had been little application of new technology in the organization, planning and management of construction process, including the aspects of prefabrication, transport research, site preparation and project development. It was increasingly recognized that research and development of new construction technology required international cooperation, especially in education and training. At the time of writing, the policy of MOC focused on the following objectives:

- encourage professional institutes to move towards modernization;
- assessment and review of the level of construction technology in the country;
- application of modern information technology;
- enhancement and modernization of standards that aims to secure joint venture (JV) with international community;
- training of young groups; and
- reform of training program for technical labour.

Prefabrication is one kind of construction technology. In Vietnam, there were 54 enterprises in the country producing about 4 million cubic metres of prefabricated and ready-mix concrete. The advantages of prefabricated structures in construction had generally been recognized. The government had pushed companies towards using them more frequently and widely. However, there were a number of problems. First, there was a great concern about low-quality production caused by cement shortage, particularly special cements, and ill competition between enterprises. Furthermore, prefabrication also tends to incur higher costs because certain advanced concrete production equipment needs to be imported. These difficulties lead to the use of fewer prefabricated products. Some factories operate at 20 per cent of their capacity. The actual difficulties were mainly overcome by individual enterprises. Many former concrete prefabricated factories in the urban regions of Hanoi, Ho Chi Minh City had joint venture agreements with foreign contractors for producing ready-mix concrete. On the other hand, LICOGI had developed a formwork system that reduces cost and speeds up construction by two to three times. Due to the diversity of projects with increasing *in situ* cast items, the shift to the popular western standards, the prefabrication technology that was well developed in the 1970s for various Soviet-style primary building structures (wall, floor, beam) had been reduced to some secondary utility products such as road partition, irrigation culverts and lighting poles.

The MOC had set the policy that, by the year 2010, the industry should raise its construction science and technology to the regional level with regard to the following areas:

- complete systems of policies, codes, research and development, and education;
- ability of SOEs to construct complex projects underground or on the sea, and increase the percentage of turn-key projects; and
- high-skill professional teams in sufficient quantities.

Table 12.20 presents some initiatives for improving the level of construction standards/methods and for replacing imported products. In addition, there were applications for the application of higher technology in the civil engineering area. These examples include the technology in building bridges with deep, long-span foundations, manufacturing 33 metres reinforced concrete beams, casting the cantilever beams of 64 and 102 metres, long-span steel beams and driving pile with diameter of over 1 metre using modern concrete pumps and mixing equipment.

The use of information technology in construction is another strategic development area. The MOC regards information and communication technology (ICT) as the engine for world economic development at the end of the twentieth century and the

Table 12.20 New technology in automation, mechanization and prefabrication

Technology and brief description	*Developer*
Cement productive equipment and technology	LICOGI, Material Research Institute
High-tech ironing	LILAMA
High-quality form work production	LICOGI
High-temperature brick from alumni ores	Material Research Institute, MOC
High-rise building construction	Ha Noi Corporation, VINACONEX
Underground construction technology	Song Da Corporation
Steel pre-stressed technology	Construction Science and Technology Institute, MOC
Ready-mix concrete	Foreign joint venture with Chem, Xuan Mai, Thinh Liet concrete factories

beginning of twenty-first century. In 1993, the government issued and implemented the government resolution on IT development of the 1996–2000 period. Vietnam had took its first step by making significant investment in communication infrastructure such as construction new high-tech centres or software parks in three big cities (Hanoi, Ho Chi Minh City and Danang).

Despite these achievements, ICT development in Vietnam had remained at a lower level in comparison with countries in the region and beyond. Vietnam was facing human resource constraints in developing communication infrastructure and nurturing the conditions for developing a knowledge-based economy. It was widely agreed that ICT development can meet the increasing demand for information transmission in construction industry. There was a need to improve the institutional and policy framework to promote the development and application of ICT. In this context, the resolution 07/CP, issued by the government on 5 June 2000 on 'Research and development of software technology for the 2000–2001 period', and after that the Instruction 58-CT/TW, issued by the Political Committee of the Party in October 2000 on 'Promotion of IT application and development for industrialization and modernization', have shown a strong commitment made by Vietnam in this direction.

Initiatives on the use of construction IT were increasingly pursued in design work and in quantity surveying in many construction consultant corporations. In consulting firms, the application is moving from PC-based to local application network (LAN) and wide application network (WAN). In the fields of architecture and civil engineering, 2D and 3D-design and documenting works were widely used. AutoCAD program is the most popular application by virtue of its diverse utilities and looser copyright. The use of Microstation program is also popular in the national surveying field. Some structural computing programs such as STAAD, SAP2000, etc. are popularly used together with the conventional design of building and large structures. However, these applications are limited to calculation of internal loads based on foreign standards and English language. MOC has developed RDW2000 program to automate the structural computing from internal load stage to drawing stage based on Vietnamese language and standards. For quantity surveying, the use of Vietnamese accounting programs is very popular because of its use of domestic language and survey standards. The remaining issue is how these programs can be customized to make them compatible with international quantity surveying practices. However, these operations are found mostly in centralized consulting firms and national education institutions. Many local construction firms are still limited to the use of word processing.

12.3.8 Environmental conservation

Vietnam had promulgated the Environmental Law since 1994 as a means to establish a statutory system of environmental conservation. This enables the government to provide guidelines on management of natural resources and occupational health, and setting national environmental standards.

The Decree 26CP 26 April 1996 stipulates that administrative fines on environmental damages, environmental impact assessment (EIA), auditing and compensation had been made compulsory in the project development procedures (Table 12.21). Environmental conflicts had increasingly been well perceived by developers, communities, media, and are often actively addressed by the administrative authorities.

Table 12.21 Ten steps of project development in Hanoi city

1	Investment intention	6	Land use license
2	Site introduction and survey	7	Compensation
3	Land compensation plan	8	Building permit
4	Pre-feasibility and EIA	9	As-built archive
5	Feasibility	10	Land and house certificate

Source: Based upon Hanoi New Town project developed in 1998.

However, at the time of writing, many problems exist in the implementation of Environmental Law. These problems include weak construction management and lack of willingness to comply with the law, especially in domestic projects. The government was trying to apply economic means, such as tax, input/output pollution fees, waste disposal permit, environmental collateral/fund, with a view to combating the environmental degradation during the construction process.

12.3.9 State investments in construction

Table 12.22 shows the amount of state investments in the field of construction in 1999. Table 12.23 illustrates the geographical distribution of the state investment in Vietnam.

Table 12.22 State investment outlays in 1999 at the then prices as of 2002 by type of component (VND billion)

Sector	*Investment outlay*
Construction and assembly works	645.0
Machinery and equipment	428.4
Others	126.6
Total	1,236.0

Source: Based upon statistics provided by National Institute for Urban and Rural Planning, Vietnam.

Table 12.23 State investment outlays of local government in 1999 at the then prices as of 2002 by region, province

	(VND billion)
Whole country	26,959.7
Red river delta	3,109.2
North East	2,715.4
North West	551.9
North central coast	2,010.1
South central coast	2,658.2
Central high lands	956.0
South East	9,622.6
Mekong river delta	5,336.3

Source: Vietnam Country Report (2001).

12.4 Construction outlook 2001–2010

12.4.1 Competitive strategy

The construction sector had suffered a considerable setback after the Asian financial crisis. The construction sector in Vietnam comprises both SOEs under the ministries and non-state-owned sector enterprises (NSOEs) (foreign, joint venture and private contractors). The scale of SOEs was much larger than the NSOEs. Since new projects were fewer, there was tougher competition between companies. Contractors, during the time of writing, had to bid for lower construction costs and higher construction quality. Many contractors have to operate at low profit margins and substantial layoffs and salary reductions have been common. To overcome the economic slowdown, three major strategies are being considered for the construction sector: (a) capital funding; (b) project management; and (c) diversification.

In the area of capital funding, the objective of the government policy is to make the most efficient use of the existing investments. One of the measures is to reform the tendering process so that project award is not entirely based upon contractor capacity and capital resources, but also on quality and timing. There is also a closer monitor of the progress of the then on-going construction projects (particularly compensation process) in order to ensure timely operation for the economy (Nghi, 2000). Many projects that were identified as expensive or unable to directly promote production or consumption (such as government offices) would be cancelled or reduced in scale. Other projects abandoned by foreign investors were changed into domestic or new foreign investment partners. Project implementation now favours the use of alternative domestic resources (consultant, material). More infrastructure projects were sought by BOT or decentralized to local governments. The state-owned housing was privatized to recover the capital for new projects as well as to reduce the operation costs that the government is difficult to bear.

In terms of project management, the government strategy is to regroup construction enterprises into larger corporations to be supervised by the Central Ministries such as MOC, Ministry of Transport (MOT), Ministry of Agriculture and Rural Development (MARD), Ministry of Industry (MOI) and so on. Redundant workers in the SOEs are allowed to stay home with basic wages with medical insurance. More flexibility is also introduced into the labour contracts. For example, in the private sector, contracts can now be signed for shorter periods like 3–6 months or 1 year. Contract extension is only offered to competent labour until completion of project.

Another important strategy is to diversify and to expand construction activities into national and foreign markets. Many construction companies, during the time of writing, diversify part of their activities into material production and trading, land development for housing, office, entertainment and so on. Some construction companies make a noticeable business shift from civil construction to the area of infrastructure. The state also supports exporting a part of construction activities to the international construction markets.

12.4.2 Business opportunities and major projects

As a developing country, Vietnam has a great demand on new infrastructure to support its economic growth. Thus, construction continues to be one of its major growth industries. The economic prospect of the construction sector in Vietnam is fairly good. A major contributing factor to such optimism comes from the state policy. The State's plans for both 1996–2000 and 2001–2010 continue to put emphasis on industrialization and modernization of the country. Infrastructure building remains high in the policy agenda, with the government allocating 3 per cent of the GDP specifically for its development.

Vietnam has set to grow its economy at an annual growth rate of 7.2–7.5 per cent with a view to doubling its GDP value by 2010. The government has pledged to focus on upgrading the manufacturing technology of selective industries, and developing high-tech and defence industries. Labour-intensive industrial sectors and some high-tech ones, such as informatics, biotechnology, new materials and automation, could become key pillars in the domestic manufacturing sector.

Industry and construction growth rate is set to be about 10–10.5 per cent per annum. By 2010, industry and construction sector is expected to contribute about 40 per cent of the GDP and provide employment for about 24 per cent of the domestic labour force. On the other hand, the service industries are estimated to grow at about 7–8 per cent annually, making up about 43 per cent of the GDP and 27 per cent of the labour force. The agricultural sector is not neglected in the process of modernization. Vietnam will develop cash crops and fruit cultivation, and strengthen food-processing industries so as to industrialize the rural areas and ensure food sufficiency. Agriculture, forestry and fishery sectors are expected to grow at about 4–4.5 per cent per year, contributing some 17 per cent of the GDP by 2010. Forest coverage is estimated to be about 43 per cent by then.

Despite current economic difficulties, the national reconstruction program, which forms part of the construction demand stimulus policy, is still underway in Vietnam. Table 12.24 shows the major construction and infrastructure projects in the short and

Table 12.24 Business opportunities and major projects in the pipeline

Type	*Project no.*	*Brief description*	*Capital distribution*		*Foreign participation needed*
			USD million	*VND billion* (USD million $)	
Short term 2000–2005					Capital, consulting and construction
Infrastructure	18	Water supply, power, road	1,251	14,613 (1,043)	
Industry	5	Industrial processing	121	903 (65)	
Commerce	8	Serviced housing	177	501 (36)	
Institutional and residential	5	State facilities	0	732 (52)	
Total	36		1,549	16,749 (1,196)	
Medium term 2000–2005					
Infrastructure	25	Electricity, road, wastewater, etc.	2,200	24,189 (1,728)	
Industry	5	Industrial processing zone	1,205	3,622 (259)	
Commerce	6	Leisure	88	814 (58)	
Institutional and residential	4	State office, housing	1,333	4,040 (289)	
Total	40				

medium terms. Infrastructure projects make up the majority part of the programme and they are related to the provision of urban infrastructure and industrial zones in major cities, rational utilization of land and protection of the environment.

12.4.3 State policy on liberalization and foreign investment

Creating favourable physical, legal and economic environments for all businesses is a stated policy under the Vietnamese economic reform strategy. To encourage investors especially the small and medium enterprises to invest in industrial production, new infrastructure and supportive facilities were set in place. These include the several export processing zones (EPZs) established since 1991, and a number of industrial zones set up since 1994. At the time of writing, some 56 industrial zones, three EPZs and one high-tech zone (HTZ) had been established throughout the country. These zones were planned and built in areas where there were highly favourable conditions for industrial development. These conditions include proximity to large cities and consumption markets, accessibility to material resources and convenient connections to transport routes and newly developed infrastructure (Vietnam Industrial Zones Authority, 2004).

The revised Law on Foreign Investment in Vietnam (LFI) and the Government Decree No. 36/CP, at the time of writing, provided a legal framework and offered a wide range of incentives for foreign investors to invest in the industrial zones. Foreign investors was expected to be granted investment licenses for operating in the industrial zones. The approval process for investment license was decentralized and notably simplified for foreign companies (Vietnam Industrial Zones Authority, 2004).

An enterprise investing in the industrial zones is entitled to (Vietnam Industrial Zones Authority, 2004):

- organize production and business in accordance with the investment license;
- sub-lease land in the industrial zones;
- carry out direct import and export;
- start building the projects without construction permits once the technical design has been approved;
- recruit and appoint foreigners for positions that require a level of technical and management expertise not found in the local Vietnamese workforce;
- open banking accounts at the foreign banks for borrowing if it is so required by the foreign lender and approved by State Bank of Vietnam; and
- enjoy simple and prompt entry-exit procedures.

Business enterprises investing in the industrial zones normally enjoy more administrative convenience and investment incentives than those outside the zones. For instance, they can file their applications for an investment license directly with the Provincial Industrial Zones Management Boards. Investment license can be granted within 15 days rather than 60 days under other circumstances. They can also register their plans for exports and for selling products in the domestic market directly with the Provincial Industrial Zones Management Boards. For companies classified as export processing enterprises, they are not subject to the turnover tax or VAT tax (from

1 January 1999). Goods and foreign exchanges sent from abroad in and out of the export processing zones are free from import–export duties and are subject to convenient and efficient customs clearance (Vietnam Industrial Zones Authority, 2004).

The ninth party conference had resolved to continue market liberalization and participation as the government policy for the coming years. The Vietnamese government would encourage further investments in the following sectors and regions.

The economic sectors include:

- export production industries;
- projects in animal husbandry farming and processing of agricultural products and afforesting;
- projects that utilize high technology and modern techniques, protect the ecological environment and enhance research and development;
- projects that are labour intensive, involve raw material and efficiently utilize the natural resources of Vietnam; and
- construction of infrastructure facilities and important industrial production establishments.

The favoured investment regions include:

- mountainous and remote regions
- regions with difficult economic and social conditions.

However, there is a list of sectors in which investment will not be licensed as follows:

- projects which are prejudicial to the national security defence and public interest;
- projects which are detrimental to the historical and cultural relics, fine customs and tradition of Vietnam;
- projects which are prejudicial to the ecological environment, such as projects for treatment of imported toxic waste from foreign countries;
- projects for the production of toxic chemicals and utilization of toxic agents which are prohibited under an international treaty; and
- nuclear power, radioactive materials, industrial explosives and leathers dyeing.

Insofar as the construction and engineering sector is concerned, Vietnam and the US signed a bilateral trade agreement on 13 July 2000 to operate in accordance with the WTO rules. Construction service was the main service trade that Vietnam had agreed to follow the principles and regulations stipulated in the General Agreement on Trade of Services (GATS) under the WTO legal framework. This agreement was likely to lead to rationalization of the organization of the existing architectural and engineering service. Small organizations in this field seem to be reducing in number and they were replaced by larger or joint-venture corporations. Service costs and prices were also increasing.

To attract participation of foreign contractors and professionals in the market, the Vietnamese government had took a number of reform measures as follows.

- Introduction of the new Enterprise Law, which gives private citizens the right to establish joint stock companies in a field of their choice.

- Signing of a comprehensive bilateral trade agreement with the US. Although this agreement still needs to be ratified, it commits the Vietnamese government to provide more transparency in its law reforming and formulation process, opening up key economic sectors and reducing barriers to trade.
- Opening of Vietnamese securities exchange that is still small scale and overly regulated, but nonetheless has expanded the availability of capital sources in Vietnam.
- Amendments to the LFI that has been approved in July 2000.

There are a number of key changes in the new LFI. First, it removes the restrictions on foreign lenders in accepting mortgages of land use rights in return for loans. The new LFI permits all banks in Vietnam to take the land use rights as a security. This marks a significant first step in freeing up project financing. Second, it introduces new tax incentives for foreign investors. These new incentives include:

- reduction in the profit remittance tax rates from 5, 7 and 10 to 3, 5 and 7 per cent, depending on the particular projects;
- allowing the parties to business cooperation contracts (BCCs) to carry forward tax losses for 5 successive years. Under the old law, losses carried forwards were available only to the other forms of foreign enterprises, thereby making BCCs unpopular;
- extension of the 5-year import duty exemption for raw materials used for investment projects especially in favoured activities; and
- a 20 per cent reduction in corporate income tax.

Furthermore, the new LFI allowed the government to enter into written agreements with the foreign investors to provide investment guarantees. It also improved the time frame for investment license approval, and created a new class of projects that must merely be registered in order to obtain their investment licenses. Foreign enterprises providing accommodation for their workers, and those operating in the fields of disease treatment, training and education, or scientific research, were entitled to pay the lowest scale of land rents and be exempt from or pay the lowest rate of all types of taxes. In sum, the new LFI had improved the legal framework and made it more favourable for foreign investment. It had enhanced the attractiveness and competitiveness of the investment environment in Vietnam (Table 12.25).

There were signs of intensified cooperation among the Asian countries, especially in the context of the ASEAN. For instance, in an informal summit in Singapore, all ASEAN member countries had committed themselves to speeding up e-commerce and increase investment towards human development. To enhance international cooperation in its construction market, the Vietnamese government had adopted the following strategies:

- encourage and create favourable conditions for the operation, participation and promotion of institutes, enterprises and professional associations in the Asian region and abroad; and
- expand and diversify international cooperation in the areas of scientific research, technology transfer, training, conference, professional development and information sharing.

Table 12.25 Top-ten sources of FDI inflows to Vietnam in 2000

State/territory	*Project number*	*Total value* (US$ billion)
Taiwan	281	358.3
Japan	96	334.6
Russia	9	253.1
Hong Kong	57	203.3
British Virgin Islands	40	144.1
Korea	75	122.8
Malaysia	26	114.9
France	32	111.2
United States	14	79.1
United Kingdom	14	51.5

Source: MPI (2000).

12.4.4 Strategic areas for foreign investments

Foreign direct investments are badly needed to support the development of Vietnam. Official plans of the government have estimated the transport and communication sector require an investment of at least US$23 billion during the period 2000–2010. On the other hand, the power industry is expected to require about US$1.2 billion over the next 5 years since the time of writing, whereas the development of the oil and gas industry will require another US$5–7 billion. Besides, the government has taken an official perspective of encouraging the import of equipment and machinery. The investment law has been amended to attract continued foreign investment, reduce the red-tape procedures and time involved in project scrutiny.

The power industry is one of the key strategic development sectors. About US$29 billion are to be invested for the generation and network expansion between 1999 and 2010. With the expansion of the power generating capacity, the power transmission (500, 220, 110 kV) and distribution (22, 35 kV) networks will be further developed as well. The existing facilities will be gradually upgraded and modernized, particularly in the congested urban areas (CP Exhibition, 2004).

Two scenarios have been prepared in the Power Demand Forecast up to the year 2010 (Table 12.26). To satisfy the demand, Electricity of Vietnam (EVN) has worked out an attractive investment plan that includes the construction of large power plants, development of BOT projects, and transmission and distribution networks in the country. There are some 21 plans for expansion of major power generating plants for the period from 2005 to 2010, including the Mekong Delta and Nhon Trach CCGT

Table 12.26 Power demand forecast upto the year 2010 (in million kWh)

Year	*Base case*	*High case*
2000	27,500	30,000
2005	47,737	53,650
2010	78,486	87,323

Source: http://www.cpexhibition.com/electric/elect_info.html.

(1200–2400 MW), the Purchase from Laos (1500–2000 MW) and the Son La hydro (3600 MW) (CP Exhibition, 2004).

Another strategic growth sector is the construction material industry. For this industry, the Vietnamese government had earmarked an investment need for some 51,500 billion Vietnamese dong (about US$3.5 billion). It is expected to recruit 180,000 workers to make it capable of meeting the domestic demand for construction and have a surplus for export in 2010. Such investment will include about 22,323 billion Vietnamese dong for the years till 2005. The decision, coded 115/2001/QD-TTG, approves the master plan for the development of Vietnam's construction materials industry till 2010. Under the plan, the industry will export at least 20 per cent of its annual output of tiles and ceramics toilet articles, and 30 per cent of its construction glass by 2005. On the other hand, it was expected that, as from 2003 onwards, all cement joint ventures will have to meet their export targets stipulated in their joint venture contracts and their licenses. As from 2004, domestic cement producers will have to export at least 1 million tonnes per year. The industry had required to reorganize its conventional small-sized construction material production in order to minimize environmental pollution and to put an end to hand-built brick making kilns in suburban areas before 2005 and in other areas before 2010. Investment will be injected into the building tunnel brick production systems using domestically manufactured equipment, generating between 7 and 10 million bricks a year (Vietnamese news agency, 2001).

12.4.5 Looking ahead

The Renovation Policy (Doi Moi) since 1986 had brought about substantial economic progress in Vietnam. During this period, the construction industry had made remarkable contributions to the economic growth of the country, especially in building up its physical infrastructure. From 2002, the government was expected to concentrate on capital investment in key projects that promote modernization and industrialization of the country. These projects include airport expansion (of Noi Bai, Tan son Nhat and Chu Lai), deep-water ports (at Cai Lan, Thi Vai, Dung Quat, Van Phong, Lien Chieu and Nghi Son), building materials plants, resource and energy projects, water supply and sewerage networks, highways and so on. According to the proceedings of the Ninth National Party Congress, Vietnam had determined to upgrade its construction industry to advanced standards in order to meet the domestic demand and possibly expand its service abroad. To achieve these objectives, the government had recognized the need to enhance application of modern technology, raise planning quality and effectiveness, improve architectural aesthetics and design, develop consulting activities especially in strategic infrastructure projects, and to strengthen the state administration in project planning and construction management.

Specifically, it is planned that the whole construction industry will have to undergo reforms in the following directions:

- Investing in technology and modern construction equipment to promote industrialized production. In this respect, the priority would be given to the areas of foundation engineering, underground construction works, concrete works, transportation and housing, equipment erection works, steel structure fabrication, decoration and finishing instruments.

- Reorganizing existing industry into 4–6 strong construction groups and units that are capable of implementing large-scale construction projects with complicated techniques in the local economic development regions and industrial zones, competing with the foreign countries in Vietnam and engaging in competitive bidding for works contracts abroad.
- Strengthening the abilities of the construction organizations in order to perform different kinds of construction business.
- Setting up management institutions, and strengthening the legal environment to create favourable conditions for the contractors to carry out their business activities, and to support domestic and foreign construction companies to participate in foreign investment projects.

The then Prime Minister Khai had confirmed that economic development would form the core of the government policy for the period between 2002 and 2007. The construction industry was expected to continue to constitute a key pillar in contributing to this policy goal.

References

ADB (1995) *Country Operational Strategy Study: Vietnam*, http://www.adb.org/Documents/COSSs/vie.pdf.
ADB (2000) *Asian Development Outlook*.
Asia Business Network (2002) http://www.businessasia.net/.
Bureau of East Asia and Pacific Affairs, US Department of State (1998) *Background Notes: Vietnam*, http://dosfan.lib.uic.edu/ERC/bgnotes/eap/vietnam9810.html.
CIA World Factbook (2001) http://www.ciafactbook.com.
CIEM (2000, 2001) *Vietnam's Economy in 1999, 2000*, Statistical Publishing House: Hanoi.
Ernst & Young International (2000) www.eyi.com.
Nghi, D.Q. (2000) 'Viet Nam 1997–1998' in 6th AsiaConstruct Conference, October, Malaysia.
SPH (Statistical Publishing House) 'Viet Nam Infrastructure Renovation in 1985–1995 period, 1996–2000'.
UNIDO (2004) http://www.un.org.vn/unido/frame3/text3.htm.
MPI (Ministry of Planning and Investment) (2000) 'General Report on Foreign Investment Situation in Vietnam in 2000', MPI: Hanoi.
The Government Decree No. 61CP dated 5/7/1994
Vietnam Country Report (2001) *Proceedings of AsiaConstruct 7*, New Delhi, India.
Vietnam Investment Review (2002) http://www.saigonnet.vn/english/re/jan02/.
Vietnamese News Agency (2001) various issues.
Vietnam Investment Zones Authority (2004) http://www.vnn.vn/province/Zones.htm.
Xay Dung Newspaper, No. 3l, 5–11 August 1999.
CP Exhibition, http://www.cpexhibition.com/electric/elect_info.html.

Index

Australia
- Asian financial crisis, impacts of 46–7
- development process 31
- economic outlook 48–9
- economic reforms 36
- economic resilience 34
- economy 32–9
- ESA 62
- foreign direct investment (FDI) 63
- GST 36–7, 42
- land area 31
- NSW government 55, 60–2
- Olympics, impacts of 37
- population 30
- property prices 43–6
- TNTS 36
- Whitlam government 32
- WST 36

Australia, construction of
- ABCB 53
- buildability 61
- business opportunities 61
- costs 50
- investments 39–42
  - engineering 42
  - non residential 42
  - residential 42
- labour productivity 49–50
- labour skills 56
- quality 52–3
- research and development (R&D) 56–7
- role 38
- safety 54–5

China Hong Kong
- development process 100–1
- economic policies 109–10
- history 98–9
- land price 106–7
- political structure 99–100

China Hong Kong, construction of
- average material price 112–14
- business opportunities 128
- China's accession to WTO 128–9
- construction investment 105–6
- environmental assessment method 126
- environmental corrective measures 124–5
- environmental impact assessment 125–6
- environmental planning 126
- prefabrication technology 120–2
- property price 107–9
- quality 118–19
- rent 107–9
- research 119–20
- review 101–3
- role of construction/property market 103–5
- safety 114–19
- salaries and wages 114
- unit cost 112
- use of IT 122–4
- value added per employee 111–12

China mainland
- Asian financial crisis, impacts of 80–1
- development process 67
- economic outlook 73, 82
- economic structure 68
- economy 69–73
- Eighth Five Year Plan 85, 89
- foreign direct investment (FDI) 71, 77, 80, 82, 92–3
- foreign investment 94–6
- land area 67
- Olympics 94, 95
- population 66
- property prices 77–80
- Shanghai index 81
- WTO accession 78, 82–3

China mainland, construction of
business opportunities 93–4
costs 86
equipment 84–5
investments 75–6, 81
labour productivity 83–4
quality 86–7
research and development (R&D) 92–3
role 74–5
safety 91
skills 91–2
supervision 89–90
wages 86

India
Asian financial crisis, impacts of 142–3
construction and property markets 137–41
demographic outlook 131
development prospect 143–8
economic development history 132–4
economic outlook 135–7
economic reforms 142
economic structure 136
economy 134–48
foreign trade and investment (FDI) 135–7
geographic outlook 131
land area 131
land price 140–1
location 131
population 131
rents 141
India, construction of
business opportunities 157–9
costs 149–50
investments 138–41
labour productivity 148–9
labour skills 151
labour wages 149
project supervision 151–2
quality 150–2
research and development (R&D) 153
safety 152
state investment 135, 138–41
Indonesia
Asian financial crisis, impacts of 164, 166–9, 172, 176
East Timor 164
economic development history 166–7
economic outlook 179–80
fiscal sector 176
geography 164–5
languages 164
political structure 165–6
population 164
separatist movements 164
Indonesia, construction of
bidding practices 171
business opportunities 186–7
costs 182–3
government sector 174–5
growth rates 173
investments 172
private sector 172–4
productivity 181–2
quality 183–4
R&D 185–6
recovery 177–9

Japan
agriculture 193
construction and property markets 223–4
economic outlook 197–8, 224–6
economic reforms 195–8
economy 193
foreign direct investment (FDI) 316–18, 319–21
government 192–3
inflation 329–31
labour 194
land prices 236–42
land transactions 241–2
population 225–6
property
market 234–45
prices and rents 234–9, 242–3
transport sector 194
Japan, construction of
access for foreign firms 295–7, 319–21, 322–4
Asian crisis 199–202, 218–23, 270–1, 273–4
companies 201–3, 245–51, 270–2
competition
bidding system 210–11, 283, 289–95
consultants 252–3
costs 233–5
environmental protection 205–8
firms *see* construction companies
industry structure 208–13
infrastructure 213–14, 298–311
international exchange and cooperation 214–18, 311–16
investment 198–9
labour
costs 234
labour supply 254–7, 264–7

Japan, construction of (Continued)
market review and outlook 225–35
materials 267–9
overseas work 273–8
prices 234–6
productivity 264–5
research and development (R&D) 282–3
safety 260–1
technologies 204–7, 284–8
see also labour costs

Malaysia
Asian financial crisis 328, 331, 334, 337, 342, 357
development process 326
East Asian 328, 342
economic development 342
economic outlook 337, 341, 345
economic policies 341
economy, export 327, 330
economy, global 341, 343, 345
economy, national 328, 329, 330, 331, 334, 343, 352, 362
foreign investment 329, 344, 346, 356–8
land 325–6, 328, 340–1, 357, 360
law 351, 352
political 326–8, 346
population 325, 328, 333, 345
prices, house and residential 339
property market 337, 338, 343
regulatory 352
renovation 332
rents 337
Malaysia, construction of
barriers 345, 357
business opportunities 356
competition 331, 351
Construction Industry Development Board 334, 349, 352, 355
cost 344, 347, 351
investments 328, 329, 331, 342, 357–8
labour 327, 333, 347
prefabrication 347
private sector 332–4, 358
productivity 346–7, 351, 355
public sector 331, 335–6, 343–4
quality 343–4, 349–52, 356, 358
R&D 344, 356, 356–7
regulation 350, 352–4
safety 344, 350, 352–4
skills 347, 355–6
supervision 355
supply 331, 334, 336, 338–40, 343–4

regional overview
access to markets 25–6
Asian financial crisis, impacts of 2–5
business opportunities 19–21
construction outlook 22–5
construction sector 8–9
cooperation 19–21
economic outlook 5–6
globalization 9–11
investment 12–14
productivity 12
property cycles 6–8
quality 1
research and development (R&D) 17–18
safety 18–19

Singapore
development history 361–2
economic growth rates 362–3
economic policies 374
geography 360–1
political structure 361
population composition 360
population policy 360
Singapore, construction of
business opportunities 388–9
costs 375–9
environmental conservation 387
outlook in 2002 368–72
productivity 374–5
property prices 372–4
quality 380–2
research and development (R&D) 385–6
role in economy 364–6
safety record 382–3
skill level, workforce 383–5
South Korea
agriculture 397
Asian financial crisis, impacts of 422–6
balance of payments 408
construction and property markets 422–5
economic development history 394–9
economic outlook 403–13
economic structure 403–13
economy 394–9, 403–13
exports and imports 406–7
foreign direct investment (FDI) 411
government 393–4
history 392–3
inflation 412
land prices 425
population 393

property
market 423–5
prices and rents 425
transport sector 397–9
unemployment 412–13
South Korea, construction of
Asian crisis 422–9
business opportunities 429–30
companies 426–7
competition 429
costs 417
and development 413–14
environmental protection 421
infrastructure 430–2
investment 422–5
labour 399
costs 417
labour skills 420
labour supply 426–8
prices 417–18
productivity 415–17
prospects 429–38
quality 419
research and development (R&D) 420–1
safety 420
see also labour costs
Sri Lanka
Asian financial crisis, impacts of 440, 444–7
construction and property markets 443–4
demographic outlook 439
development prospect 447–8
economic development history 440
economic outlook 440–3
economic reforms 440, 447–8
economic structure 442
economy 440–9
foreign trade and investment 441
geographic outlook 439
land area 439
land price 443
location 439
population 439
rents 444
Sri Lanka, construction of
business opportunities 455–6
costs 450–1
investments 445
labour productivity 449–50
labour skills 453
labour wages 451
prefabrication 454
project supervision 455–6
quality 451–3
state investment 443
Vietnam
Asian financial crisis, impacts of 463, 467–8, 469, 471, 472, 480
categories of land 468
construction and property markets 467–70
development prospect 486–7
Doi Moi (Renovation) 460, 462–3, 486
economic development history 461–2
economic outlook 470
economic reforms 460–1, 462, 464, 470, 473, 482
economic structure 463
economy 462–70
foreign trade and investment 462, 466–7, 483–7
land area 459
land rents 468
location 459
population 459
property prices 472–3
Vietnam, construction of
business opportunities 480–2
competition 480
costs 472–3
investments 467
labour productivity 470–1
labour skills 474–5
labour supply 475
prefabrication 476–7
project supervision 475–6
quality 473–4
R&D 476–8
safety 474
state investment 479